APPLIED
MATHEMATICS

APPLIED MATHEMATICS

A CONTEMPORARY APPROACH

J. David Logan

Department of Mathematics and Statistics
University of Nebraska-Lincoln
Lincoln, Nebraska

A WILEY-INTERSCIENCE PUBLICATION

JOHN WILEY & SONS

New York · Chichester · Brisbane · Toronto · Singapore

Library of Congress Cataloging-in-Publication Data:

Logan, J. David (John David)
   Applied mathematics.

   "A Wiley-Interscience Publication."
   Bibliography: p.
   1. Mathematics—1961–   .I. Title.

QA37.2.L64 1988   510   87-13318
ISBN 0-471-85083-7

Printed in the United States of America

10 9 8 7 6 5 4

To
Tess
and our little Aquarian
David R.

PREFACE

This is a textbook designed to introduce applied mathematics to seniors and beginning graduate students majoring in mathematics, engineering, and the physical sciences. Prerequisites include a good command of calculus, elementary linear algebra (matrices), and postcalculus differential equations, as well as familiarity with a few of the concepts presented in an elementary physics course. It differs from other books in that an attempt is made to present some of the more current topics in applied mathematics in an elementary format. These include singular perturbation, nonlinear waves, similarity methods, and bifurcation phenomena. An effort was made to write in a style that makes the topics accessible to students with widely varying backgrounds and interests but who have a common need to know the rudiments of these subjects. Some of the more standard topics are covered as well, such as the calculus of variations, dimensional analysis, Fourier methods, integral equations, and the numerical solution of partial differential equations. Because many of the chapters are independent, there is considerable flexibility for the instructor in using this book as a text for either a one-year or one-semester course.

The text was spawned from a two-semester three-credit-hour sequence in applied mathematics at the University of Nebraska. The course tries to strike a balance between the mathematical aspects of a subject and its origins in empirics, and to teach a way of thinking about problems that emphasizes the interplay between mathematics and science. The insight gained can be of benefit later when students try to apply the methods or explore new concepts on their own.

I took the task of writing this book because of a belief that there is a need for a survey of applied mathematics at this level, particularly one that incorporates the current topics just mentioned. Just as in algebra, for example, where an introductory course includes the study of groups, rings, etc., a beginning course in applied mathematics should introduce some of the basic areas of study. Such a course can be of tremendous benefit to students in a terminal Masters Degree program or to seniors who may be considering graduate study. At the University of Nebraska this course is followed by specialized courses covering each of the topics at a more advanced level. The guiding principle in the exposition in the text was to take a classical approach that would be accessible to students with a wide range of interests and previous mathematical training.

Applied mathematics is a broad field of study and every applied mathematician will view its role and content differently. This presentation is one practitioner's view. The topics covered herein are ones classically associated with mathematics applied to physical sciences and do not include important topics like control theory, optimization, combinatorics, or such. In this sense the scope of the text is limited.

Scaling and dimensional analysis are topics usually ignored in treatments of applied mathematics. These subjects are often left to the folklore in which the student is supposed to pick up as needed. Yet a good understanding of scaling is essential for perturbation calculations, and dimensional analysis is required in the mathematical modeling of physical phenomena. In Chapter 1 a short introduction to the basic concepts is presented, including an elementary proof of the Buckingham Pi theorem.

In Chapter 2 the underlying ideas of regular and singular perturbation theory are offered in the context of ordinary differential equations. Chapter 3 introduces the classical techniques of the calculus of variations in a functional analytic setting.

In Chapter 4 begins a study of the fundamental equations of applied mathematics, partial differential equations and integral equations. Classical techniques involving Fourier series and transform methods are illustrated on the diffusion equation. In Chapter 5 the study of evolution equations continues with emphasis on wave propagation. An approach is taken in which model equations are developed to illustrate basic physical processes such as convection, diffusion, dispersion, distortion, and so on. The differences between linear and nonlinear phenomena become apparent.

Fluid dynamics in one and three dimensions is discussed in Chapter 5 as well. Wave propagation in continuous media is a nonsterile example of wave phenomena and provides the correct context for developing the wave equation and for understanding some of the origins of singular perturbation and bifurcation theory. For these reasons and others there seems to be an increasing demand among mathematicians to learn fluid dynamics. What was once a standard part of the applied mathematics curriculum appears to be undergoing a renaissance; it offers a rich context for illustrating mathematical modeling and analysis.

Chapter 6 contains an introduction to stability and bifurcation. The latter has become a popular area of research in applied analysis. One dimensional problems are presented in a fair amount of detail within the context of singularity theory. Nonlinear systems and phase plane phenomena are discussed as well as hydrodynamic stability.

Chapter 7 on similarity methods shows how one can take advantage of the symmetry or invariance properties of a problem to obtain a significant simplification or a solution. This is carried out in problems in the calculus of variations where the famous E. Noether theorem on conservation laws is presented and in partial differential equations where it is shown how symmetries permit a reduction to an ordinary differential equation. Similarity methods are not often taught in elementary partial differential equation courses as a technique, but their wide applicability is firmly established and there is indication that the method should be introduced in elementary contexts.

The final chapter on finite difference methods for partial differential equations contains some of the basic numerical algorithms for solving the diffusion equation, the wave equation, Laplace's equation, and hyperbolic systems. Concepts of convergence and stability are introduced and programs in BASIC are presented for some of the algorithms. The idea is to indicate the logical structure and the ease with which the calculations can be performed, even on a microcomputer. The student is encouraged to make the programs more efficient and applicable to more general problems.

The exercises form an essential element of the text and the course. They range from routine problems designed to build confidence and test basic technique to more challenging problems that build technique.

The bibliography has been selected to suit the needs of an introductory text. At the end of each chapter are listed a few standard, and in most cases classical, references to the material. In these the reader can find parallel discussions or extended coverage of the topics.

Equations are numbered consecutively starting anew at the beginning of each section. Theorems, definitions, and examples are numbered within each section as well. For example, Theorem 3.2 refers to the second theorem in Section 3 of the current chapter.

This work was influenced either directly or indirectly by many individuals. The University of Nebraska supported my efforts during the summers of 1984 through 1987 by relieving me of some of my duties as Chairman of the Department. Special thanks go to Mr. Kevin TeBeest, who carefully proofread the manuscript and made many suggestions and corrections, and to my colleague Dr. Steven Dunbar who was frequently a sounding board. The comments of Professors Ivar Stakgold at Delaware, Bernard J. Matkowsky at Northwestern, and Gunter H. Meyer at Georgia Tech also led to many improvements. Maria Taylor, the editor at Wiley-Interscience, shared my enthusiasm for this project and skillfully managed its development and production. Several versions of the manuscript were typed with skill and dedication by Rhonda Bordeaux at the University. Three colleagues with whom I have worked during the last ten years deserve a strong acknowledgment for their influence; these are Dr. John Bdzil at Los Alamos, Dr. Robert Krueger at Iowa State Ames Research Laboratory, and Dr. Kane Yee at Livermore.

Finally, it is rare that one gets to thank in such a permanent, public form those who have made the quality of one's life so high. On this occasion I thank my mother Dorothy for her ideals, devotion, and for having a vision for herself and for me. To my wife Tess goes my deepest gratitude for her support; without her steady encouragement this book may never have been completed.

J. DAVID LOGAN

Lincoln, Nebraska
September 1987.

Suggestions for Use of the Text. The following diagram shows the dependence of the Chapters. The dashed lines indicate only a weak dependence and the earlier material can be referred to only as needed.

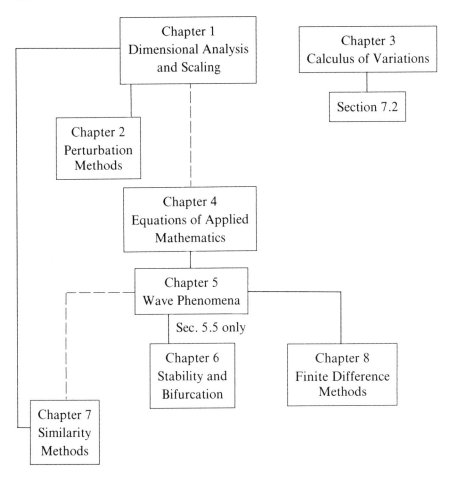

For a one-year course: First Semester (Chapters 1, 2, 3)
Second Semester (Chapters 4, 5, 6)

Chapters 1, 4 (except integral equations), 5, and 8 have been used for a one-semester introduction to partial differential equations, which includes numerical methods; Chapter 7 can be substituted for Chapter 8. If the section on hydrodynamic stability is omitted, Chapter 6 is independent of the remaining parts of the book.

CONTENTS

1. DIMENSIONAL ANALYSIS AND SCALING   **1**

  1.1   Dimensional Analysis / 1
         The Program of Applied Mathematics / *1*
         Dimensional Methods / *3*
         Exercises / *4*

  1.2   The Buckingham Pi Theorem / 5
         Formulation / *5*
         Application to a Diffusion Problem / *7*
         Proof of the Pi Theorem / *10*
         Exercises / *14*

  1.3   Scaling / 16
         Characteristic Scales / *16*
         Heat Conduction / *18*
         The Projectile Problem / *21*
         Scaling Known Functions / *26*
         Exercises / *30*
         References / *33*

2. PERTURBATION METHODS   **34**

  2.1   Regular Perturbation / 34
         The Perturbation Method / *34*
         Motion in a Nonlinear Resistive Medium / *36*

A Nonlinear Oscillator / 39
The Poincaré-Lindstedt Method / 42
Asymptotics / 44
Exercises / 49

2.2 Singular Perturbation / 53
Failure of Regular Perturbation / 53
Inner and Outer Approximations / 54
Algebraic Equations and Balancing / 58
Exercises / 59

2.3 Boundary Layer Analysis / 60
The Inner Approximation / 60
Matching / 62
Uniform Approximations / 63
A Worked Example / 64
Boundary Layer Phenomena / 66
Exercises / 69

2.4 Two Applications / 70
Damped Harmonic Oscillator / 71
A Chemical Kinetics Problem / 76
Exercises / 83
References / 84

3. CALCULUS OF VARIATIONS   **85**

3.1 Variational Problems / 85
Functionals / 85
Examples / 87
Exercises / 90

3.2 Necessary Conditions for Extrema / 90
Normed Linear Spaces / 90
Derivatives of Functionals / 95
Necessary Conditions / 98
Exercises / 99

3.3 The Simplest Problem / 101
The Euler Equation / 101
Solved Examples / 104
First Integrals / 106
Exercises / 108

3.4 Generalizations / 110
Higher Derivatives / 110

Several Functions / 112
Multiple Integral Problems / 114
Natural Boundary Conditions / 118
Exercises / 122

3.5 Hamiltonian Theory / 126
Hamilton's Principle / 126
Hamilton's Principle Versus Newton's Law / 132
The Canonical Formalism / 133
The Inverse Problem / 137
Exercises / 140

3.6 Isoperimetric Problems / 143
Necessary Conditions / 143
Exercises / 148
References / 150

4. EQUATIONS OF APPLIED MATHEMATICS　　　　　　**151**

4.1 Partial Differential Equations / 151
Definitions / 151
Linearity versus Nonlinearity / 155
Superposition / 157
Exercises / 158

4.2 The Diffusion Equation / 159
Heat Conduction / 159
Well-Posed Problems / 162
Diffusion in Higher Dimensions / 166
Exercises / 172

4.3 Classical Techniques / 174
Separation of Variables / 174
Fourier Series / 178
Sturm-Liouville Problems / 182
Integral Transforms / 188
Exercises / 195

4.4 Integral Equations / 199
Classification and Origins / 199
Relationship to Differential Equations / 204
Fredholm Equations / 208
Symmetric Kernels / 212
Volterra Equations / 216
Exercises / 224
References / 229

5. WAVE PHENOMENA IN CONTINUOUS SYSTEMS   230

5.1   Wave Propagation / 230
Waves / 230
Linear Waves / 234
Nonlinear Waves / 237
Burgers' Equation / 242
The Korteweg-deVries Equation / 245
Conservation Laws / 248
Quasi-Linear Equations / 253
Exercises / 260

5.2   Mathematical Models of Continua / 264
Kinematics / 264
Mass Conservation / 270
Momentum Conservation / 272
Thermodynamics and Energy Conservation / 276
The Acoustic Approximation / 282
Stress Waves in Solids / 284
Exercises / 288

5.3   The Wave Equation / 292
D'Alembert's Solution / 292
Scattering and Inverse Problems / 298
Exercises / 301

5.4   Gasdynamics / 305
Conservation Laws / 305
Riemann's Method / 307
The Rankine-Hugoniot Conditions / 313
Exercises / 315

5.5   Fluid Motions in R^3 / 316
Kinematics / 316
Dynamics / 323
Energy / 331
Exercises / 335
References / 340

6. STABILITY AND BIFURCATION   341

6.1   Intuitive Ideas / 342
Stability and Population Dynamics / 342
Bifurcation of a Bead on a Hoop / 346

Stability and Chemotaxis of Amoebae / *351*
Exercises / *357*

6.2   One Dimensional Problems / 357
Stability / *357*
Classification of Bifurcation Points / *364*
Exchange of Stability / *368*
Continuously Stirred Tank Reactor / *373*
Exercises / *377*

6.3   Two Dimensional Problems / 378
Phase Plane Phenomena / *378*
Linear Systems / *385*
Nonlinear Systems / *391*
Bifurcation / *394*
Exercises / *404*

6.4   Hydrodynamic Stability / 406
A Layered Fluid / *406*
Perturbation Equations / *408*
Rayleigh's Example / *410*
Exercises / *414*
References / *415*

7. SIMILARITY METHODS   **416**

7.1   Invariant Variational Problems / 418
Local Lie Groups / *418*
Invariance of Functionals / *423*
The Noether Theorem / *429*
Conservation Laws in Mechanics / *435*
Exercises / *437*

7.2   Invariant Partial Differential Equations / 441
Self-Similar Solutions / *441*
Similarity and Dimensional Analysis / *444*
The Method of Stretchings / *447*
The Lie Plane / *457*
Exercises / *461*

7.3   The General Similarity Method / 464
Local Lie Groups on R^3 / *464*
Invariant Partial Differential Equations / *469*
The Determination of Symmetries / *474*
Self-Similar Solutions / *476*

Exercises / 486
References / 488

**8. DIFFERENCE METHODS FOR PARTIAL
   DIFFERENTIAL EQUATIONS**   **489**

8.1   Finite Difference Methods / 490
      Discretization / 490
      Discrete Approximations for Derivatives / 493
      Exercises / 496

8.2   The Diffusion Equation / 497
      An Explicit Scheme / 497
      Truncation Error and Convergence / 500
      Implicit Scheme / 503
      General Boundary Conditions / 509
      Stability / 510
      Matrix Stability Analysis / 513
      Exercises / 514

8.3   The Laplace Equation / 517
      Finite Difference Approximation / 517
      Iterative Methods for Linear Systems / 521
      The Neumann Problem / 533
      Accelerating Iterative Schemes / 540
      Exercises / 543

8.4   Hyperbolic Equations / 547
      The Wave Equation / 547
      Hyperbolic Systems / 552
      Characteristic Methods / 558
      Conservation Laws / 562
      Exercises / 564
      References / 566

INDEX   **567**

1

DIMENSIONAL ANALYSIS
AND SCALING

The techniques of dimensional analysis and scaling are basic in the theory and practice of mathematical modeling. In every physical setting a good grasp of the possible relationships and comparative magnitudes among the various dimensioned parameters nearly always leads to a better understanding of the problem and sometimes points the way toward approximations and solutions. In this chapter we briefly introduce some of the basic concepts from these two topics. Along with several examples, a statement and proof of the fundamental result in dimensional analysis, the Buckingham Pi theorem, is presented, and scaling is discussed in the context of reducing problems to dimensionless form. The notion of scaling also points the way toward a proper treatment of perturbation methods, especially boundary layer phenomena in singular perturbation theory.

1.1 DIMENSIONAL ANALYSIS

The Program of Applied Mathematics

There are many phases to the solution of a problem that arises in a physical context and that requires careful mathematical analysis. One way to view the attack on such a physical problem is as follows. When a problem arises in empirics, the first stage is to formulate a mathematical model of the situation. This step includes defining the relevant quantities and formulating a set of governing equations that describe the process involved in detail. We can regard the mathematical problem represented by these model equations as a

1

pure mathematics problem. Its solution by some mathematical technique is the second stage of analysis. Once the solution is obtained, the third stage is to go back and verify that the analytical results are consistent with the experimental observations in the original physical problem. If indeed there is consistency, and if the solution is predictive of other similar physical results, then we can conclude that the devised mathematical equations do in fact represent a realistic model.

It would be a limited view, in fact an incorrect one, to believe that applied mathematics consists only of developing techniques and algorithms to solve problems that arise in a physical or applied context. Applied mathematics deals with all these stages, not merely the formal solution as represented in stage two. It is true that an important aspect of applied mathematics consists of studying, investigating, and developing procedures that are useful in solving such mathematical problems: these include analytic and approximation techniques, numerical analysis, and methods for solving differential and integral equations. It is more the case, however, that applied mathematics deals with every phase of the problem. Formulating the model and understanding its origin in empirics are crucial steps. Because there is a constant interplay between the various stages, the scientist, engineer, or mathematician must understand each phase. For example, in the second stage the solution to a problem sometimes involves making approximations that lead to a simplification. The approximations often come from a careful examination of the physical reality, which in turn suggests what terms may be neglected, what quantities (if any) are small, and so on. Finally, inaccurate predictions may suggest refinements in the model that lead to even better descriptions of reality. All of this is the practice of applied mathematics; heuristic reasoning, manipulative skills, and physical insight are all essential elements.

In this chapter our aim is to focus upon the first stage, or modeling process. We carry this out by formulating models for various physical systems while emphasizing the interdependence of mathematics and the physical world. Through study of the modeling process we gain insight into the equations themselves. For example, it is possible to study the diffusion equation, a partial differential equation of the form

$$u_t(x, t) - u_{xx}(x, t) = 0$$

without regard to its origin. We can investigate it mathematically by asking questions regarding the existence of solutions, methods of solution, and so on. Such an endeavor, however, is sterile from the point of view of applied mathematics; the origins and analysis are equally important. Indeed, physical insight forces us toward the right questions and at times leads us to the theorems and their proofs.

In addition to presenting some concrete examples of modeling, we also discuss two techniques that are useful in developing and interpreting the model equations. One technique is dimensional analysis and the other is scaling. The former permits us to understand the dimensional (meaning length, time, mass, etc.) relationships of the quantities in the equations and the resulting implications of dimensional homogeneity. Scaling is a technique that helps us understand the magnitude of the terms that appear in the model equations by comparing the quantities to intrinsic reference quantities that appear naturally in the physical situation.

Dimensional Methods

One of the basic techniques that is useful in the initial or modeling stage of a problem is the analysis of the relevant quantities and how they must relate to each other in a dimensional way. Simply put, apples cannot equal oranges; equations must have a consistency to them that precludes every possible relationship among the variables. Stated still differently, equations must be dimensionally homogeneous. These simple observations form the basis of the subject known as *dimensional analysis*. The methods of dimensional analysis developed over the last century or so have led to important results in determining the nature of physical phenomena, even when the governing equations were not known. This has been especially true in continuum mechanics, out of which the general methods of dimensional analysis evolved.

The cornerstone result in dimensional analysis is known as the *Pi theorem*. Roughly, the Pi theorem states that if there is a physical law that gives a relation among a certain number of physical quantities, then there is an equivalent law that can be expressed as a relation among certain dimensionless quantities (often noted by $\pi_1, \pi_2, \ldots$, and hence the name). The Pi theorem appears to have been first stated by A. Vashy in 1892. Later, in 1914, E. Buckingham gave a proof of the Pi theorem for special cases, and now the theorem often carries his name. Birkhoff [1] can be consulted for a bibliography and history.

To communicate the flavor and power of this classical result, let us consider an example. This calculation was made by G. I. Taylor in the late 1940s to compute the yield of the first atomic explosion after viewing photographs of the spread of the fireball. In such an explosion a large amount of energy e is released in a short time (essentially instantaneously) in a region small enough to be considered a point. From the center of the explosion a strong shock wave spreads outward; the pressure behind it is on the order of hundreds of thousands of atmospheres, far greater than the ambient air pressure whose magnitude can be accordingly neglected in the early stages of the explosion. It is reasonable that there should exist a relation between the radius of the blast wave front r, time t, the initial air density ρ, and the energy released e. Hence we *assume* there is a physical law

$$g(t, r, \rho, e) = 0 \qquad (1)$$

which provides a relationship among these quantities. The Pi theorem of Buckingham states that there is an equivalent physical law between the independent dimensionless quantities that can be formed from t, r, e, and ρ. We note that t has dimensions of time, r has dimensions of length, e has dimensions of mass $\cdot$ length^2 $\cdot$ time^{-2}, and ρ has dimensions of mass $\cdot$ length^{-3}. Hence, the quantity $r^5\rho/t^2e$ is dimensionless, since all of the dimensions cancel out of this quantity (this is easy to check). It is not difficult to observe that no other independent dimensionless quantities can be formed from t, r, e, and ρ. The Pi theorem then guarantees that the physical law (1) is equivalent to a physical law involving only the dimensionless quantities; in this case

$$f\left(\frac{r^5\rho}{t^2e}\right) = 0 \tag{2}$$

since there is only one such quantity. Here, f is some function. From (2) it follows that the physical law must take the form

$$\frac{r^5\rho}{t^2e} = \text{constant}$$

or

$$r = C\left(\frac{et^2}{\rho}\right)^{1/5} \tag{3}$$

where C is a constant. Therefore, just from dimensional reasoning it has been shown that the radius of the wave front depends on the two-fifths power of time. Experiments and photographs of explosions confirm this dependence. The constant C depends upon the ratio of specific heat at constant pressure to the specific heat at constant volume, and it can be determined by invoking the governing equations of gasdynamics. By fitting the curve (3) to experimental data of r versus t, the initial energy yield e can be computed, since C and ρ are known quantities.

   Although the previous calculation is only a simple version of the original argument given by Taylor, we can infer that heuristic dimensional reasoning can give real insight into the nature of a physical process and is an invaluable tool for the applied mathematican, physicist, or engineer.

   There is another aspect of dimensional analysis that is important in engineering, namely the design of small-scale models (say of an airplane or ship) that behave like their real counterparts. Discussions of this important topic are not treated here, but can be found in the references listed at the end of the chapter.

EXERCISES

1.1   In the blast wave problem discussed in this section let us assume instead of (1), a physical law of the form

$$g(t, r, \rho, e, P) = 0 \tag{4}$$

where P is the ambient pressure. By inspection, how many dimensionless parameters can be formed from t, r, ρ, e, and P? (Two) Calling the two dimensionless parameters π_1 and π_2 and assuming (4) is equivalent to

$$f(\pi_1, \pi_2) = 0$$

does it still follow that r varies like the two-fifths power of t?

1.2 The law governing how far an object falls in a field of constant gravitational acceleration g with no air resistance is

$$x = \tfrac{1}{2}gt^2 \tag{5}$$

How many dimensionless quantities can be formed from t, x, and g? Rewrite the physical law (5) in terms of dimensionless quantities $\pi_1, \pi_2, \ldots$. Can the distance a body falls depend on the mass m as well? That is, can there be a physical law of the form $f(t, x, g, m) = 0$?

1.2   THE BUCKINGHAM PI THEOREM

Formulation

As mentioned in the last section, it is generally true that a physical law

$$f(q_1, q_2, q_3, \ldots, q_m) = 0 \tag{1}$$

relating m quantities $q_1, q_2, \ldots, q_m$ is equivalent to a physical law that relates the dimensionless quantities that can be formed from $q_1, q_2, \ldots, q_m$. This is the content of the Pi theorem. Before making a formal statement, however, we formulate some basic ideas.

First, the m quantities $q_1, q_2, \ldots, q_m$, which are like the quantities t, r, ρ, and e in the blast wave example in the preceding section, are dimensioned quantities. This means that they can be expressed in a natural way in terms of certain selected fundamental dimensions $L_1, L_2, \ldots, L_n$ ($n < m$), appropriate to the problem being studied. In the blast wave problem, time T, length L, and mass M can be taken to be the fundamental dimensions, since each quantity t, r, ρ, and e can be expressed in terms of T, L, and M. For example, the dimensions of the energy e are ML^2T^{-2}. In general, the dimensions of q_i, denoted by $[q_i]$, can be written in terms of the fundamental dimensions as

$$[q_i] = L_1^{a_{1i}} L_2^{a_{2i}} \cdots L_n^{a_{ni}} \tag{2}$$

for some choice of exponents $a_{1i}, \ldots, a_{ni}$. If $[q_i] = 1$, then q_i is said to be *dimensionless*. The $n \times m$ matrix

$$A = \begin{bmatrix} a_{11} & \cdots & a_{1m} \\ a_{21} & \cdots & a_{2m} \\ \vdots & & \vdots \\ a_{n1} & \cdots & a_{nm} \end{bmatrix}$$

is called the *dimension matrix*. The elements in the ith column give the exponents for q_i in terms of the powers of $L_1, \ldots, L_n$.

The fundamental assumption regarding the physical law (1) goes back to the simple statement that apples cannot equal oranges. We assume that (1) is *unit free* in the sense that it is independent of the particular set of units chosen to express the quantities $q_1, q_2, \ldots, q_m$. We are distinguishing the word *unit* from the word *dimension*. By units we mean specific physical units like seconds, hours, days, and years; all of these units have dimensions of time. Similarly, grams, kilograms, slugs, etc., are units of the dimension mass. Any fundamental dimension L_i has the property that its units can be changed upon multiplication by the appropriate conversion factor $\lambda_i > 0$ to obtain $\bar{L}_i$ in a new system of units. We write

$$\bar{L}_i = \lambda_i L_i, \qquad i = 1, \ldots, n$$

The units of derived quantities q can be changed in a similar fashion. If

$$[q] = L_1^{b_1} L_2^{b_2} \cdots L_n^{b_n}$$

then

$$\bar{q} = \lambda_1^{b_1} \lambda_2^{b_2} \cdots \lambda_n^{b_n} q$$

gives q in the new system of units. The physical law (1) is said to be independent of the units chosen to express the dimensional quantities $q_1, q_2, \ldots, q_m$, or unit free, if $f(q_1, \ldots, q_m) = 0$ and $f(\bar{q}_1, \ldots, \bar{q}_m) = 0$ are equivalent physical laws. More formally:

Definition 2.1   *The physical law (1) is unit free if for all choices of real numbers* $\lambda_1, \ldots, \lambda_n$, *with* $\lambda_i > 0$, $i = 1, \ldots, n$, *we have* $f(\bar{q}_1, \ldots, \bar{q}_m) = 0$ *if, and only if,* $f(q_1, \ldots, q_m) = 0$.

Example 2.1   The physical law

$$f(x, t, g) \equiv x - \tfrac{1}{2}gt^2 = 0 \tag{3}$$

relates the distance x a body falls in a constant gravitational field g to the time t. In a certain system of units, let x be given in centimeters (cm), t in seconds, and g in cm/sec^2. If we change units for the fundamental quantities x and t to inches and minutes, then in the new system of units

$$\bar{x} = \lambda_1 x \quad \text{and} \quad \bar{t} = \lambda_2 t$$

where $\lambda_1 = \frac{1}{2.54}$ (in./cm) and $\lambda_2 = \frac{1}{60}$ (min/sec). Since $g = $ (length)(time)$^{-2}$, we have

$$\bar{g} = \lambda_1 \lambda_2^{-2} g$$

Then

$$f(\bar{x}, \bar{t}, \bar{g}) = \bar{x} - \tfrac{1}{2}\bar{g}\bar{t}^2 = \lambda_1 x - \tfrac{1}{2}\left(\lambda_1 \lambda_2^{-2} g\right)\left(\lambda_2 t\right)^2 = \lambda_1\left(x - \tfrac{1}{2}gt^2\right)$$

Hence (3) is unit free.

We are now able to give a precise statement of the Pi theorem.

Theorem 2.1   (*Pi Theorem*)   *Let*

$$f(q_1, q_2, \ldots, q_m) = 0 \tag{4}$$

be a unit free physical law that relates the dimensional quantities $q_1, q_2, \ldots, q_m$. Let $L_1, \ldots, L_n$ ($n < m$) be fundamental dimensions with

$$[q_i] = L_1^{a_{1i}} L_2^{a_{2i}} \cdots L_n^{a_{ni}}, \qquad i = 1, \ldots, m$$

and let $r = \text{rank } A$, where A is the dimension matrix (2). Then there exists $m - r$ independent dimensionless quantities $\pi_1, \pi_2, \ldots, \pi_{m-r}$, which can be formed from $q_1, \ldots, q_m$, and the physical law (4) is equivalent to an equation

$$F(\pi_1, \pi_2, \ldots, \pi_{m-r}) = 0 \tag{5}$$

expressed only in terms of the dimensionless quantities.

The proof of the Pi theorem is presented later. For the present we analyze an example. Before continuing, however, we note that the existence of a physical law (4) is an assumption. In practice one must conjecture which are the relevant variables in a problem and then apply the machinery of the theorem. The resulting dimensionless physical law (5) must be checked by experiment, or whatever, in an effort to determine the validity of the original assumptions.

Application to a Diffusion Problem

Example 2.2   At time $t = 0$ an amount of heat energy e, concentrated at a point in space, is allowed to diffuse outward into a region with temperature

the region has temperature zero.

zero. If r denotes the radial distance from the source and t is time, the problem is to determine the temperature u as a function of r and t. Of course, this problem can be formulated as a boundary value problem for a partial differential equation (the heat or diffusion equation), but let us see what can be learned from a careful dimensional analysis of the situation. The first step is to make an assumption regarding which quantities affect the temperature u. Clearly t, r, and e are relevant quantities. It also seems reasonable that the heat capacity c, with dimensions of energy per degree per volume, of the region will play a role, as well as the rate at which heat diffuses outward. The latter is characterized by the thermal diffusivity k of dimensions length-squared per unit time; k is the thermal conductivity divided by the heat capacity or the amount of heat energy flowing across a unit length per unit time at a given temperature, *per* unit of heat capacity. Therefore, we conjecture a physical law of the form

$$f(t, r, u, e, k, c) = 0$$

which relates the six quantities, t, r, u, e, k, c. The next step is to determine independent fundamental dimensions $L_1, \ldots, L_n$ by which the six dimensional quantities can be expressed. A suitable selection would be the four quantities T (time), L (length), Θ (temperature), and E (energy). Then

$$[t] = T \qquad [e] = E$$
$$[r] = L \qquad [k] = L^2 T^{-1}$$
$$[u] = \Theta \qquad [c] = E\Theta^{-1}L^{-3}$$

and the dimension matrix A is given by

$$
\begin{array}{c c c c c c c}
 & t & r & u & e & k & c \\
T & 1 & 0 & 0 & 0 & -1 & 0 \\
L & 0 & 1 & 0 & 0 & 2 & -3 \\
\Theta & 0 & 0 & 1 & 0 & 0 & -1 \\
E & 0 & 0 & 0 & 1 & 0 & 1
\end{array}
$$

Here $m = 6$, $n = 4$, and the rank of the dimension matrix is clearly $r = 4$. Consequently there are $m - r = 2$ dimensionless quantities that can be formed from t, r, u, e, k, and c. To find them we proceed as follows: If π is dimensionless, then for some choice of $\alpha_1, \ldots, \alpha_6$

$$1 = [\pi] = [t^{\alpha_1} r^{\alpha_2} u^{\alpha_3} e^{\alpha_4} k^{\alpha_5} c^{\alpha_6}]$$
$$= T^{\alpha_1} L^{\alpha_2} \Theta^{\alpha_3} E^{\alpha_4} (L^2 T^{-1})^{\alpha_5} (E\Theta^{-1}L^{-3})^{\alpha_6}$$
$$= T^{\alpha_1 - \alpha_5} L^{\alpha_2 + 2\alpha_5 - 3\alpha_6} \Theta^{\alpha_3 - \alpha_6} E^{\alpha_4 + \alpha_6}$$

Therefore the exponents must vanish and we obtain four homogeneous linear equations for $\alpha_1, \ldots, \alpha_6$, namely

$$\alpha_1 - \alpha_5 = 0$$
$$\alpha_2 + 2\alpha_5 - 3\alpha_6 = 0$$
$$\alpha_3 - \alpha_6 = 0$$
$$\alpha_4 + \alpha_6 = 0$$

The coefficient matrix of this homogeneous linear system is just the dimension matrix A. From elementary matrix theory the number of independent solutions equals the number of unknowns minus the rank of A. Each independent solution will give rise to a dimensionless variable. Now the method unfolds and we can see the origin of the rank condition in the statement of the Pi theorem.

By standard methods for solving linear systems we find that two linearly independent solutions are

$$\alpha_1 = -\tfrac{1}{2}, \qquad \alpha_2 = 1, \qquad \alpha_3 = \alpha_4 = 0, \qquad \alpha_5 = -\tfrac{1}{2}, \qquad \alpha_6 = 0$$

and

$$\alpha_1 = \tfrac{3}{2}, \qquad \alpha_2 = 0, \qquad \alpha_3 = 1, \qquad \alpha_4 = -1, \qquad \alpha_5 = \tfrac{3}{2}, \qquad \alpha_6 = 1$$

These give rise to the two dimensionless quantities

$$\pi_1 = \frac{r}{\sqrt{kt}}$$

and

$$\pi_2 = \frac{uc}{e}(kt)^{3/2}$$

Therefore the Pi theorem guarantees that the original physical law $f(t, r, u, e, k, c) = 0$ is equivalent to a physical law of the form

$$F(\pi_1, \pi_2) = 0$$

Solving for π_2 we get

$$\pi_2 = g(\pi_1)$$

for some g, or

$$u = \frac{e}{c}(kt)^{-3/2}g\left(\frac{r}{\sqrt{kt}}\right) \tag{6}$$

Again, without solving any partial differential equation governing diffusion processes we have been able via dimensional analysis to argue that the temperature of the region varies according to (6). For example, the temperature near the source $r = 0$ falls off like $t^{-3/2}$.

Proof of the Pi Theorem

To prove the Pi theorem we must demonstrate two propositions.

(i) Among the quantities $q_1, \ldots, q_m$ there are $m - r$ independent dimensionless variables that can be formed, where r is the rank of the dimension matrix A.

(ii) If $\pi_1, \ldots, \pi_{m-r}$ are the $m - r$ dimensionless variables, then (4) is equivalent to a physical law of the form $F(\pi_1, \ldots, \pi_{m-r}) = 0$.

$$(4) \quad f(q_1, q_2; \cdots; q_m) = 0$$

The proof of (i) is straightforward; the general argument proceeds exactly like the construction of the dimensionless variables in Example 2.2. It makes use of a familiar result from linear algebra, namely that the number of linearly independent solutions of a set of n homogeneous equations in m unknowns is $m - r$, where r is the rank of the coefficient matrix. For, let π be a dimensionless quantity. Then

$$\pi = q_1^{\alpha_1} q_2^{\alpha_2} \cdots q_m^{\alpha_m} \tag{7}$$

for some $\alpha_1, \alpha_2, \ldots, \alpha_m$. In terms of the fundamental dimensions $L_1, \ldots, L_n$,

$$\pi = \left(L_1^{a_{11}} L_2^{a_{21}} \cdots L_n^{a_{n1}}\right)^{\alpha_1} \left(L_1^{a_{12}} L_2^{a_{22}} \cdots L_n^{a_{n2}}\right)^{\alpha_2} \cdots \left(L_1^{a_{1m}} L_2^{a_{2m}} \cdots L_n^{a_{nm}}\right)^{\alpha_m}$$

$$= L_1^{a_{11}\alpha_1 + a_{12}\alpha_2 + \cdots + a_{1m}\alpha_m} \cdots L_n^{a_{n1}\alpha_1 + a_{n2}\alpha_2 + \cdots + a_{nm}\alpha_m}$$

Since $[\pi] = 1$, the exponents vanish, or

$$a_{11}\alpha_1 + a_{12}\alpha_2 + \cdots + a_{1m}\alpha_m = 0$$

$$\vdots \tag{8}$$

$$a_{n1}\alpha_1 + a_{n2}\alpha_2 + \cdots + a_{nm}\alpha_m = 0$$

By the aforementioned theorem in linear algebra the system (8) has exactly $m - r$ independent solutions $[\alpha_1, \ldots, \alpha_m]$. Each solution gives rise to a dimensionless variable via (7), and this completes the proof of (i). The independence of the dimensionless variables is in the sense of linear algebraic independence.

The proof of (ii) makes strong use of the hypothesis that the law is unit free. The argument is subtle but it can be made almost transparent if we examine a particular example.

Example 2.3   Consider the unit free law

$$f(x, t, g) = x - \tfrac{1}{2} g t^2 = 0 \tag{9}$$

from Example 2.1. If length and time are chosen as fundamental dimensions a straightforward calculation shows there is a single dimensionless variable given by

$$\pi_1 = t^2 \frac{g}{x}$$

The remaining variable g can be expressed as $g = x\pi_1/t^2$, and we can define a function G by

$$G(x, t, \pi_1) \equiv f\left(x, t, x\pi_1/t^2\right)$$

Clearly the law $G(x, t, \pi_1) = 0$ or

$$x - \frac{1}{2}\left(\frac{x\pi_1}{t^2}\right)t^2 = 0 \tag{10}$$

is equivalent to (9) and is unit free since f is. Then F is defined by

$$F(\pi_1) \equiv G(1, 1, \pi_1) = 1 - \tfrac{1}{2}\pi_1 = 0$$

which is obviously equivalent to (10) and (9).

Rather than present a general proof of (ii), we present a special case when $m = 4$, $n = 2$, and $r = 2$. The notation will be easier, and the general argument for arbitrary m, n, and r proceeds in exactly the same manner. Therefore we consider a unit-free physical law

$$f(q_1, q_2, q_3, q_4) = 0 \tag{11}$$

with

$$[q_j] = L_1^{a_{1j}} L_2^{a_{2j}}, \qquad j = 1, \dots, 4$$

where L_1 and L_2 are fundamental dimensions. The dimension matrix

$$A = \begin{pmatrix} a_{11} & a_{12} & a_{13} & a_{14} \\ a_{21} & a_{22} & a_{23} & a_{24} \end{pmatrix}$$

is assumed to have rank $r = 2$. If π is a dimensionless quantity, then

$$\pi = q_1^{\alpha_1} q_2^{\alpha_2} q_3^{\alpha_3} q_4^{\alpha_4} \tag{12}$$

where the exponents $\alpha_1, \alpha_2, \alpha_3, \alpha_4$ satisfy the homogeneous system (8), which in this case is, written in vector form,

$$\begin{pmatrix} a_{11} \\ a_{21} \end{pmatrix}\alpha_1 + \begin{pmatrix} a_{12} \\ a_{22} \end{pmatrix}\alpha_2 + \begin{pmatrix} a_{13} \\ a_{23} \end{pmatrix}\alpha_3 + \begin{pmatrix} a_{14} \\ a_{24} \end{pmatrix}\alpha_4 = \begin{pmatrix} 0 \\ 0 \end{pmatrix} \tag{13}$$

We wish to determine α_1, α_2, α_3, and α_4, and the form of the two dimensionless variables. Without loss of any generality, we can assume the first two columns of A are linearly independent. This is because we can rearrange the indices on the q_j so that the two independent columns appear as the first two. Then columns three and four can be written as linear combinations of the first two, or

$$
\begin{pmatrix} a_{13} \\ a_{23} \end{pmatrix} = c_{31} \begin{pmatrix} a_{11} \\ a_{21} \end{pmatrix} + c_{32} \begin{pmatrix} a_{12} \\ a_{22} \end{pmatrix}
$$
$$
\begin{pmatrix} a_{14} \\ a_{24} \end{pmatrix} = c_{41} \begin{pmatrix} a_{11} \\ a_{21} \end{pmatrix} + c_{42} \begin{pmatrix} a_{12} \\ a_{22} \end{pmatrix}
$$

(14)

for some constants c_{31}, c_{32}, c_{41}, and c_{42}. Substituting (14) into (13) gives

$$
(\alpha_1 + c_{31}\alpha_3 + c_{41}\alpha_4) \begin{pmatrix} a_{11} \\ a_{21} \end{pmatrix} + (\alpha_2 + c_{32}\alpha_3 + c_{42}\alpha_4) \begin{pmatrix} a_{12} \\ a_{22} \end{pmatrix} = \begin{pmatrix} 0 \\ 0 \end{pmatrix}
$$

This is a zero linear combination of linearly independent vectors, and therefore the coefficients must vanish

$$
\alpha_1 + c_{31}\alpha_3 + c_{41}\alpha_4 = 0
$$
$$
\alpha_2 + c_{32}\alpha_3 + c_{42}\alpha_4 = 0
$$

Hence we can solve for α_1 and α_2 in terms of α_3 and α_4 and we can write

$$
\begin{pmatrix} \alpha_1 \\ \alpha_2 \\ \alpha_3 \\ \alpha_4 \end{pmatrix} = \alpha_3 \begin{pmatrix} -c_{31} \\ -c_{32} \\ 1 \\ 0 \end{pmatrix} + \alpha_4 \begin{pmatrix} -c_{41} \\ -c_{42} \\ 0 \\ 1 \end{pmatrix}
$$

The two vectors on the right represent two linearly independent solutions of (13); hence the two dimensionless quantities are

$$
\pi_1 = q_1^{-c_{31}} q_2^{-c_{32}} q_3
$$
$$
\pi_2 = q_1^{-c_{41}} q_2^{-c_{42}} q_4
$$

Now define a function G by

$$
G(q_1, q_2, \pi_1, \pi_2) \equiv f(q_1, q_2, \pi_1 q_1^{c_{31}} q_2^{c_{32}}, \pi_2 q_1^{c_{41}} q_2^{c_{42}}).
$$

The physical law

$$
G(q_1, q_2, \pi_1, \pi_2) = 0
$$

(15)

holds if, and only if, (11) holds, and therefore (15) is an equivalent physical law. Since $f = 0$ is unit free, it easily follows that (15) is unit free (note that $\bar{\pi}_1 = \pi_1$, $\bar{\pi}_2 = \pi_2$ under any change of units; i.e., dimensionless variables have the same value in all systems of units). For the final stage of the argument we show that (15) is equivalent to the physical law

$$G(1, 1, \pi_1, \pi_2) = 0 \tag{16}$$

which will give the result we are seeking, for (16) implies $F(\pi_1, \pi_2) = 0$, where $F(\pi_1, \pi_2) \equiv G(1, 1, \pi_1, \pi_2)$. Since (15) is a unit free law, we must have

$$G(\bar{q}_1, \bar{q}_2, \pi_1, \pi_2) = 0 \tag{17}$$

where

$$\bar{q}_1 = \lambda_1^{a_{11}} \lambda_2^{a_{21}} q_1, \qquad \bar{q}_2 = \lambda_1^{a_{12}} \lambda_2^{a_{22}} q_2$$

for *every* choice of the conversion factors $\lambda_1, \lambda_2 > 0$. We select λ_1 and λ_2 so that $\bar{q}_1 = \bar{q}_2 = 1$. We are able to make this choice because

$$\lambda_1^{a_{11}} \lambda_2^{a_{21}} q_1 = 1, \qquad \lambda_1^{a_{12}} \lambda_2^{a_{22}} q_2 = 1 \tag{18}$$

implies

$$\begin{aligned} a_{11} \ln \lambda_1 + a_{21} \ln \lambda_2 &= -\ln q_1 \\ a_{12} \ln \lambda_1 + a_{22} \ln \lambda_2 &= -\ln q_2 \end{aligned} \tag{19}$$

And, since the coefficient matrix

$$\begin{pmatrix} a_{11} & a_{21} \\ a_{12} & a_{22} \end{pmatrix}$$

in (19) is nonsingular (recall the assumption that the first two columns of the dimension matrix A are linearly independent), the system (19) has a unique solution $(\ln \lambda_1, \ln \lambda_2)$ from which λ_1 and λ_2 can be determined to satisfy (18). Thus (16) is an equivalent physical law and the argument is complete.

The general argument for arbitrary m, n, and r can be found in Birkhoff [1], from which the last proof was adapted. This classic book also provides additional examples and many comments of a historical natural.

In performing a dimensional analysis on a given problem, two judgements are required at the beginning.

1. The selection of the pertinent variables.
2. The choice of the fundamental dimensions.

The first is a matter of experience and may be based on intuition or experiments. There is, of course, no guarantee that the selection will lead to a useful formula after the procedure is applied. Second, the choice of fundamental dimensions may involve tacit assumptions that may not be valid in a given problem. For example, including mass, length, and time but not force in a given problem assumes there is some relation (Newton's second law) that plays an important role and causes force not to be an independent dimension. As a specific example, a small sphere falling under gravity in a viscous fluid is observed to fall, after a short time, at constant velocity. Since the motion is unaccelerated we need not make use of the proportionality of force to acceleration, and so force can be treated as a separate, independent fundamental dimension. In summary, intuition and experience are important ingredients in applying the dimensional analysis formalism to a specific physical problem.

EXERCISES

2.1   The speed v of a wave in deep water is determined by its wavelength Λ and the acceleration g due to gravity. What does dimensional analysis imply regarding the relationship between v, Λ, and g?

2.2   A small sphere of radius r and density ρ is falling at constant velocity v under the influence of gravity g in a liquid of density ρ_l and viscosity μ (μ is given in mass per length per unit time). It is observed experimentally that

$$v = \frac{2}{9} r^2 \rho g \mu^{-1} \left(1 - \frac{\rho_l}{\rho}\right)$$

Is this law unit free?

2.3   A physical phenomenon is described by the quantities P, l, m, t, and ρ, representing pressure, length, mass, time, and density, respectively. If there is a physical law

$$f(P, l, m, t, \rho) = 0$$

relating these quantities, show that there is an equivalent physical law of the form $G(l^3\rho/m, t^6 P^3/m^2\rho) = 0$.

2.4   A physical system is described by a law $f(E, P, A) = 0$, where E, P, and A are energy, pressure, and area, respectively. Find the equivalent physical law in terms of dimensionless variables.
Answer: $PA^{3/2}/E$ = constant.

2.5   A thin rectangular *flyer* of mass per unit volume m_f is placed upon a block of explosive of mass per unit volume m_e that is backed by a tamper of essentially infinite mass. When the explosive detonates, the flyer is driven off vertically upward at velocity v_f. If E_g is the Gurney energy (joules/kg) of the explosive, that is, the energy available in the explosive to do work, determine as far as possible, using dimensional analysis methods, the velocity of the flyer in terms of m_e, m_f, and E_g. The correct relation is

$$v_f = \sqrt{2E_g} \left(\frac{m_f}{m_e} + \frac{1}{3} \right)^{-1/2}$$

2.6   Using dimensional analysis methods prove Pythagoras's theorem, which states that in a right triangle the sum of the squares of the legs equals the square of the hypotenuse. *Hint:* The area A of the right triangle is determined by its hypotenuse c and the smallest acute angle ϕ. Apply the same principle to the two similar right triangles found by dropping a perpendicular to the hypotenuse c.

2.7   A pendulum executing small vibrations has period T, length l, and m is the mass of the bob. Can T depend only on l and m? If we assume T depends on l, m, and the acceleration g due to gravity, then show that $T = $ constant $\cdot (L/g)^{1/2}$.

2.8   Suppose one wishes to determine the power P that must be applied to keep a ship of length l moving at a constant speed V. If it is the case, as seems reasonable, that P depends on the density of water ρ, the acceleration due to gravity g, and the viscosity of water ν (in length squared per unit time), as well as l and V, then show that

$$\frac{P}{\rho l^2 V^3} = f(\mathrm{Fr}, \mathrm{Re})$$

where Fr is the Froude number and Re is the Reynolds number defined by

$$\mathrm{Fr} \equiv \frac{V}{\sqrt{lg}}, \qquad \mathrm{Re} \equiv \frac{Vl}{\nu}$$

2.9   A spherical gas bubble with ratio of specific heats γ is surrounded by an infinite sea of liquid of density ρ_l. The bubble oscillates with growth and contraction periodically with small amplitude at a well-defined frequency ω. Assuming a physical law

$$f(P, R, \rho_l, \omega, \gamma) = 0$$

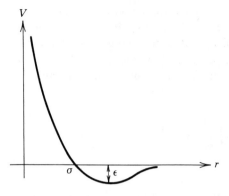

Figure 1.1. Lennard-Jones potential.

where P is the mean pressure inside the bubble and R is the mean radius, show that the frequency must vary inversely with the mean radius R. The constant γ is dimensionless.

2.10 The Lennard-Jones potential is defined by

$$V(r) = 4\varepsilon\left[\left(\frac{\sigma}{r}\right)^{12} - \left(\frac{\sigma}{r}\right)^{6}\right]$$

It represents for suitable choice of ε and σ the approximate potential energy between two molecules separated by a distance r (see Fig. 1.1). In general it is known that the viscosity μ (mass per length per time) of a gas depends only on the temperature and molecular properties. Show that it is not possible, however, for

$$\mu = f(m, \varepsilon, \sigma, T)$$

for some function f, where m is the molecular mass and T is temperature. On the other hand, by including Boltzmann's constant k (1.38 $\times$ 10^{-6} erg/deg) show that

$$\frac{\mu\sigma^2}{\sqrt{m\varepsilon}} = f\left(\frac{kT}{\varepsilon}\right)$$

for some function f.

1.3 SCALING

Characteristic Scales

Another procedure useful in formulating a mathematical model of a physical situation is that of *scaling*. Roughly, scaling means selecting new, usually

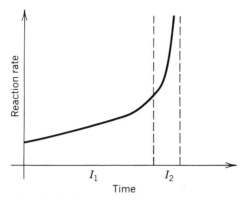

Figure 1.2. Physical process with a multiple time scale.

dimensionless variables and reformulating the problem in terms of those variables. Not only is the procedure useful but it often is a necessity, especially when comparisons of the magnitudes of various terms in an equation must be made in order to neglect small terms, for example. These ideas are of particular importance in the application of perturbation methods to a problem to identify small and large parameters.

For motivation let us suppose that time t is a variable in a given problem, measured in units of seconds. If the problem involved the motion of a glacier, clearly the unit of seconds is too fast because significant changes in the glacier could not be observed on the order of seconds. On the other hand, if the problem involved a nuclear reaction then the unit of seconds is too slow; all of the important action would be over before the first second ticked. Evidently, every problem has an *intrinsic time scale*, or *characteristic time* t_c, which is appropriate to the given problem. This is the shortest time for discernible changes to be observed in the physical quantities. For example, the characteristic time for glacier motion would be of the order of years, whereas the characteristic time for a nuclear reaction would be of the order of microseconds. Some problems may have multiple time scales; a chemical reaction, for example, may begin slowly and then rapidly go to completion. Figure 1.2 depicts such a reaction. Obviously, a time scale appropriate for the time interval I_1 would not be appropriate for the interval I_2 where rapid changes are occurring. This basic observation is at the heart of singular perturbation or boundary layer theory.

Once the characteristic time is known, then a new dimensionless variable $\bar{t}$ can be defined by

$$\bar{t} = \frac{t}{t_c}$$

If t_c is chosen correctly, then the *dimensionless time* $\bar{t}$ is neither too large nor too small, but rather of order unity. The question remains to determine the time scale t_c for a particular problem. The same question applies to lengths or to any other variables in the problem. The general rule is that the characteristic quantities are formed by taking combinations of the various dimensional constants in the problem and should be roughly the same order of magnitude of the quantity itself. This general principle is illustrated by the following examples.

Heat Conduction

We examine the problem of heat flow in a rod of finite length with known physical parameters. Using an energy balance law we obtain a partial differential equation whose solution gives the temperature $u(x, t)$ at any location x in the rod at any time t. After formulating the governing equation, we select characteristic scales for the dependent and independent variables in the problem and reduce it to a form in which all of the variables are dimensionless. Scaling and reduction to dimensionless form are standard procedures that should be followed in modeling most physical problems.

Example 3.1   Consider heat condition in a homogeneous rod extending from $x = 0$ to $x = l$ with constant cross-sectional area A (see Fig. 1.3). We assume that the rod is insulated laterally, so that heat flows only in the x direction and that the temperature is constant in any cross section. Initially the bar is at zero degrees, and both ends are held at a constant T_0 degrees for $t > 0$. The quantity $u(x, t)$ that we wish to determine denotes the temperature in the rod at position x and at the instant t of time. We seek an equation governing $u(x, t)$. To this end let c_v denote the *specific heat at constant volume* of the material from which the rod is composed; that is, c_v is the amount of heat energy required to raise one unit mass of the material one degree. In *cgs* units c_v is measured in calories/(gram · degree C). Then the amount of heat in a segment of the rod from x to $x + \Delta x$ is approximately given by

$(\rho A \Delta x)$ is the mass of the segment

$$c_v \rho u(\xi, t) A \Delta x$$

where ρ is the density and $x \le \xi \le x + \Delta x$. If $\phi(x, t)$ denotes the *heat flux*, or the amount of heat energy per unit time flowing through the face at x, then

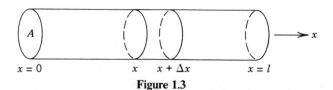

$$x = 0 \qquad x \quad x + \Delta x \qquad x = l$$

Figure 1.3

we may write an energy balance equation for the segment $[x, x + \Delta x]$ as

$$\frac{\partial}{\partial t}(c_v\rho u(\xi, t) A \Delta x) = \phi(x, t) - \phi(x + \Delta x, t) \tag{1}$$

In words this equation states that the time rate of change of energy in the segment must equal the rate at which heat flows in at x minus the rate at which heat flows out at $x + \Delta x$. By convention, the heat flux is positive when heat is flowing to the right. Dividing by Δx in the last equation and then taking the limit as $\Delta x \to 0$ gives the partial differential equation

$$c_v\rho A u_t(x, t) = -\phi_x(x, t) \tag{2}$$

which is a local differential form of the energy balance equation (1). Here it has been noted that $\xi \to x$ as $\Delta x \to 0$.

There are two unknowns in (2), both u and ϕ. Up to this point no assumption has been made concerning the rate at which heat flows. Clearly the rate is material dependent; for example, metals conduct heat faster than ceramics. Therefore we introduce a *constitutive relation*, which for heat flow is an assumption regarding the conductivity of the material. This relation, in its simplest form, states that flux is proportional to the cross-sectional area A and the temperature gradient u_x, the latter measuring the steepness of the temperature profile in the bar at a given instant. In symbols,

$$\phi(x, t) = -KAu_x(x, t) \tag{3}$$

where the proportionality constant K is the *thermal conductivity*, or the amount of heat per unit time flowing across a unit length, per unit degree. In cgs units K is measured in calories/(cm · sec · degree C), and it depends upon the material. Equation (3) is known as *Fourier's heat law*, and it expresses experimental observations. Therefore it is a different type of equation from (2), the latter expressing a fundamental law of nature, namely conservation of energy.

Substituting (3) into (2) gives a single partial differential equation for the temperature $u(x, t)$

$$c_v\rho u_t(x, t) = Ku_{xx}(x, t)$$

If we introduce the *thermal diffusivity* k defined by

$$k = \frac{K}{c_v\rho}, \quad \text{in} \quad \frac{\text{length}^2}{\text{time}}$$

then the partial differential equation can be written

$$u_t(x, t) - ku_{xx}(x, t) = 0 \tag{4}$$

which is known as the *heat* or *diffusion equation*. It holds for $t > 0$ and $0 < x < l$.

From the other stated conditions it is clear that

$$u(x,0) = 0 \qquad \text{for} \quad 0 < x < l \tag{5}$$

and

$$u(0, t) = u(l, t) = T_0 \qquad \text{for} \quad t > 0 \tag{6}$$

The differential equation (4) along with the two supplementary conditions (5) and (6) are known as a *boundary value problem* for the unknown temperature distribution $u(x, t)$. Condition (5) is an *initial condition*, since it holds at $t = 0$; conditions (6) are *boundary conditions*, since they hold at the boundaries of the rod for all $t > 0$. This boundary value problem is a mathematical model of heat flow in a cylindrical rod, and as for any model, it must be tested against experimental observations. In Chapter 4 we discuss some of its deficiencies and its extensions.

We now scale this mathematical problem by selecting characteristic values for the dependent and independent variables. These include the characteristic time t_c, characteristic length l_c, and characterisitic temperature θ_c. The analysis showed that the constants that occur in the model are l, T_0, and k. It seems obvious that $l_c = l$ and $\theta_c = T_0$; that is, length x should be measured relative to the length of the rod and temperature should be measured relative to the temperature at which the ends of the bar are maintained. But what about time? Since the only combination of the constants l, T_0, and k that has time units is l^2/k, it seems reasonable to select

$$t_c = \frac{l^2}{k}$$

Consequently, we define dimensionless independent and dependent variables by

$$\bar{x} = \frac{x}{l}, \qquad \bar{t} = \frac{t}{l^2/k}, \qquad \bar{u} = \frac{u}{T_0} \tag{7}$$

Reformulating the boundary value problem (4), (5), and (6) in terms of these scaled variables easily gives the *scaled problem*

$$\begin{aligned}
\bar{u}_{\bar{t}}(\bar{x}, \bar{t}) - \bar{u}_{\bar{x}\bar{x}}(\bar{x}, \bar{t}) = 0, & \qquad \bar{t} > 0, \quad 0 < \bar{x} < 1 \\
\bar{u}(\bar{x},0) = 0, & \qquad 0 < \bar{x} < 1 \\
\bar{u}(0, \bar{t}) = \bar{u}(1, \bar{t}) = 1, & \qquad \bar{t} > 0
\end{aligned}$$

This process is known as *reducing a problem to dimensionless form*. It involves

TABLE 1.1. Thermal Parameters for Various Materials

Material	ρ (g/cm^3)	c_v (cal/g · deg)	K (cal/cm · deg · sec)	k (cm^2/sec)
Copper	8.9	0.093	1.09	1.32
Cast iron	7.4	0.136	0.12	0.119
Granite	2.6	0.210	0.006	0.011
Glass	2.5	0.198	0.002	0.004
Wood	0.41	0.30	0.0006	0.004
Lucite	1.18	0.35	0.0006	0.001
Cork	0.15	0.48	0.0001	0.001

selecting new, dimensionless independent and dependent variables and then changing variables in the equations via transformations like (7). The latter involves application of the chain rule for partial differentiation (see Exercises 3.2 and 3.3). One advantage in working with the dimensionless model is that once the problem is solved for $\bar{u}(\bar{x}, \bar{t})$, then the solution is known for any choice of the physical constants l, T_0, and k by merely using the conditions (7).

We conclude with some heuristic remarks of a physical nature concerning the diffusivity k that occurs in the heat equation. In heat conduction problems it is the diffusivity rather than the thermal conductivity K that is important, since k depends upon the specific heat c_v of the material. The process of conduction involves getting heat in and out of the various bodies, and thus depends intimately upon c_v. Table 1.1 gives some of the parameters associated with various materials. Roughly, in decreasing order of diffusivity, materials can be ordered as follows: good metals, poor metals, crystalline materials, amorphous materials, and plastics. The computation of t_c, the characteristic time, for a given problem using the formula $t_c = l^2/k$ will yield a practical time scale for the problem and give some notion of when discernible changes will occur. For example, if the heat flow in a copper rod of length 2.54 cm is to be investigated, then the characteristic time for the problem is $t_c = (2.54)^2/1.32 = 4.88$ sec. Hence it would be senseless to make observations on the order of an hour or develop a numerical procedure that takes time steps every minute, for example.

The Projectile Problem

The following example, as first pointed out by Lin and Segel [2], is a good illustration of the importance of choosing correct scales in a problem, particularly when it is desired to make a simplification by neglecting small quantities. Terms in an equation that appear small are not always as they seem, and proper scaling is essential in determining the orders of magnitude of the terms.

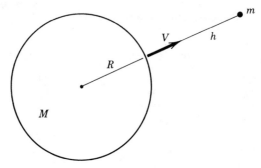

Figure 1.4

Example 3.2 (The Projectile Problem) In this example the problem of describing the motion of a projectile thrust vertically upward from the surface of the earth is analyzed in detail. The steps are similar to those taken in the preceding example, namely, the development of a mathematical model and then the determination of appropriate scales that are used to cast the problem in dimensionless form.

At time $t = 0$ on the surface of the earth, with radius R and mass M, an object of mass m is given a vertical upward velocity of magnitude V. To be determined is the height h above the earth's surface that the mass reached at time t (see Fig. 1.4). The forces on the object are the force due to gravity and the force due to air resistance. In an initial effort to formulate a governing equation we assume the force due to air resistance is negligible in the particular physical situation we are considering. In general, as a first approximation it is common to neglect what are believed to be small effects, since in that case the equations are more tractable for analysis. Should the analytic results compare unfavorably with experiment or should a more detailed description be required, then additional effects can be included.

In the present problem the governing equation or mathematical model comes from a physical law, Newton's universal gravitational law, which states that the force between the two objects is proportional to the product of the masses and inversely proportional to the square of the distance between them, where the mass of each object can be regarded as concentrated at its center. Therefore, since by Newton's second law the force on m must equal the mass of m times its acceleration,

$$m \frac{d^2 h}{dt^2} = -G \frac{Mm}{(h + R)^2}$$

where G is the proportionality constant in the universal gravitational law. When $h = 0$, that is, at the earth's surface, the gravitational force must equal $-mg$, and so

$$\frac{GM}{R^2} = g$$

where g is the acceleration due to gravity. Thus

$$\frac{d^2h}{dt^2} = -\frac{R^2g}{(h+R)^2} \tag{8}$$

with initial conditions

$$h(0) = 0, \qquad \frac{dh}{dt}(0) = V \tag{9}$$

The initial value problem (8) and (9) represents the mathematical model for the problem.

At this point we can undertake a dimensional analysis of the problem and gain considerable insight without actually attempting a solution. From our model the relevant dimensional quantities are t, h, R, V, and g having dimensions

$$[t] = \text{time } (T) \qquad [V] = \text{velocity } (LT^{-1})$$
$$[h] = \text{length } (L) \qquad [g] = \text{acceleration } (LT^{-2})$$
$$[R] = \text{length } (L)$$

We are using T (time) and L (length) as fundamental units. Following the procedure described in Section 1.2, if π is a dimensionless combination of t, h, R, V, and g, then

$$[\pi] = [t^{\alpha_1}h^{\alpha_2}R^{\alpha_3}V^{\alpha_4}g^{\alpha_5}]$$
$$= T^{\alpha_1 - \alpha_4 - 2\alpha_5}L^{\alpha_2 + \alpha_3 + \alpha_4 + \alpha_5}$$
$$= 1$$

Therefore,

$$\alpha_1 - \alpha_4 - 2\alpha_5 = 0$$
$$\alpha_2 + \alpha_3 + \alpha_4 + \alpha_5 = 0 \tag{10}$$

This system has rank two and so there are three independent dimensionless variables. Either by inspection or solving (10) we find that these quantities are

$$\pi_1 = \frac{h}{R}, \qquad \pi_2 = \frac{t}{R/V}, \qquad \pi_3 = V/\sqrt{gR} \tag{11}$$

By the Pi theorem, if there is a physical law relating t, h, R, V, and g (and we assume there must be, since in theory we could solve (8) and (9) to obtain that law), then there is an equivalent law that can be expressed as

$$\frac{h}{R} = f\left(\frac{t}{R/V}, \frac{V}{\sqrt{gR}}\right) \tag{12}$$

for some function $f(\pi_2, \pi_3)$.

Actually there is considerable information in (12). For example suppose we are interested in finding the time t_{max} that is required for the object to reach its maximum height for a given velocity V. Then differentiating (12) with respect to t and setting $h'(t)$ equal to zero gives

$$\frac{\partial f}{\partial \pi_2}\left(\frac{t_{max}}{R/V}, \frac{V}{\sqrt{gR}}\right) = 0$$

or

$$\frac{t_{max}}{R/V} = F\left(\frac{V}{\sqrt{gR}}\right) \tag{13}$$

for some function F. Remarkably, with little analysis beyond dimensional reasoning we have found that the time to maximum height depends only on the single quantity $V/\sqrt{gR}$. The value in knowing this kind of information lies in the efficiency of (13); a *single* graph of $t_{max}/(R/V)$ versus $V/\sqrt{Rg}$ contains all of the data of the graphs of t_{max} versus V for *all* choices of g and R. For example, an experimenter making measurements on different planets of t_{max} versus V would not need a separate plot of data for each planet.

The next step in the analysis is to choose characteristic time and length scales and recast the problem represented by (8) and (9) into dimensionless form. For the present problem this is more subtle than it originally appears. The general method requires us to select a new dimensionless dependent variable $\bar{h}$ and independent variable $\bar{t}$ by

$$\bar{t} = \frac{t}{t_c}, \qquad \bar{h} = \frac{h}{h_c} \tag{14}$$

where t_c is an intrinsic time scale and h_c is an intrinsic length scale; the values of t_c and h_c should be chosen by taking combinations of the constants in the problem, which in this case are R, V, and g. Unlike the heat conduction problem in Example 3.1, where the choices were fairly obvious, this problem presents several choices. For a length scale h_c we could take either R or V^2/g. Possible time scales are R/V, $\sqrt{R/g}$, and V/g. Which choice is the most appropriate? Actually, Equations (14) represent a legitimate transformation of variables for any choice of t_c and h_c; after the change of variables an equivalent problem would result. From a scaling viewpoint, however, one particular choice will be advantageous. The three choices

$$\bar{t} = \frac{t}{R/V}, \qquad \bar{h} = \frac{h}{R} \tag{15}$$

$$\bar{t} = \frac{t}{\sqrt{R/g}}, \qquad \bar{h} = \frac{h}{R} \tag{16}$$

and

$$\bar{t} = \frac{t}{Vg^{-1}}, \qquad \bar{h} = \frac{h}{V^2g^{-1}} \tag{17}$$

lead to the following three dimensionless problems, which are equivalent to (8) and (9)

$$\varepsilon \frac{d^2\bar{h}}{d\bar{t}^2} = -\frac{1}{(1+\bar{h})^2}, \qquad \bar{h}(0) = 0, \qquad \frac{d\bar{h}}{d\bar{t}}(0) = 1 \qquad (18)$$

$$\frac{d^2\bar{h}}{d\bar{t}^2} = -\frac{1}{(1+\bar{h})^2}, \qquad \bar{h}(0) = 0, \qquad \frac{d\bar{h}}{d\bar{t}}(0) = \sqrt{\varepsilon} \qquad (19)$$

and

$$\frac{d^2\bar{h}}{d\bar{t}^2} = -\frac{1}{(1+\varepsilon\bar{h})^2}, \qquad \bar{h}(0) = 0, \qquad \frac{d\bar{h}}{d\bar{t}}(0) = 1 \qquad (20)$$

respectively, where ε is a dimensionless parameter defined by

$$\varepsilon = \frac{V^2}{gR}$$

To illustrate how difficulties may arise in selecting an incorrect scaling let us modify our original problem by examining the situation when ε is known to be a small quantity; that is, V^2 is much smaller than gR. Then one may be tempted, in order to make an approximation, to delete the terms involving ε in the scaled problem. Problem (18) would then become

$$(1+\bar{h})^{-2} = 0, \qquad \bar{h}(0) = 0, \qquad \frac{d\bar{h}}{d\bar{t}}(0) = 1$$

which has no solution, and problem (19) would become

$$\frac{d^2\bar{h}}{d\bar{t}^2} = -\frac{1}{(1+\bar{h})^2}, \qquad \bar{h}(0) = 0, \qquad \frac{d\bar{h}}{d\bar{t}}(0) = 0$$

which has no physically valid solution. In the latter case note that the graph of $\bar{h}(\bar{t})$ would pass though the origin with zero slope and be concave downward, thereby making $\bar{h}$ negative. Therefore it appears that terms involving small parameters cannot be neglected. This is indeed unfortunate, since this kind of technique is a common practice in making approximations in applied problems. What went wrong was that (15) and (16) represent *incorrect* scalings; in that case, terms that appear small may in fact not be small. For example, in the term $\varepsilon d^2\bar{h}/d\bar{t}^2$ the parameter ε may be small but $d^2\bar{h}/d\bar{t}^2$ may be large, and hence the term may not be negligible compared to other terms in the equation.

If, on the other hand, the term $\varepsilon\bar{h}$ is neglected in (20), then $d^2\bar{h}/d\bar{t}^2 = -1$, or $\bar{h} = \bar{t} - \bar{t}^2/2$, after applying the initial conditions. Therefore

$$h = -\tfrac{1}{2}gt^2 + Vt$$

and we have obtained an approximate solution that is consistent with our experience with falling bodies. In this case we are able to neglect the small term and obtain a valid approximation because the scaling was correct. That (17) gives the correct time and length scales can be argued physically. If V is small, then the body will be acted upon by a constant gravitational field; hence, launched with speed V, it will uniformly decelerate and reach its maximum height in V/g units of time, which is the characteristic time. It will travel a distance of about (V/g) times its average velocity $\frac{1}{2}(V + 0)$, or $V^2/2g$. Hence V^2/g is a good selection for the length scale.

In general, if a correct scaling is chosen, then terms in the equations that appear small are indeed small and may be safely neglected. In fact, one goal of scaling is to select intrinsic, characteristic reference quantities so that each term in the dimensional equation transforms into a term which the dimensionless coefficient in the transformed term represents the order of magnitude or approximate size of that term. Pictorially

$$
\begin{bmatrix} \text{Dimensional} \\ \text{term} \end{bmatrix} \rightarrow \begin{bmatrix} \text{coefficient} \\ \text{representing} \\ \text{the order of} \\ \text{magnitude} \\ \text{of the term} \end{bmatrix} \cdot \begin{bmatrix} \text{dimensionless} \\ \text{factor} \\ \text{of order unity} \end{bmatrix}
$$

By *order of unity* we mean a term that is neither extremely large nor small. In the next section this notion is made precise.

Scaling Known Functions

In Example 3.1 where we considered heat conduction in a cylindrical rod, we scaled the temperature u by the constant temperature T_0 which was applied at both ends of the bar for $t > 0$; we recall that the initial temperature of the bar was $u = 0$, so we had little choice in the selection of the temperature scale. Now let us change the initial and boundary conditions. Assume that both ends at $x = 0$ and $x = l$ are held at zero degrees and that at $t = 0$ there is an initial temperature distribution in the bar defined by a given function $f(x), 0 < x < l$. These new auxiliary conditions lead to the boundary value problem

$$
\begin{aligned}
u_t(x, t) - ku_{xx}(x, t) &= 0, & 0 < x < l, \quad t > 0 \\
u(x, 0) &= f(x), & 0 < x < l \\
u(0, t) = u(l, t) &= 0, & t > 0
\end{aligned}
$$

consisting of the heat equation, an initial condition, and boundary conditions. By the same arguments as in Example 3.1 we infer that the length and time scales are l and l^2/k, respectively. Several possibilities exist for scaling the temperature u. One, for example, is to choose the average value of initial temperature $f(x)$ over the interval $0 < x < l$. Another, which is simpler and more common, is to scale u by the maximum value of $f(x)$ over the interval $0 < x < l$. We adopt this latter approach for the general definition.

Therefore, let $F(x)$ be a function that is bounded on an interval I. By the *magnitude* of F we mean the number M where

$$M = \sup_{x \in I} |F(x)|$$

If F is continuous and I is a closed interval, then the supremum, or least upper bound, may be replaced by maximum. The general principle is to scale a function by its magnitude. Frequently it is helpful to have the concept of the *order of the magnitude* of a function F. For example, if the maximum of a function is 937, then its order of magnitude is 10^3. Quite generally, a bounded function has order of magnitude 10^n over the interval I if

$$n - \tfrac{1}{2} < \log_{10} M \le n + \tfrac{1}{2}$$

To gain further insight into the concepts of scaling and characteristic quantities let us consider the following situation. Suppose $f(t)$ represents the time evolution of some quantity over an interval of time $0 \le t \le t_1$. To fix the idea let us suppose f has dimensions of velocity. By the previous discussion the function f can be scaled by M, the maximum of $|f|$ over the interval $[0, t_1]$. Is there any way that the time scale can be discovered for this physical process? To answer this question let us assume that the phenomenon is governed by a first order ordinary differential equation of the form

$$G(f, f') = 0, \qquad 0 \le t \le t_1$$

To scale this problem requires introducing a characteristic time t_c and a characteristic velocity (which must be M) and then defining dimensionless variables by

$$\bar{t} = \frac{t}{t_c}, \qquad \bar{f} = \frac{f}{M}$$

The derivative df/dt that occurs in the differential equation would then become

$$\frac{df}{dt} = \frac{M}{t_c} \frac{d\bar{f}}{d\bar{t}}$$

This derivative term, a function in its own right, should be scaled by $\max |df/dt|$. Therefore, for correct scaling

$$\frac{M}{t_c} = \max_{[0, t_1]} \left| \frac{df}{dt} \right|$$

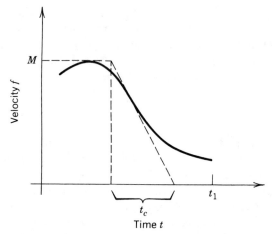

Figure 1.5. Time scale t_c as the ratio of the maximum value M to the maximum slope.

since M/t_c should represent the magnitude of the term df/dt. This forces upon us the time scale

$$t_c = \frac{\max |f|}{\max |df/dt|} \tag{21}$$

where both maxima are taken over $[0, t_1]$. Geometrically, since $\max |df/dt|$ is the maximum slope of the curve over $[0, t_1]$ and M is the maximum of $|f|$ over $[0, t_1]$, then t_c can be regarded as the base of the right triangle with height M and hypotenuse with slope $\max |df/dt|$ (see Fig. 1.5). This interpretation is consistent with the earlier remark that the time scale for a problem should be the shortest time in which discernible changes can be observed.

In some cases the previous discussion is inadequate, especially when there is radically different behavior on different subintervals of $[0, t_1]$. The following example illustrates the point.

Example 3.3   Let

$$f(t) = t + e^{-10000t}, \qquad 0 \le t \le 1$$

The derivative is

$$f'(t) = 1 - 10000e^{-10000t}, \qquad 0 \le t \le 1$$

and the graph of f is shown in Fig. 1.6. Its minimum occurs at $t = 0.0004 \ln 10 \cong 0.00092$. The derivative is maximized at $t = 0$ with an absolute value of about 10^4. By the preceding discussion the time scale, according to (21), is approximately 10^{-4}. This value clearly does not represent an ap-

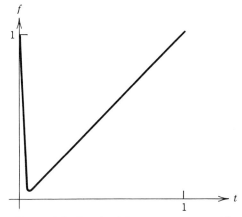

Figure 1.6. Graph of $f(t) = t + \exp(-10^4 t)$.

propriate characteristic time over the whole interval of interest. Rather, it represents a time scale only in the region near $t = 0$ where rapid changes are taking place. In the remainder of the interval the time scale is much larger because no significant discernible changes in f can be observed on the order of 10^{-4} units of time. The characteristic time is approximately unity for the portion of the interval $t \geq 10^{-4}$.

The preceding example illustrates a case where multiple time scales are required. Such situations are not uncommon in nature, and it is the modeling of these types of phenomena that gives rise to singular perturbation problems.

Historically, these problems became of interest in the early part of the century in examining the fluid flow over a wing, or airfoil. The correct model of this phenomenon, which includes lift and drag, involves introduction of a

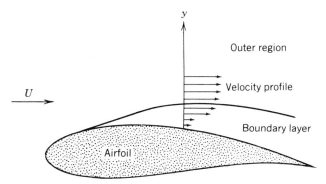

Figure 1.7. Boundary layer over an airfoil showing the horizontal velocity profile as a function of the vertical distance.

narrow *boundary layer* adjacent to the wing where rapid changes in the velocity of the fluid are taking place. At the wing the horizontal velocity component is zero. As the vertical distance y increases only slightly, there is a significant increase in this horizontal velocity component. In the outer region away from this narrow boundary layer the velocity is nearly that of the incoming uniform velocity U of the fluid. In Fig. 1.7 the arrows indicate the horizontal component of the fluid velocity. To describe this velocity component as a function of the vertical distance y it is clear that two characteristic lengths are appropriate. When scaling this problem one must take into account this behavior and introduce two length scales, one appropriate for each region.

EXERCISES

▷ **3.1**   For each of the following cases assume the given function f describes the evolution of a physical system governed by a first order differential equation. Find the scale for f and the appropriate time scale(s).

(a) $f(t) = A \cos \lambda t$, $0 \le t < \infty$, $A, \lambda > 0$

(b) $f(t) = e^{-at}$, $0 \le t < \infty$, $a > 0$

(c) $f(t) = 100 e^{(1-t)/1000}$, $0 \le t \le 1$

(d) $f(t) = e^{(t-1)/\varepsilon} + e^{-t/\sqrt{\varepsilon}}$, $0 \le t \le 1$, $0 < \varepsilon \ll 1$

3.2   Let $u = u(t)$ be a given function and let

$$\bar{t} = \frac{t}{t_c}, \qquad \bar{u} = \frac{u}{u_c}$$

represent a change of variables, where t_c and u_c are constants. Use the chain rule to show that

$$\frac{du}{dt} = \frac{u_c}{t_c} \frac{d\bar{u}}{d\bar{t}} \qquad \text{and} \qquad \frac{d^2u}{dt^2} = \frac{u_c}{t_c^2} \frac{d^2\bar{u}}{d\bar{t}^2}$$

▷ **3.3**   Let $u = u(x, t)$ be a given function and let

$$\bar{x} = \frac{x}{x_c}, \qquad \bar{t} = \frac{t}{t_c}, \qquad \bar{u} = \frac{u}{u_c}$$

represent a change of variables, where x_c, t_c, and u_c are constants. Use the chain rule to show that

$$\frac{\partial u}{\partial x} = \frac{u_c}{x_c}\frac{\partial \bar{u}}{\partial \bar{x}}, \qquad \frac{\partial^2 u}{\partial t^2} = \frac{u_c}{t_c^2}\frac{\partial^2 \bar{u}}{\partial \bar{t}^2}$$

and

$$\frac{\partial^2 u}{\partial x\,\partial t} = \frac{u_c}{x_c t_c}\frac{\partial^2 \bar{u}}{\partial \bar{x}\,\partial \bar{t}}$$

3.4   A cast iron bar one meter long is initially at 100°C. Both ends are then put next to reservoirs at 0°. What is the characteristic time scale or heat flow in the bar?

▷ **3.5**   The state s of a physical system evolves in time t according to the law $s(t) = at\exp(-bt)$, $0 \le t \le T$, where $T = 2/b$ and a and b are positive constants with appropriate units. What is a scale for s? What is a reasonable time scale for the process?

▷ **3.6**   Consider heat flow in a bar of length π and diffusivity k, where the ends are held at zero degrees and the initial temperature is $f(x) = \sin x$. From Example 3.1 the temperature $u(x, t)$ satisfies

$$u_t - k u_{xx} = 0, \qquad\qquad t > 0, \quad 0 < x < \pi$$
$$u(x,0) = \sin x, \qquad\qquad 0 < x < \pi$$
$$u(0, t) = u(\pi, t) = 0, \qquad t > 0$$

In the interval $0 \le t \le t_c$, where t_c is the characteristic time, what percentage decrease occurs in the initial temperature distribution (solve the problem exactly by assuming $u(x, t) = \phi(t)\psi(x)$ for some ϕ and ψ)?

3.7   Generalize the characterization of the time scale to the case where the physical process   9-15   Hwk   ferential equation

1. $\{1,2\}$
2. $\{1,2,3,6,8,9,10\}$
where $f = f(t)$.
3. $\{1,3,5,6,9,10\}$
Answer:

$$t_c$$

$$\left(\;\underset{I}{\max}\;|\,(\cdot)_I\,|\;\;\vee\;\;\underset{I}{\max}\;|\,(t)|\;\right)$$

where

$$M = \max_{I} |f(t)|$$

3.8   A ball of mass m is tossed upward with initial velocity V. Assuming that the force due to air resistance is proportional to the velocity of the ball and the gravitational field is constant, formulate an initial value problem whose solution would give the height of the ball at any time t. Determine characteristic length and time scales and recast the problem into dimensionless form.

▷ **3.9**   A pendulum executing angular displacements $\theta = \theta(t)$ has length l, the bob has mass m, and $\theta = 0$ when the pendulum hangs vertically downward. Derive the equation of motion for the pendulum by applying Newton's second law to the bob. (Note the acceleration is d^2s/dt^2 where s is the length of the circular arc traveled by the bob, and $s = l\theta$. The force mg on the bob can be resolved into a tangential component to the arc and one normal to the arc. Only the tangential component affects the motion. See Fig. 1.8.)
Answer:

$$\frac{d^2\theta}{dt^2} + \frac{g}{l}\sin\theta = 0, \qquad t > 0$$

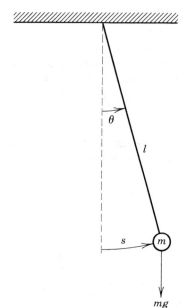

mg **Figure 1.8.** Simple pendulum.

If at $t = 0$ the bob is released from a small positive angle A, formulate an initial value problem whose solution will give $\theta = \theta(t)$. Introduce appropriate dimensionless variables Θ and τ for this problem and rewrite it in dimensionless form.

Answer:

$$\frac{d^2\Theta}{d\tau^2} + \frac{\sin(A\Theta)}{A} = 0, \qquad \tau > 0$$

$$\Theta(0) = 1, \qquad \frac{d\Theta}{d\tau}(0) = 0$$

▷ **3.10** Rework Exercise 3.9 when the initial conditions are changed to

$$\theta(0) = 0, \qquad \frac{d\theta}{dt}(0) = \Omega$$

that is, the pendulum is given an initial angular velocity Ω from the equilibrium position. If Ω is small, what are the characteristic time and angular deflection in this case. Recast the initial value problem

$$\frac{d^2\theta}{dt^2} + \frac{g}{l}\sin\theta = 0, \qquad \theta(0) = 0, \qquad \frac{d\theta}{dt}(0) = \Omega$$

into dimensionless form.

REFERENCES

1. G. Birkhoff, *Hydrodynamics: A Study in Logic, Fact, and Similitude*, Princeton University Press, Princeton, N.J. (1950).
2. C. C. Lin and L. A. Segel, *Mathematics Applied to Deterministic Problems in the Natural Sciences*, Macmillan Publishing Co., New York (1974).

2

PERTURBATION METHODS

When a mathematical model is formulated for a physical problem it is often represented by equations that are not solvable exactly by analytic techniques. Therefore one must resort to approximation and numerical methods. Foremost among approximation techniques are the so-called perturbation methods. Roughly, a perturbation method allows us to obtain an approximate solution to a problem when the model equations have terms that are small. These terms arise because the physical process has small effects. For example, in a fluid flow problem the viscosity may be small or in the problem of the motion of a projectile the force due to air resistance may be small. These low order effects are represented by terms in the model equations that, when compared to the other terms, are small. When scaled properly the order of magnitude of these terms is represented by a coefficient parameter, say ε, that is small. By a perturbation solution we mean an approximate solution that is the first few terms of a Taylor-like expansion in the parameter ε.

2.1 REGULAR PERTURBATION

The Perturbation Method

To fix the idea let us consider a differential equation of second order that we symbolically write as

$$F(t, y, \dot{y}, \ddot{y}; \varepsilon) = 0, \qquad t \in I \tag{1}$$

where t is the independent variable, I is an interval, and y is the dependent variable. The *overdot* denotes the time derivative, and the appearance of the parameter ε is shown explicitly. In general, initial or boundary conditions may accompany the equation, but for the present we ignore auxiliary conditions. To denote that ε is a small parameter we usually write

$$\varepsilon \ll 1$$

This means that ε is *small compared to unity*, a glib phrase in which no specific cutoff is explicitly defined; certainly $0.001 \ll 1$, but 0.75 is not small compared to unity. The previous remarks also apply to equations with a large parameter λ, since in that case we may introduce $\varepsilon = 1/\lambda$, which is small.

By a *perturbation series* we understand a power series in ε of the form

$$y_0(t) + \varepsilon y_1(t) + \varepsilon^2 y_2(t) + \cdots \tag{2}$$

The basis of the *regular perturbation method* is to assume a solution of (1) of the form (2) where the functions $y_0, y_1, y_2, \ldots$ are to be determined by substitution of (2) into (1). The first few terms of such a series form an approximate solution, a so-called *perturbation solution* to the problem; usually no more than two or three terms are taken. Generally the method will be successful if the approximation is uniform; that is, the difference between the approximate solution and the exact solution converges to zero at some well-defined rate as ε approaches zero uniformly on I. These ideas are made precise later. We emphasize in this context that there is no advantage in taking a priori a specific value for ε; rather we consider ε to be an arbitrary small number, so that the analysis will be valid for any choice of ε. Of particular interest in many problems is the behavior of the solutions as $\varepsilon \to 0$. In some problems the perturbation expansion in ε may be a good approximation even for ε close to one. Generally all that can be said is that the perturbation series is a valid representation for $\varepsilon < \varepsilon_0$ for some undetermined ε_0.

The term y_0 in perturbation series is called the *leading order* term. The terms $\varepsilon y_1, \varepsilon^2 y_2, \ldots$ are regarded as higher order correction terms that are expected to be small. If the method is successful, y_0 will be the solution of the *unperturbed problem*

$$F(t, y, \dot{y}, \ddot{y}; 0) = 0, \qquad t \in I$$

in which ε is set to zero. In this context (1) is called the *perturbed problem*.

The origins of this terminology are easily illustrated by a brief, simplistic discussion of planetary motion. We consider the planar motion of the earth about the sun. This two-body problem can be solved exactly to determine the trajectory $y_0(t)$ of the earth. The relevant equation is Newton's second law

$$m\ddot{\mathbf{y}} = \mathbf{F}_{\text{sun}}, \qquad \mathbf{y} = (y_1, y_2) \tag{3}$$

where y_1 and y_2 are the rectangular coordinates of the position of the earth, m

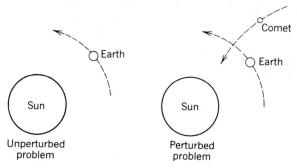

Figure 2.1. Unperturbed and perturbed system.

is the mass of the earth, and $\mathbf{F}_{sun}$ is the gravitational force on the earth caused by the sun. Now suppose a small comet passes close to the earth (see Fig. 2.1). What is the effect on the earth's path due to the presence of this *small perturbation*? The differential equation governing the motion of the earth now becomes

$$m\ddot{\mathbf{y}} = \mathbf{F}_{sun} + \varepsilon \mathbf{F}_{comet} \tag{4}$$

where $\varepsilon \mathbf{F}_{comet}$ is the force on the earth due to the comet and $\varepsilon \ll 1$. When ε is set to zero in the perturbed problem (4) we obtain the unperturbed problem (3) without the comet. It seems reasonable that the solution $\mathbf{y}(t)$ of the perturbed problem should be represented as the solution of the unperturbed problem plus small corrections, that is,

$$\mathbf{y}(t) = \mathbf{y}_0(t) + \varepsilon \mathbf{y}_1(t) + \varepsilon^2 \mathbf{y}_2(t) + \cdots$$

where the correction terms $\varepsilon \mathbf{y}_1, \varepsilon^2 \mathbf{y}_2, \ldots$ arise from the presence of the comet.

Therefore when an equation is encountered in which a small parameter appears it is often regarded as a perturbed equation where the term(s) containing the small parameter represent small perturbations or changes from some basic unperturbed problem. It is implicit that the unperturbed equation should always be solvable so that the leading order behavior of the solution is known. One frequently encounters problems of this sort whose solutions are facilitated by the fact that the governing equation, the boundary condition, or the shape of the region in higher dimensional problems, are not too different from those of a simpler problem.

Motion in a Nonlinear Resistive Medium

Example 1.1 A body of mass m initially with velocity V_0 moves in a straight line in a medium that offers a resistive force of magnitude $av - bv^2$, where

$v = v(\tau)$ the velocity of the object as a function of time τ and a and b are positive constants with $b \ll a$. Therefore the nonlinear part of the force is assumed to be small compared to the linear part. The constants a and b have units of force per unit velocity and force per unit velocity squared, respectively. By Newton's second law the equation of motion is

$$m\frac{dv}{d\tau} = -av + bv^2, \qquad v(0) = V_0$$

The velocity scale for this problem is the maximum velocity, which is clearly V_0, since the object must slow down in the resistive medium. If the small nonlinear term bv^2 were not present, then the velocity would decay like $\exp[(-a/m)\tau]$; therefore, the characteristic time is m/a. Introducing dimensionless variables

$$y = \frac{v}{V_0}, \qquad t = \frac{\tau}{m/a}$$

the problem becomes

$$\frac{dy}{dt} = -y + \varepsilon y^2, \qquad t > 0 \tag{5}$$

$$y(0) = 1$$

where

$$\varepsilon \equiv \frac{bV_0}{a} \ll 1$$

Equation (5) is a Bernoulli equation and can be solved exactly (see, for example, Boyce and DiPrima [1]) by making the substitution $w = y^{-1}$ and then integrating the resulting linear equation to obtain

$$y_{\text{exact}} = \frac{e^{-t}}{1 + \varepsilon(e^{-t} - 1)} \tag{6}$$

Equation (5) is a slightly altered or perturbed form of the linear equation

$$\frac{dy}{dt} = -y, \qquad y(0) = 1 \tag{7}$$

which is easily solved by $y = \exp(-t)$. Since the nonlinear term εy^2 is small the function $\exp(-t)$ appears to be a good approximate solution to the problem, provided the scaling was performed correctly. The exact solution (6) can be expanded in a Taylor series in powers of ε as

$$y_{\text{exact}} = e^{-t} + \varepsilon(e^{-t} - e^{-2t}) + \varepsilon^2(e^{-t} - 2e^{-2t} + e^{-3t}) + \cdots \qquad (8)$$

If ε is small, $\exp(-t)$ is a reasonable approximation. The leading order approximation $\exp(-t)$, however, does not include any effects of the nonlinear term in the original equation.

To obtain a solution by the perturbation method we employ a perturbation series. That is, we assume that the solution of (5) is representable as

$$y = y_0(t) + \varepsilon y_1(t) + \varepsilon^2 y_2(t) + \cdots \qquad (9)$$

which is a series in powers of ε. The functions $y_0, y_1, y_2, \ldots$ are to be determined by substituting (9) into both the differential equation (5) and initial condition, and then equating coefficients of like powers of ε. When this is done the differential equation becomes

$$\dot{y}_0 + \varepsilon \dot{y}_1 + \varepsilon^2 \dot{y}_2 + \cdots = -\left(y_0 + \varepsilon y_1 + \varepsilon^2 y_2 + \cdots \right)$$
$$+ \varepsilon \left(y_0 + \varepsilon y_1 + \varepsilon^2 y_2 + \cdots \right)^2$$

which, when coefficients are collected, gives a sequence of linear differential equations

$$\dot{y}_0 = -y_0$$
$$\dot{y}_1 = -y_1 + y_0^2$$
$$\dot{y}_2 = -y_2 + 2y_0 y_1, \ldots$$

The initial condition gives

$$y_0(0) + \varepsilon y_1(0) + \varepsilon^2 y_2(0) + \cdots = 1$$

or the sequence of initial conditions

$$y_0(0) = 1, \qquad y_1(0) = y_2(0) = \cdots = 0$$

Therefore we have obtained a set of linear initial value problems for $y_0, y_1, y_2, \ldots$. These are easily solved in sequence to obtain

$$y_0 = e^{-t}$$
$$y_1 = e^{-t} - e^{-2t}$$
$$y_2 = e^{-t} - 2e^{-t} + e^{-3t}, \ldots$$

Notice that y_1 and y_2 are the first and second order correction terms to the leading order approximation $y_0 = \exp(-t)$, which is in agreement with (8). Therefore we have obtained a three-term perturbation solution

$$y_{\text{approx}} = e^{-t} + \varepsilon(e^{-t} - e^{-2t}) + \varepsilon^2(e^{-t} - 2e^{-2t} + e^{-3t}) \qquad (10)$$

which is an approximation of y_{exact} and includes nonlinear effects due to the term εy^2 in the original differential equation. We notice that the approximate solution is the first three terms of the Taylor expansion of the exact solution. The error in the approximation (10) is given by the difference

$$y_{\text{exact}} - y_{\text{approx}} = m_1(t)\varepsilon^3 + m_2(t)\varepsilon^4 + \cdots, \qquad t > 0$$

for some bounded functions $m_1, m_2, \ldots$, which can be computed. For a fixed positive t the error approaches zero as $\varepsilon \to 0$ at the same rate as ε^3 goes to zero. In fact one can show that the convergence is uniform as $\varepsilon \to 0$ in the interval $0 \le t < \infty$. Convergence notions are explored further after an additional example.

A Nonlinear Oscillator

In the previous example application of the perturbation method led to a satisfactory result that was consistent with our intuition and that compared favorably with the exact solution. In the following example the procedure is the same, but the result does not turn out completely satisfactory. The perturbation approximation will be valid only if certain restrictions are placed on the interval of time that the solution evolves.

Example 1.2   Let us consider a spring–mass system as shown in Fig. 2.2. Such a system is a model of a typical nonlinear oscillator that occurs in many physical contexts (e.g., electrical circuits or atomic potentials). The mass is connected to a spring whose restoring force has magnitude $ky + ay^3$, where y is the displacement of the mass m measured positively from equilibrium, and k and a are positive constants that characterize the stiffness properties of the spring. We assume that the nonlinear portion of the restoring force is small in magnitude compared to the linear part, that is, $a \ll k$. If initially the mass is released from a positive displacement A, then the motion of the mass is given

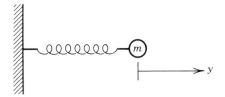

Figure 2.2. Spring–mass oscillator.

by a function $y = y(\tau)$ of time τ, which satisfies Newton's second law

$$m\frac{d^2y}{d\tau^2} = -ky - ay^3, \qquad \tau > 0 \tag{11}$$

and the initial conditions

$$y(0) = A, \qquad \frac{dy}{d\tau}(0) = 0 \tag{12}$$

Because of the presence of the nonlinear term ay^3 this problem cannot be solved exactly. Since $a \ll k$, however, a perturbation method is suggested.

In the first chapter the reader should have gained enough skepticism not to just neglect the term ay^3 and proceed without rescaling. Therefore, to properly analyze the problem with an eye upon attempting a perturbation method, we seek appropriate time and length scales that will reduce the problem to dimensionless form. The constants in the problem are k, a, m, and A having dimensions

$$[k] = \frac{\text{mass}}{\text{time}^2}, \qquad [a] = \frac{\text{mass}}{\text{length}^2 \cdot \text{time}^2}, \qquad [m] = \text{mass}, \qquad [A] = \text{length}$$

An obvious choice is to scale y by the amplitude A of the initial displacement. To scale time τ we argue as follows. If the small term ay^3 is neglected, then the differential equation is $my'' = -ky$, which has periodic solutions of the form $\cos\sqrt{k/m}\,t$ with a period proportional to $\sqrt{m/k}$. Hence we choose $\sqrt{m/k}$ to be the characteristic time and introduce dimensionless variables t and u via

$$t = \frac{\tau}{\sqrt{m/k}}, \qquad u = \frac{y}{A}$$

Under this change of variables the differential equation (11) and initial conditions (12) become

$$\ddot{u} + u + \varepsilon u^3 = 0, \qquad t > 0$$
$$u(0) = 1, \qquad \dot{u}(0) = 0 \tag{13}$$

where

$$\varepsilon \equiv \frac{aA^2}{k} \ll 1$$

is a dimensionless parameter. It now becomes clear exactly what the assumption of a small nonlinear restoring force means; our precise assumption is $aA^2 \ll k$. The differential equation (13) is known as the *Duffing equation*.

As in Example 1.1 we attempt a solution of the form

$$u(t) = u_0(t) + \varepsilon u_1(t) + \varepsilon^2 u_2(t) + \cdots$$

where $u_0, u_1, \ldots$ are to be determined. Substituting this expression into the differential equation and initial conditions and equating coefficients of like powers of ε gives a sequence of linear initial value problems

$$\ddot{u}_0 + u_0 = 0, \qquad u_0(0) = 1, \qquad \dot{u}_0(0) = 0 \qquad (14a)$$

$$\ddot{u}_1 + u_1 = -u_0^3, \qquad u_1(0) = 0, \qquad \dot{u}_0(0) = 0, \ldots \qquad (14b)$$

The solution to (14a) is easily found to be

$$u_0(t) = \cos t$$

Then (14b) becomes

$$\ddot{u}_1 + u_1 = -\cos^3 t, \qquad u_1(0) = \dot{u}_1(0) = 0$$

To solve this initial value problem for the first correction u_1 we employ the trigonometric identity $\cos 3t = 4\cos^3 t - 3\cos t$ to write

$$\ddot{u}_1 + u_1 = -\tfrac{1}{4}(3\cos t + \cos 3t) \qquad (15)$$

The general solution of the homogeneous equation $\ddot{u}_1 + u_1 = 0$ is $c_1\cos t + c_2\sin t$. A particular solution of (15) can be found by the method of undetermined coefficients and will be of the form

$$u_p = C\cos 3t + Dt\cos t + Et\sin t$$

Substituting u_p into (15) and equating like terms gives $C = \tfrac{1}{32}$, $D = 0$, and $E = -\tfrac{3}{8}$. Thus the general solution of (15) is

$$u_1 = c_1\cos t + c_2\sin t + \tfrac{1}{32}\cos 3t - \tfrac{3}{8}t\sin t$$

Applying the initial conditions on u_1 gives

$$u_1 = \tfrac{1}{32}(\cos 3t - \cos t) - \tfrac{3}{8}t\sin t$$

Therefore in scaled variables a two-term approximate solution takes the form

$$u_{\text{approx}} = \cos t + \varepsilon\left[\tfrac{1}{32}(\cos 3t - \cos t) - \tfrac{3}{8}t\sin t\right] \qquad (16)$$

The leading order behavior of the approximate solution is $\cos t$. The second term, or the correction term, however, is not necessarily small. For a fixed value of time t the term goes to zero as $\varepsilon \to 0$, but if t itself is of the order of ε^{-1} or larger as $\varepsilon \to 0$, then the term $-\frac{3}{8}t \sin t$ will be large. Such a term is called a *secular term*. Therefore we expect that the amplitude of the approximate solution (16) to grow with time, which is not consistent with the physical situation or with the exact solution. Indeed, it is easy to show that the latter is bounded for all $t > 0$ (see Exercise 1.2). This example is in sharp contrast to Example 1.1 where no secular terms arose and where the correction terms could be made arbitrarily small for t in $(0, \infty)$ by choosing ε small enough. And, it is not possible to improve upon (16) by calculating additional higher order terms; they too will contain secular terms that will not cancel the effects of the lower order terms. One may legitimately ask if there is any value at all in the approximate solution (16). In a limited sense the answer is yes. If we restrict the independent variable t to a *finite* interval $[0, T]$, then the correction term $\varepsilon[\frac{1}{2}(\cos 3t - \cos t) - \frac{3}{8}t \sin t]$ can be made arbitrarily small by choosing ε sufficiently small, for any $t \in [0, T]$. So, as long as the coefficient $3\varepsilon t/8$ is kept small by limiting t and taking ε small, the leading order term $\cos t$ is a reasonable approximate solution.

The Poincaré-Lindstedt Method

A straightforward application of the regular perturbation method on the initial value problem (13) for the Duffing equation led to a secular term in the first correction that ruined the approximation unless t was restricted to small values. In this section we present a method to remedy this type of singular behavior that occurs in all cases of periodic motion. The key to the analysis is to recognize that not only does the correction term grow in amplitude but it also does not correct for the difference between the exact period of oscillation and the approximate period 2π of the leading order term $\cos t$. Over several oscillations the error in the period increases until the approximation and the actual solution are completely out of phase. The idea of the Poincaré-Lindstedt method is to introduce a distorted time scale in the perturbation series. In particular we let

$$u(\tau) = u_0(\tau) + \varepsilon u_1(\tau) + \varepsilon^2 u_2(\tau) + \cdots \tag{17}$$

where

$$\tau = \omega t \tag{18}$$

and

$$\omega = 1 + \varepsilon \omega_1 + \varepsilon^2 \omega_2 + \cdots \tag{19}$$

Here ω_0 has been chosen to be unity, the frequency of the solution of the unperturbed problem. Under the scale transformation (18) the initial value problem (13) becomes

$$\omega^2 u'' + u + \varepsilon u^3 = 0, \qquad \tau > 0 \tag{20}$$

$$u(0) = 1, \qquad u'(0) = 0 \tag{21}$$

where $u = u(\tau)$ and prime denotes differentiation with respect to τ. Substituting (17) and (19) into (20) and (21) gives

$$(1 + 2\varepsilon\omega_1\omega_0 + \cdots)(u_0'' + \varepsilon u_1'' + \cdots) + (u_0 + \varepsilon u_1 + \cdots)$$
$$+ \varepsilon(u_0^3 + 3\varepsilon u_0^2 u_1 + \cdots) = 0$$

and

$$u_0(0) + \varepsilon u_1(0) + \cdots = 1, \qquad u_0'(0) + \varepsilon u_1'(0) + \cdots = 0$$

Collecting powers of ε gives

$$u_0'' + u_0 = 0, \qquad u_0(0) = 1, \qquad u_0'(0) = 0 \tag{22}$$

$$u_1'' + u_1 = -2\omega_1 u_0'' - u_0^3, \qquad u_1(0) = u_1'(0) = 0 \ldots \tag{23}$$

The solution of (22) is

$$u_0(\tau) = \cos\tau$$

Then the differential equation in (23) becomes

$$u_1'' + u_1 = 2\omega_1\cos\tau - \cos^3\tau$$
$$= \left(2\omega_1 - \tfrac{3}{4}\right)\cos\tau - \tfrac{1}{4}\cos 3\tau \tag{24}$$

We observe that we can avoid a secular term if we choose

$$\omega_1 = \tfrac{3}{8}$$

Then (24) becomes

$$u_1'' + u_1 = -\tfrac{1}{4}\cos 3\tau$$

which has general solution

$$u_1(\tau) = c_1\cos\tau + c_2\sin\tau + \tfrac{1}{32}\cos 3\tau$$

The initial conditions on u_1 lead to

$$u_1(\tau) = \tfrac{1}{32}(\cos 3\tau - \cos\tau)$$

Therefore a first order perturbation solution of (13) is

$$u(\tau) = \cos \tau + \tfrac{1}{32}\varepsilon(\cos 3\tau - \cos \tau) + \cdots$$

where

$$\tau = t + \tfrac{3}{8}\varepsilon t + \cdots$$

This method is successful on a number of similar problems and it is one of a general class of multiple scale methods (see, for example, Jordan and Smith [2]).

Asymptotics

With insight from the two previous examples we now define some notions regarding convergence and uniformity. It has been observed that substitution of a perturbation series into a differential equation does not always lead to a valid approximate solution. Ideally we would like to say that a few terms in a truncated perturbation series provides, for a given ε, an approximate solution for the entire range of the independent variable t. Unfortunately, as we have seen, this is not always the case. Failure of this regular perturbation method is the rule rather than the exception.

To aid in the analysis of approximate solutions we introduce some basic notation and terminology that permits the comparison of two functions as their common argument approaches some fixed value.

Definition 1.1 *Let $f(\varepsilon)$ and $g(\varepsilon)$ be defined in some neighborhood (or punctured neighborhood) of $\varepsilon = 0$. We write*

$$f(\varepsilon) = o(g(\varepsilon)) \qquad as \quad \varepsilon \to 0 \tag{25}$$

if

$$\lim_{\varepsilon \to 0} \left| \frac{f(\varepsilon)}{g(\varepsilon)} \right| = 0$$

and we write

$$f(\varepsilon) = O(g(\varepsilon)) \qquad as \quad \varepsilon \to 0 \tag{26}$$

if there exists a positive constant M such that

$$|f(\varepsilon)| \le M|g(\varepsilon)|$$

for all ε in some neighborhood (punctured neighborhood) of zero.

In this definition $\varepsilon \to 0$ may be replaced by a one-sided limit or by $\varepsilon \to \varepsilon_0$, where ε_0 is any finite or infinite number, with the domain of f and g defined appropriately. If (25) holds, we say f is *little oh* of g as $\varepsilon \to 0$, and if (26) holds we say f is *big oh* of g as $\varepsilon \to 0$. A common comparison function is $g(\varepsilon) = \varepsilon^n$ for some exponent n; another comparison function is $g(\varepsilon) = \varepsilon^n \ln^m \varepsilon$ for exponents m and n.

The statement $f(\varepsilon) = O(1)$ means f is bounded in a neighborhood of $\varepsilon = 0$ and $f(\varepsilon) = o(1)$ means $f(\varepsilon) \to 0$ as $\varepsilon \to 0$. If $f = o(g)$, then f goes to zero faster than g goes to zero as $\varepsilon \to 0$.

Example 1.3   Verify $\varepsilon^2 \ln \varepsilon = o(\varepsilon)$ as $\varepsilon \to 0^+$. By L'Hôpital's rule

$$\lim_{\varepsilon \to 0^+} \frac{\varepsilon^2 \ln \varepsilon}{\varepsilon} = \lim_{\varepsilon \to 0^+} \frac{\ln \varepsilon}{(1/\varepsilon)} = \lim_{\varepsilon \to 0^+} \frac{(1/\varepsilon)}{(-1/\varepsilon^2)} = 0$$

Example 1.4   Verify $\sin \varepsilon = O(\varepsilon)$ as $\varepsilon \to 0^+$. By the mean value theorem in calculus there is a number c between 0 and ε such that

$$\frac{\sin \varepsilon - \sin 0}{\varepsilon - 0} = \cos c$$

Hence $|\sin \varepsilon| = |\varepsilon \cos c| \le |\varepsilon|$, since $|\cos c| \le 1$. An alternate argument is to note that $(\sin \varepsilon)/\varepsilon \to 1$ as $\varepsilon \to 0^+$. Since the limit exists the function $(\sin \varepsilon)/\varepsilon$ must be bounded for $0 < \varepsilon < \varepsilon_0$, for some ε_0. Therefore $|(\sin \varepsilon)/\varepsilon| \le M$, for some constant M and $\sin \varepsilon = O(\varepsilon)$.

Definition 1.1 may be extended to functions of ε and another variable t lying in an interval I. First we review the notion of uniform convergence. Let $h(t, \varepsilon)$ be a function defined for ε in a neighborhood of $\varepsilon = 0$, possibly not including the value $\varepsilon = 0$ itself, and for t in some interval I, either finite or infinite. We say

$$\lim_{\varepsilon \to 0} h(t, \varepsilon) = 0 \ \textit{uniformly on } I$$

if the convergence to zero is at the same rate for each $t \in I$; that is, if for any positive number η there can be chosen a positive number ε_0 independent of t such that $|h(t, \varepsilon)| < \eta$ for all $t \in I$, whenever $|\varepsilon| < \varepsilon_0$. In other words, if $h(t, \varepsilon)$ can be made arbitrarily small over the entire interval I by choosing ε small enough, then the convergence is uniform. If merely $\lim_{\varepsilon \to 0} h(t_0, \varepsilon) = 0$ for each fixed $t_0 \in I$, then we say that the convergence is *pointwise* on I.

One method of proving $\lim_{\varepsilon \to 0} h(t, \varepsilon) = 0$ uniformly on I is to find a function $H(\varepsilon)$ such that the inequality $|h(t, \varepsilon)| \le H(\varepsilon)$ holds for all $t \in I$, and having the property $H(\varepsilon) \to 0$ as $\varepsilon \to 0$. To prove that convergence is not uniform on I it is sufficient to produce a $\bar{t} \in I$ such that $|h(\bar{t}, \varepsilon)| \geq \eta$ for some positive η, regardless of how small ε is chosen.

Definition 1.2   *Let $f(t, \varepsilon)$ and $g(t, \varepsilon)$ be defined for all $t \in I$ and all ε in a (punctured) neighborhood of $\varepsilon = 0$. We write*

$$f(t, \varepsilon) = o(g(t, \varepsilon)) \qquad as \quad \varepsilon \to 0$$

if

$$\lim_{\varepsilon \to 0} \left| \frac{f(t, \varepsilon)}{g(t, \varepsilon)} \right| = 0 \tag{27}$$

pointwise on I. If the limit in (27) is uniform on I, we write $f(t, \varepsilon) = o(g(t, \varepsilon))$ as $\varepsilon \to 0$ uniformly on I. If there exists a positive function $M(t)$ on I such that

$$|f(t, \varepsilon)| \leq M(t)|g(t, \varepsilon)|$$

for all $t \in I$ and ε in some neighborhood of zero, then we write

$$f(t, \varepsilon) = O(g(t, \varepsilon)) \qquad as \quad \varepsilon \to 0, \quad t \in I$$

If $M(t)$ is a bounded function on I, then we write

$$f(t, \varepsilon) = O(g(t, \varepsilon)) \qquad as \quad \varepsilon \to 0, \; uniformly \; on \; I$$

The big oh and little oh notation permit us to make quantitative statements about the error in a given approximation. In general, the following definition holds.

Definition 1.3   *A function $y_a(t, \varepsilon)$ is a uniformly valid asymptotic approximation to a function $y(t, \varepsilon)$ on an interval I as $\varepsilon \to 0$ if the error $E(t, \varepsilon)$ defined by*

$$E(t, \varepsilon) \equiv y(t, \varepsilon) - y_a(t, \varepsilon)$$

converges to zero as $\varepsilon \to 0$ uniformly for $t \in I$.

We often express the fact that $E(t, \varepsilon)$ is little oh or big oh of ε^n (for some n) as $\varepsilon \to 0$ to make an explicit statement regarding the rate that the error goes to zero and whether or not the convergence is uniform.

Example 1.5   Let

$$y(t, \varepsilon) = e^{-t\varepsilon}, \qquad t > 0, \quad \varepsilon \ll 1$$

The first three terms of the Taylor expansion in powers of ε provide an

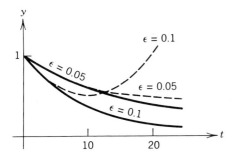

Figure 2.3. Comparison of $y(t, \varepsilon)$ (solid) and $y_a(t, \varepsilon)$ (dashed) for $\varepsilon = 0.1, 0.05$.

approximation

$$y_a(t, \varepsilon) = 1 - t\varepsilon + \tfrac{1}{2}t^2\varepsilon^2$$

The error is

$$E(t, \varepsilon) = e^{-t\varepsilon} - 1 + t\varepsilon - \frac{1}{2}t^2\varepsilon^2 = -\frac{1}{3!}t^3\varepsilon^3 + \cdots$$

For a fixed t the error can be made as small as desired by choosing ε small enough. Thus $E(t, \varepsilon) = o(\varepsilon^2)$ as $\varepsilon \to 0$. If ε is fixed, however, regardless how small, t may be chosen large enough so that the approximation is totally invalid. This phenomenon is illustrated in Fig. 2.3. Thus the approximation is not uniform on $I = [0,\infty)$. Clearly, by choosing $t = 1/\varepsilon$ we have $E(1/\varepsilon, \varepsilon) = e^{-1} - \frac{1}{2}$, which is not small. We may *not* write $E(t, \varepsilon) = o(\varepsilon^2)$ as $\varepsilon \to 0$, uniformly on $[0, \infty)$.

The difficulty of these definitions with regard to differential equations is that the exact solution to the equation is seldom known and thus a direct error estimate cannot be made. Therefore we require some notion of how well an approximate solution satisfies the differential equation and the auxiliary conditions. For definiteness consider the differential equation in (1). We say that an approximate solution $y_a(t, \varepsilon)$ *satisfies the differential* (1) *uniformly for* $t \in I$ *as* $\varepsilon \to 0$ if

$$r(t, \varepsilon) \equiv F\big(t, y_a(t, \varepsilon), \dot{y}_a(t, \varepsilon), \ddot{y}_a(t, \varepsilon), \varepsilon\big) \to 0$$

uniformly on I as $\varepsilon \to 0$. We can regard $r(t, \varepsilon)$ as the residual error, that is, it measures how well the approximate solution $y_a(t, \varepsilon)$ satisfies the equation.

Example 1.6 Consider the initial value problem

$$\ddot{y} + \dot{y}^2 + \varepsilon y = 0, \qquad t > 0, \quad 0 < \varepsilon \ll 1$$
$$y(0) = 0, \qquad \dot{y}(0) = 1$$

Substituting the perturbation series $y = y_0 + \varepsilon y_1 + \cdots$ gives the initial value problem

$$\ddot{y}_0 + \dot{y}_0^2 = 0, \qquad t > 0$$
$$y_0(0) = 0, \qquad \dot{y}_0(0) = 1$$

for the leading order term y_0. It is easily found that $y_0(t) = \ln(t + 1)$ and hence

$$r(t, \varepsilon) \equiv \ddot{y}_0 + \dot{y}_0^2 + \varepsilon y_0 = \varepsilon \ln(t + 1)$$

Thus $r(t, \varepsilon) = O(\varepsilon)$ as $\varepsilon \to 0$, but not uniformly on $[0, \infty)$. On any finite interval $[0, T]$, however, we have $|\varepsilon \ln(t + 1)| \leq \varepsilon \ln(T + 1)$, and so $r(t, \varepsilon) = O(\varepsilon)$ as $\varepsilon \to 0$ uniformly on $[0, T]$.

Specifically, the regular perturbation method produces an expansion

$$y_0(t) + y_1(t)\varepsilon + y_2(t)\varepsilon^2 + \cdots$$

for which an approximate solution can be obtained by taking the first few terms. Such an expansion in the integral powers of ε, that is, $1, \varepsilon, \varepsilon^2, \ldots$, is called an *asymptotic power series*. In some problems the expansion is taken to be

$$y_0(t) + y_1(t)\sqrt{\varepsilon} + y_2(t)\varepsilon + y_3(t)\varepsilon^{3/2} + \cdots$$

in terms of the sequence $1, \varepsilon^{1/2}, \varepsilon, \varepsilon^{3/2}, \ldots$. In yet other problems we assume an expansion

$$y_0(t)\ln \varepsilon + y_1(t)\varepsilon \ln \varepsilon + y_2(t)\varepsilon + y_3(t)\varepsilon^2 \ln^2\varepsilon$$
$$+ y_4(t)\varepsilon^2\ln \varepsilon + y_5(t)\varepsilon^2 + \cdots$$

The type of expansion depends upon the problem. In general we say a sequence $\{g_n(t, \varepsilon)\}$ is an *asymptotic sequence* as $\varepsilon \to 0$, $t \in I$, if

$$g_{n+1}(t, \varepsilon) = o(g_n(t, \varepsilon)) \text{ as } \varepsilon \to 0$$

for $n = 0, 1, 2, \ldots$. That is, each term in the sequence tends to zero faster than its predecessor as $\varepsilon \to 0$. Given a function $y(t, \varepsilon)$ and an asymptotic sequence $\{g_n(t, \varepsilon)\}$ as $\varepsilon \to 0$, the formal series

$$\sum_{n=0}^{\infty} a_n g_n(t, \varepsilon), \qquad a_n \text{ constants} \tag{28}$$

is said to be an *asymptotic expansion* of $y(t, \varepsilon)$ as $\varepsilon \to 0$, if

$$y(t, \varepsilon) - \sum_{n=0}^{N} a_n g_n(t, \varepsilon) = o(g_N(t, \varepsilon)), \qquad \text{as} \quad \varepsilon \to 0$$

for every N. In other words, for any partial sum the remainder is little oh of the last term. If the limits just cited are uniform for $t \in I$ then we speak of a uniform asymptotic sequence and uniform asymptotic expansion. In most cases the sequence $\{g_n(t, \varepsilon)\}$ is of the form of a product $g_n(t, \varepsilon) = y_n(t)\phi_n(\varepsilon)$ as in the previous examples.

The formal series (28) need not converge. The value of such expansions, although perhaps divergent, is that often only a few terms are required to obtain an accurate approximation, whereas a convergent Taylor series may yield an accurate approximation only if many terms are calculated.

A rather obvious question arises at this point. If a given approximate solution $y_a(t, \varepsilon)$ satisfies the differential equation uniformly for $t \in I$, is it in fact a uniformly valid approximation to the exact solution $y(t, \varepsilon)$? A complete discussion of this question is beyond the scope of this book, but a few remarks are appropriate in order to caution the reader regarding the nature of this problem. Probably more familiar is the situation in linear algebra where we consider a linear system of equations

$$A\mathbf{x} = \mathbf{b}$$

Let $\mathbf{x}_a$ be an approximate solution. A measure of how well it satisfies the system is the magnitude $|\mathbf{r}|$ of the *residual vector* $\mathbf{r}$ defined by

$$\mathbf{r} = A\mathbf{x}_a - \mathbf{b}$$

If $\mathbf{r} = \mathbf{0}$, then $\mathbf{x}_a$ must be the exact solution $\bar{\mathbf{x}}$. But if $|\mathbf{r}|$ is small, it does not necessarily follow that the magnitude $|\mathbf{e}|$ is small, where $\mathbf{e} = \bar{\mathbf{x}} - \mathbf{x}_a$ is the error, or difference between the exact solution and the approximate solution. In *ill-conditioned systems* where $det\, A$ is close to zero, a small residual may not imply a small error. A similar state of affairs exists for differential equations. Therefore one must proceed cautiously in interpreting the validity of a perturbation solution. Numerical calculations or the computation of additional correction terms may aid in the interpretation. Often a favorable comparison with experiment leads one to conclude that an approximation is valid.

EXERCISES

1.1   In a spring–mass problem (see Fig. 2.2) assume that the restoring force is $-ky$ and that there is also a resistive force numerically equal to $a\dot{y}^2$, where k and a are constants with appropriate units. With initial conditions $y(0) = A$, $\dot{y}(0) = 0$ determine the correct time and displacement scales for small damping and show that the problem can be written in dimensionless form as

$$\bar{y}'' + \varepsilon(\bar{y}')^2 + \bar{y} = 0$$
$$\bar{y}(0) = 1, \qquad \bar{y}'(0) = 0$$

where $\varepsilon \equiv aA/m$ is a dimensionless parameter and prime denotes the derivative with respect to the scaled time $\bar{t}$.

1.2   Show that the solution of the initial value problem

$$\ddot{u} + u + \varepsilon u^3 = 0, \qquad t > 0$$
$$u(0) = 1, \qquad \dot{u}(0) = 0$$

is bounded for all $t > 0$. *Hint:* Multiply the equation by $\dot{u}$ and integrate to get

$$\tfrac{1}{2}\dot{u}^2 + \tfrac{1}{2}u^2 + \tfrac{1}{4}\varepsilon u^4 = C$$

Find C from the initial conditions and note that $\tfrac{1}{2}u^2 \le C$ for all $t > 0$. Sketch the solution path in uv space, where $v = \dot{u}$. Is the solution periodic?

1.3   Verify the following order relations:

(a) $\varepsilon^2 \tanh \varepsilon = O(\varepsilon^2)$ as $\varepsilon \to \infty$

(b) $\exp(-\varepsilon) = o(1)$ as $\varepsilon \to \infty$

(c) $\sqrt{\varepsilon(1 - \varepsilon)} = O(\sqrt{\varepsilon})$ as $\varepsilon \to 0^+$

(d) $\dfrac{\sqrt{\varepsilon}}{1 - \cos \varepsilon} = 0(\varepsilon^{-3/2})$ as $\varepsilon \to 0^+$

(e) $\varepsilon = O(\varepsilon^2)$ as $\varepsilon \to \infty$

(f) $\exp(\varepsilon) - 1 = O(\varepsilon)$ as $\varepsilon \to 0$

(g) $\displaystyle\int_0^\varepsilon \exp(-x^2)\, dx = O(\varepsilon)$ as $\varepsilon \to 0^+$

(h) $\exp(\tan \varepsilon) = O(1)$ as $\varepsilon \to 0$

(i) $e^{-\varepsilon} = O(\varepsilon^{-p})$ as $\varepsilon \to \infty$ for all $p > 0$

(j) $\ln \varepsilon = o(\varepsilon^{-p})$ as $\varepsilon \to 0^+$ for all $p > 0$

1.4   Find the first two terms of the perturbation series solution to the initial value problem

$$\frac{dy}{dt} = 1 + (1 + \varepsilon)y^2, \qquad y(0) = 1, \qquad t > 0$$

where $0 < \varepsilon \ll 1$. Find the exact solution and compare the approximation. Is it uniform?

1.5   Consider the algebraic equation

$$x^2 + \varepsilon x - 1 = 0, \qquad 0 < \varepsilon \ll 1$$

Determine the first three terms in a perturbation series solution

$$x = x_0 + \varepsilon x_1 + \varepsilon^2 x_2 + \cdots$$

for each root. Compare with the exact roots.

1.6   Show that the sequence $\ln \varepsilon$, $\varepsilon \ln \varepsilon$, ε, $\varepsilon^2 \ln^2 \varepsilon$, $\varepsilon^2 \ln \varepsilon$, $\varepsilon^2, \ldots$ is an asymptotic sequence as $\varepsilon \to 0^+$.

1.7   Find a three-term perturbation expansion for the root of

$$x = 1 + \varepsilon x^2, \qquad 0 < \varepsilon \ll 1$$

near $x = 1$. Compare it to the exact solution for $\varepsilon = 0.1$ and $\varepsilon = 0.001$.

1.8   Find a two-term perturbation expansion for the solution of

$$y'' - \varepsilon y = 0, \qquad 0 < t < 1, \quad 0 < \varepsilon \ll 1$$
$$y(0) = 0, \qquad y(1) = 1$$

Compare to the exact solution and show that the two-term approximate solution satisfies the differential equation up to an $o(\varepsilon)$ term. How well does it satisfy the boundary conditions? Is the approximation uniform?

1.9   Consider the initial value problem

$$\ddot{y} + (1 + \varepsilon) y = 0, \qquad t > 0, \quad 0 < \varepsilon \ll 1$$
$$y(0) = 1, \qquad \dot{y}(0) = 0$$

Find the exact solution. Find a two-term perturbation approximation and show that the correction term is a secular term. Compare the exact solution to the perturbation approximation for large t.

1.10   To find approximations to the roots of the cubic equation

$$x^3 - 4.001x + 0.002 = 0$$

why is it easier to examine the equation

$$x^3 - (4 + \varepsilon)x + 2\varepsilon = 0?$$

1.11   Consider the algebraic system

$$0.01x + y = 0.1$$
$$x + 101y = 11$$

In the first equation ignore the apparently small first term $0.01x$ and obtain an approximate solution for the system. Is the approximation a good one? Analyze what went wrong by examining the solution of the system

$$\varepsilon x + y = 0.1$$
$$x + 101y = 11$$

1.12 In Section 1.3 we obtained the initial value problem

$$\frac{d^2h}{dt^2} = -(1 + \varepsilon h)^{-2}, \qquad h(0) = 0, \quad h'(0) = 1, \quad 0 < \varepsilon \ll 1$$

governing the motion of a projectile. Use regular perturbation theory to obtain a three-term perturbation approximation. Up to the accuracy of ε^2 terms determine the value t_m when h is maximum. Find $h_{max} \equiv h(t_m)$ up through ε^2 terms.

1.13 In Exercise 3.9 of Chapter 1 the pendulum problem was scaled to obtain the initial value problem

$$\frac{d^2\theta}{d\tau^2} + \frac{\sin A\theta}{A} = 0, \qquad \tau > 0, \quad 0 < A \ll 1$$

$$\theta(0) = 1, \qquad \frac{d\theta}{d\tau}(0) = 0$$

Apply the regular perturbation method to find a two-term expansion. Show that the correction term is secular and comment on the validity of the approximation.

1.14 Consider the initial value problem $y'' = \varepsilon ty$, $0 < \varepsilon \ll 1$, $y(0) = 0$, $y'(0) = 1$. Using regular perturbation theory obtain a three-term approximate solution on $t \geq 0$. Does the approximation satisfy the differential equation uniformly on $t \geq 0$ as $\varepsilon \to 0^+$?

1.15 Consider the boundary value problem

$$Ly \equiv t^2\ddot{y} + \varepsilon t^2\dot{y} + \tfrac{1}{4}y = 0, \qquad 1 \leq t \leq e, \quad 0 < \varepsilon \ll 1$$

$$y(1) = 1, \qquad y(e) = 0$$

(a) Using regular perturbation find the leading order behavior $y_0(t)$.

(b) Compute an upper bound for $|Ly_0|$ on $1 \leq t \leq e$ when $\varepsilon = 0.01$. Can you conclude that y_0 is a good approximation to the exact solution?

1.16 Show that the three-term expansion $\sin t + \varepsilon \cos t - \varepsilon^2/2 \sin t$ is a uniformly valid approximation of $\sin(t + \varepsilon)$ on $-\infty < t < \infty$.

1.17 Show that the three-term expansion

$$\sqrt{t} + \frac{\varepsilon}{2\sqrt{t}} - \frac{\varepsilon^2}{16t^{3/2}}$$

is not a uniformly valid approximation of $\sqrt{t + \varepsilon}$ in any interval I containing $t = 0$.

1.18 Work Exercise 1.1 when the initial conditions are $y(0) = 0$, $\dot{y}(0) = V$.

1.19 Apply the Poincaré-Lindstedt method to the scaled pendulum problem in Exercise 1.13 to obtain a two-term perturbation solution.

2.2 SINGULAR PERTURBATION

Failure of Regular Perturbation

There are many instances when a straightforward application of the regular perturbation method fails. We do not mean failure for large t due to the appearance of secular terms in the perturbation series as occurred in Example 1.2, but rather a different failure. Often, with a perturbation series some problems do not even permit the calculation of the leading order behavior because the perturbed problem is of totally different character than the unperturbed problem. The origins of this type of singular behavior will be investigated in the sequel.

The exposition in this section is deliberate and tutorial in nature, and some of the points are subtle and warrant a careful examination. The material herein does not lend itself well to a definition-theorem-proof format, but is instead a discussion that is motivational and intuitive. With these remarks in mind we begin with an example that illustrates the salient points and the inadequacy of a regular perturbation calculation on a finite interval.

Example 2.1 Consider the boundary problem

$$\varepsilon y'' + (1 + \varepsilon) y' + y = 0, \qquad 0 < t < 1, \quad 0 < \varepsilon \ll 1$$
$$y(0) = 0, \qquad y(1) = 1 \tag{1}$$

Let us assume a perturbation series of the form

$$y = y_0(t) + \varepsilon y_1(t) + \varepsilon^2 y_2(t) + \cdots$$

Substitution into the differential equation gives

$$\varepsilon\left(y_0'' + \varepsilon y_1'' + \varepsilon^2 y_2'' + \cdots \right) + \left(y_0' + \varepsilon y_1' + \varepsilon^2 y_2' + \cdots \right)$$
$$+ \left(\varepsilon y_0' + \varepsilon^2 y_1' + \varepsilon^3 y_2' + \cdots \right) + \left(y_0 + \varepsilon y_1 + \varepsilon^2 y_2 + \cdots \right) = 0$$

and equating to zero the coefficients of like powers of ε gives

$$y_0' + y_0 = 0$$
$$y_1' + y_1 = -y_0'' - y_0', \ldots$$

The boundary conditions force

$$y_0(0) = 0, \qquad y_0(1) = 1$$
$$y_1(0) = 0, \qquad y_1(1) = 0, \ldots$$

As before we have obtained a sequence of boundary value problems for $y_0, y_1, \ldots$; the zeroth order problem is

$$y_0' + y_0 = 0, \qquad y_0(0) = 0, \qquad y_0(1) = 1 \tag{2}$$

Already we see a difficulty. The differential equation is first order, yet there are two conditions to be satisfied. The general solution of the equation is

$$y_0(t) = ce^{-t} \tag{3}$$

Application of the boundary condition $y_0(0) = 0$ gives $c = 0$, and so $y_0(t) = 0$. This function cannot satisfy the boundary condition at $t = 1$. Conversely, application of the boundary condition $y_0(1) = 1$ in (3) gives $c = e$, and so

$$y_0(t) = e^{1-t} \tag{4}$$

This function cannot satisfy the condition at $t = 0$. Therefore we are at an impasse; regular perturbation has failed at the first step.

Inner and Outer Approximations

A careful examination of this problem will show what went wrong and point the way toward a correct, systematic method for obtaining an approximate solution. First of all, we should have been suspicious from the beginning. The unperturbed problem found by setting $\varepsilon = 0$ is

$$y' + y = 0, \qquad y(0) = 0, \qquad y(1) = 1 \tag{5}$$

This problem is of a different character than the perturbed problem (1) in that it is first order rather than second. Because the small parameter multiplied the

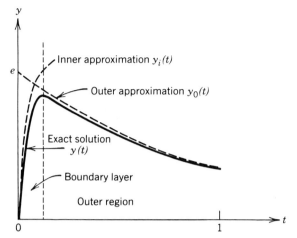

Figure 2.4. Schematic of exact solution compared to inner and outer approximation for a fixed value of ε.

highest derivative in (1), the second derivative term disappeared when ε was set to zero. This type of phenomenon signals in almost all cases the failure of the regular perturbation method.

Second, a look at the graph of the solution of (1) will reveal the main reason for the difficulty. Fortunately equation (1) can be solved exactly since it is a linear equation with constant coefficients. Its solution is

$$y(t) = \frac{1}{e^{-1} - e^{-1/\varepsilon}}\left[e^{-t} - e^{-t/\varepsilon}\right] \tag{6}$$

The graph of $y(t)$ is plotted in Fig. 2.4. We observe that $y(t)$ is changing very rapidly in a narrow interval called the *boundary layer* near the origin, and more slowly in the larger interval called the *outer region* away from the origin. Two time scales are indicated, one for each interval. It is instructive to compute the derivatives of y and estimate the size of the terms in the equation. An easy calculation shows

$$y' = \frac{1}{e^{-1} - e^{-1/\varepsilon}}\left(-e^{-t} + \frac{1}{\varepsilon}e^{-t/\varepsilon}\right)$$

$$y'' = \frac{1}{e^{-1} - e^{-1/\varepsilon}}\left(e^{-t} - \frac{1}{\varepsilon^2}e^{-t/\varepsilon}\right)$$

First we examine the second derivative. Suppose ε is small and t is in the narrow boundary layer near $t = 0$. For definiteness assume $t = \varepsilon$. Then

$$y''(\varepsilon) = \frac{1}{e^{-1} - e^{-1/\varepsilon}}\left(e^{-\varepsilon} - \frac{1}{\varepsilon^2}e^{-1}\right) = O(\varepsilon^{-2})$$

Hence y'' is very large inside this narrow band, and therefore the term $\varepsilon y''$ is not small as would be anticipated in a regular perturbation calculation; in fact $\varepsilon y'' = O(\varepsilon^{-1})$. We see again that terms that appear small in a differential equation are not always necessarily small. The answer to this dilemma is that the differential equation (1) as it stands is not scaled properly for small t and a rescaling must occur in the boundary layer if correct conclusions are to be drawn. For values of t that are away from the boundary layer the term $\varepsilon y''$ is indeed small. For example, if $t = \frac{1}{2}$, then

$$y''\left(\frac{1}{2}\right) = \frac{1}{e^{-1} - e^{-1/\varepsilon}}\left(e^{-1/2} - \frac{1}{\varepsilon^2}e^{-1/2\varepsilon}\right) = O(1)$$

Consequently, in the outer region $\varepsilon y''$ is small and may be safely neglected. The reader may deduce the conclusions regarding the first derivative terms.

This calculation suggests that in the outer region, the region away from the boundary layer, the zeroth order problem (2) or (5) obtained by setting $\varepsilon = 0$ in the original problem is a valid approximation provided we take only the right ($t = 1$) boundary condition. As we recall from (4), this gave

$$y_0(t) = e^{1-t} \tag{7}$$

This is consistent with the exact solution (6). For, if ε is small, then $e^{-1} - e^{-1/\varepsilon} \approx e^{-1}$ and y may be approximated by

$$y(t) \approx e^{1-t} - e^{1-t/\varepsilon} \tag{8}$$

For t of order one we have $e^{1-t/\varepsilon} \approx 0$, and so

$$y(t) \approx e^{1-t}, \qquad t = O(1)$$

The approximate solution (7), which is valid in the outer region, is called the *outer approximation* and it is plotted in Fig. 2.4.

What about an approximate solution in the narrow boundary layer near $t = 0$? From (8), if t is small, then

$$y(t) \approx e - e^{1-t/\varepsilon}, \qquad t \text{ small} \tag{9}$$

This approximate solution, which is valid only in the boundary layer where rapid changes are taking place, is called the *inner approximation* and is denoted by $y_i(t)$: it is also plotted in Fig. 2.4. Here we have had the advantage of knowing the exact solution, and as yet we have no clue how to determine the inner solution if the exact solution is unknown. The key to the analysis in the boundary layer is rescaling. There the term $\varepsilon y''$ in the differential equation was not a coefficient, which indicated the order of magnitude of the term times an order one quantity, as required by proper scaling, but rather a small quantity times a large one. The product was not negligible. Consequently we must rescale the independent variable t in the boundary layer by selecting a time scale that will reflect rapid and abrupt changes and will force each term in the equation into its proper form in the rescaled variables, namely as the product of a coefficient representing the magnitude of the term and a term of order unity.

Before embarking upon an analysis of the boundary layer we summarize our observations so far and make some general remarks. We have noted that the regular perturbation method introduced in Section 2.1 does not always produce an approximate solution. In fact, there are several indicators that often suggest its failure.

1. When the small parameter multiplies the highest derivative in the problem as in Example 1.1.
2. When setting the parameter equal to zero completely changes the character of the problem, as in the case of a partial differential equation changing type (elliptic to parabolic, for example) or an algebraic equation changing degree. In other words, the solution for $\varepsilon = 0$ is fundamentally different in character from the solutions for ε close to zero.
3. When problems occur on infinite domains.
4. When singular points are present in the interval of interest.
5. When the equations model physical processes that have multiple time or length scales.

Such perturbation problems fall in the general category of singular perturbation problems. For ordinary differential equations problems involving boundary layers are common. The procedure is to determine whether there is a boundary layer and where it is located. If there is a boundary layer, then the zeroth order perturbation term found by setting $\varepsilon = 0$ in the equation will often provide a valid approximation in the outer region. The inner approximation in the boundary layer is found by rescaling, which we shall discuss. Finally, the inner and outer approximations are *matched* in order to obtain a uniformly valid approximation over the entire interval of interest. For this reason the singular perturbation method applied in this context is sometimes called the method of *matched asymptotic expansions*. It also goes by the name of *boundary layer theory*.

Algebraic Equations and Balancing

Before returning to the differential equation we digress to the simpler example of an algebraic equation. The latter will give insight into the balancing of terms required in the rescaling of a problem in differential equations.

Example 2.2   Using singular perturbation methods we solve the quadratic equation

$$\varepsilon x^2 + 2x + 1 = 0, \qquad 0 < \varepsilon \ll 1$$

This equation is a slight alteration of the unperturbed equation

$$2x + 1 = 0$$

which has solution $x = -\frac{1}{2}$. Here, singular perturbation methods are indicated, since the unperturbed problem (linear) is fundamentally different from the original problem (quadratic). If we attempt a regular perturbation method by substituting in the series

$$x = x_0 + x_1\varepsilon + x_2\varepsilon^2 + \cdots$$

then we obtain, after setting the coefficients of like powers of ε equal to zero, the sequence of equations

$$2x_0 + 1 = 0$$
$$x_0^2 + 2x_1 = 0$$
$$2x_1x_0 + 2x_2 = 0,\ldots$$

Hence, $x_0 = -\frac{1}{2}$, $x_1 = -\frac{1}{8}$, $x_2 = -\frac{1}{16},\ldots$, and we obtain the solution

$$x = -\frac{1}{2} - \frac{1}{8}\varepsilon - \frac{1}{16}\varepsilon^2 - \cdots$$

Regular perturbation methods give only one root. To find the second root we examine the three terms εx^2, $2x$, and 1 of the equation more closely. Discarding the εx^2 gave the root close to $x = -\frac{1}{2}$, and in that case, the term εx^2 is small compared to $2x$ and 1. We suspect that to find the second root we must assume εx^2 is not small. Two cases are possible if one term is to be neglected from the equation to make a simplification.

(i) εx^2 and 1 are the same order and $2x \ll 1$.

(ii) εx^2 and $2x$ are the same order and both are large compared to 1.

In Case (i) we have $x = O(1/\sqrt{\varepsilon})$; but then $2x \ll 1$ could not hold since $2x$ would be large. In Case (ii) we have $x = O(1/\varepsilon)$; hence εx^2 and $2x$ both are of order $1/\varepsilon$ and are large compared to unity. Therefore Case (ii) is consistent and the second root is of order $1/\varepsilon$. This provides a clue to the scaling. To find the root let us rescale the equation by choosing a new variable $\bar{x}$ defined by

$$\bar{x} = \frac{x}{1/\varepsilon}$$

Note that $\bar{x}$ is order unity. Under this change of variables the original equation becomes

$$\bar{x}^2 + 2\bar{x} + \varepsilon = 0$$

Now the principle of proper scaling holds—each term has a magnitude defined by its coefficient. Moreover the unperturbed problem is still quadratic so the problem is no longer singular. Assuming a perturbation series

$$\bar{x} = \bar{x}_0 + \bar{x}_1\varepsilon + \bar{x}_2\varepsilon^2 + \cdots$$

and substituting into the last equation, we obtain

$$\bar{x}_0^2 + 2\bar{x}_0 = 0$$
$$2\bar{x}_0\bar{x}_1 + 2\bar{x}_1 + 1 = 0, \ldots$$

Hence $\bar{x}_0 = -2$, $\bar{x}_1 = \frac{1}{2}, \ldots$, giving

$$\bar{x} = -2 + \tfrac{1}{2}\varepsilon + \cdots$$

or

$$x = -\tfrac{2}{\varepsilon} + \tfrac{1}{2} + \cdots$$

as the second root.

EXERCISES

2.1 Use the balancing method of Example 2.2 to determine the leading order behavior of the roots of the following algebraic equations. In all cases $0 < \varepsilon \ll 1$.

(a) $\varepsilon x^4 + \varepsilon x^3 - x^2 + 2x - 1 = 0$ (c) $\varepsilon^2 x^6 - \varepsilon x^4 - x^3 + 8 = 0$

(b) $\varepsilon x^3 + x - 2 = 0$ (d) $\varepsilon x^5 + x^3 - 1 = 0$

In (b) find a first order correction for the leading behavior (*Hint*: For the scaled equation assume $\bar{x} = x_0 + x_1\sqrt{\varepsilon} + x_2\varepsilon + \cdots$).

2.2 Show that regular perturbation fails on the boundary value problem

$$\varepsilon y'' + y' + y = 0, \qquad 0 < t < 1, \quad 0 < \varepsilon \ll 1$$
$$y(0) = 0, \qquad y(1) = 1$$

Find the exact solution and sketch it for $\varepsilon = 0.05$ and $\varepsilon = 0.005$. If $t = O(\varepsilon)$, show that $\varepsilon y''(t)$ is large; if $t = O(1)$, show that $\varepsilon y''(t) = O(1)$. Find the inner and outer approximations from the exact solution.

2.3 BOUNDARY LAYER ANALYSIS

The Inner Approximation

We return to the boundary value problem of Example 2.1

$$\varepsilon y'' + (1 + \varepsilon) y' + y = 0, \qquad 0 < t < 1, \quad 0 < \varepsilon \ll 1$$
$$y(0) = 0, \qquad y(1) = 1 \tag{1}$$

It was discovered by examining the exact solution near $t = 0$ that rapid and abrupt changes were taking place in y, y', and y'', suggesting a short characteristic time; the term $\varepsilon y''$ was not small as it appears to be in the equation. Away from this boundary layer, in the region where $t = O(1)$, it was found that $\varepsilon y''$ and $\varepsilon y'$ were small compared to y' and y, and so the solution could be approximated accurately by setting $\varepsilon = 0$ in the equation to obtain

$$y' + y = 0$$

and solving, along with the boundary condition $y(1) = 1$, to get the outer solution

$$y_0(t) = e^{1-t} \tag{2}$$

To analyze the behavior in the boundary layer we notice that significant changes in y take place in a very short time, which suggests a time scale on the order of some function of ε, say $\delta(\varepsilon)$. Then, if we change variables via

$$\tau = \frac{t}{\delta(\varepsilon)}, \qquad Y = y \tag{3}$$

the differential equation (1) becomes

$$\frac{\varepsilon}{\delta(\varepsilon)^2} Y''(\tau) + \frac{(1 + \varepsilon)}{\delta(\varepsilon)} Y'(\tau) + Y(\tau) = 0 \tag{4}$$

where $Y(\tau) = y(\tau\delta(\varepsilon))$ and prime denotes the derivative with respect to τ. Another way of looking at this rescaling is to regard (3) as a scale transformation that permits us to examine this boundary layer region close up, as under a microscope.

There are four terms in (4) with coefficients

$$\frac{\varepsilon}{\delta(\varepsilon)^2}, \qquad \frac{1}{\delta(\varepsilon)}, \qquad \frac{\varepsilon}{\delta(\varepsilon)}, \qquad 1 \tag{5}$$

If the scaling is correct each will represent the order of magnitude of the term in which it appears. To determine the scale factor $\delta(\varepsilon)$ we estimate these magnitudes by considering all possible dominant balances between pairs of terms in (5). In the pairs we include the first term because it was deleted in the outer region and it is known that it plays a significant role in the boundary layer. Since the goal is to make a simplification in the problem, we do not at this time consider dominant balancing of three terms. If all four terms are equally important, no simplification can be made at all. Therefore there are three cases to consider.

(i) The terms $\varepsilon/\delta(\varepsilon)^2$ and $1/\delta(\varepsilon)$ are of the same magnitude and $\varepsilon/\delta(\varepsilon)$ and 1 are small in comparison.

(ii) $\varepsilon/\delta(\varepsilon)^2$ and 1 are of the same magnitude and $1/\delta(\varepsilon)$ and $\varepsilon/\delta(\varepsilon)$ are small in comparison.

(iii) The terms $\varepsilon/\delta(\varepsilon)^2$ and $\varepsilon/\delta(\varepsilon)$ are of the same magnitude and $1/\delta(\varepsilon)$ and 1 are small in comparison.

We will see that only Case (i) is possible. For, in Case (ii) $\varepsilon/\delta(\varepsilon)^2 \sim 1$ implies $\delta(\varepsilon) = O(\sqrt{\varepsilon})$, but then $1/\delta(\varepsilon)$ is not small compared to 1. In Case (iii) $\varepsilon/\delta(\varepsilon)^2 \sim \varepsilon/\delta(\varepsilon)$ implies $\delta(\varepsilon) = O(1)$, which leads to the outer approximation. In Case (i) $\varepsilon/\delta(\varepsilon)^2 \sim 1/\delta(\varepsilon)$ forces $\delta(\varepsilon) = O(\varepsilon)$; then $\varepsilon/\delta(\varepsilon)^2$ and $1/\delta(\varepsilon)$ are both order $1/\varepsilon$, which is large compared to $\varepsilon/\delta(\varepsilon)$ and 1. Therefore a consistent scaling is possible if we select $\delta(\varepsilon) = O(\varepsilon)$; hence we take

$$\delta(\varepsilon) = \varepsilon \tag{6}$$

and the scaled differential equation (4) becomes

$$Y''(\tau) + Y'(\tau) + \varepsilon Y'(\tau) + \varepsilon Y(\tau) = 0 \tag{7}$$

At this point (7) is amenable to a regular perturbation analysis. Since we are now interested only in the leading-term behavior, however, we set $\varepsilon = 0$ in (7) to obtain

$$Y''(\tau) + Y'(\tau) = 0 \tag{8}$$

which has general solution

$$Y(\tau) = C_1 + C_2 e^{-\tau}$$

Since the boundary layer is located near $t = 0$, we apply the boundary condition $y(0) = 0$, or $Y(0) = 0$. This yields $C_2 = -C_1$, and so

$$Y(\tau) = C_1(1 - e^{-\tau}) \tag{9}$$

or in terms of y and t,

$$y(t) = C_1(1 - e^{-t/\varepsilon}) \tag{10}$$

This is the inner solution for $t = 0(\varepsilon)$ and it is in agreement with (9) of Section 2, which was obtained from the exact solution. We denote this inner solution by $y_i(t)$.

In summary, we have an approximate solution

$$
\begin{aligned}
y_0(t) &= e^{1-t}, & t &= 0(1) \\
y_i(t) &= C_1(1 - e^{-t/\varepsilon}), & t &= 0(\varepsilon)
\end{aligned}
\tag{11}
$$

each valid for an appropriate range of t. There remains to determine the constant C_1. This is accomplished by the process of matching, which we now discuss.

Matching

First of all, by matching we do *not* mean selecting a specific value of ε, say ε_0, and requiring that $y_0(\varepsilon_0) = y_i(\varepsilon_0)$ to obtain the constant C_1. This process is nothing more than patching the two approximations together so as to be continuous at some fixed $t = \varepsilon_0$. Rather, our goal is to construct a single composite expansion in ε that is uniformly valid on the entire interval $[0, 1]$ as $\varepsilon \to 0$. Thus there is little gain in pinpointing the edge of the boundary layer, since it becomes narrower as $\varepsilon \to 0$. We figuratively state that the boundary layer for this problem has *width* ε. In general it is the scaling factor $\delta(\varepsilon)$ that defines the width of the boundary layer.

One can proceed as follows, however. It seems reasonable that the inner and outer expansions should agree to some order in an overlap domain that is intermediate between the boundary layer and outer region (see Fig. 2.5). If $t = O(\varepsilon)$, then t is in the boundary layer, and if $t = O(1)$, then t is in the

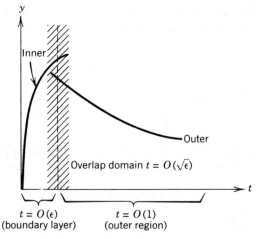

Figure 2.5. Overlap domain.

outer region; therefore, this overlap domain could be characterized as values of t for which $t = O(\sqrt{\varepsilon})$, for example, since orderwise $\sqrt{\varepsilon}$ is between ε and 1. This is because $\sqrt{\varepsilon}$ goes to zero slower than ε does. This intermediate scale allows the introduction of a new scaled independent variable η in the overlap domain defined by

$$\eta = \frac{t}{\sqrt{\varepsilon}} \tag{12}$$

The condition for matching will be that the inner approximation written in terms of the *intermediate variable* η should agree in the limit as $\varepsilon \to 0^+$, with the outer approximation written in terms of the intermediate variable. In symbols, for matching we require that for fixed η

$$\lim_{\varepsilon \to 0^+} y_0(\sqrt{\varepsilon}\,\eta) = \lim_{\varepsilon \to 0^+} y_i(\sqrt{\varepsilon}\,\eta) \tag{13}$$

In the present problem

$$\lim_{\varepsilon \to 0^+} y_0(\sqrt{\varepsilon}\,\eta) = \lim_{\varepsilon \to 0^+} \exp(1 - \sqrt{\varepsilon}\,\eta) = e$$

and

$$\lim_{\varepsilon \to 0^+} y_i(\sqrt{\varepsilon}\,\eta) = \lim_{\varepsilon \to 0^+} C_1(1 - \exp(-\eta/\sqrt{\varepsilon})) = C_1$$

Therefore matching requires $C_1 = e$ and the inner approximation becomes

$$y_i(t) = e(1 - e^{-t/\varepsilon}) \tag{14}$$

Uniform Approximations

To obtain a composite expansion that is uniformly valid throughout $[0, 1]$ we note the sum of the outer and inner approximations is

$$y_0(t) + y_i(t) = e^{1-t} + e - e^{1-t/\varepsilon}$$

$$\approx \begin{cases} e^{1-t} + e, & t = O(1) \\ 2e - e^{1-t/\varepsilon}, & t = O(\varepsilon) \end{cases}$$

Hence, by subtracting the common limit (13), which has value e, we have

$$y_u(t) \equiv y_0(t) + y_i(t) - e = e^{1-t} - e^{1-t/\varepsilon} \tag{15}$$

When t is in the outer region the second term is small and $y_u(t)$ is approximately e^{1-t}, which is the outer approximation. When t is in the boundary layer the first term is nearly e and $y_u(t)$ is approximately $e(1 - e^{-t/\varepsilon})$, which is the inner approximation. In the intermediate or overlap region both the inner and outer approximations are approximately equal to e. Therefore in the overlap domain the sum of $y_0(t)$ and $y_i(t)$ gives $2e$, or twice the contribution. This is why we must subtract the common limit from the sum. In summary, $y_u(t)$ defined by (15) provides a uniform approximate solution throughout the interval $[0, 1]$. Substituting $y_u(t)$ into the differential equation gives

$$\varepsilon y_u'' + (1 + \varepsilon) y_u' + y_u = 0$$

so $y_u(t)$ satisfies the differential equation exactly on $(0, 1)$. Checking the boundary conditions

$$y_u(0) = 0, \qquad y_u(1) = 1 - e^{1 - 1/\varepsilon}$$

The left boundary condition is satisfied exactly and the right boundary condition holds up to $O(\varepsilon^n)$ for any $n > 0$, since

$$\lim_{\varepsilon \to 0^+} \frac{e^{1 - 1/\varepsilon}}{\varepsilon^n} = 0, \qquad \text{for any } n > 0$$

Thus y_u is a uniformly valid approximation on $[0, 1]$.

A Worked Example

Now we work through another simple example without the detailed exposition of the preceding paragraphs. We determine an approximate solution of the boundary value problem

$$\varepsilon y'' + y' = 2t, \qquad 0 < t < 1, \quad 0 < \varepsilon \ll 1$$
$$y(0) = 1, \qquad y(1) = 1$$

using singular perturbation methods. Clearly, regular perturbation will fail since the unperturbed problem is

$$y' = 2t$$

which has the general solution

$$y(t) = t^2 + C$$

and such a function cannot satisfy both boundary conditions. Consequently we assume a boundary layer at $t = 0$ and impose the boundary condition $y(1) = 1$ (since $t = 1$ is in the outer region) to get the outer approximation

$$y_0(t) = t^2, \qquad t = O(1)$$

To determine the width $\delta(\varepsilon)$ of the boundary layer we rescale near $t = 0$ via

$$\tau = \frac{t}{\delta(\varepsilon)}, \qquad Y = y$$

In scaled variables the differential equation becomes

$$\frac{\varepsilon}{\delta(\varepsilon)^2} Y''(\tau) + \frac{1}{\delta(\varepsilon)} Y'(\tau) = 2\delta(\varepsilon)\tau$$

If $\varepsilon/\delta(\varepsilon)^2 \sim 2\delta(\varepsilon)$ is the dominant balance, then $\delta(\varepsilon) = O(\varepsilon^{1/3})$ and the second term $1/\delta(\varepsilon)$ would be $O(\varepsilon^{-1/3})$, which is not small compared to the assumed dominant terms. Therefore assume $\varepsilon/\delta(\varepsilon)^2 \sim 1/\delta(\varepsilon)$ is the dominant balance. In that case $\delta(\varepsilon) = O(\varepsilon)$ and the term $2\delta(\varepsilon)$ has order $O(\varepsilon)$, which is small compared to $\varepsilon/\delta(\varepsilon)^2$ and $1/\delta(\varepsilon)$, both of which are $O(\varepsilon^{-1})$. Therefore $\delta(\varepsilon) = \varepsilon$ is consistent and the scaled differential equation is

$$Y''(\tau) + Y'(\tau) = 2\varepsilon^2 \tau$$

The inner approximation to first order satisfies

$$Y''(\tau) + Y'(\tau) = 0$$

whose general solution is

$$Y(\tau) = C_1 + C_2 e^{-\tau}$$

In terms of y and t

$$y(t) = C_1 + C_2 e^{-t/\varepsilon}$$

Applying the boundary condition $y(0) = 1$ in the boundary layer gives $C_1 = 1 - C_2$, and therefore the inner approximation is

$$y_i(t) = (1 - C_2) + C_2 e^{-t/\varepsilon}$$

To find the constant C_2 we introduce an overlap domain of order $\sqrt{\varepsilon}$ and an appropriate intermediate scaled variable

$$\eta = t/\sqrt{\varepsilon}$$

Then $t = \sqrt{\varepsilon}\,\eta$ and the matching condition becomes

$$\lim_{\substack{\varepsilon \to 0^+ \\ \eta \text{ fixed}}} y_0(\sqrt{\varepsilon}\,\eta) = \lim_{\substack{\varepsilon \to 0^+ \\ \eta \text{ fixed}}} y_i(\sqrt{\varepsilon}\,\eta)$$

or

$$\lim_{\substack{\varepsilon \to 0^+ \\ \eta \text{ fixed}}} \varepsilon\eta^2 = \lim_{\substack{\varepsilon \to 0^+ \\ \eta \text{ fixed}}} \left[(1 - C_2) + C_2 e^{-\eta/\sqrt{\varepsilon}} \right]$$

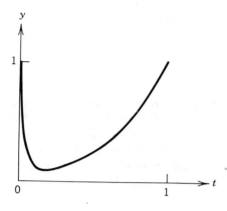

Figure 2.6. Graph of $y_u = t^2 + \exp(-t/\varepsilon)$.

This gives

$$O = 1 - C_2 \quad \text{or} \quad C_2 = 1$$

Therefore the inner approximation is

$$y_i(t) = e^{-t/\varepsilon}$$

A uniform composite approximation $y_u(t)$ is found by adding the inner and outer approximations and subtracting the common limit in the overlap domain, which is zero in this case. Hence

$$y_u(t) = t^2 + e^{-t/\varepsilon}$$

is the desired approximation and its graph is shown in Fig. 2.6.

Boundary Layer Phenomena

In the two preceding examples the boundary layer occurred at $t = 0$, or at the left endpoint. In the general case boundary layers can occur at any point in the interval, at the right endpoint or an interior point; in fact, multiple boundary layers can occur in the same problem. Boundary layers can also appear in initial value problems. The exercises will point out different boundary layer phenomena. When solving a problem one should assume a boundary layer at $t = 0$ (or the left endpoint) and then proceed. If this assumption is in error, then the procedure will break down when trying to match the inner and outer approximations. At that time one can make an assumption of a boundary layer at the right endpoint. The analysis is exactly the same, but the scale transformation in the boundary layer becomes

$$\bar{t} = \frac{t_0 - t}{\delta(\varepsilon)}$$

where t_0 is the right endpoint. Furthermore, even though both of the earlier examples exhibited a boundary layer width of $\delta(\varepsilon) = \varepsilon$ it is by no means true that this is always the case. Again the exercises contain other examples.

Here we have only attempted to match the leading order behavior of the inner and outer approximations. It is possible to refine the matching procedure to include higher order terms, and we refer the reader to Nayfeh [3] or Bender and Orszag [4] for a discussion of the topic. Finally we point out that this method is not a universal technique. For certain classes of problems it works well, but for other problems significant modifications must be made. At the present time singular perturbation theory is an active area of research in applied mathematics and well-developed rigorous theory is only available for restricted classes of differential equations. The references at the end of the chapter contain a wealth of examples and information on the subject.

We conclude this section with a general singular perturbation theorem for linear equations with variable coefficients. For this class of problems the boundary layer can be completely characterized, provided suitable restrictions are placed on the coefficients.

Theorem 3.1   *Consider the boundary value problem*

$$\varepsilon y'' + p(t)y' + q(t)y = 0, \qquad 0 < t < 1, \quad 0 < \varepsilon \ll 1$$
$$y(0) = a, \qquad y(1) = b \tag{16}$$

where p and q are continuous functions on $0 \le t \le 1$ and $p(t) > 0$ for $0 \le t \le 1$. Then there exists a boundary layer at $t = 0$ with inner and outer approximations given by

$$y_i(t) = C_1 + (a - C_1)e^{-p(0)t/\varepsilon} \tag{17}$$

$$y_0(t) = b \exp\left(\int_t^1 \frac{q(s)}{p(s)} \, ds \right) \tag{18}$$

where

$$C_1 = b \exp\left(\int_0^1 \frac{q(s)}{p(s)} \, ds \right) \tag{19}$$

Proof   We show that the assumption of a boundary layer at $t = 0$ is consistent and leads to (17), (18), and (19). If the boundary layer is at $t = 0$, then the outer solution $y_0(t)$ will satisfy

$$p(t)y_0' + q(t)y_0 = 0 \tag{20}$$

and the condition $y_0(1) = b$. Separating variables in (20), integrating, and then applying the condition at $t = 1$ gives (18) directly. In the boundary layer we introduce a scaled variable τ defined by $\tau = t/\delta(\varepsilon)$, where $\delta(\varepsilon)$ is to be determined. If $Y(\tau) = y(\delta(\varepsilon)\tau)$, then the differential equation becomes

$$\frac{\varepsilon}{\delta(\varepsilon)^2} Y''(\tau) + \frac{p(\delta(\varepsilon)\tau)}{\delta(\varepsilon)} Y'(\tau) + q(\delta(\varepsilon)\tau)Y(\tau) = 0 \qquad (21)$$

As $\varepsilon \to 0^+$ the coefficients behave like

$$\frac{\varepsilon}{\delta(\varepsilon)^2}, \qquad \frac{p(0)}{\delta(\varepsilon)}, \qquad q(0)$$

It is easy to see that the dominant balance is $\varepsilon/\delta(\varepsilon)^2 \sim p(0)/\delta(\varepsilon)$, and therefore the boundary layer has thickness $\delta(\varepsilon) = O(\varepsilon)$. For definiteness take $\delta(\varepsilon) = \varepsilon$. Equation (21) then becomes

$$Y''(\tau) + p(\varepsilon\tau)Y'(\tau) + \varepsilon q(\varepsilon\tau)Y(\tau) = 0$$

which to first order is

$$Y''(\tau) + p(0)Y'(\tau) = 0 \qquad (22)$$

The general solution of (22) is

$$Y(\tau) = C_1 + C_2 e^{-p(0)\tau}$$

Applying the boundary condition $Y(0) = a$ yields $C_2 = a - C_1$, and thus the inner approximation is

$$y_i(t) = C_1 + (a - C_1)e^{-p(0)t/\varepsilon} \qquad (23)$$

To match we introduce the intermediate variable $\eta = t/\sqrt{\varepsilon}$ and require that for fixed η

$$\lim_{\varepsilon \to 0^+} y_i(\sqrt{\varepsilon}\,\eta) = \lim_{\varepsilon \to 0^+} y_0(\sqrt{\varepsilon}\,\eta)$$

In this case the matching condition becomes

$$\lim_{\varepsilon \to 0^+} \left\{ C_1 + (a - C_1)e^{-p(0)\eta/\sqrt{\varepsilon}} \right\} = \lim_{\varepsilon \to 0^+} \left\{ b \exp\left(\int_{\sqrt{\varepsilon}\eta}^1 \frac{q(s)}{p(s)} ds \right) \right\}$$

which forces

$$C_1 = b \exp\left(\int_0^1 \frac{q(s)}{p(s)} ds \right)$$

Consequently the inner approximation is given as stated in (17) and (19) and the proof is complete.

A uniform composite approximation $y_u(t)$ is given by

$$y_u(t) = y_0(t) + y_i(t) - C_1$$

It can be shown that $y_u(t) - y(t) = O(\varepsilon)$ as $\varepsilon \to 0^+$ uniformly on $[0, 1]$, where $y(t)$ is the exact solution to (16). If $p(t) < 0$ on $0 \le t \le 1$, then no match is possible, since in that case $y_i(t)$ would grow exponentially unless $C_1 = a$. On the other hand, a match is possible in this case if the boundary layer is at $t = 1$. In summary one can show that the boundary layer is at $t = 0$ if $p(t) > 0$ and at $t = 1$ if $p(t) < 0$. As a final observation, there can be no boundary layer at an interior point $0 < t < 1$ in either case. We leave the demonstration of these facts to the exercises.

EXERCISES

3.1 Derive (6) in Section 2.2 and sketch graphs of $y(t)$ for $\varepsilon = 0.1, 0.01$.

3.2 Use singular perturbation methods to obtain a uniform approximate solution to the following problems. In each case assume $0 < \varepsilon \ll 1$ and $0 < t < 1$.

(a) $\varepsilon y'' + 2y' + y = 0$
    $y(0) = 0, \qquad y(1) = 1$

(b) $\varepsilon y'' + y' + y^2 = 0$
    $y(0) = \frac{1}{4}, \qquad y(1) = \frac{1}{2}$

(c) $\varepsilon y'' - y = 0$
    $y(0) = 1, \qquad y(1) = 2$

(d) $\varepsilon y'' + (t + 1)y' + y = 0$
    $y(0) = 0, \qquad y(1) = 1$

(e) $\varepsilon y'' + t^{1/3}y' + y = 0$
    $y(0) = 0, \qquad y(1) = \exp(-\frac{3}{2})$

(f) $\varepsilon y'' + ty' - ty = 0$
    $y(0) = 0, \qquad y(1) = e$

(g) $\varepsilon y'' + 2y' + e^y = 0$
    $y(0) = y(1) = 0$

(h) $\varepsilon y'' - (2 - t^2)y = -1$
    $y'(0) = 0, \qquad y(1) = 1$

3.3 By examining the exact solution shown why singular perturbation methods fail on the boundary value problem

$$\varepsilon y'' + y = 0, \qquad 0 < t < 1, \quad 0 < \varepsilon \ll 1$$
$$y(0) = 1, \qquad y(1) = 2$$

3.4 Obtain a uniform approximation to the problem

$$\varepsilon y'' - y' = 2t, \qquad 0 < t < 1, \quad 0 < \varepsilon \ll 1$$
$$y(0) = y(1) = 1$$

▷ **3.5** Obtain a uniform approximation to the problem

$$\varepsilon y'' - (2t + 1)y' + 2y = 0, \qquad 0 < t < 1, \quad 0 < \varepsilon \ll 1$$
$$y(0) = 1, \qquad y(1) = 0$$

3.6 Is the problem

$$\varepsilon y'' + \frac{1}{t}y' + y = 0, \qquad t > 0, \quad 0 < \varepsilon \ll 1$$
$$y(0) = 1, \qquad y'(0) = 0$$

a singular perturbation problem? Discuss.

☑ **3.7** Try singular perturbation methods on the boundary value problem

$$\varepsilon y'' + \left(t - \tfrac{1}{2}\right)y = 0, \qquad 0 < t < 1, \quad 0 < \varepsilon \ll 1$$
$$y(0) = 1, \qquad y(1) = 2$$

Discuss.

3.8 **(a)** Under the hypotheses of Theorem 3.1 show that a boundary layer cannot exist at $t = 1$ or at an interior point t_0 where $0 < t_0 < 1$.

(b) In the case that $p(t) < 0$ on $0 \le t \le 1$ show that a boundary layer exists at $t = 1$ and find the inner and outer approximations.

3.9 Find uniform asymptotic approximations to the solutions of the following problems on $0 \le t \le 1$ with $0 < \varepsilon \ll 1$.

(a) $\varepsilon y'' + (t^2 + 1)y' - t^3 y = 0$ ▷**(d)** $\varepsilon(y'' + y') - y = 0$
    $y(0) = y(1) = 1$ $y(0) = y(1) = 0$

(b) $\varepsilon y'' + (\cosh t)y' - y = 0$ **(e)** $\varepsilon y'' + (1 + t)^{-1}y' + \varepsilon y = 0$
    $y(0) = y(1) = 1$ $y(0) = 0, \qquad y(1) = 1$

▷**(c)** $\varepsilon y'' + 2\varepsilon t^{-1}y' - y = 0$
    $y(0) = 0, \qquad y'(1) = 1$

2.4 TWO APPLICATIONS

In this section we apply singular perturbation techniques to two problems that have their origins in a physical setting. We often let the physical situation be an insightful guide to their solution. Both examples are initial value problems as opposed to the boundary value problems considered in the last section.

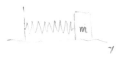

Damped Harmonic Oscillator

In Section 2.1 we obtained a model equation for a spring–mass system for a spring that exerted a nonlinear restoring force. In this section we study a simpler linear problem where the spring satisfies Hooke's law and there is a linear damping term. Let y denote the positive displacement (to the right) of the mass m from equilibrium and suppose the two forces on the spring are given by

$$F_{\text{spring}} = -ky, \qquad F_{\text{damping}} = -a\dot{y}, \qquad (k, a > 0)$$

where k and a are the spring constant and damping constant, respectively. By Newton's second law the governing differential equation is

$$m\ddot{y} + a\dot{y} + ky = 0, \qquad t > 0 \tag{1}$$

Initially, we assume the displacement is zero and that the mass is put into motion by imparting to it (say, with a hammer blow) a positive impulse I. Therefore the initial conditions are given by

$$y(0) = 0, \qquad m\dot{y}(0) = I \tag{2}$$

We are interested in determining the leading order behavior of the system in the case that the mass is very small in magnitude.

Of course, this problem can be solved exactly since (1) is a linear, second order equation with constant coefficients. This is why the problem is a good textbook example; comparisons can eventually be made to the exact solution to gain additional insight into the validity of the approximation procedures. Moreover, treated as a problem in singular perturbation theory it requires careful analysis not unlike the analysis that may be required on a *real* problem that cannot be solved exactly.

The first step is to subject the problem to a dimensional analysis in order to gain information about dimensionless variables, possible length and time scales, and so on. The independent and dependent variables are t and y having dimensions of time T and length L, respectively. The constants are m, a, k, and I have dimensions

$$[m] = M, \qquad [a] = MT^{-1}, \qquad [k] = MT^{-2}, \qquad [I] = MLT^{-1}$$

where M is a mass dimension. The dimension matrix is

$$
\begin{array}{c}
\begin{array}{cccccc}
t & y & m & a & k & I
\end{array} \\
\begin{array}{c} T \\ L \\ M \end{array}
\begin{pmatrix}
1 & 0 & 0 & -1 & -2 & -1 \\
0 & 1 & 0 & 0 & 0 & 1 \\
0 & 0 & 1 & 1 & 1 & 1
\end{pmatrix}
\end{array}
$$

which has rank three. Thus there are six minus three, or three dimensionless variables that can be formed from t, y, m, a, k, and I. If π is a dimensionless variable, then

$$\pi = t^{\alpha_1} y^{\alpha_2} m^{\alpha_3} a^{\alpha_4} k^{\alpha_5} I^{\alpha_6}$$

for some choice of $\alpha_1, \ldots, \alpha_6$. Hence

$$T^{\alpha_1} L^{\alpha_2} M^{\alpha_3} (MT^{-1})^{\alpha_4} (MT^{-2})^{\alpha_5} (MLT^{-1})^{\alpha_6} = 1$$

or, equating the powers of T, L, and M to zero,

$$\alpha_1 - \alpha_4 - 2\alpha_5 - \alpha_6 = 0$$
$$\alpha_2 + \alpha_6 = 0$$
$$\alpha_3 + \alpha_4 + \alpha_5 + \alpha_6 = 0$$

Clearly we can choose α_4, α_5, and α_6 arbitrarily and write

$$\begin{pmatrix} \alpha_1 \\ \alpha_2 \\ \alpha_3 \\ \alpha_4 \\ \alpha_5 \\ \alpha_6 \end{pmatrix} = \alpha_4 \begin{pmatrix} 1 \\ 0 \\ -1 \\ 1 \\ 0 \\ 0 \end{pmatrix} + \alpha_5 \begin{pmatrix} 2 \\ 0 \\ -1 \\ 0 \\ 1 \\ 0 \end{pmatrix} + \alpha_6 \begin{pmatrix} 1 \\ -1 \\ -1 \\ 0 \\ 0 \\ 1 \end{pmatrix}$$

which exhibits the three independent solutions. Hence three dimensionless variables are

$$\pi_1 = \frac{ta}{m}, \qquad \pi_2 = \frac{t^2 k}{m}, \qquad \pi_3 = \frac{tI}{ym} \tag{3}$$

From (3) it is easy to observe that possible *time scales* are

$$\frac{m}{a}, \qquad \sqrt{\frac{m}{k}}, \qquad \frac{a}{k} \tag{4}$$

The first two were obtained from π_1 and π_2 and the third was obtained from π_2/π_1. Upon substitution of these quantities into π_3 we obtain possible *length scales*

$$\frac{I}{a}, \qquad \frac{I}{\sqrt{km}}, \qquad \frac{aI}{km} \tag{5}$$

By assumption $m \ll 1$, and so $m/a \ll 1$, $\sqrt{m/k} \ll 1$, $I/\sqrt{km} \gg 1$, and $aI/km \gg 1$. Such relations are often important in determining appropriate time and length scales, since scales should be chosen so that the dimensionless dependent and independent variables are of order one.

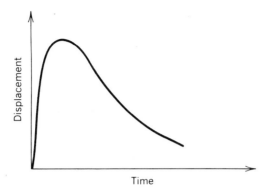

Figure 2.7. Displacement of mass versus time.

Before proceeding further let us use our intuition regarding the motion of the mass. The positive impulse given to the mass will cause a rapid displacement to some maximum value, at which time the force due to the spring will attempt to restore it to its equilibrium position. Since the mass is small there will be very little inertia and therefore it will probably not oscillate about equilibrium; the system is strongly overdamped. Consequently we expect the graph of $y = y(t)$ to exhibit the features shown in Fig. 2.7—a quick rise and then a gradual falling. Such a function is indicative of a multiple time-scale description. In the region near $t = 0$ where there is an abrupt change a small time scale seems appropriate, and away from this boundary layer an order of one time scale seems appropriate. In all, the problem has the characteristics of a singular perturbation problem, namely multiple time scales and an apparent small quantity m multiplying the highest derivative term.

Of the length scales given in (5) it appears that only I/a is suitable. The remaining two are large, which violates our intuition regarding the maximum displacement. Of the time scales only a/k in (4) is first order. The remaining two are small, and one of them should be suitable for the assumed boundary layer near $t = 0$. As a guess, since m/a depends on the mass and the damping, and $\sqrt{m/k}$ depends on the mass and the spring, we predict the former. The reason lies in our intuition concerning the dominant processes during the early stages of motion. The high initial velocity should influence the damping force more than the force due to the spring. Therefore the terms $m\ddot{y}$ and $a\dot{y}$ should dominate during the initial phase, whereas $m\ddot{y}$ should be low order in the later phase.

With these remarks in mind we introduce scaled variables

$$\bar{t} = \frac{t}{a/k}, \qquad \bar{y} = \frac{y}{I/a} \qquad (6)$$

which seem appropriate to the outer region, away from the boundary layer at

$t = 0$. Note that both $\bar{t}$ and $\bar{y}$ are of order unity. In terms of these variables the initial value problem (1) and (2) becomes

$$\varepsilon\bar{y}'' + \bar{y}' + \bar{y} = 0, \qquad \bar{t} > 0$$
$$\bar{y}(0) = 0, \qquad \varepsilon\bar{y}'(0) = 1 \tag{7}$$

where prime denotes the derivative with respect to $\bar{t}$ and the dimensionless constant ε is given by

$$\varepsilon = \frac{mk}{a^2} \ll 1 \tag{8}$$

Using singular perturbation techniques we can obtain an approximate solution to (7). To obtain the leading order approximation in the outer region we set $\varepsilon = 0$ to obtain

$$\bar{y}' + \bar{y} = 0$$

which gives

$$\bar{y}_0(\bar{t}) = Ce^{-\bar{t}}, \qquad C \text{ constant} \tag{9}$$

as the outer approximation. Neither initial condition in (7) is appropriate to apply, since both are at $\bar{t} = 0$, which is in the assumed boundary layer. To obtain an inner approximation we rescale according to

$$\tau = \frac{\bar{t}}{\delta(\varepsilon)}, \qquad Y = \bar{y} \tag{10}$$

Then (7) becomes

$$\frac{\varepsilon}{\delta(\varepsilon)^2}Y''(\tau) + \frac{1}{\delta(\varepsilon)}Y'(\tau) + Y(\tau) = 0 \tag{11}$$

The dominant balance is $\varepsilon/\delta(\varepsilon)^2 \sim 1/\delta(\varepsilon)$, which gives $\delta(\varepsilon) = \varepsilon$. (The balance $\varepsilon/\delta(\varepsilon)^2 \sim 1$ implies $\delta(\varepsilon) = O(\sqrt{\varepsilon})$, but then $1/\delta(\varepsilon)$ is not small, giving a contradiction.) Consequently the boundary layer has width ε, or $O(m)$. The scale transformation (10) is then

$$\tau = \frac{\bar{t}}{\varepsilon}, \qquad Y = \bar{y}$$

and the differential equation (11) becomes

$$Y''(\tau) + Y'(\tau) + \varepsilon Y(\tau) = 0, \qquad \tau > 0$$

Setting $\varepsilon = 0$ and solving the resulting equation gives

$$Y(\tau) = A + Be^{-\tau}$$

or

$$\bar{y}_i(\bar{t}) = A + Be^{-\bar{t}/\varepsilon}$$

as the leading order behavior of the inner solution. The initial condition $\bar{y}_i(0) = 0$ forces $B = -A$ and the condition $\varepsilon \bar{y}_i'(0) = 1$ forces $A = 1$. Therefore the inner approximation is completely determined by

$$\bar{y}_i(\bar{t}) = 1 - e^{-\bar{t}/\varepsilon} \tag{12}$$

It must now be matched with the outer solution (9) to determine the constant C in (9). Introducing an overlap domain with time scale $\sqrt{\varepsilon}$ and scaled dimensionless variable

$$\eta = \frac{\bar{t}}{\sqrt{\varepsilon}} \tag{13}$$

the matching condition can be written

$$\lim_{\substack{\varepsilon \to 0^+ \\ \eta \text{ fixed}}} \bar{y}_0(\sqrt{\varepsilon}\,\eta) = \lim_{\substack{\varepsilon \to 0^+ \\ \eta \text{ fixed}}} \bar{y}_i(\sqrt{\varepsilon}\,\eta)$$

or

$$\lim_{\substack{\varepsilon \to 0^+ \\ \eta \text{ fixed}}} Ce^{-\sqrt{\varepsilon}\,\eta} = \lim_{\substack{\varepsilon \to 0^+ \\ \eta \text{ fixed}}} \left(1 - e^{-\eta/\sqrt{\varepsilon}}\right)$$

Consequently

$$C = 1$$

A uniformly valid approximation is therefore

$$\bar{y}_u(\bar{t}) = \bar{y}_0(\bar{t}) + \bar{y}_i(\bar{t}) - \lim_{\substack{\varepsilon \to 0^+ \\ \eta \text{ fixed}}} y_0(\bar{t})$$

or

$$\bar{y}_u(\bar{t}) = e^{-\bar{t}} - e^{-\bar{t}/\varepsilon}$$

In terms of the original dimensioned variables t and y

$$y_u(t) = \frac{I}{a}\left(e^{-kt/a} - e^{-at/m}\right) \tag{14}$$

Our intuition proved to be correct. For small t, say $t = O(m)$, we have

$$y_u(t) \cong \frac{I}{a}(1 - e^{-at/m})$$

which describes a rapidly rising function with time scale m/a. For larger t, say $t = O(1)$, the first term in (14) dominates and

$$y_u(t) \cong \frac{I}{a}e^{-kt/a}$$

This represents a function that is decaying exponentially with characteristic time a/k. In Exercise 4.1 the reader is asked to compare this approximate solution with the exact solution.

In summary, this example shows how physical reasoning can complement mathematical analysis in solving problems arising out of empirics. The constant interplay between the physics and the mathematics can provide good understanding of a problem and give deep insight into its structure and methods of solution.

A Chemical Kinetics Problem

In the following paragraphs we formulate and solve a problem arising in the determination of concentrations of chemical species when certain reactions take place. The process will be modeled by a system of first order differential equations for the various concentrations with initial values given. Unlike the preceding example this problem is not solvable exactly. It has its origins in biochemistry where enzyme chemistry is used to enhance certain biochemical reactions. Detailed references can be found in Lin and Segel [5], who discuss this problem in its entirety and some of the present discussion is adapted from their treatment.

In certain chemical processes it is important to know the concentrations of the chemical species when certain reactions take place. To model a certain class of chemical reactions we consider an idealized reaction

$$A + B \rightleftarrows C \rightarrow P + B \tag{15}$$

where reactant molecules A and B combine to form a complex molecule C. The molecule C in turn breaks up to form a product molecule P and the original molecule B. We also consider the possibility of the complex breaking up into the original reactants A and B, which is denoted by the double arrow in (15); in this case the reaction is said to be *reversible*. The goal is to formulate a set of equations that govern the concentrations a, b, c, and p of A, B, C, and P as functions of time. We measure concentration in number of molecules per unit volume.

Before setting up the governing equations we first analyze some simpler reactions. We consider a simple one-way reaction

$$A + B \rightarrow C$$

By holding the number of molecules of B fixed it is reasonable to assume that the rate at which molecules of C are formed is proportional to the number of molecules of A present, since the formation of a molecule C results from an effective collision between A and B. Hence

$$\frac{dc}{dt} \propto a$$

Similarly

$$\frac{dc}{dt} \propto b$$

Therefore

$$\frac{dc}{dt} = k_1 ab$$

for some constant k_1 called the *rate constant*. Clearly

$$\frac{da}{dt} = -\frac{dc}{dt} = -k_1 ab$$

since the rate at which A disappears must be the negative of the rate at which C is created. Similarly

$$\frac{db}{dt} = -k_1 ab$$

For a breakup reaction

$$C \rightarrow A + B \tag{16}$$

the rate at which C is depleted is obviously proportional to the number of molecules of C present, whereas the rate at which A or B is created is $-dc/dt$. Hence

$$\frac{dc}{dt} = -k_2 c, \qquad \frac{da}{dt} = k_2 c, \qquad \frac{db}{dt} = k_2 c$$

where k_2 is the rate constant for reaction (16). Putting these results together, we conclude that for the reversible reaction

$$A + B \underset{k_2}{\overset{k_1}{\rightleftarrows}} C \tag{17}$$

we have

$$\frac{dc}{dt} = k_1 ab - k_2 c$$

while

$$\frac{da}{dt} = k_2 c - k_1 ab$$

and

$$\frac{db}{dt} = k_2 c - k_1 ab$$

In (17) we have adhered to the customary notation of writing the rate constants for the reactions next to their indicated directions.

Based on these ideas we can write the governing reaction rate equations for the reaction

$$A + B \underset{k_2}{\overset{k_1}{\rightleftarrows}} C \overset{k_3}{\to} P + B$$

as

$$\frac{da}{dt} = -k_1 ab + k_2 c \tag{18a}$$

$$\frac{db}{dt} = -k_1 ab + k_2 c + k_3 c \tag{18b}$$

$$\frac{dc}{dt} = k_1 ab - k_2 c - k_3 c \tag{18c}$$

$$\frac{dp}{dt} = k_3 c \tag{18d}$$

At $t = 0$ the initial concentrations are

$$a(0) = \hat{a}, \qquad b(0) = \hat{b}, \qquad c(0) = 0, \qquad p(0) = 0 \tag{19}$$

where we assume $\hat{b}$ *is small compared to* $\hat{a}$. A substantial reduction can be made by adding (18b) and (18c) to obtain $b + c = $ constant $= \hat{b}$. Furthermore the first three equations do not involve p. Therefore the system of four equations (18) can be reduced to two, namely

$$\frac{da}{dt} = -k_1 a(\hat{b} - c) + k_2 c \tag{20a}$$

$$\frac{dc}{dt} = k_1 a(\hat{b} - c) - (k_2 + k_3)c \tag{20b}$$

Once these are integrated for a and c then p can be found from (18d) by simple quadrature and b is given by $\hat{b} - c$.

In summary we must solve the differential equations (20) subject to the initial conditions (19) where the initial concentration $\hat{b}$ of B is assumed to be small compared to the initial concentration $\hat{a}$ of A.

Before tackling the problem directly let us use intuition as we did in the spring–mass oscillator problem to discover some relevant features of the reaction. In the early stages we expect A and B to combine rapidly to produce molecules of C. This is because it is relatively easy for the few molecules of B to bond with the abundant molecule A. In the long run we expect C to decrease to zero, since at any time some of it goes to form the product P through an irreversible reaction. Under these assumptions the concentration of A would decrease quickly at first, then ultimately level off at some positive value. Thus at first instant it appears that the problem has singular perturbation character with a narrow boundary layer near $t = 0$ where possibly abrupt changes are occurring.

We select dimensionless variables consistent with the scaling principle, that is, so that the magnitude of each term is correctly estimated by the coefficient that precedes it. It is clear that a should be scaled by its value at $t = 0$, namely $\hat{a}$. To scale c we choose the concentration $\hat{b}$, since eventually all of B will go into forming the complex C. Although there are two time scales for the process, it is not clear at this point what either scale is. Therefore, for the moment we choose a time scale $\hat{t}$ that is still unknown. Consequently, under the scale transformation

$$\bar{t} = \frac{t}{\hat{t}}, \qquad \bar{a} = \frac{a}{\hat{a}}, \qquad \bar{c} = \frac{c}{\hat{b}} \tag{21}$$

the differential equations (20) become

$$\frac{\hat{a}}{\hat{t}} \frac{d\bar{a}}{d\bar{t}} = -k_1 \hat{a} \hat{b} \bar{a}(1 - \bar{c}) + k_2 \hat{b} \bar{c} \tag{22a}$$

$$\frac{\hat{b}}{\hat{t}} \frac{d\bar{c}}{d\bar{t}} = k_1 \hat{a} \hat{b} \bar{a}(1 - \bar{c}) - (k_2 + k_3) \hat{b} \bar{c} \tag{22b}$$

In (22a) the term $-k_1 \hat{a} \hat{b} \bar{a}(1 - \bar{c})$ defines the rate at which A is depleted, while the term $k_2 \hat{b} \bar{c}$ defines the rate at which A is formed from the breakup of C. If the back reaction $C \rightarrow A + B$ dominated the forward reaction $A + B \rightarrow C$, then little would be accomplished in the total reaction. Hence we assume the dominant balance in (22a) is between the two terms

$$\frac{\hat{a}}{\hat{t}} \frac{d\bar{a}}{d\bar{t}} \quad \text{and} \quad -k_1 \hat{a} \hat{b} \bar{a}(1 - \bar{c})$$

which forces the time scale to be

$$\hat{t} = 1/k_1\hat{b} \tag{23}$$

With this choice the dimensionless equations (22) become

$$\frac{d\bar{a}}{d\bar{t}} = -\bar{a} + (\bar{a} + \lambda)\bar{c} \tag{24a}$$

$$\varepsilon\frac{d\bar{c}}{d\bar{t}} = \bar{a} - (\bar{a} + \mu)\bar{c} \tag{24b}$$

where the parameters λ and μ are given by

$$\lambda = \frac{k_2}{\hat{a}k_1}, \qquad \mu = \frac{k_2 + k_3}{\hat{a}k_1}$$

and $\varepsilon = \hat{b}/\hat{a} \ll 1$. The initial conditions are

$$\bar{a}(0) = 1, \qquad \bar{c}(0) = 0$$

The outer approximation can be found by setting $\varepsilon = 0$ in (24). If $\bar{a}_0(\bar{t})$ and $\bar{c}_0(\bar{t})$ denote these outer approximations, then

$$\frac{d\bar{a}_0}{d\bar{t}} = -\bar{a}_0 + (\bar{a}_0 + \lambda)\bar{c}_0$$

$$0 = \bar{a}_0 - (\bar{a}_0 + \mu)\bar{c}_0$$

The second equation is an algebraic equation that yields

$$\bar{c}_0 = \frac{\bar{a}_0}{\bar{a}_0 + \mu} \tag{25}$$

Substituting into the first equation then gives a single first order equation for $\bar{a}_0$ that when variables are separated, gives

$$\left(1 + \frac{\mu}{\bar{a}_0}\right) d\bar{a}_0 = -(\mu - \lambda)\, d\bar{t}$$

Integrating gives

$$\bar{a}_0 + \mu \ln \bar{a}_0 = -(\mu - \lambda)\bar{t} + K \tag{26}$$

where K is a constant of integration and $\mu - \lambda > 0$. Equation (26) defines $\bar{a}_0(\bar{t})$ implicitly. Since it is assumed to hold in the outer region neither initial condition can be applied; the constant K will have to be determined by matching. In summary, (26) and (25) determine the outer approximation.

To find the inner approximation in the boundary layer near $t = 0$ we define a change of scale via

$$\tau = \frac{\bar{t}}{\delta(\varepsilon)}, \qquad A = \bar{a}, \qquad C = \bar{c}$$

Equations (24) become

$$\frac{1}{\delta(\varepsilon)} \frac{dA}{d\tau} = -A + (A + \lambda)C \tag{27a}$$

$$\frac{\varepsilon}{\delta(\varepsilon)} \frac{dC}{d\tau} = A - (A + \mu)C \tag{27b}$$

where $A = A(\tau)$ and $C = C(\tau)$. In the boundary layer we expect the term involving $dC/d\tau$ to be important. Thus from (27b) we note that we must choose $\delta(\varepsilon) = \varepsilon$, which turns (27) into

$$\frac{dA}{d\tau} = [-A + (A + \lambda)C]\varepsilon \tag{28a}$$

$$\frac{dC}{d\tau} = A - (A + \mu)C \tag{28b}$$

Upon setting $\varepsilon = 0$, we find $A(\tau) =$ constant. From the initial condition $A(0) = 1$ we find $A(\tau) = 1$, and hence

$$\bar{a}_i(\bar{t}) = 1 \tag{29}$$

Equation (28b) then becomes

$$\frac{dC}{d\tau} = 1 - (1 + \mu)C$$

which is a linear first order equation for C. Solving, we obtain

$$C = \frac{1}{\mu + 1} + Me^{-(\mu+1)\tau}$$

where M is a constant of integration. Using $C(0) = 0$ gives $M = -1/(\mu + 1)$, and so the inner approximation for $\bar{c}$ is completely determined by

$$\bar{c}_i(\bar{t}) = \frac{1}{\mu + 1}(1 - e^{-(\mu+1)\bar{t}/\varepsilon}) \tag{30}$$

To match the outer approximation (25) and (26) with the inner approximation (29) and (30) we postulate an overlap domain characterized by the dimensionless time variable

$$\eta = \bar{t}/\sqrt{\varepsilon}$$

Then the matching conditions are, for fixed η,

$$\lim_{\varepsilon \to 0^+} \bar{a}_0(\sqrt{\varepsilon}\eta) = \lim_{\varepsilon \to 0^+} \bar{a}_i(\sqrt{\varepsilon}\eta) \tag{31}$$

and

$$\lim_{\varepsilon \to 0^+} \bar{c}_0(\sqrt{\varepsilon}\eta) = \lim_{\varepsilon \to 0^+} \bar{c}_i(\sqrt{\varepsilon}\eta) \tag{32}$$

The left-hand side of (31) cannot be evaluated directly, since $\bar{a}_0(\bar{t})$ is defined only implicitly by (26). Rewriting (26) in terms of η, however, gives

$$\bar{a}_0(\sqrt{\varepsilon}\eta) + \mu \ln \bar{a}_0(\sqrt{\varepsilon}\eta) = -(\mu - \lambda)\sqrt{\varepsilon}\eta + K$$

Taking the limit as $\varepsilon \to 0^+$ then gives

$$\lim_{\varepsilon \to 0^+} \bar{a}_0(\sqrt{\varepsilon}\eta) + \mu \lim_{\varepsilon \to 0^+} \ln \bar{a}_0(\sqrt{\varepsilon}\eta) = K$$

or

$$1 + \mu \ln(1) = K$$

whence

$$K = 1$$

The limit on the right side of (31) is also unity, so (31) is satisfied identically. From (30) it is easy to see that

$$\lim_{\varepsilon \to 0^+} \bar{c}_i(\sqrt{\varepsilon}\eta) = \frac{1}{\mu + 1}$$

To compute the limit on the left side of (32) we use (25) to obtain

$$\lim_{\varepsilon \to 0^+} \bar{c}_0(\sqrt{\varepsilon}\eta) = \lim_{\varepsilon \to 0^+} \frac{\bar{a}_0(\sqrt{\varepsilon}\eta)}{\bar{a}_0(\sqrt{\varepsilon}\eta) + \mu} = \frac{1}{1 + \mu}$$

Therefore (32) gives no new information, but it is consistent.

In summary, uniform leading order approximations for $\bar{a}$ and $\bar{c}$ are

$$\bar{a}_u(\bar{t}) = \bar{a}_0(\bar{t}) + \bar{a}_i(\bar{t}) - 1$$

$$\bar{c}_u(\bar{t}) = \bar{c}_0(\bar{t}) + \bar{c}_i(\bar{t}) - \frac{1}{1 + \mu}$$

or

$$\bar{a}_u(\bar{t}) = \bar{a}_0(\bar{t}) \tag{33}$$

$$\bar{c}_u(\bar{t}) = \bar{c}_0(\bar{t}) + \bar{c}_i(\bar{t}) - \frac{1}{1+\mu} \tag{34}$$

where $\bar{a}_0(\bar{t})$ is defined implicitly by the algebraic equation

$$\bar{a}_0(\bar{t}) + \mu \ln \bar{a}_0(\bar{t}) = -(\mu - \lambda)\bar{t} + 1$$

Further calculations require numerical procedures and are requested in the exercises.

EXERCISES

▷ **4.1** Solve the differential equation (1) with initial conditions (2) exactly and compare the approximate solution given by (14) with the exact solution.

4.2 In the damped harmonic oscillator problem assume an *incorrect* scaling

$$\bar{t} = \frac{t}{\sqrt{m/k}}, \qquad \bar{y} = \frac{y}{I/a}$$

in the outer region. Write the problem in dimensionless form. What does this scaling imply regarding the dominant processes in the outer region? Find the outer approximation in this case and determine whether there are times when this approximation holds.

4.3 Formulate the differential equations governing the concentrations a, b, c, and p for the reaction

$$A + B \underset{k_2}{\overset{k_1}{\rightleftarrows}} C \underset{k_4}{\overset{k_3}{\rightleftarrows}} P + B$$

▷ **4.4** Choosing $\mu = 1$, $\lambda = 0.625$, and $\varepsilon = 0.1$, sketch graphs of the uniform approximations $\bar{a}_u(\bar{t})$ and $\bar{c}_u(\bar{t})$ given by (33) and (34).

4.5 Show that for large times $\bar{t}$

$$\bar{a}_0(\bar{t}) \approx \exp\left(-\frac{\mu - \lambda}{\mu}\bar{t}\right)$$

How long after $\bar{t} = 0$ will this approximation become valid?

REFERENCES

1. W. E. Boyce and R. DiPrima, *Elementary Differential Equations and Boundary Value Problems*, 4th ed., John Wiley & Sons, New York (1986).
2. D. W. Jordan and P. Smith, *Nonlinear Ordinary Differential Equations*, Oxford University Press, London/New York (1977).
3. A. Nayfeh, *Perturbation Methods*, Wiley–Interscience, New York (1973).
4. C. M. Bender and S. A. Orszag, *Advanced Mathematical Methods for Scientists and Engineers*, McGraw-Hill Book company, New York (1978).
5. C. C. Lin and L. A. Segel, *Mathematics Applied to Deterministic Problems in the Natural Sciences*, Macmillan, New York (1974).

3

CALCULUS OF VARIATIONS

Generally, the calculus of variations is concerned with the optimization of variable quantities called functionals over some admissible class of competing objects. Many of its methods were developed over two hundred years ago by Euler (1701–1783), Lagrange (1736–1813), and others. It continues to the present day to bring important techniques to many branches of engineering and physics, however, and as an area of mathematics it introduces many important methods to applied analysis.

3.1 VARIATIONAL PROBLEMS

Functionals

To motivate the basic concepts of the calculus of variations we first review some simple notions in elementary differential calculus. One of the central problems in the calculus is to maximize or minimize a given real valued function of a single variable. If f is a given function defined in an open interval I, then f has a *local* (or *relative*) *minimum* at a point x_0 in I if $f(x_0) \leq f(x)$ for all x, satisfying $|x - x_0| < \delta$ for some δ. If f has a local minimum at x_0 in I and f is differentiable in I, then it is well known that

$$f'(x_0) = 0 \tag{1}$$

where the prime denotes the ordinary derivative of f. Similar statements can be made if f has a local maximum at x_0.

85

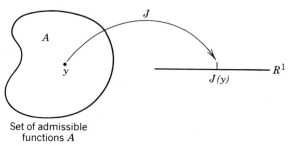

Set of admissible
functions A

Figure 3.1. A functional $J: A \to R^1$.

Condition (1) is called a *necessary condition* for a local minimum; that is, if f has a local minimum at x_0, then (1) necessarily follows. Equation (1) is not sufficient for a local minimum, however; that is, if (1) holds, it does not guarantee that x_0 provides an actual minimum. The following conditions are *sufficient conditions* for f to have a local minimum at x_0

$$f'(x_0) = 0 \quad \text{and} \quad f''(x_0) > 0$$

provided f'' exists. Again, similar conditions can be formulated for local maxima. If (1) holds, we say f is *stationary* at x_0 and that x_0 is an *extreme point* for f.

The calculus of variations is a subject that deals with generalizations of this problem from the calculus. Rather than find conditions under which functions have extreme values, the calculus of variations deals with extremizing general quantities called functionals. A *functional* is a rule that assigns a real number to each function $y(t)$ in a well-defined class. Like a function, a functional is a rule or association, but its domain is some set of functions rather than a subset of real numbers. To be more precise let A be a set of functions $y, z, \ldots$; then a functional J on A is a rule that associates to each $y \in A$ a real number denoted by $J(y)$. Figure 3.1 schematically illustrates this notion and it is symbolically represented as $J: A \to R^1$. We say J *maps* A into R^1. A fundamental problem of the calculus of variations can be stated as follows: Given a functional J and a well-defined set of functions A, determine which functions in A afford a minimum (or maximum) value to J. The word minimum can be interpreted as a local minimum (if A is equipped with some device to measure closeness of its elements) or an absolute minimum, that is, a minimum relative to all elements in A. The well-defined set A is called the set of *admissible functions*. It is those functions that are the competing functions for extremizing J. For example, the set of admissible functions might be the set of all continuous functions on an interval $[a, b]$, the set of all continuously differentiable functions on $[a, b]$ satisfying the condition $f(a) = 0$, or whatever, as long as the set is well defined. Later we shall impose some general conditions on A. For the most part the classical calculus of variations restricts itself to functionals that are defined by certain integrals and to the determination of both necessary and sufficient conditions for extrema. The problem of extremizing a functional J over the set A is called a *variational problem*. Several examples are presented in the next section.

To a certain degree the calculus of variations could be characterized as the calculus of functionals. In the present discussion we restrict ourselves to an analysis of necessary conditions for extrema. An elementary treatment of sufficient conditions can be found in Gelfand and Fomin [1]. Applications of the calculus of variations are found in geometry, physics, engineering, and optimal control theory. An elementary introduction to the theory is given in Troutman [4].

Examples

In this section we formulate several variational problems and illustrate the idea of a functional and the set of corresponding admissible functions.

Example 1.1   Let A be the set of all continuously differentiable functions on an interval $a \le x \le b$, which satisfy the boundary conditions $y(a) = y_0$, $y(b) = y_1$. Let J be the arclength functional on A defined by

$$J(y) = \int_a^b \sqrt{1 + y'(x)^2} \, dx$$

To each y in A the functional J associates a real number that is the arclength of the curve $y = y(x)$ between the two fixed points P: (a, y_0) and Q: (b, y_1). The associated calculus of variations problem is to minimize J, that is, the arclength, over the set A. Clearly the minimizing arc is the straight line between P and Q given by

$$y(x) = \frac{y_1 - y_0}{b - a}(x - a) + y_0$$

The general methods of the calculus of variations will allow us to prove that this is the case and we will therefore have a careful demonstration of the well known geometric fact that the shortest distance between two points is a straight line.

Example 1.2   (The Brachistochrone Problem)   A bead of mass m with initial velocity zero slides with no friction under the force of gravity g from a point (x_1, y_1) to a point (x_2, y_2) along a wire defined by a curve $y = y(x)$ in the xy plane $(x_1 < x_2, y_1 > y_2)$. Which curve leads to the fastest time of descent? Historically this problem was important in the development of the calculus of variations. Johann Bernoulli posed it in 1696 and several well-known mathematicians of that day proposed solutions; Euler's solution led to general

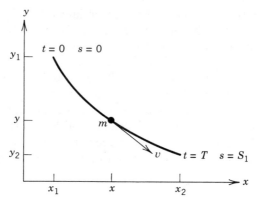

Figure 3.2. The brachistochrone curve.

methods that were useful in solving a wide variety of such problems. The solution curve itself is an arc of a cycloid, the curve traced out by a point on the rim of a rolling wheel; so the *brachistochrone*, or the curve of shortest time, is a cycloid.

To formulate this problem analytically we compute the time of descent T for a fixed curve $y = y(x)$ connecting the points (x_1, y_1) and (x_2, y_2) (see Fig. 3.2). Letting s denote the arclength along the curve measured from the initial point (x_1, y_1) we have

$$T = \int_0^T dt = \int_0^{s_1} \frac{dt}{ds}\, ds = \int_0^{s_1} \frac{1}{v}\, ds$$

where s_1 is the total arclength of the curve and v denotes the velocity of the mass. Since $ds = (1 + y'(x)^2)^{1/2}\, dx$ we have

$$T = \int_{x_1}^{x_2} \frac{\sqrt{1 + y'(x)^2}}{v}\, dx$$

To obtain an expression for v we use the fact that energy is conserved through the motion; that is,

(Kinetic energy at $t > 0$) + (potential energy at $t > 0$)

= (Kinetic energy at $t = 0$) + (potential energy at $t = 0$)

Expressed in terms of our notation

$$\tfrac{1}{2}mv^2 + mgy = 0 + mgy_1$$

Solving for v gives

$$v = \sqrt{2g(y_1 - y(x))}$$

Therefore the time required for the bead to descend is

$$T(y) = \int_{x_1}^{x_2} \frac{\sqrt{1 + y'(x)^2}}{\sqrt{2g(y_1 - y(x))}} \, dx \qquad (2)$$

where we have explicitly noted that T depends on the curve y. Here the set of admissible functions A is the set of all continuously differentiable functions y defined on $x_1 \le x \le x_2$ that satisfy the boundary conditions $y(x_1) = y_1$, $y(x_2) = y_2$, and the conditions

$$y \le y_1, \qquad \int_{x_1}^{x_2} (y_1 - y)^{-1/2} \, dx < +\infty$$

Equation (2) defines a functional on A. A real number is associated to each curve, namely the time required for the bead to descend the curve. The associated variational problem is to minimize $T(y)$, where $y \in A$, that is, to find which curve y in A minimizes the time. Because of the fixed boundary conditions this problem is called a *fixed endpoint problem*. Example 1.1 is another example of a fixed endpoint problem.

Example 1.3   Let A be the set of all nonnegative continuous functions on an interval $x_1 \le x \le x_2$. Then a functional on A can be defined by

$$J(y) = \int_{x_1}^{x_2} y(x) \, dx$$

Here J gives the area under the curve $y = y(x)$. Clearly $y(x) = 0$ provides an absolute minimum for J and there is no maximum.

Example 1.4   Although not directly of interest here, functionals need not be defined by integrals. For example $J(y) = \exp y(0)$ is a functional defined on the set of all functions that have $x = 0$ in their domain.

The types of functionals that are of interest in the classical calculus of variations are ones defined by expressions of the form

$$J(y) = \int_a^b L(x, y(x), y'(x)) \, dx, \qquad y \in A \qquad (3)$$

where $L = L(x, y, y')$ is some given function and A is a well-defined class of functions. In Examples 1.1, 1.2, and 1.3 the functionals are each of the form

(3) with

$$L = \sqrt{1 + y'^2} \tag{4}$$

$$L = \sqrt{1 + y'^2} / \sqrt{2g(y_1 - y)} \tag{5}$$

$$L = y \tag{6}$$

respectively. The function L is called the *Lagrangian* (after Lagrange), and in general it is assumed to be twice continuously differentiable in each of its three arguments. Notationally it is customary to drop the independent variable x in parts of the integrand in (3) and just write

$$J(y) = \int_a^b L(x, y, y') \, dx, \quad y \in A$$

In problems where time is an independent variable we write

$$J(y) = \int_a^b L(t, y, \dot{y}) \, dt, \quad y \in A$$

where the overdot on y denotes differentiation with respect to t.

EXERCISE

1.1   Show that the functional

$$J(y) = \int_0^1 \left\{ \dot{y} \sin \pi y - (t + y)^2 \right\} dt$$

where y is continuously differentiable on $0 \le t \le 1$, actually assumes its maximum value $2/\pi$ for the function $y(t) = -t$.

3.2   NECESSARY CONDITIONS FOR EXTREMA

Normed Linear Spaces

As stated in Section 3.1, a necessary condition for a function f to have a local minimum at some point x_0 is $f'(x_0) = 0$. We might conjecture that a similar condition should hold true for functionals provided the concept of derivative can be defined. In elementary calculus the derivative of a function f at x_0 is defined by a limit process, that is

$$f'(x_0) = \lim_{\Delta x \to 0} \frac{f(x_0 + \Delta x) - f(x_0)}{\Delta x}$$

provided the limit exists. If we are given a functional $J: A \rightarrow R^1$, the formula just noted makes no sense if f is replaced by J, and x_0 and Δx are replaced by functions in A. The essence of the derivative, namely the limit process, however, can be carried over to functionals in a restricted manner that we now describe.

To define the derivative of a functional J at some element y_0 in its domain A we require a notion of closeness in the set A so that the value $J(y_0)$ can be compared to $J(y)$ for y *close to* y_0. To carry out this idea in a fairly general manner we assume that the set of admissible functions A is a subset of a *normed linear space*. Such a space is a set of objects in which addition and multiplication by scalars (real numbers) are defined, and there is some measuring device to give definiteness to the idea of the *size* of an object. The objects in the normed linear space may be vectors in the geometric sense, functions, numbers, tensors, matrices, etc. We make precise definition in two parts—first we define a linear space, and then we define a norm.

A collection V of objects $u, v, w, \ldots$ is called *real linear space* if the following conditions are satisfied.

1. There is a binary operation $+$ on V such that $u + v \in V$ whenever $u, v \in V$. That is, V is *closed* under the operation $+$, which is called *addition*.

2. Addition is *commutative* and *associative*; that is, $u + v = v + u$ and $u + (v + w) = (u + v) + w$.

3. There is *zero element* 0 in V that has the property that $u + 0 = u$ for all $u \in V$. The zero element is also called the identity under addition and it is unique.

4. For each $u \in V$ there is a $v \in V$ such that $u + v = 0$. The object v is called the *inverse* of u and is denoted by $-u$.

5. For each $u \in V$ and $\alpha \in R^1$ the *scalar multiplication* αu is defined and $\alpha u \in V$.

6. Scalar multiplication is associative and distributive; that is,

$$\alpha(\beta u) = (\alpha\beta)u$$
$$(\alpha + \beta)u = \alpha u + \beta u$$
$$\alpha(u + v) = \alpha u + \alpha v$$

   for all $\alpha, \beta \in R^1$, $u, v \in V$.

7. $1u = u$ for all $u \in V$.

The previous conditions or axioms put an algebraic structure on the set V. The first four conditions state that there is an addition defined on V that satisfies the usual rules of arithmetic; the last three conditions state that the objects of V can be multiplied by real numbers and that reasonable rules governing that multiplication hold true. The reader is already familiar with some examples of linear spaces.

Example 2.1   The set R^n. Here an object is an n-tuple of real numbers $(x_1, x_2, \ldots, x_n)$, where $x_i \in R^1$. Addition and scalar multiplication are defined by

$$(x_1, x_2, \ldots, x_n) + (y_1, y_2, \ldots, y_n) = (x_1 + y_1, x_2 + y_2, \ldots, x_n + y_n)$$

and

$$\alpha(x_1, x_2, \ldots, x_n) = (\alpha x_1, \alpha x_2, \ldots, \alpha x_n)$$

The zero element is $(0, 0, \ldots, 0)$ and the inverse of $(x_1, \ldots, x_n)$ is $(-x_1, \ldots, -x_n)$. It is easily checked that the remaining properties hold true. The objects of R^n are called vectors.

Example 2.2   The set $C[a, b]$. The objects in this set are the continuous functions $f, g, h, \ldots$ defined on $a \le x \le b$. Addition and scalar multiplication are defined pointwise by

$$(f + g)(x) = f(x) + g(x)$$
$$(\alpha f)(x) = \alpha f(x), \qquad \alpha \in R^1$$

From elementary calculus it is known that the sum of two continuous functions is a continuous function, and a continuous function multiplied by a constant is also a continuous function. The zero element is the 0 function, that is, the function that is constantly zero on $[a, b]$. The inverse of f is $-f$, which is the reflection of $f(x)$ through the x axis. Again it is easily checked that the other conditions are satisfied. Geometrically the sum of the two functions f and g is the graphical sum, and multiplication of f by α stretches the graph of f vertically by a factor α.

Example 2.3 $C^2[a, b]$ is the set of all continuous functions on an interval $[a, b]$ whose second derivative is also continuous. Here again the objects are functions. Addition and scalar multiplication are defined as in Example 2.2. It is clear that $C^2[a, b]$ satisfies the other conditions for a linear space. If $y \in C^2[a, b]$, we say y is a function *of class* C^2 on $[a, b]$.

Example 2.4   Let V be the set of all singular (having determinant zero) 2-by-2 matrices with real entries. Here the elements of V are of the form

$$\begin{pmatrix} a & b \\ c & d \end{pmatrix}$$

with $ad - bc = 0$. If addition is defined in the usual manner, that is,

$$\begin{pmatrix} a_1 & b_1 \\ c_1 & d_1 \end{pmatrix} + \begin{pmatrix} a_1 & b_2 \\ c_2 & d_2 \end{pmatrix} = \begin{pmatrix} a_1 + a_2 & b_1 + b_2 \\ c_1 + c_2 & d_1 + d_2 \end{pmatrix}$$

then it easy to see that V is *not* a linear space because it is not closed under addition. For example,

$$\begin{pmatrix} 2 & 8 \\ 1 & 4 \end{pmatrix} \quad \text{and} \quad \begin{pmatrix} 3 & 6 \\ 2 & 4 \end{pmatrix}$$

are both elements of V, but their sum

$$\begin{pmatrix} 5 & 14 \\ 3 & 8 \end{pmatrix}$$

is not in V.

The linear spaces in Examples 2.2 and 2.3 are typical of those occurring in the calculus of variations; they are spaces whose objects are functions and therefore are often called *function spaces*. The most common function spaces occurring in applied analysis are the spaces $C^n[a, b]$, that is, the set of all continuous functions on the interval $a \leq x \leq b$ whose nth derivatives exist and are continuous. The operations in $C^n[a, b]$ are those defined in Example 2.2.

A linear space is said to have a norm on it if there is a rule that uniquely determines the size of a given element in the space. In particular if V is a linear space, then a *norm* on V is a mapping that associates to each $y \in V$ a nonnegative real number denoted by $\|y\|$, called the norm of y, and that satisfies the following conditions:

8. $\|y\| = 0$ if, and only if, $y = 0$.
9. $\|\alpha y\| = |\alpha| \, \|y\|$ for all $\alpha \in R^1$, $y \in V$.
10. $\|y_1 + y_2\| \leq \|y_1\| + \|y_2\|$ for all $y_1, y_2 \in V$.

A *normed linear space* is a linear space V on which there is defined a norm $\|\cdot\|$. The number $\|y\|$ is interpreted as the magnitude or size of y. Thus a norm puts geometric structure on V.

Example 2.5 The linear space R^n is a normed linear space if we define a norm by

$$\|\mathbf{y}\| = \sqrt{y_1^2 + \cdots + y_n^2}, \qquad \mathbf{y} = (y_1, y_2, \ldots, y_n)$$

It is not difficult to show that this norm satisfies the conditions 8 through 10.

A norm on a linear space is not unique. In R^n the quantity

$$\|\mathbf{y}\| = \max\{|y_1|, |y_2|, \ldots, |y_n|\}$$

also defines a norm.

Example 2.6 $C[a, b]$, the linear space of all continuous functions on $[a, b]$, is a normed linear space if we define a norm by

$$\|y\|_M = \max_{a \le x \le b} |y(x)| \tag{1}$$

The quantity

$$\|y\|_1 = \int_a^b |y(x)| \, dx \tag{2}$$

also defines a norm on $C[a, b]$. Either (1) or (2) provides a way to define the magnitude of a continuous function on $[a, b]$ and both satisfy the rules 8 through 10.

Example 2.7   In the linear space $C^1[a, b]$ the norm, given by

$$\|y\|_w = \max_{a \le x \le b} |y(x)| + \max_{a \le x \le b} |y'(x)| \tag{3}$$

is called the *weak norm* and it plays an important role in the calculus of variations. The norm is also defined in $C^n[a, b]$, $n \ge 2$.

If V is a normed linear space with norm $\|\cdot\|$, then the *distance* between the elements y_1 and y_2 in V is defined by

$$\|y_1 - y_2\|$$

The motivation for this definition comes from the set of real numbers R^1, where the absolute value $|x|$ is a norm. The distance between two real numbers a and b is $|a - b|$. As another example the distance between two functions y_1 and y_2 in $C^1[a, b]$ measured in the weak norm (3) is

$$\|y_1 - y_2\|_w = \max_{a \le x \le b} |y_1(x) - y_2(x)| + \max_{a \le x \le b} |y_1'(x) - y_2'(x)| \tag{4}$$

Having the notion of distance allows us to talk about closeness. Two elements of a normed linear space are close if the distance between them is small. Two elements y_1 and y_2 in $C^1[a, b]$ are close in the weak norm (3) if the quantity $\|y_1 - y_2\|_w$ defined by (4) is small; this means that y_1 and y_2 have values that do not differ greatly as well as derivatives that do not differ greatly. The two functions y_1 and y_2 shown in Fig. 3.3 are not close in the

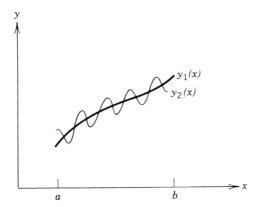

Figure 3.3. Two functions close in the strong norm but not close in the weak norm.

weak norm. In the *maximum norm* (also called the *strong* norm) however, defined by (1), the two functions are close, since $\max|y_1(x) - y_2(x)|$ is small. Hence closeness in one norm does not imply closeness in another.

Derivatives of Functionals

If we assume the set of admissible functions A is a subset of a normed linear space V, then the set A can inherit the norm and algebraic properties of the space V. Hence the notions of *local* maxima and minima make sense. If J: $A \to R^1$ is a functional on A, where $A \subseteq V$ and V a normed linear space with norm $\| \cdot \|$, then J has a *local minimum* at $y_0 \in A$, provided $J(y_0) \leq J(y)$ for all $y \in A$ with $\|y - y_0\| < d$, for some positive number d. In the special case that A is a set of functions and $A \subseteq C^1[a, b]$ with the weak norm defined by (3) we say J has a *weak relative minimum* at y_0. If the norm on $C^1[a, b]$ is the strong norm (1), then we speak of a *strong relative minimum*. Generally the norm comes into play only in the theory of sufficient conditions for extrema.

To motivate the definition of a derivative of a functional let us rewrite the limit definition of derivative of a function f at x_0 at the beginning of this section as

$$f(x_0 + \Delta x) - f(x_0) = f'(x_0)\,\Delta x + o(\Delta x)$$

where $o(\Delta x)$ denotes higher order terms in Δx having the property that

$$\lim_{\Delta x \to 0} \frac{o(\Delta x)}{\Delta x} = 0$$

that is, they go to zero faster than Δx. Thus we observe that the *differential* of f at x_0, defined by $df(x_0, \Delta x) = f'(x_0)\Delta x$, is the linear part in the increment

Δx of the total change $\Delta f \equiv f(x_0 + \Delta x) - f(x_0)$. That is

$$\Delta f = df(x_0, \Delta x) + \text{higher order terms in } \Delta x$$

With these notions in mind let us consider a functional $J\colon A \to R^1$, where A is a subset of a normed linear space V. Let $y_0 \in A$. To increment y_0 we fix an element $h \in V$ such that $y_0 + \varepsilon h$ is in A for all real numbers ε sufficiently small. The increment εh is called the *variation of the function* y_0 and is often denoted by δy_0, that is, $\delta y_0 \equiv \varepsilon h$. Then we define the total change in the functional J due to the change εh in y_0 by

$$\Delta J = J(y_0 + \varepsilon h) - J(y_0)$$

Our goal is to calculate the linear part of this change. To accomplish this we define the function

$$\mathcal{J}(\varepsilon) \equiv J(y_0 + \varepsilon h)$$

So, $\mathcal{J}$ is a function of a real variable ε defined by evaluating the functional J on the one parameter family of functions $y_0 + \varepsilon h$. Then, assuming $\mathcal{J}$ is sufficiently differentiable we can expand $\mathcal{J}(\varepsilon)$ by Taylor's theorem to get

$$\mathcal{J}(\varepsilon) = \mathcal{J}(0) + \mathcal{J}'(0)\varepsilon + \frac{\mathcal{J}''(0)}{2!}\varepsilon^2 + \cdots$$

or

$$J(y_0 + \varepsilon h) - J(y_0) = \mathcal{J}'(0)\varepsilon + \frac{\mathcal{J}''(0)}{2!}\varepsilon^2 + \cdots$$

Therefore

$$\Delta J = \mathcal{J}'(0)\varepsilon + \text{higher order terms in } \varepsilon$$

Thus we are led to the following definition.

Definition 2.1   *Let $J\colon A \to R^1$ be a functional on A, where $A \subseteq V$, V a normed linear space. Let $y_0 \in A$ and $h \in V$ such that $y_0 + \varepsilon h \in A$ for all ε sufficiently small. Then the first variation (or Gâteaux variation) of J at y_0 in the direction of h is defined by*

$$\delta J(y_0, h) \equiv \mathcal{J}'(0) \equiv \frac{d}{d\varepsilon}J(y_0 + \varepsilon h)|_{\varepsilon=0} \tag{5}$$

provided the derivative exists. Such a direction h for which (5) exists is called an admissible variation at y_0.

Thus it is clear that δJ is like a differential; the quantity $\varepsilon\,\delta J(y_0, h)$ is the lowest order term in the increment ΔJ. There is still a closer analogy of δJ with a directional derivative. Let $f(\mathbf{X})$ be a real valued function defined for $\mathbf{X} = (x, y)$ in R^2, and let $\mathbf{n} = (n_1, n_2)$ be a unit vector. We recall from calculus that the directional derivative of f at $\mathbf{X}_0 = (x_0, y_0)$ in the direction $\mathbf{n}$ is defined by

$$D_n f(\mathbf{X}_0) \equiv \lim_{\varepsilon \to 0} \frac{f(\mathbf{X}_0 + \varepsilon\mathbf{n}) - f(\mathbf{X}_0)}{\varepsilon}$$

When (5) is written out in limit form it becomes

$$\delta J(y_0, h) = \lim_{\varepsilon \to 0} \frac{J(y_0 + \varepsilon h) - J(y_0)}{\varepsilon}$$

and the analogy stands out clearly.

In summary, to calculate the first variation $\delta J(y_0, h)$ we evaluate J at $y_0 + \varepsilon h$, where h is chosen such that $y_0 + \varepsilon h \in A$. Then we take the derivative with respect to ε, afterwards setting $\varepsilon = 0$.

In some special cases it is easy to provide a geometrical interpretation. Let $J: A \to R^1$, where $A = C^1[0, 1]$ and let $y_0 \in C^1[0, 1]$. If $h \in C^1[0, 1]$, then $y_0 + \varepsilon h$ belongs to $C^1[0, 1]$ and represents a one parameter family of curves close to y_0 (see Fig. 3.4). It is along this family of curves that we take a limit (as $\varepsilon \to 0$, $y_0 + \varepsilon h \to y_0$) in order to define the differential $\delta J(y_0, h)$. Choosing a different h gives a different set of *varied curves* and hence a different differential. Hence $\delta J(y_0, h)$ depends on which function h is chosen to define the increment δy_0 and this dependence is explicitly shown in the notation.

Example 2.8 Let $J(y) = \exp(\int_0^1 y'(x)^2\,dx)$, $y \in C^1[0, 1]$. In order to calculate $\delta J(y_0, h)$ we choose $h \in C^2[0, 1]$ and form

$$J(y_0 + \varepsilon h) = \exp\left(\int_0^1 [y_0'(x) + \varepsilon h'(x)]^2\,dx\right)$$

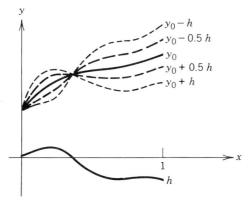

Figure 3.4. A one parameter family of varied curves $y_0(x) + \varepsilon h(x)$ for values $\varepsilon = -0.5, 0, 0.5$.

Then

$$\delta J(y_0, h) = \frac{d}{d\varepsilon} \left[\exp \int_0^1 (y_0'(x) + \varepsilon h'(x))^2 \, dx \right]_{\varepsilon=0}$$

$$= \exp\left(\int_0^1 y_0'(x)^2 \, dx \right) \int_0^1 2 y'(x) h'(x) \, dx$$

Example 2.9   Let $J(y) = \exp(\int_0^1 y'(x)^2 \, dx)$, where $y \in C^1[0, 1]$ with $y(0) = 1$ and $y(1) = 2$. Now the admissible functions are class C^1 functions, which are required to satisfy boundary conditions at $x = 0$ and $x = 1$. Let $y_0 \in A$. In order to have $y_0 + \varepsilon h$ an admissible function we must choose $h \in C^1[0, 1]$ *and* $h(0) = h(1) = 0$. Fixing h at the endpoints will have no effect on the calculation of δJ made in Example 2.8. It will have an effect, however, when we further manipulate the expression for δJ to make it more tractable for analysis.

Necessary Conditions

Let $J: A \to R^1$ be a given functional on A, where A is a subset of some normed linear space V and let $y_0 \in A$ provide a local minimum for J relative to the norm $\| \cdot \|$ in V. Guided by what we know from the calculus we might expect that the first variation δJ of J should vanish at y_0. It is not difficult to see that this is indeed the case. For, let $h \in V$ be such that $y_0 + \varepsilon h$ is in A for sufficiently small values of ε. Then the function $\mathscr{I}(\varepsilon)$ defined by

$$\mathscr{I}(\varepsilon) = J(y_0 + \varepsilon h)$$

has a local minimum at $\varepsilon = 0$. Hence from ordinary calculus we must have $\mathscr{I}'(0) = 0$ or $\delta J(y_0, h) = 0$. This follows no matter which $h \in V$ is chosen. Hence we have proved the following theorem that gives a necessary condition for a local minimum.

Theorem 2.1   *Let $J: A \to R^1$ be a given functional, $A \subseteq V$, where V is a normed linear space. If $y_0 \in A$ provides a local minimum for J relative to the norm $\| \cdot \|$, then*

$$\delta J(y_0, h) = 0 \tag{6}$$

for all admissible variations h.

   The fact that the condition (6) holds for all *admissible variations h* often allows us to eliminate h from the condition and obtain an equation just in terms of y_0, which can then be solved for y_0. Generally the equation for y_0 is a differential equation. Since (6) is a necessary condition we are not guaranteed

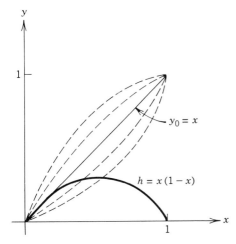

Figure 3.5. The one parameter family of curves $x + \varepsilon x(1 - x)$ for different values of ε.

that solutions y_0 actually will provide a minimum. Therefore the solutions y_0 to (6) are called (*local*) *extremals* or *stationary functions*, and are the candidates for maxima and minima. If $\delta J(y_0, h) = 0$, we say J is *stationary* at y_0 in the direction h.

Example 2.10 Consider the functional $J(y) = \int_0^1 (1 + y'(x)^2)\, dx$, where $y \in C^1[0, 1]$ and $y(0) = 0$, $y(1) = 1$. Let $y_0(x) = x$ and $h(x) = x(1 - x)$. The family of curves $y_0 + \varepsilon h$ is given by $x + \varepsilon x(1 - x)$ and a few members are sketched in Fig. 3.5. We evaluate J on the family $y_0 + \varepsilon h$ to get

$$\mathscr{J}(\varepsilon) = J(y_0 + \varepsilon h)$$

$$= \int_0^1 \left[1 + (y_0'(x) + \varepsilon h'(x))^2\right] dx$$

$$= \int_0^1 \left[1 + (1 + \varepsilon(1 - 2x))^2\right] dx = 2 + \frac{\varepsilon^2}{3}$$

Then $\mathscr{J}'(\varepsilon) = \frac{2}{3}\varepsilon$ and $\mathscr{J}'(0) = 0$. Hence we conclude that $\delta J(y_0, h) = 0$ and J is stationary at $y_0 = x$ in the direction $h = x(1 - x)$.

EXERCISES

2.1   Determine whether the given set constitutes a real linear space under the usual operations associated with elements of the set.

    (a) The set of elements in R^3 with first component 0.

    (b) The set of elements in R^3 with first component 1.

(c) The set of all polynominals of degree ≤ 2.

(d) The set of all nonsingular (determinant nonzero) 2 by 2 matrices.

(e) The set of all continuous functions on $[0, 1]$ satisfying $f(0) = 0$.

(f) The set of all continuous functions on $[0, 1]$ satisfying $f(1) = 1$.

2.2   Prove that (1) and (2) both satisfy properties (8) through (10) for a norm on $C[a, b]$.

2.3   In $C^1[0, 1]$ compute the distances between the two functions $y_1(x) = 0$ and $y_2(x) = \frac{1}{100} \sin(1000x)$ in both the weak and strong norms.

2.4   Show that the first variation $\delta J(y_0, h)$ satisfies the homogeneity condition

$$\delta J(y_0, \alpha h) = \alpha \, \delta J(y_0, h), \quad \alpha \in R^1$$

2.5   A functional $J: V \to R^1$, where V is a normed linear space, is *linear* if

$$J(y_1 + y_2) = J(y_1) + J(y_2), \qquad y_1, y_2 \in V$$

and

$$J(\alpha y_1) = \alpha J(y_1), \qquad \alpha \in R^1, \quad y_1 \in V$$

Which of the following functionals on $C^1[a, b]$ are linear?

(a) $J(y) = \int_a^b yy' \, dx$

(b) $J(y) = \int_a^b (y'^2 + 2y) \, dx$

(c) $J(y) = \exp(y(a))$

(d) $J(y) = \int_a^b \int_a^b K(x, t) y(x) y(t) \, dx dt$

(e) $J(y) = \int_a^b y \sin x \, dx$

(f) $J(y) = \int_a^b y'^2 \, dx + G(y(b))$
where G is a given differentiable function.

2.6   A functional $J: V \to R^1$, where V is a normed linear space, is *continuous* at $y_0 \in V$ if for any $\varepsilon > 0$ there is a $\delta > 0$ such that $|J(y) - J(y_0)| < \varepsilon$ whenever $\|y - y_0\| < \delta$.

(a) Let $V = C^1[a, b]$. Prove that if J is continuous in the strong norm, then it is continuous in the weak norm.

(b) Give an example to show that continuity in the weak norm does not imply continuity in the strong norm. *Hint*: Arclength functional.

2.7   Show that the functional $J(y) = \int_a^b (\sin^3 x + y^2) \, dx$ is continuous in the strong norm at any $y_0 \in C[a, b]$.

2.8   Compute the first variation of the functionals on $C^1[a, b]$ given in Exercise 2.5a–2.5f.

2.9   Let $J: V \to R^1$ be a linear functional and V a normed linear space. Prove that if J is continuous at $y_0 = 0$, then it is continuous at each $y \in V$.

2.10   The second variation of a functional $J: A \to R^1$ at $y_0 \in A$ in the direction h is defined by

$$\delta^2 J(y_0, h) \equiv \frac{d^2}{d\varepsilon^2} J(y_0 + \varepsilon h)|_{\varepsilon = 0}$$

Find the second variation of the functional

$$J(y) = \int_0^1 (xy'^2 + y \sin y') \, dx$$

where $y \in C^2[0, 1]$.

2.11   If $J(y) = \int_0^1 (x^2 - y^2 + y'^2) \, dx$, $y \in C^2[0, 1]$, calculate ΔJ and $\delta J(y, h)$ when $y(x) = x$ and $h(x) = x^2$.

2.12   Let V be the set of all continuous functions on R^1 that fail to have a first derivative at $x = 0$. With the usual operations of addition and scalar multiplication, is V a linear space?

2.13   Consider the functional $J(y) = \int_0^{2\pi} y'^2 \, dx$. Sketch the function $y_0(x) = x$ and the varied family $y_0(x) + \varepsilon h(x)$, where $h(x) = \sin x$. Compute $\mathcal{J}(\varepsilon) \equiv J(y_0 + \varepsilon h)$ and show that $\mathcal{J}'(0) = 0$. Deduce that J is stationary at x in the direction $\sin x$.

2.14   Let $J(y) = \int_0^1 (3y^2 + x) \, dx + y(0)^2$, where y is a continuous function on $0 \le x \le 1$. Let $y = x$ and $h = x + 1, 0 \le x \le 1$. Compute $\delta J(x, h)$. *Answer:* $\delta J(x, h) = 5$.

3.3 THE SIMPLEST PROBLEM

The Euler Equation

What is often called the *simplest problem* in the calculus of variations is to find a local minimum for the functional

$$J(y) = \int_a^b L(x, y, y') \, dx \tag{1}$$

where $y \in C^2[a, b]$ and $y(a) = y_0$, $y(b) = y_1$ Here L is a given function that is twice continuously differentiable on $[a, b] \times R^2$, and $\| \cdot \|$ is some norm in $C^2[a, b]$. Actually, our assumption that y is of class C^2 is too strong; it is possible to obtain the same results for $y \in C^1[a, b]$. The increased smoothness assumption, however, will make our work a little easier, so here and in the sequel we shall assume that the competing functions are class C^2. The references at the end of the chapter can be consulted to see how the analysis changes if smoothness conditions are relaxed.

We seek a necessary condition. Guided by the last two sections we compute the first variation of J. Let y be a local minimum and h a twice continuously differentiable function satisfying $h(a) = h(b) = 0$. Then $y + \varepsilon h$ is an admissible function and

$$J(y + \varepsilon h) = \int_a^b L(x, y + \varepsilon h, y' + \varepsilon h') \, dx$$

Therefore

$$\frac{d}{d\varepsilon} J(y + \varepsilon h) = \int_a^b \frac{\partial}{\partial \varepsilon} L(x, y + \varepsilon h, y' + \varepsilon h') \, dx$$

$$= \int_a^b \left\{ L_y(x, y + \varepsilon h, y' + \varepsilon h')h + L_{y'}(x, y + \varepsilon h, y' + \varepsilon h')h' \right\} dx$$

where L_y denotes $\partial L/\partial y$ and $L_{y'}$ denotes $\partial L/\partial y'$. Hence

$$\delta J(y, h) = \frac{d}{d\varepsilon} J(y + \varepsilon h)\big|_{\varepsilon=0} = \int_a^b \left\{ L_y(x, y, y')h + L_{y'}(x, y, y')h' \right\} dx$$

So a necessary condition for $y(x)$ to be a local minimum is

$$\int_a^b \left\{ L_y(x, y, y')h + L_{y'}(x, y, y')h' \right\} dx = 0 \tag{2}$$

for all $h \in C^2[a, b]$ with $h(a) = h(b) = 0$.

Condition (2) is not useful as it stands for determining $y(x)$. Using the fact that it must hold for all h, however, we can thus eliminate h and thereby obtain a condition for y alone. First we integrate the second term in (2) by parts to obtain

$$\int_a^b \left(L_y(x, y, y') - \frac{d}{dx} L_{y'}(x, y, y') \right) h \, dx + L_{y'}(x, y, y')h \big|_{x=a}^{x=b} = 0 \tag{3}$$

Since h vanishes at a and b, the boundary terms vanish and the necessary condition becomes

$$\int_a^b \left(L_y(x, y, y') - \frac{d}{dx} L_{y'}(x, y, y') \right) h \, dx = 0 \tag{4}$$

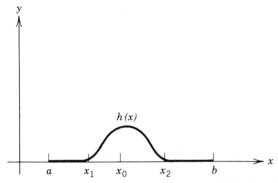

Figure 3.6. Graph of the twice continuously differentiable function $h(x)$.

for all $h \in C^2[a, b]$ with $h(a) = h(b) = 0$. The following lemma due to Lagrange provides the final step in our calculation by effectively allowing us to eliminate h from condition (4). This lemma is one version of what is called the fundamental lemma of the calculus of variations.

Lemma 3.1 *If $f(x)$ is continuous on $[a, b]$ and if*

$$\int_a^b f(x)h(x)\, dx = 0 \tag{5}$$

for every twice continuously differentiable function h with $h(a) = h(b) = 0$, then $f(x) = 0$ for $x \in [a, b]$.

Proof In order to obtain a contradiction assume for some x_0 in (a, b) that $f(x_0) > 0$. Since f is continuous, $f(x) > 0$ for all x in some interval (x_1, x_2) containing x_0. For $h(x)$ choose

$$h(x) = \begin{cases} (x - x_1)^3(x_2 - x)^3, & x_1 \leq x \leq x_2 \\ 0, & \text{otherwise} \end{cases}$$

a graph of which is shown in Fig. 3.6. The cube of the factors $(x - x_1)$ and $(x_2 - x)$ appears so that h is smoothed out at x_1 and x_2 and is therefore of class C^2. Then

$$\int_a^b f(x)h(x)\, dx = \int_{x_1}^{x_2} f(x)(x - x_1)^3(x_2 - x)^3\, dx > 0$$

since f is positive on (x_1, x_2). This contradicts (5) and the lemma is proved.

Applying Lemma 3.1 to (4) with $f(x) = L_y(x, y, y') - (d/dx)L_{y'}(x, y, y')$, which is continuous since L is twice continuously differentiable, we obtain the following theorem.

Theorem 3.1 *If a function y provides a local minimum to the functional*

$$J(y) = \int_a^b L(x, y, y') \, dx$$

where $y \in C^2[a, b]$ and

$$y(a) = y_0, \qquad y(b) = y_1$$

then y must satisfy the equation

$$L_y(x, y, y') - \frac{d}{dx} L_{y'}(x, y, y') = 0, \qquad x \in [a, b] \qquad (6)$$

Equation (6) is called the *Euler equation* or *Euler-Lagrange equation*. It is a second order ordinary differential equation that is in general nonlinear. It represents a *necessary* condition for a local minimum and it is analogous to the derivative condition $F'(x) = 0$ in differential calculus. Therefore its solutions are not necessarily local minima. In general the solutions of (6) are called (*local*) *extremals*. If y is an extremal, then $\delta J(y, h) = 0$ for all h and we say that J is *stationary* at y. An extremal y is sometimes called a *stationary function* for J. By writing out the total derivative in (6) the Euler equation becomes

$$L_y - L_{y'x}(x, y, y') - L_{y'y}(x, y, y')y' - L_{y'y'}(x, y, y')y'' = 0$$

Thus it is evident that the Euler equation is second order provided $L_{y'y'} \neq 0$.

Solved Examples

We now determine the extremals for some specific variational problems.

Example 3.1 Find the extremals of the functional

$$J(y) = \int_0^1 (y'^2 + 3y + 2x) \, dx, \qquad y(0) = 0, \quad y(1) = 1$$

The Lagrangian is $L = y'^2 + 3y + 2x$. Then $L_y = 3$, $L_{y'} = 2y'$ and the Euler equation becomes

$$3 - \frac{d}{dx}(2y') = 0 \quad \text{or} \quad y'' = \frac{3}{2}$$

Integrating twice gives the extremals

$$y = \frac{3}{4}x^2 + C_1 x + C_2$$

The boundary conditions readily give $C_2 = 0$ and $C_1 = \frac{1}{4}$. Therefore the extremal of J satisfying the boundary conditions is

$$y = \frac{3}{4}x^2 + \frac{1}{4}x$$

Further calculations, which are not discussed in this work, are required to determine whether this represents a maximum or minimum.

Example 3.2 We now show that extremals for the arclength functional are straight lines. To this end let

$$J(y) = \int_a^b \sqrt{1 + y'^2} \, dx$$

where $y \in C^2[a, b]$, $y(a) = y_0$ and $y(b) = y_1$. We are allowing all C^2 functions passing through the fixed points (a, y_0) and (b, y_1) to compete. A necessary condition for J to be a local minimum at y is that y satisfy the Euler equation

$$L_y - \frac{d}{dx}L_{y'} = 0$$

or

$$\frac{d}{dx}\left(\frac{y'}{\sqrt{1 + y'^2}} \right) = 0$$

Hence

$$\frac{y'}{\sqrt{1 + y'^2}} = C$$

where C is a constant. Solving for y' gives

$$y' = K$$

where K is the constant $C^2/(1 - C^2)$. Integrating the last equation gives

$$y = Kx + M, \qquad M \quad \text{constant}$$

Therefore the extremals are straight lines. Now we apply the boundary conditions to determine K and M. The condition $y(a) = y_0$ implies

$$y_0 = Ka + M$$

The condition $y(b) = y_1$ implies

$$y_1 = Kb + M$$

Solving the last two equations for K and M gives

$$K = \frac{y_1 - y_0}{b - a}, \qquad M = \frac{by_0 - ay_1}{b - a}$$

Hence the unique extremal is

$$y = \frac{y_1 - y_0}{b - a}x + \frac{by_0 - ay_1}{b - a}$$

which is the equation of the straight line between (a, y_0) and (b, y_1).

First Integrals

The Euler equation is a second order ordinary differential equation and therefore the solution will generally contain two arbitrary constants. In special cases when the Lagrangian does not depend explicitly upon one of its variables x, y, or y', it is possible to make an immediate simplification of the Euler equation. We consider three cases:

1. If $L = L(x, y)$, then the Euler equation is $L_y(x, y) = 0$, which is an algebraic equation.
2. If $L = L(x, y')$, then the Euler equation is

$$L_{y'}(x, y') = \text{constant} \tag{7}$$

3. If $L = L(y, y')$, then the Euler equation is

$$L(y, y') - y'L_{y'}(y, y') = \text{constant} \tag{8}$$

Propositions 1 and 2 are clear. To prove 3 we note that

$$\frac{d}{dx}\left[L - y'L_{y'}\right] = \frac{dL}{dx} - y'\frac{d}{dx}L_{y'} - L_{y'}y''$$

$$= L_y y' + L_{y'}y'' - y'\frac{d}{dx}L_{y'} - L_{y'}y''$$

$$= y'\left(L_y - \frac{d}{dx}L_{y'}\right)$$

$$= 0$$

from which (8) follows.

Equations (7) and (8) are first integrals of the Euler equation. A *first integral* of a second order differential equation $F(x, y, y', y'') = 0$ is an expression of the form $g(x, y, y')$, involving one lower derivative, which is constant whenever y is a solution of the original equation $F(x, y, y', y'') = 0$. Hence

$$g(x, y, y') = \text{constant}$$

represents an integration of the second order equation. In the calculus of variations if L is independent of x, then $L - y'L_{y'}$ is a first integral of the Euler equation and we say $L - y'L_{y'}$ is constant on the extremals.

Example 3.3   We now find the extremals of the brachistochrone problem that was posed in Example 1.2. The functional is

$$J(y) = \int_{x_1}^{x_2} \frac{\sqrt{1 + y'^2}}{\sqrt{2g(y_1 - y)}} \, dx$$

subject to the boundary conditions $y(x_1) = y_1$, $y(x_2) = y_2$, $y_1 > y_2$. The Lagrangian is independent of x and therefore a first integral of the Euler equation is given by (8). Then

$$\frac{\sqrt{1 + y'^2}}{\sqrt{y_1 - y}} - (y')^2 \frac{(1 + y'^2)^{-1/2}}{\sqrt{y_1 - y}} = C$$

which simplifies to

$$\left(\frac{dy}{dx} \right)^2 = \frac{1 - C^2(y_1 - y)}{C^2(y_1 - y)}$$

Taking the square root of both sides and separating variables gives

$$dx = - \frac{\sqrt{y_1 - y}}{\sqrt{C_1 - (y_1 - y)}} \, dy, \qquad C_1 = C^{-2}$$

where the minus sign is taken since $dy/dx < 0$. The last equation can be integrated by making the trigonometric substitution

$$y_1 - y = C_1 \sin^2 \frac{\phi}{2} \tag{9}$$

One obtains

$$dx = C_1 \sin^2 \frac{\phi}{2} \, d\phi = \frac{C_1}{2} (1 - \cos \phi) \, d\phi$$

Integrating gives

$$x = \frac{C_1}{2}(\phi - \sin \phi) + C_2 \tag{10}$$

Equations (9) and (10) are parametric equations for a cycloid. Here, in contrast to the problem of finding the curve of shortest length between two points, it is not clear that the cycloids just obtained minimize the given functional. Further calculations would be required in this case.

Example 3.4 (Fermat's Principle) In the limit of geometrical optics the principle of Fermat (1601–1665) states that the time elapsed in the passage of light between two fixed points in a medium is an extremum with respect to all possible paths connecting the points. For simplicity we consider only light rays that lie in the xy plane. Let $c = c(x, y)$ be a positive continuously differentiable function representing the velocity of light in the medium. Its reciprocal $n = c^{-1}$ is called the index of refraction of the medium. If $P: (x_1, y_1)$ and $Q: (x_2, y_2)$ are two fixed points in the plane, then the time required for light to travel along a given path $y = y(x)$ connecting the two points is

$$T(y) = \int_P^Q \frac{ds}{c} = \int_{x_1}^{x_2} n(x, y)\sqrt{1 + y'^2}\, dx \tag{11}$$

Therefore, the actual light path connecting P and Q is the one that extremizes the integral (11). The differential equation for the extremal path is the Euler equation, which in this case is

$$n_y(x, y)\sqrt{1 + y'^2} - \frac{d}{dx}\left(\frac{n(x, y)y'}{\sqrt{1 + y'^2}}\right) = 0$$

If the index of refraction does not depend on x, then a first integral is given by (8), or

$$\frac{n(y)}{\sqrt{1 + y'^2}} = \text{constant}$$

EXERCISES

3.1   Find the extremals

(a) $J(y) = \int_0^1 y'\, dx,$ $y \in C^2[0, 1],$ $y(0) = 0,$ $y(1) = 1$

(b) $J(y) = \int_0^1 yy'\, dx,$ $y \in C^2[0, 1],$ $y(0) = 0,$ $y(1) = 1$

(c) $J(y) = \int_0^1 xyy'\, dx,$ $y \in C^2[0, 1],$ $y(0) = 0,$ $y(1) = 1$

3.2   Find extremals for the following functionals:

(a) $\int_a^b \dfrac{y'^2}{x^3} dx$

(b) $\int_a^b (y^2 + y'^2 + 2 y e^x) dx$

3.3   Show that the Euler equation for the functional

$$J(y) = \int_a^b f(x, y)\sqrt{1 + y'^2}\, dx$$

has the form

$$f_y - f_x y' - \frac{fy''}{1 + y'^2} = 0$$

3.4   Find an extremal for

$$J(y) = \int_1^2 \frac{\sqrt{1 + y'^2}}{x}\, dx, \qquad y(1) = 0, \quad y(2) = 1$$

3.5   Calculate the second variation of the functional

$$J(y) = \int_a^b L(x, y, y')\, dx, \qquad y \in C^2[a, b]$$

3.6   Obtain a necessary condition for a function $y \in C[a, b]$ to be a local minimum of the functional

$$J(y) = \iint_R K(s, t)\, y(s)\, y(t)\, ds\, dt + \int_a^b y(t)^2\, dt - 2\int_a^b y(t) f(t)\, dt$$

where $K(s, t)$ is a given continuous function of s and t on the square $R: \{(s, t)|a \le s \le b, a \le t \le b\}$, $K(s, t) = K(t, s)$, and $f \in C[a, b]$. *Answer*: The condition is a Fredholm integral equation.

3.7   Consider the functional $J(x) = \int_a^b L(x, y, y')\, dx$, where $y \in C^2[a, b]$. Let $h, g \in C^2[a, b]$ and define the *modified* variation of J at y in the directions h and g by

$$\hat{\delta}J(y; h, g) \equiv \frac{d}{d\varepsilon} \int_a^b L(x, y_\varepsilon, z_\varepsilon)\, dt \bigg|_{\varepsilon = 0}$$

provided the limit exists, where

$$y_\varepsilon = y + \varepsilon h$$
$$z_\varepsilon = y' + \varepsilon g$$

for ε sufficiently small. (Note that here y and y' are being *varied* independently, for we have not assumed that $g = h'$). If $y \in C^2[a, b]$ is a local minimum, does it follow that $\delta J(y; h, g) = 0$ for all $h, g \in C^2[a, b]$?

3.8   Describe the paths of light rays in the plane where the medium has index of refraction given by

(a) $n = ky$ (b) $n = kx$

(c) $n = ky^{-1}$ (d) $n = ky^{1/2}$,

(e) $n = ky^{-1/2}$,   where $k > 0$

3.9   Show that the Euler equation for the problem

$$J(y) = \int L(t, y, \dot{y}) \, dt$$

can be written in the form

$$L_t - \frac{d}{dt}(L - \dot{y}L_{\dot{y}}) = 0$$

3.10   Show that the minimal area of a surface of revolution is a catenoid, that is, the surface found by revolving a catenary

$$y = c_1 \cosh\left(\frac{x + c_1}{c_1}\right)$$

about the x axis.

3.11   Find the extremals

$$J(y) = \int_a^b (x^2 y'^2 + y^2) \, dx, \qquad y \in C^2[a, b]$$

3.4   GENERALIZATIONS

Higher Derivatives

In the simplest variational problem discussed in Section 3.3 the Lagrangian depended upon x, y, and y'. An obvious generalization is to include higher derivatives in the Lagrangian. Thus we consider the *second order problem*

$$J(y) = \int_a^b L(x, y, y', y'') \, dx, \qquad y \in A \qquad (1)$$

where A is the set of all functions in $C^4[a, b]$ that satisfy the boundary conditions

$$y(a) = A_1, \qquad y'(a) = A_2, \qquad y(b) = B_1, \qquad y'(b) = B_2 \qquad (2)$$

The function L is assumed to be twice continuously differentiable in each of its four arguments.

Again we seek a necessary condition. Let $y \in A$ provide a local minimum for J with respect to some norm in $C^4[a, b]$ and choose $h \in C^4[a, b]$ satisfying

$$h(a) = h'(a) = h(b) = h'(b) = 0 \qquad (3)$$

To compute the first variation we form the function $J(y + \varepsilon h)$. Then

$$\delta J(y, h) = \frac{d}{d\varepsilon} J(y + \varepsilon h)|_{\varepsilon = 0}$$

$$= \frac{d}{d\varepsilon} \int_a^b L(x, y + \varepsilon h, y' + \varepsilon h', y'' + \varepsilon h'') \, dx|_{\varepsilon = 0}$$

$$= \int_a^b \left[L_y h + L_{y'} h' + L_{y''} h'' \right] dx$$

where L_y, $L_{y'}$, and $L_{y''}$ are evaluated at (x, y, y', y''). In this calculation two integrations by parts are required. Thus

$$L_{y'} h' = \frac{d}{dx}(L_{y'} h) - h \frac{d}{dx} L_{y'}$$

and

$$L_{y''} h'' = h \frac{d^2}{dx^2} L_{y''} + \frac{d}{dx}\left(L_{y''} h' - \frac{d}{dx} L_{y''} h \right)$$

and therefore

$$\delta J(y, h) = \int_a^b \left[L_y - \frac{d}{dx} L_{y'} + \frac{d^2}{dx^2} L_{y''} \right] h \, dx$$

$$+ \left[L_{y''} h' - \frac{d}{dx} L_{y''} h + h L_{y'} \right]_{x=a}^{x=b} \qquad (4)$$

By (3) the terms on the boundary at $x = a$ and $x = b$ vanish. Therefore, since y provides a relative minimum,

$$\int_a^b \left[L_y - \frac{d}{dx} L_{y'} + \frac{d^2}{dx^2} L_{y''} \right] h \, dx = 0 \qquad (5)$$

for all $h \in C^4[a, b]$ satisfying (3). To complete the argument another version of the lemma of Lagrange is needed (see Lemma 3.1).

Lemma 4.1   *If f is a continuous function on* $[a, b]$ *and if*

$$\int_a^b f(x)h(x)\, dx = 0$$

for all $h \in C^4[a, b]$ *satisfying* (3), *then* $f(x) \equiv 0$ *for* $x \in [a, b]$.

The proof is nearly the same as before and is left as an exercise (Exercise 4.1). Applying this lemma to (5) shows that a necessary condition for y to be a local minimum is

$$L_y - \frac{d}{dx}L_{y'} + \frac{d^2}{dx^2}L_{y''} = 0, \qquad a \le x \le b \tag{6}$$

which is called the *Euler equation* for the second order problem defined by (1). Equation (6) is a fourth order ordinary differential equation for y. Its general solution will involve four arbitrary constants that can be determined from the boundary conditions (2).

The *nth order variational problem* is defined by the functional

$$J(y) = \int_a^b L(x, y, y', y'', \ldots, y^{(n)})\, dx, \quad y \in A$$

where A is the set of all functions in $C^{2n}[a, b]$ satisfying the $2n$ boundary conditions

$$y(a) = A_1, \qquad y'(a) = A_2, \ldots, y^{(n-1)}(a) = A_n$$
$$y(b) = B_1, \qquad y'(b) = B_2, \ldots, y^{(n-1)}(b) = B_n$$

It is not difficult to show that the Euler equation is

$$L_y - \frac{d}{dx}L_{y'} + \frac{d^2}{dx^2}L_{y''} - \cdots + (-1)^n \frac{d^n}{dx^n}L_{y^{(n)}} = 0$$

which is an ordinary differential equation of order $2n$.

Several Functions

A generalization in a different direction is to have the functional J depend on several functions $y_1, \ldots, y_n$. To fix the notion let $n = 2$ and consider the functional

$$J(y_1, y_2) = \int_a^b L(x, y_1, y_2, y_1', y_2')\, dx \tag{7}$$

where y_1, $y_2 \in C^2[a, b]$ with boundary conditions given by

$$y_1(a) = A_1, \qquad y_1(b) = B_1$$
$$y_2(a) = A_2, \qquad y_2(b) = B_2 \qquad (8)$$

In this case suppose y_1 and y_2 provide a local minimum for J. (Note that a minimum now consists of a pair of functions.) We vary each independently by choosing h_1 and h_2 in $C^2[a, b]$, satisfying

$$h_1(a) = h_2(a) = h_1(b) = h_2(b) = 0 \qquad (9)$$

and forming a one parameter admissible pair of functions $y_1 + \varepsilon h_1$ and $y_2 + \varepsilon h_2$. Then

$$\mathcal{J}(\varepsilon) = \int_a^b L(x, y_1 + \varepsilon h_1, y_2 + \varepsilon h_2, y_1' + \varepsilon h_1', y_2' + \varepsilon h_2') \, dx$$

and $\mathcal{J}$ has a local minimum at $\varepsilon = 0$. Now

$$\mathcal{J}'(0) = \int_a^b (L_{y_1} h_1 + L_{y_2} h_2 + L_{y_1'} h' + L_{y_2'} h_2') \, dx$$

Integrating the last two terms in the integrand by parts gives

$$\mathcal{J}'(0) = \int_a^b \left\{ \left(L_{y_1} - \frac{d}{dx} L_{y_1'} \right) h_1 + \left(L_{y_2} - \frac{d}{dx} L_{y_2'} \right) h_2 \right\} dx \qquad (10)$$
$$+ (L_{y_1'} h_1 + L_{y_2'} h_2) \Big|_{x=a}^{x=b}$$

From (9) the boundary terms vanish in (10) and we conclude that

$$\int_a^b \left\{ \left(L_{y_1} - \frac{d}{dx} L_{y_1'} \right) h_1 + \left(L_{y_2} - \frac{d}{dx} L_{y_2'} \right) h_2 \right\} dx = 0 \qquad (11)$$

for all h_1, $h_2 \in C^2[a, b]$. Now picking $h_2 = 0$ on $[a, b]$, it follows that

$$\int_a^b \left(L_{y_1} - \frac{d}{dx} L_{y_1'} \right) h_1 \, dx = 0$$

for all $h_1 \in C^2[a, b]$. By the fundamental lemma, Lemma 3.1,

$$L_{y_1} - \frac{d}{dx} L_{y_1'} = 0 \qquad (12)$$

Now in (11) pick $h_1 = 0$ on $[a, b]$. Then

$$\int_a^b \left(L_{y_2} - \frac{d}{dx} L_{y_2'} \right) h_2 \, dx = 0$$

for all $h_2 \in C^2[a, b]$. Applying Lemma 3.1 again yields

$$L_{y_2} - \frac{d}{dx} L_{y_2'} = 0 \tag{13}$$

Consequently, if the pair y_1, y_2 provides a local minimum for J, then y_1 and y_2 must satisfy the system of two ordinary differential equations (12) and (13). These are known as the Euler equations for the problem. The four arbitrary constants appearing in the general solution can be determined from the boundary data (8).

In general if J depends on n functions, that is, it has the form

$$J(y_1, \ldots, y_n) = \int_a^b L(x, y_1, \ldots, y_n, y_1', \ldots, y_n') \, dx \tag{14}$$

where $y_1 \in C^2[a, b]$, and

$$y_i(a) = A_i, \qquad y_i(b) = B_i, \qquad i = 1, \ldots, n$$

then a necessary condition for $y_1, \ldots, y_n$ to provide a local minimum for J is that $y_1, \ldots, y_n$ satisfy the system of n ordinary differential equations

$$L_{y_i} - \frac{d}{dx} L_{y_i'} = 0, \qquad i = 1, \ldots, n \tag{15}$$

known as the Euler equations for the variational problem (14). If the Lagrangian L in (14) is independent of explicit dependence upon x, that is, $L_x = 0$, then

$$L - \sum_{i=1}^{n} y_i' L_{y_i'} = \text{constant}$$

is a first integral of (15). This is a generalization of (8) in Section 3.3 and its proof is easily demonstrated.

Multiple Integral Problems

Now we consider variational problems where the functionals are defined on a set of admissible functions whose elements are functions of several variables. To fix the notion we first analyze the case of functions of two variables.

Let R be a closed region in the xy plane. By $C^2(R)$ we mean the set of all continuous functions $u = u(x, y)$ defined on R having continuous second partial derivatives on the interior of R. Geometrically $u = u(x, y)$ represents a smooth surface over the region R. For the set of admissible functions A we take the set of functions in $C^2(R)$ whose values are fixed on the curve C, which bounds the region R in the xy plane. Hence $u \in A$ if $u \in C^2(R)$ and

$$u(x, y) = f(x, y), \qquad (x, y) \quad \text{on} \quad C \tag{16}$$

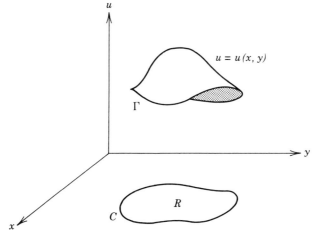

Figure 3.7. Admissible surface $u = u(x, y)$ with boundary Γ defined over the domain R.

where $f(x, y)$ is a given function defined over C and whose values trace out a fixed curve Γ, which forms the boundary of the surface u (see Fig. 3.7).

The variational problem is to minimize

$$J(u) = \iint_R L\Big(x, y, u(x, y), u_x(x, y), u_y(x, y)\Big)\, dx\, dy \qquad (17)$$

where $u \in A$. As before we seek a necessary condition for a minimum. Let $u(x, y)$ provide a local minimum for the functional J and consider the family of admissible functions

$$u(x, y) + \varepsilon h(x, y)$$

where $h \in C^2(R)$ and $h(x, y) = 0$ for (x, y) on C. Then

$$\delta J(u, h) = \frac{d}{d\varepsilon} J(u + \varepsilon h)\Big|_{\varepsilon = 0}$$

$$= \frac{d}{d\varepsilon} \iint_R L\big(x, y, u + \varepsilon h, u_x + \varepsilon h_x, u_y + \varepsilon h_y\big)\, dx\, dy\Big|_{\varepsilon = 0}$$

$$= \iint_R \Big[L_u h + L_{u_x} h_x + L_{u_y} h_y\Big]\, dx\, dy$$

$$= \iint_R \Big[L_u - \frac{\partial}{\partial x} L_{u_x} - \frac{\partial}{\partial y} L_{u_y}\Big] h\, dx\, dy$$

$$+ \iint_R \Big[\frac{\partial}{\partial x}\big(h L_{u_x}\big) + \frac{\partial}{\partial y}\big(h L_{u_y}\big)\Big]\, dx\, dy$$

The second integral can be transformed to a line integral over C using Green's theorem, which states that if P and Q are functions in $C^1(R)$, then

$$\iint_R \left(\frac{\partial Q}{\partial x} - \frac{\partial P}{\partial y} \right) dx\, dy = \int_C P\, dx + Q\, dy$$

Consequently

$$\delta J(u, h) = \iint_R \left[L_u - \frac{\partial}{\partial x} L_{u_x} - \frac{\partial}{\partial y} L_{u_y} \right] h\, dx\, dy$$

$$+ \int_C \left(-hL_{u_y}\, dx + hL_{u_x}\, dy \right) \tag{18}$$

But since $h = 0$ on C, the line integral vanishes and

$$\delta J(u, h) = \iint_R \left[L_u - \frac{\partial}{\partial x} L_{u_x} - \frac{\partial}{\partial y} L_{u_y} \right] h\, dx\, dy \tag{19}$$

Since u is a local minimum it follows that $\delta J(u, h) = 0$ for every $h \in C^2(R)$ with $h(x, y) = 0$ on C. The following generalized version of the fundamental lemma of the calculus of variations permits us to obtain the required necessary condition (compare Lemmas 3.1 and 4.1).

Lemma 4.2   *If $g \in C(R)$ and $\iint_R g(x, y)h(x, y)\, dx\, dy = 0$ for every h in $C^2(R)$ with $h(x, y) = 0$ on C, then $g(x, y) = 0$ for $(x, y) \in R$.*

The proof is left to Exercise 4.7. Applying this lemma to the condition $\delta J(u, h) = 0$, where δJ is given by (19) we obtain the necessary condition

$$L_u - \frac{\partial}{\partial x} L_{u_x} - \frac{\partial}{\partial y} L_{u_y} = 0 \tag{20}$$

which is the Euler equation for the problem (17). Equation (20) is a second order partial differential equation for $u = u(x, y)$. Its solutions are called *extremals* or *extremal surfaces*.

Example 4.1   Find the Euler equation for the functional

$$J(u) = \iint_R \left[\tfrac{1}{2} u_x^2 + \tfrac{1}{2} u_y^2 + \rho(x, y)u \right] dx\, dy$$

where $\rho(x, y)$ is a given function on R. In this problem

$$L_u = \rho, \qquad L_{u_x} = u_x, \qquad L_{u_y} = u_y$$

and so the Euler equation is

$$\rho - \frac{\partial}{\partial x} u_x - \frac{\partial}{\partial y} u_y = 0$$

or

$$u_{xx} + u_{yy} = \rho(x, y)$$

which is Poisson's equation.

Example 4.2 (Plateau's Problem)   Given a fixed closed curve Γ in space (see Fig. 3.7) find the surface $u = u(x, y)$ with boundary Γ whose surface area is least. In this case the problem is to minimize the surface area functional

$$J(u) = \iint_R \sqrt{1 + u_x^2 + u_y^2} \, dx \, dy$$

where R is the region enclosed by the curve C, which is the projection of Γ onto the xy plane. The function u should satisfy the boundary condition

$$u(x, y) = f(x, y), \qquad (x, y) \in C$$

where f is the function defining Γ. Here

$$L_u = 0, \qquad L_{u_x} = \frac{u_x}{\sqrt{1 + u_x^2 + u_y^2}}, \qquad L_{u_y} = \frac{u_y}{\sqrt{1 + u_x^2 + u_y^2}}$$

and the Euler equation is

$$-\frac{\partial}{\partial x} \frac{u_x}{\sqrt{1 + u_x^2 + u_y^2}} - \frac{\partial}{\partial y} \frac{u_y}{\sqrt{1 + u_x^2 + u_y^2}} = 0$$

After simplification, which we leave for the reader, we obtain

$$\left(1 + u_y^2\right) u_{xx} - 2 u_x u_y u_{xy} + \left(1 + u_x^2\right) u_{yy} = 0$$

which is the equation for the minimal surface. This nonlinear partial differential equation is difficult or impossible to solve for a given boundary Γ.

The generalization of these ideas to triple integrals and higher order integrals is straightforward. In the most general case let $u = u(x_1, \ldots, x_m)$ be a function of m variables. Then we consider the integral functional

$$J(u) = \int \cdots \int_{R_m} L\left(x_1, \ldots, x_m, u, \frac{\partial u}{\partial x_1}, \ldots, \frac{\partial u}{\partial x_m}\right) dx_1 \, dx_2 \, \ldots \, dx_m$$

where R_m is a closed region in m dimensional Euclidean space. The Euler equation is in this case

$$L_u - \frac{\partial}{\partial x_1} \frac{\partial L}{\partial u_{x_1}} - \frac{\partial}{\partial x_2} \frac{\partial L}{\partial u_{x_2}} - \cdots - \frac{\partial}{\partial x_m} \frac{\partial L}{\partial u_{x_m}} = 0$$

where $u_{x_i} = \partial u / \partial x_i$. If several functions $u^1(x_1, \ldots, x_m), \ldots, u^n(x_1, \ldots, x_m)$ are involved and J is defined by

$$J(u^1, \ldots, u^n)$$
$$= \int \cdots \int_{R_m} L\left(x_1, \ldots, x_m, u^1, \ldots, u^n,\right.$$
$$\left. u^1_{x_1}, \ldots, u^n_{x_1}, \ldots, u^1_{x_m}, \ldots, u^n_{x_m}\right) dx^1, \ldots, dx^m$$

then the n Euler equations are

$$L_{u^i} - \sum_{j=1}^{m} \frac{\partial}{\partial x_j} \frac{\partial L}{\partial u^i_{x_j}} = 0, \qquad i = 1, \ldots, n$$

Natural Boundary Conditions

Let us consider the following problem: A river with parallel straight banks b units apart has stream velocity given by $\mathbf{v}(x, y) = v(x)\mathbf{j}$, where $\mathbf{j}$ is the unit vector in the y direction (see Fig. 3.8). Assuming that one of the banks is the y axis and that the point $(0, 0)$ is the point of departure, what path should a boat take to reach the opposite bank in the shortest possible time? Assume that the speed of the boat in still water is c, where $c^2 > v^2$.

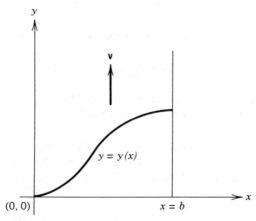

Figure 3.8. Path $y = y(x)$ across a river with velocity $\mathbf{v} = v(x)\mathbf{j}$.

This problem differs from problems in earlier sections in that the right-hand endpoint, the point of arrival on the line $x = b$, is not specified; it must be determined as part of the solution to the problem. It is shown in Exercise 4.24 that the time required for the boat to cross the river along a given path $y = y(x)$ is

$$J(y) = \int_0^b \frac{\sqrt{c^2(1 + y'^2) - v^2} - vy'}{c^2 - v^2}\, dx \qquad (21)$$

Hence the variational problem is to minimize $J(y)$ subject to the conditions

$$y(0) = 0, \qquad y(b) \text{ unspecified}$$

Such a problem is called a *free endpoint problem* and if $y(x)$ is an extremal, then a certain condition must hold at $x = b$. Conditions of these types, called *natural boundary conditions*, are the subject of this section. Just as common are problems where both endpoints are unspecified.

To fix the notion we consider the problem

$$J(y) = \int_a^b L(x, y, y')\, dx \qquad (22)$$

where $y \in C^2[a, b]$ and

$$y(a) = y_0, \qquad y(b) \text{ unspecified} \qquad (23)$$

Let y be a local minimum. Then the variations h must be C^2 functions that satisfy the single condition

$$h(a) = 0 \qquad (24)$$

Figure 3.9 shows several variations $y + \varepsilon h$ of the extremal y. No condition on h at $x = b$ is required, since the admissible functions are unspecified at the right endpoint.

The Gâteaux variation is

$$\delta J(y, h) = \frac{d}{d\varepsilon} \int_a^b L(x, y + \varepsilon h, y' + \varepsilon h')\, dx \Big|_{\varepsilon = 0}$$

$$= \int_a^b \left(L_y - \frac{d}{dx} L_{y'} \right) h\, dx + L_{y'} h \Big|_{x=a}^{x=b}$$

where we have used the same calculation as in Section 3.3. Since $h(a) = 0$, a necessary condition for y to be a local minimum is

$$\int_a^b \left(L_y - \frac{d}{dx} L_{y'} \right) h\, dx + L_{y'}(b, y(b), y'(b)) h(b) = 0 \qquad (25)$$

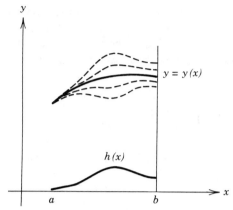

Figure 3.9. Variations $y(x) + \varepsilon h(x)$ of $y(x)$ satisfying $h(a) = 0$.

for every h in $C^2(a, b)$ satisfying $h(a) = 0$. Since (25) holds for these h, it must hold for h also satisfying the condition $h(b) = 0$. Hence

$$\int_a^b \left(L_y - \frac{d}{dx} L_{y'} \right) h \, dx = 0 \qquad (26)$$

and by the fundamental lemma, y must satisfy the Euler equation

$$L_y - \frac{d}{dx} L_{y'} = 0 \qquad (27)$$

Substituting (27) into (25) then gives

$$L_{y'}(b, y(b), y'(b)) h(b) = 0 \qquad (28)$$

Since (28) holds for all choices of $h(b)$, we must have

$$L_{y'}(b, y(b), y'(b)) = 0 \qquad (29)$$

which is a condition on the extremal y at $x = b$. Equation (29) is called a *natural boundary condition*. The Euler equation (27), the fixed boundary condition $y(a) = A$, and the natural boundary condition (29) are enough to determine the extremal for the variational problem (22) and (23). By similar arguments if the left endpoint $y(a)$ is unspecified, then the natural boundary condition on an extremal y at $x = a$ is

$$L_{y'}(a, y(a), y'(a)) = 0 \qquad (30)$$

Example 4.3   Find the differential equation and boundary conditions for the extremal of the variational problem

$$J(y) = \int_0^1 \left(p(x) y'^2 - q(x) y^2 \right) dx$$

$$y(0) = 0, \qquad y(1) \text{ unspecified}$$

where p and q are positive smooth functions on $[0, 1]$. In this case

$$L_y = -2q(x)y \quad \text{and} \quad L_{y'} = 2p(x)y'$$

The Euler equation is

$$\frac{d}{dx}\left(p(x) y' \right) + q(x) y = 0, \qquad 0 < x < 1$$

The natural boundary condition at $x = 1$ is given by (29) or

$$2p(1) y'(1) = 0$$

Since $p > 0$ the boundary conditions are

$$y(0) = 0, \qquad y'(1) = 0$$

Example 4.4   We find the natural boundary condition at $x = b$ for the river crossing problem defined by the functional (21). In this case

$$L_{y'} = \frac{1}{c^2 - v^2}\left(\frac{c^2 y'}{\left(c^2(1 + y'^2) - v^2 \right)^{1/2}} - v \right)$$

The boundary condition (29) becomes

$$\frac{c^2 y'(b)}{\left(c^2(1 + y'(b)^2) - v(b)^2 \right)^{1/2}} - v(b) = 0$$

This can be simplified to

$$y'(b) = \frac{v(b)}{c}$$

Thus the slope that the boat enters the bank at $x = b$ is the ratio of the water speed at the bank to the boat velocity in still water.

EXERCISES

4.1   Prove Lemma 4.1.

4.2   Solve the following variational problems by finding extremals satisfying the given conditions.

(a) $J(y_1, y_2) = \int_0^{\pi/4} [4y_1^2 + y_2^2 + y_1'y_2'] \, dx$

$y_1(0) = 1, \qquad y_1(\frac{\pi}{4}) = 0, \quad y_2(0) = 0, \quad y_2(\frac{\pi}{4}) = 1$

(b) $J(y) = \int_0^1 (1 + y''^2) \, dx$

$y(0) = 0, \qquad y'(0) = 1, \quad y(1) = 1, \quad y'(1) = 1$

4.3   Prove that the Euler equation for $J(y) = \int_a^b L(x, y, y', y'') \, dx$ has first integral $L_{y'} - (d/dx)L_{y''} = C$ when L is independent of y. If L is independent of x, show that

$$L - y'\left(L_{y'} - \frac{d}{dx}L_{y''}\right) - y''L_{y''} = C$$

is a first integral.

4.4   Write down the Euler equations for the functional

$$J(y, z) = \int_a^b \left[y''z' + xyz'' + z'''\, y^2\right] dx$$

where y and z are sufficiently smooth functions.

4.5   Find the extremals

(a) $J(y) = \int_0^1 (yy' + y''^2) \, dx, \qquad y(0) = 0, \quad y'(0) = 1, \quad y(1) = 2,$
$y'(1) = 4$

(b) $J(y) = \int_0^\infty [y^2 + y'^2 + (y'' + y')^2] \, dx, \qquad y(0) = 1, \quad y'(0) = 2,$
$y(\infty) = 0, \quad y'(\infty) = 0$

4.6   Find the Euler equations corresponding to the following functionals

(a) $J(u) = \iint_R (x^2u_x^2 + y^2u_y^2) \, dx \, dy, \qquad u = u(x, y)$

(b) $J(u) = \iint_R (u_t^2 - c^2u_x^2) \, dt \, dx, \qquad u = u(x, t), \quad c$ constant

(c) $J(u) = \iiiint_R \frac{1}{2}\sum_{\alpha=1}^4 \sum_{\beta=1}^4 g^{\alpha\beta} \frac{\partial u}{\partial x_\alpha} \frac{\partial u}{\partial x_\beta} \, dx_1 \, dx_2 \, dx_3 \, dx_4,$ where $u = u(x_1, x_2, x_3, x_4)$ and where $g^{\alpha\beta} = 1$ if $\alpha = \beta \neq 4$, $g^{\alpha\beta} = -1$ if $\alpha = \beta = 4$ and $g^{\alpha\beta} = 0$ if $\alpha \neq \beta$.

4.7   Prove Lemma 4.2.

4.8   Show that the Euler equation for the surface area functional $J(u) = \iint_R \sqrt{1 + u_x^2 + u_y^2}\, dx\, dy$ is $(1 + u_y^2)u_{xx} - 2u_x u_y u_{xy} + (1 + u_x^2)u_{yy} = 0$.

4.9   Derive the Euler equation for the functional

$$J(u) = \iint_R L(x, y, u, u_x, u_y, u_{xx}, u_{xy}, u_{yy})\, dx\, dy, \text{ where } u(x, y) \in C^4(R)$$

4.10   Find the Euler equation for

$$J(u) = \iint_R \left\{ \tfrac{1}{2}\left(u_{xx}^2 + u_{yy}^2\right) + \mu u_{xx} u_{yy} + (1 - \mu)u_{xy}^2 \right\} dx\, dy, \quad \mu \text{ constant}$$

4.11   Find the Euler equation for

$$J(u) = \iiiint_R \tfrac{1}{2}\left(u_t^2 - u_x^2 - u_y^2 - u_z^2 - m^2 u^2\right) dt\, dx\, dy\, dz, \qquad m \text{ constant}$$

4.12   Find extremals for the functional $J(y) = \int_a^b (y^2 + 2\dot{y}^2 + \ddot{y}^2)\, dt$.

4.13   Find extremals for a functional of the form

$$J(y, z) = \int_{t_0}^{t_1} L(\dot{y}, \dot{z})\, dt$$

given that L satisfies the condition

$$L_{\dot{y}\dot{y}} L_{\dot{z}\dot{z}} - L_{\dot{y}\dot{z}}^2 \neq 0$$

4.14   Show that the Euler equations corresponding to the functional

$$J = \iiiint_{R_4} \left(\tfrac{1}{2}\mathbf{E} \cdot \mathbf{E} - \tfrac{1}{2}\mathbf{B} \cdot \mathbf{B} \right) dt\, dx\, dy\, dz$$

are Maxwell's equations *in vacuo* in electrodynamics given by

$$\text{curl } \mathbf{E} + \frac{\partial \mathbf{B}}{\partial t} = 0, \qquad \text{curl } \mathbf{B} - \frac{\partial \mathbf{E}}{\partial t} = 0$$

where $\mathbf{E} = \mathbf{E}(t, x, y, z)$ and $\mathbf{B} = \mathbf{B}(t, x, y, z)$ are the electric and magnetic field vectors, respectively. To make this calculation introduce the four-potential (A_0, A_1, A_2, A_3) defined by the equations curl $\mathbf{A} = \mathbf{B}$ and $\mathbf{E} + \partial \mathbf{A}/\partial t = \mathbf{grad}\ A_0$, where $\mathbf{A} = (A_1, A_2, A_3)$, and subject to the Lorentz condition $\partial A_0/\partial t - \operatorname{div} \mathbf{A} = 0$ (see Logan [2]).

4.15 Find the extremals for

(a) $J(y) = \int_0^1 (y'^2 + y^2)\, dx,$ $y(0) = 1,$ $y(1)$ unspecified

(b) $J(y) = \int_0^1 (\frac{1}{2} y'^2 + y'y + y' + y)\, dx,$ $y(0) = \frac{1}{2},$
$y(1)$ unspecified

4.16 Find an extremal for

$$ J(y) = \int_1^e \left(\tfrac{1}{2} x^2 y'^2 - \tfrac{1}{8} y^2 \right) dx, \qquad y(1) = 1, \quad y(e) \text{ unspecified} $$

4.17 Determine the natural boundary condition for the multiple integral problem

$$ J(u) = \iint_R L(x, y, u, u_x, u_y)\, dx\, dy, $$

$$ u \in C^2(R), \quad u \text{ unspecified on Bd } R $$

4.18 Let

$$ J(u) = \iint_R \left\{ a\left(u_x^2 + u_y^2 \right) - bu^2 \right\} dx\, dy $$

where a and b are functions of x and y. Find the natural boundary condition on Bd R.
Answer: $a(\partial u/\partial n) = 0$, where $\partial u/\partial n$ denotes the derivative of u in the direction of the outer unit normal $\mathbf{n}$, that is, $\partial u/\partial n = \mathbf{n} \cdot \mathbf{grad}\ u$.

4.19 Determine the natural boundary condition at $x = b$ for the variational problem defined by

$$ J(y) = \int_a^b L(x, y, y')\, dx + G(y(b)), \qquad y \in C^2[a, b] $$
$$ y(a) = y_0 $$

where g is a given differentiable function on R^1.

4.20 Find an extremal for

$$ J(y) = \int_0^1 y'^2\, dx + y(1)^2, \quad y(0) = 1, \quad y(1) \text{ unspecified} $$

Answer: $y = -x/2 + 1$

4.21 Determine the natural boundary conditions at $x = a$ and $x = b$ for the functional

$$J(y) = \int_a^b L(x, y, y') \, dx - \beta y(b) + \alpha y(a)$$

where α and β are constants.

4.22 Consider the problem of minimizing

$$J(y) = \int_a^b L(x, y, y', y'') \, dx, \qquad y \in C^4[a, b]$$

subject to the conditions $y(a) = A$, $y'(a) = B$.

(a) Find the natural boundary condition(s) at $x = b$.

(b) Write the answer to part (a) for the case $L = y + y' + xy''^2$.

4.23 According to the theory of elasticity, a thin homogeneous rod compressed by a constant longitudinal force P at both ends is in equilibrium if its potential energy given by

$$\frac{1}{2} \int_0^L (EIy''^2 - Py'^2) \, dx \tag{31}$$

is a minimum where $y = y(x)$ is the lateral displacement, E is Young's modulus, and I is the moment of inertia of a cross section (Fig. 3.10).

(a) Find the function $y(x)$ that extremizes (31) and satisfies the *clamped* boundary conditions

$$y(0) = y(L) = y'(0) = y'(L) = 0$$

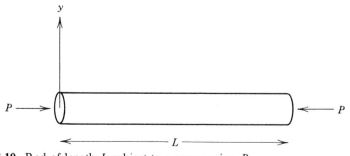

Figure 3.10. Rod of length L subject to a compression P.

(b) Suppose both ends of the bar are hinged so that they are free to bend but their position is still fixed, that is, $y(0) = y(L) = 0$. Find the appropriate natural boundary conditions. Solve the resulting problem and show that as P/EI increases a critical value is reached at which a nontrivial solution is possible. Interpret the result.

(c) What are the boundary conditions if the end of the bar at $x = L$ is completely free?

4.24 Derive (21).

3.5 HAMILTONIAN THEORY

Hamilton's Principle

Most of the applications of the calculus of variations examined so far have been geometrical in nature. Beginning with classical mechanics we now explore applications to problems in science and engineering.

According to the doctrine of classical dynamics, one associates with the system being described a set of quantities or dynamical variables, each of which is a well-defined value at each instant of time, and which define the state of the dynamical system at that instant. Further it is assumed that the time evolution of the system is completely determined if its state is known at some given instant. Analytically this doctrine is expressed by the fact that the dynamical variables satisfy a set of differential equations (the equations of motion of the system) as functions of time along with initial conditions. The program of classical dynamics, therefore, consists of listing the dynamical variables and discovering the equations of motion that predict the system's evolution in time.

One method of obtaining the equations of motion is from a variational principle. This method is based upon the idea that a system should evolve along the path of least resistance. Principles of this sort have a long history in physical theories dating back to the Greeks when Hero of Alexanderia stated a minimum principle concerning the path of reflected light rays. In the seventeenth century, Fermat's principle that light rays travel along the path of shortest time was announced. For mechanical systems Maupertuis's principle of least action stated that a system should evolve from one state to another in such a way that the *action* (a vaguely defined term with the units energy × time) is smallest. Lagrange and Gauss were also advocates of similar principles. In the early part of the nineteenth century, however, W. R. Hamilton stated what has become an encompassing, aesthetic principle that can be generalized to embrace many areas of physics.

Hamilton's principle states that the evolution in time of a mechanical system takes place in such a manner that the integral of the difference between kinetic and potential energy is stationary. To be more precise let $y_1, \ldots, y_n$

denote a set of *generalized coordinates* of a given dynamical system. That is, regarded as functions of time, we assume that $y_1, \ldots, y_n$ completely specify the state of the system at any instant. Further we assume that there are no relations among the y_i so that they may be regarded as independent. In general the y_i may be lengths, angles, or whatever. The time derivatives $\dot{y}_1, \ldots, \dot{y}_n$ (the overdot denotes d/dt) are called the *generalized velocities*. The kinetic energy T is in the most general case a quadratic form in the $\dot{y}_i$, that is

$$T = \sum_{i=1}^{n} \sum_{j=1}^{n} a_{ij}(y_1, \ldots, y_n) \dot{y}_i \dot{y}_j \tag{1}$$

where the a_{ij} are known functions of the coordinates $y_1, \ldots, y_n$. The potential energy V may be a function of t, y_i, and $\dot{y}_i$, that is

$$V = V(t, y_1, \ldots, y_n, \dot{y}_1, \ldots, \dot{y}_n) \tag{2}$$

For systems where such a scalar potential energy exists we define the *Lagrangian* by

$$L = (t, y_1, \ldots, y_n, \dot{y}_1, \ldots, \dot{y}_n) \equiv T - V \tag{3}$$

Hamilton's Principle for these systems may then be stated as follows.

Hamilton's Principle *Consider a mechanical system described by generalized coordinates $y_1, \ldots, y_n$ with Lagrangian given by (3). Then the motion of the system from time t_0 to t_1 is such that the functional*

$$J(y_1, \ldots, y_n) = \int_{t_0}^{t_1} L(t, y_1, \ldots, y_n, \dot{y}_1, \ldots, \dot{y}_n) \, dt \tag{4}$$

is stationary for the functions $y_1(t), \ldots, y_n(t)$ which describe the actual time evolution of the system.

There is a suggestive geometrical language that is helpful in restating Hamilton's principle. If we regard the set of coordinates $y_1, \ldots, y_n$ as coordinates in an n dimensional space, the so-called *configuration space*, then the equations

$$y_i = y_i(t), \quad i = 1, \ldots, n, \quad t_0 \le t \le t_1$$

can be regarded as parametric equations of a curve or path C (see Fig. 3.11 for the case $n = 3$) in that space that joins two states S_0: $(y_1(t_0), \ldots, y_n(t_0))$ and S_1: $(y_1(t_1), \ldots, y_n(t_1))$. Hamilton's principle then states that among all paths in configuration space connecting the initial state S_0 to the final state S_1 the actual motion will take place along the path that affords an extreme value to

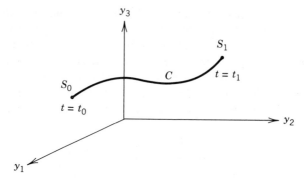

Figure 3.11. Path in configuration space.

the integral (4). Concisely, Hamilton's principle is often stated as

$$\delta \int_{t_0}^{t_1} L(t, y_1, \ldots, y_n, \dot{y}_1, \ldots, \dot{y}_n) \, dt = 0 \tag{5}$$

By (5), however, we mean the precise statement given previously. The integral functional $\int L \, dt$ is often called the *action integral* for the system. We have not said so explicitly, but we have assumed that L is twice continuously differentiable and the y_i are in $C^2[t_0, t_1]$.

Since the curve $y_i = y_i(t)$, $i = 1, \ldots, n$, along which the motion occurs must make the functional (4) stationary, it follows from the calculus of variations that $y_i(t)$ must satisfy the Euler equations

$$L_{y_i} - \frac{d}{dt} L_{\dot{y}_i} = 0, \qquad i = 1, \ldots, n \tag{6}$$

In mechanics equations (6) are known as *Lagrange's equations* rather than Euler's equations. They form the equations of motion or governing equations for the system. We say that the governing equations follow from a *variational principle* if we can find an L such that $\delta \int L \, dt = 0$ gives those governing equations as necessary conditions for an extremum.

In this context, if the Langrangian L is independent of time t, that is, $L_t = 0$, then the first integral obtained in Section 3.4, namely

$$L - \sum_{i=1}^{n} \dot{y}_i L_{\dot{y}_i} = \text{constant}$$

is called a *conservation law*. The quantity $-L + \sum_{i=1}^{n} \dot{y}_i L_{\dot{y}_i}$ is called the *Hamiltonian* of the system and it frequently represents the total energy. Thus if L is independent of explicit dependence on time, then energy is conserved.

Example 5.1 (Harmonic Oscillator) Consider the motion of a mass m attached to a spring obeying Hooke's law, that is, the restoring force is $F = -ky$, where y is a single generalized coordinate representing the positive displacement of the mass from equilibrium and $k > 0$ is the spring constant. The kinetic energy is

$$T = \tfrac{1}{2}m\dot{y}^2$$

and the potential energy is

$$V = \tfrac{1}{2}ky^2$$

Recall that $F = -dV/dy$, so $V = -\int(-ky)\,dy = (k/2)y^2$. The Langrangian is therefore

$$L = T - V = \tfrac{1}{2}m\dot{y}^2 - \tfrac{1}{2}ky^2$$

and Hamilton's principle states that the motion takes place in such a way that

$$J(y) = \int_{t_0}^{t_1}\left(\tfrac{1}{2}m\dot{y}^2 - \tfrac{1}{2}ky^2\right)dt$$

is stationary. The Langrange equation is

$$L_y - \frac{d}{dt}L_{\dot{y}} = -ky - \frac{d}{dt}(m\dot{y}) = 0$$

or

$$\ddot{y} + \frac{k}{m}y = 0$$

This equation, which expresses Newton's second law that force equals mass times acceleration, is the governing equation for the system. Its solution is

$$y(t) = C_1 \cos\sqrt{\frac{k}{m}}\,t + C_2 \sin\sqrt{\frac{k}{m}}\,t$$

where C_1 and C_2 are constants that can be determined from boundary conditions.

Example 5.2 (Simple Pendulum) Consider a simple pendulum of length l and bob mass m suspended from a frictionless support as shown in Fig. 3.12. To describe the state at any time t we choose the generalized coordinate θ measuring the angle displaced from the vertical equilibrium position. If s denotes the actual displacement of the bob along a circular arc measured

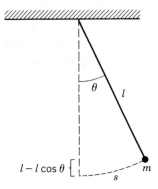

Figure 3.12. Simple pendulum.

from equilibrium, then the kinetic energy is

$$T = \tfrac{1}{2}m\dot{s}^2 = \tfrac{1}{2}ml^2\dot{\theta}^2$$

since $s = l\theta$. The potential energy of the bob is mg times the height above its equilibrium position, or

$$V = mg(l - l\cos\theta)$$

By Hamilton's principle the motion takes place so that

$$J(\theta) = \int_{t_0}^{t_1}\left\{\tfrac{1}{2}ml^2\dot{\theta}^2 - mg(l - l\cos\theta)\right\}dt$$

is stationary. Hence Lagrange's equation is

$$L_\theta - \frac{d}{dt}L_{\dot{\theta}} = -mgl\sin\theta - \frac{d}{dt}(ml^2\dot{\theta}) = 0$$

or

$$\ddot{\theta} + \frac{g}{l}\sin\theta = 0$$

which is the governing equation for the system. For small displacements $\sin\theta \approx \theta$ and the equation becomes $\ddot{\theta} + (g/l)\theta = 0$, whose solution exhibits simple harmonic motion.

Example 5.3 (Motion in a Central Force Field)   Consider the planar motion of a mass m that is attracted to the origin with a force inversely proportional to the square of the distance from the origin (see Fig. 3.13). For generalized coordinates we take the polar coordinates r and θ of the mass. The kinetic

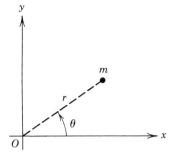

Figure 3.13. Polar coordinates for the motion of a mass m in the plane.

energy is

$$T = \frac{1}{2}mv^2 = \frac{1}{2}m(\dot{x}^2 + \dot{y}^2)$$

$$= \frac{1}{2}m\left[\left(\frac{d}{dt}(r\cos\theta)\right)^2 + \left(\frac{d}{dt}(r\sin\theta)\right)^2\right]$$

$$= \frac{1}{2}m\left[(\dot{r}\cos\theta - r\sin\theta\dot{\theta})^2 + (\dot{r}\sin\theta + r\cos\theta\dot{\theta})^2\right]$$

$$= \frac{1}{2}m(\dot{r}^2 + r^2\dot{\theta}^2)$$

Since the force is $-k/r^2$ for some constant k, the potential energy is given by

$$V = -\int -\frac{k}{r^2}\,dr$$

$$= -\frac{k}{r}$$

Thus Hamilton's principle requires that

$$J(r,\theta) = \int_{t_0}^{t_1}\left\{\frac{m}{2}(\dot{r}^2 + r^2\dot{\theta}^2) + \frac{k}{r}\right\}dt$$

be stationary. Lagrange's equations

$$L_\theta - \frac{d}{dt}L_{\dot{\theta}} = 0, \qquad L_r - \frac{d}{dt}L_{\dot{r}} = 0$$

give the equations of motion for the particle. They are

$$-\frac{d}{dt}(mr^2\dot{\theta}) = 0, \qquad mr\dot{\theta}^2 - \frac{k}{r^2} - \frac{d}{dt}(m\dot{r}) = 0$$

or

$$mr^2\dot{\theta} = \text{constant}, \qquad m\ddot{r} - mr\dot{\theta}^2 + \frac{k}{r^2} = 0$$

This coupled pair of ordinary differential equations can be solved exactly to determine the path of the motion (see Exercise 5.13).

Hamilton's Principle Versus Newton's Law

In summary, Hamilton's principle gives us a procedure for finding the equations of motion of a system if we can write down the kinetic and potential energies. This offers an alternative approach to writing down Newton's second law for a system, which requires that we know the forces. We might at this time point out, however, that the practical value of Hamilton's principle may be questionable. After all it only results in writing down Lagrange's equations. Why not just directly determine the governing equations and forgo the variational principle altogether? Actually this is a legitimate objection, particularly in view of the fact that the variational principle is often times derived a posteriori, that is, from the known equations of motion and not conversely, as would be relevant from the point of view of the calculus of variations. Moreover if a variational principle is given as the basic principle for the system, then there are complicated sufficient conditions for extrema that must be considered and they seem to have little or no role in physical problems. And finally, although variational principles do to some extent represent a unifying concept for physical theories, the extent is by no means universal; it is impossible to state such principles for some systems with constraints or dissipative forces.

On the other hand, aside from the aesthetic view, the *ab initio* formulation of the governing law for a physical system by a variational principle also has arguments on its side. The action integral plays a fundamental role in the development of numerical methods for solving differential equations (Rayliegh-Ritz method, Galerkin methods, etc.); it also plays a decisive role in the definition of Hamilton's characteristic function, the basis for the Hamilton-Jacobi canonical theory. Furthermore many variational problems occur in geometry and other areas apart from physics; in these problems the action or fundamental integral *is* an a priori notion. Thus in spite of the objections just mentioned, it appears that the calculus of variations provides a general context in which to study wide classes of problems of interest in many areas of science, engineering, and mathematics.

The Canonical Formalism

The Euler equations for the variational problem

$$J(y_1, \ldots, y_n) = \int_{t_0}^{t_1} L(t, y_1, \ldots, y_n, \dot{y}_1, \ldots, \dot{y}_n) \, dt \tag{7}$$

form a system of n second order ordinary differential equations. We now introduce a canonical method for reducing these equations to a system of $2n$ first order differential equations. For simplicity we examine the case $n = 1$; thus

$$J(y) = \int_{t_0}^{t_1} L(t, y, \dot{y}) \, dt \tag{8}$$

and the Euler equation is

$$L_y - \frac{d}{dt} L_{\dot{y}} = 0 \tag{9}$$

We define a new variable p called the *canonical momentum* by

$$p \equiv L_{\dot{y}}(t, y, \dot{y}) \tag{10}$$

If $L_{\dot{y}\dot{y}} \neq 0$, then the implicit function theorem guarantees that (10) can be solved for $\dot{y}$ in terms of t, y, and p to get

$$\dot{y} = \phi(t, y, p) \tag{11}$$

Then we define the *Hamiltonian H* by

$$H(t, y, p) \equiv -L(t, y, \phi(t, y, p)) + \phi(t, y, p)p \tag{12}$$

Notice that H is just $-L + \dot{y}L_{\dot{y}}$, where $\dot{y}$ is given by (11). In many systems H is the total energy.

Example 5.4 For a particle of mass m moving in one dimension with potential energy $V(y)$ the Lagrangian is given by $L = \frac{1}{2}m\dot{y}^2 - V(y)$. Hence $p = m\dot{y}$, which is the momentum of the particle, and

$$\begin{aligned} H &= -\frac{1}{2}m\left(\frac{p}{m}\right)^2 + V(y) + \frac{p^2}{m} \\ &= \frac{1}{2}\frac{p^2}{m} + V(y) \end{aligned}$$

which is the total energy of the particle written in terms of the position and momentum.

Now, from (12)

$$\frac{\partial H}{\partial p} = -\frac{\partial L}{\partial \dot{y}}\frac{\partial \phi}{\partial p} + \phi + p\frac{\partial \phi}{\partial p} = \phi = \dot{y}$$

and

$$\frac{\partial H}{\partial y} = -\frac{\partial L}{\partial y} - \frac{\partial L}{\partial \dot{y}}\frac{\partial \phi}{\partial y} + p\frac{\partial \phi}{\partial y} = -L_y = -\frac{d}{dt}L_{\dot{y}} = -\dot{p}$$

We have shown that the Euler equation (9) can be written as an equivalent system of equations

$$\dot{y} = \frac{\partial H}{\partial p}(t, y, p), \qquad \dot{p} = -\frac{\partial H}{\partial y}(t, y, p) \tag{13}$$

Equations (13) are called *Hamilton's equations* or the *canonical equations*. They form a system of first order differential equations for y and p.

Example 5.5   Consider the harmonic oscillator whose Lagrangian is

$$L(t, y, \dot{y}) = \tfrac{1}{2}m\dot{y}^2 - \tfrac{1}{2}ky^2 \tag{14}$$

The canonical momentum is

$$p = L_{\dot{y}} = m\dot{y}$$

Solving for $\dot{y}$ gives

$$\dot{y} = \frac{p}{m}, \qquad (= \phi(t, y, p))$$

Therefore the Hamiltonian is

$$\begin{aligned} H &= -L + \dot{y}L_{\dot{y}} \\ &= -\left(\frac{1}{2}m\left(\frac{p}{m}\right)^2 - \frac{1}{2}ky^2\right) + \left(\frac{p}{m}\right)p \\ &= \frac{1}{2}\frac{p^2}{m} + \frac{1}{2}ky^2 \end{aligned}$$

which is the sum of the kinetic and potential energy, or the total energy of the system. Now

$$\frac{\partial H}{\partial y} = ky, \qquad \frac{\partial H}{\partial p} = \frac{p}{m}$$

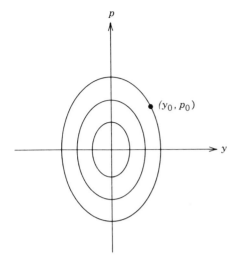

Figure 3.14. Phase trajectories $p^2 + kmy^2$ $= c$ of the system $\dot{y} = p/m$, $\dot{p} = -ky$.

so Hamilton's equations are

$$\dot{y} = \frac{p}{m}, \qquad \dot{p} = -ky \qquad\qquad (15)$$

To solve these equations in the yp plane (the so-called *phase plane*) we divide them to obtain

$$\frac{dp}{dy} = \frac{-ky}{p/m}$$

or

$$pdp + kmydy = 0$$

Integrating gives

$$p^2 + kmy^2 = c, \qquad c \text{ constant}$$

which is a family of ellipses in the py plane (Fig. 3.14). These ellipses are the integral curves of the first order system (15) and represent trajectories or paths that the system evolves along in position–momentum space, or *phase space*. Fixing initial values (y_0, p_0) at time t_0 selects out the particular trajectory that the system takes. In contrast we can compare the direct solution to the Euler equation

$$\ddot{y} + \frac{k}{m}y = 0$$

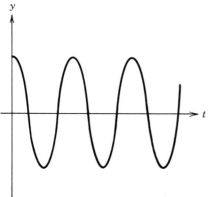

Figure 3.15. State space trajectory $y = C_1\cos(\sqrt{k/m}\,t) + C_2\sin(\sqrt{k/m}\,t)$.

which is

$$y(t) = C_1\cos\sqrt{\frac{k}{m}}\,t + C_2\sin\sqrt{\frac{k}{m}}\,t$$

The latter represents a curve in position–time coordinates (see Fig. 3.15). Consequently the canonical formulation in terms of the *phase coordinates y* and *p* gives a different geometric view of the evolution of the system.

For the more general action integral

$$J(y_1,\ldots,y_n) = \int_{t_0}^{t_1} L(t, y_1,\ldots, y_n, \dot{y}_1,\ldots, \dot{y}_n)\,dt$$

depending on *n* functions $y_1,\ldots,y_n$, we define the *n canonical momenta* by

$$p_i = L_{\dot{y}_i}(t, y_1,\ldots, y_n, \dot{y}_1,\ldots, \dot{y}_n), \qquad i = 1,\ldots, n \qquad (16)$$

Assuming that

$$\det\left(L_{\dot{y}_i\dot{y}_j}\right) \neq 0$$

we are able to solve the system of *n* equations (16) for $\dot{y}_1,\ldots, \dot{y}_n$ to obtain

$$\dot{y}_i = \phi_i(t, y_1,\ldots, y_n, p_1,\ldots, p_n), \qquad i = 1,\ldots, n$$

The Hamiltonian *H* is defined by

$$H(t, y_1,\ldots, y_n, p_1,\ldots, p_n)$$

$$= -L + \sum_{i=1}^{n} \dot{y}_i L_{\dot{y}_i} = -L(t, y_1,\ldots, y_n, \phi_1,\ldots, \phi_n)$$

$$+ \sum_{i=1}^{n} \phi_i(t, y_1,\ldots, y_n, p_1,\ldots, p_n)p_i$$

Using an argument exactly like the one we used for the case $n = 1$, we obtain the canonical form of the Euler equations

$$\dot{y}_i = \frac{\partial H}{\partial p_i}(t, y_1, \ldots, y_n, p_1, \ldots, p_n) \tag{17a}$$

$$\dot{p}_i = -\frac{\partial H}{\partial y_i}(t, y_1, \ldots, y_n, p_1, \ldots, p_n), \qquad i = 1, \ldots, n \tag{17b}$$

known as *Hamilton's equations*. These represent $2n$ first order ordinary differential equations for the $2n$ functions $y_1, \ldots, y_n, p_1, \ldots, p_n$.

A complete discussion of the role of the canonical formalism in the calculus of variations, geometry, and physics can be found in Rund [3].

The Inverse Problem

In general a variational principle exists for a given physical system if there is a Lagrangian L such that the Euler equations

$$L_{y_i} - \frac{d}{dt} L_{\dot{y}_i} = 0, \qquad i = 1, \ldots, n$$

corresponding to the action integral $\int L \, dt$ coincide with the governing equations of the system. For certain mechanical systems Hamilton's principle tells us that $L = T - V$. How can we determine the Lagrangian L for other systems if we know the equations of motion? This problem is known as the *inverse problem* of the calculus of variations. Let us state this problem precisely for $n = 1$. Given a second order ordinary differential equation

$$\ddot{y} = F(t, y, \dot{y}) \tag{18}$$

find a Lagrangian $L(t, y, \dot{y})$ such that (18) is the Euler equation

$$L_y - \frac{d}{dt} L_{\dot{y}} = 0 \tag{19}$$

Generally the problem will have infinitely many solutions. To determine L we write (19) as

$$L_y - L_{\dot{y}t} - L_{\dot{y}y}\dot{y} - L_{\dot{y}\dot{y}}\ddot{y} = 0$$

From (18)

$$L_y - L_{\dot{y}t} - L_{\dot{y}y}\dot{y} - FL_{\dot{y}\dot{y}} = 0$$

Differentiating with respect to $\dot{y}$ gives

$$L_{\dot{y}\dot{y}t} + \dot{y}L_{\dot{y}\dot{y}y} + FL_{\dot{y}\dot{y}\dot{y}} + L_{\dot{y}\dot{y}}F_{\dot{y}} = 0 \tag{20}$$

We are assuming that L is three times continuously differentiable so that the orders of differentiation can be interchanged. Letting

$$u(t, y, \dot{y}) \equiv L_{\dot{y}\dot{y}}(t, y, \dot{y}) \tag{21}$$

Equation (20) becomes

$$u_t + \dot{y}u_y + Fu_{\dot{y}} + uF_{\dot{y}} = 0 \tag{22}$$

which is a first order partial differential equation for u. If this equation is solved for u, then the Lagrangian L can be determined from (21) by integrating twice with respect to $\dot{y}$. Clearly there will be infinitely many solutions, since these integrations will involve arbitrary functions. The general solution of (22) is discussed in Chapter 5.

   If only a single Lagrangian is sought, rather than the entire class of Lagrangians, it is often simpler to proceed directly by matching terms in the Euler equation to terms in the given differential equation.

Example 5.6   Consider a damped harmonic oscillator in which a mass m is suspended from a spring with spring constant k and there is a dashpot offering a resistive force numerically equal to $a\dot{y}$, where a is a constant and y is the displacement of the mass from equilibrium (see Fig. 3.16). From Newton's law the mass times acceleration equals the restoring force of the spring plus resistive force of the dashpot. In mathematical symbols

$$m\ddot{y} = -ky - a\dot{y}$$

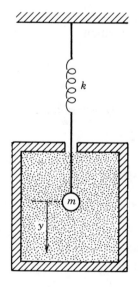

Figure 3.16. Spring–mass–dashpot system.

or

$$m\ddot{y} + a\dot{y} + ky = 0 \qquad (23)$$

Equation (23) is the damped harmonic oscillator equation. Here the system is not conservative because of the damping and there is not a scalar potential. We seek a Lagrangian $L(t, y, \dot{y})$ such that (23) is the Euler equation. Multiplying (23) by a nonnegative function $f(t)$ we obtain

$$mf(t)\ddot{y} + af(t)\dot{y} + kf(t)y = 0 \qquad (24)$$

Equation (24) must coincide with the Euler equation

$$-L_y + L_{\dot{y}t} + L_{\dot{y}y}\dot{y} + L_{\dot{y}\dot{y}}\ddot{y} = 0.$$

An obvious choice is to take

$$L_{\dot{y}\dot{y}} = mf(t)$$

so that

$$L_{\dot{y}} = mf(t)\dot{y} + M(t, y)$$

and

$$L = mf(t)\frac{\dot{y}^2}{2} + M(t, y)\dot{y} + N(t, y)$$

where M and N are arbitrary functions. It follows that

$$-L_y + L_{\dot{y}t} + L_{\dot{y}y}\dot{y} = af(t)\dot{y} + kf(t)y$$

or

$$-(M_y\dot{y} + N_y) + mf'(t)\dot{y} + M_t + M_y\dot{y} = af(t)\dot{y} + kf(t)y$$

Thus $mf'(t) = af(t)$ or $f(t) = \exp(at/m)$. Hence

$$-N_y + M_t = ky \exp\frac{at}{m}$$

Choosing $M = 0$ gives $N = -(k/2)y^2\exp(at/m)$ and therefore a Lagrangian for the damped harmonic oscillator is

$$L(t, y, \dot{y}) = \left(\frac{m}{2}\dot{y}^2 - \frac{k}{2}y^2\right)\exp\frac{at}{m}$$

EXERCISES

5.1   Consider the functional $J(y) = \int_a^b [p(t)\dot{y}^2 + q(t)y^2]\,dt$. Find the Hamiltonian $H(t, y, p)$ and write down the canonical equations for the problem.

5.2   Write down Hamilton's equations for the functional

$$J(y) = \int_a^b \sqrt{(t^2 + y^2)(1 + \dot{y}^2)}\ dt$$

Solve these equations and sketch the solution curves in the yp plane.

5.3   Derive Hamilton's equations for the functional

$$J(\theta) = \int_{t_1}^{t_2} \left[\frac{m}{2} l^2 \dot{\theta}^2 + mgl \cos\theta - mgl \right] dt$$

representing the motion of a pendulum in a plane.

5.4   Derive Hamilton's equations for the functional

$$J(\theta, \phi) = \int_{t_1}^{t_2} \left\{ \frac{m}{2} l^2 (\dot{\theta}^2 + \sin^2\theta \dot{\phi}^2) + mgl \cos\theta \right\} dt$$

representing the motion of a pendulum in space.

5.5   Give an alternate derivation of Hamilton's equations by rewriting the action integral

$$\int_{t_0}^{t_1} L(t, y, \dot{y})\,dt$$

as

$$J(p, y) = \int_{t_0}^{t_1} F(t, y, \dot{y}, p, \dot{p})\,dt$$

where $F = \dot{y}p - H(t, y, p)$, treating y and p as independent and finding the Euler equations.

5.6   A particle of mass m moves in one dimension under the influence of a force $F(y, t) = ky^{-2}\exp t$, where y is position, t is time, and k is a constant. Formulate Hamilton's principle for this system and derive the equation of motion. Determine the Hamiltonian and compare it with the total energy. Discuss conservation of energy.

5.7 A particle of mass m is falling downward under the action of gravity with resistive forces neglected. If y denotes the distance measured downward, find the Lagrangian for the system and determine the equation of motion from a variational principle.
Answer: $L = \frac{1}{2}m\dot{y}^2 + mgy$.

5.8 Consider a system of n particles where m_i is the mass of the ith particle and (x_i, y_i, z_i) is its position in Euclidean three-space. The kinetic energy of the system is

$$T = \frac{1}{2}\sum_{i=1}^{n} m_i\left(\dot{x}_i^2 + \dot{y}_i^2 + \dot{z}_i^2\right)$$

Assume that the system has a potential energy $V = V(t, x_1, \ldots, x_n, y_1, \ldots, y_n, z_1, \ldots, z_n)$ such that the force acting on the ith particle has components

$$F_i = -\frac{\partial V}{\partial x_i}, \qquad G_i = -\frac{\partial V}{\partial y_i}, \qquad H_i = -\frac{\partial V}{\partial z_i}$$

Show that Hamilton's principle applied to this system yields the equations

$$m_i\ddot{x}_i = F_i$$
$$m_i\ddot{y}_i = G_i$$
$$m_i\ddot{z}_i = H_i$$

which are Newton's equations for a system of n particles.

5.9 A particle of mass m moves in one dimension with a Lagrangian of the form

$$L = \frac{m^2}{12}\dot{y}^4 + m\dot{y}^2V(y) - V^2(y)$$

where V is a given differentiable function. Find the equation of motion and a first integral.

5.10 Use Hamilton's principle to derive the equations of motion for a mass on a spring pendulum. The spring is *linear* with spring constant k and the equilibrium length of the spring is l (see Fig. 3.17).

5.11 Consider a simple plane pendulum with a bob of mass m attached to a string of length l. After the pendulum is set in motion the string

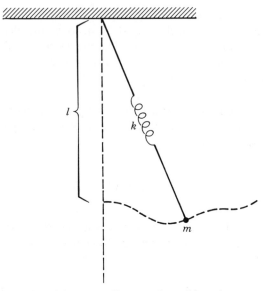

Figure 3.17. Pendulum on a linear spring with spring constant k.

is shortened by a constant rate $dl/dt = -\alpha = $ constant. Formulate Hamilton's principle and determine the equation of motion. Compare the Hamiltonian to the total energy. Is energy conserved?

5.12 If the Lagrangian L does not depend explicitly on time t, prove that $H = $ constant, and if L does not depend explicitly on a generalized coordinate y, prove that $p = $ constant.

5.13 Consider the differential equations

$$r^2\dot{\theta} = C, \qquad \ddot{r} - r\dot{\theta}^2 + \frac{k}{m}r^{-2} = 0 \tag{25}$$

governing the motion of a mass in an inverse square central force field (see Example 5.3).

(a) Show by the chain rule that

$$\dot{r} = Cr^{-2}\frac{dr}{d\theta}, \qquad \ddot{r} = C^2r^{-4}\frac{d^2r}{d\theta^2} - 2C^2r^{-5}\left(\frac{dr}{d\theta}\right)^2$$

and therefore (25) may be written

$$\frac{d^2r}{d\theta^2} - 2r^{-1}\left(\frac{dr}{d\theta}\right)^2 - r + \frac{k}{C^2m}r^2 = 0$$

(b) Let $r = u^{-1}$ and show that

$$\frac{d^2u}{d\theta^2} + u = \frac{k}{C^2m}$$

(c) Solve the differential equation in (5.13b) to obtain

$$u = r^{-1} = \frac{k}{C^2m}(1 + \varepsilon \cos(\theta - \theta_0))$$

where ε and θ_0 are constants of integration.

(d) Show that elliptical orbits are obtained when $\varepsilon < 1$.

5.14 Find a Lagrangian $L(t, y, \dot{y})$ such that the Euler equation coincides with the Emden-Fowler equation

$$\ddot{y} + \frac{2}{t}\dot{y} + y^5 = 0$$

5.15 A particle moving in one dimension in a constant external force field with frictional force proportional to its velocity has equation of motion

$$\ddot{y} + a\dot{y} + b = 0, \qquad a, b > 0$$

Find a Lagrangian $L(t, y, \dot{y})$.

5.16 Using the electrical–mechanical analogy and the result for the damped harmonic oscillator, determine a Lagrangian for an RCL electrical circuit whose differential equation is

$$L\ddot{I} + R\dot{I} + \frac{1}{C}I = 0$$

where $I(t)$ is the current, L the inductance, C the capacitance, and R the resistance.

5.17 For which Lagrangians $L(t, y, \dot{y})$ is the Euler equation satisfied identically for every y?

3.6 ISOPERIMETRIC PROBLEMS

Necessary Conditions

In differential calculus the problem of minimizing a given function $f(x, y)$ subject to a subsidiary condition $g(x, y) = constant$ is a common one. Now we consider classes of variational problems in which the competing functions are required to conform to certain restrictions that arise in addition to the normal endpoint conditions. Such restrictions are called *constraints* and they may take the form of integral relations, algebraic equations, or differential equations. In this section we limit the discussion to integral constraints.

One method for solving the constraint problem in calculus is the *Lagrange multiplier rule*. We state this necessary condition precisely since it will be used in the subsequent discussion.

Theorem 6.1 (Lagrange Multiplier Rule)   *Let f and g be differentiable functions with $f_x(x_0, y_0)$ and $f_y(x_0, y_0)$ not both zero. If (x_0, y_0) provides an extreme value to f subject to the constraint $g = C = $ constant, then there exists a constant λ such that*

$$f_x^*(x_0, y_0) = 0 \tag{1}$$
$$f_y^*(x_0, y_0) = 0 \tag{2}$$

and

$$g(x_0, y_0) = C \tag{3}$$

where

$$f^* \equiv f + \lambda g$$

A proof can be found in most calculus books. For practical calculations the conditions (1), (2), and (3) can be used to determine x_0, y_0, and the Lagrange multiplier λ.

A variational problem that has an integral constraint is known as an *isoperimetric problem*. In particular, consider the problem of minimizing the functional

$$J(y) = \int_a^b L(x, y, y') \, dx \tag{4}$$

subject to

$$W(y) \equiv \int_a^b G(x, y, y') \, dx = C \tag{5}$$

where $y \in C^2[a, b]$ and

$$y(a) = y_0, \qquad y(b) = y_1 \tag{6}$$

and C is a fixed constant. The given functions L and G are assumed to be twice continuously differentiable. The subsidiary condition (5) is called an *isoperimetric constraint*.

In essence we follow the procedure in earlier sections for problems without constraints. We embed an assumed local minimum $y(x)$ in a family of admissible functions with respect to which we carry out the extremization. A *one parameter* family $y(x) + \varepsilon h(x)$ is not, however, a suitable choice, since those curves may not maintain the constancy of W. Therefore we introduce a *two parameter* family

$$z = y(x) + \varepsilon_1 h_1(x) + \varepsilon_2 h_2(x)$$

where $h_1, h_2 \in C^2(a, b)$,

$$h_1(a) = h_1(b) = h_2(a) = h_2(b) = 0 \tag{7}$$

and ε_1 and ε_2 are real parameters ranging over intervals containing the origin. We assume that W does not have an extremum at y; then for any choice of h_1 and h_2 there will be values of ε_1 and ε_2 in the neighborhood of $(0, 0)$, for which $W(z) = C$. Evaluating J and W at z gives

$$\mathscr{J}(\varepsilon_1, \varepsilon_2) = \int_a^b L(x, z, z') \, dx$$

$$\mathscr{W}(\varepsilon_1, \varepsilon_2) = \int_a^b G(x, z, z') \, dx = C$$

Since y is the local minimum for (4) subject to the constraint (5), the point $(\varepsilon_1, \varepsilon_2) = (0, 0)$ must be a local minimum for $\mathscr{J}(\varepsilon_1, \varepsilon_2)$ subject to the constraint $\mathscr{W}(\varepsilon_1, \varepsilon_2) = C$. This is just a differential calculus problem and so the Lagrange multiplier rule may be applied. There must exist a constant λ such that

$$\frac{\partial \mathscr{J}^*}{\partial \varepsilon_1} = \frac{\partial \mathscr{J}^*}{\partial \varepsilon_2} = 0 \qquad \text{at} \quad (\varepsilon_1, \varepsilon_2) = (0, 0) \tag{8}$$

where $\mathscr{J}^*$ is defined by

$$\mathscr{J}^* = \mathscr{J} + \lambda \mathscr{W} = \int_a^b L^*(x, z, z') \, dx$$

with

$$L^* = L + \lambda G \tag{9}$$

We now calculate the derivatives in (8), afterwards setting $\varepsilon_1 = \varepsilon_2 = 0$. Accordingly

$$\frac{\partial \mathscr{J}^*}{\partial \varepsilon_i}(0, 0) = \int_a^b \left(L_y^*(x, y, y') h_i + L_{y'}^*(x, y, y') h_i' \right) dx, \qquad i = 1, 2$$

Integrating the second term by parts and applying conditions (7) give

$$\frac{\partial \mathscr{J}^*}{\partial \varepsilon_i}(0, 0) = \int_a^b \left(L_y^*(x, y, y') - \frac{d}{dx} L_{y'}^*(x, y, y') \right) h_i \, dx, \qquad i = 1, 2$$

From (8), and because of the arbitrary character of h_1 or h_2, the fundamental lemma (Lemma 3.1) implies

$$L_y^*(x, y, y') - \frac{d}{dx} L_{y'}^*(x, y, y') = 0 \tag{10}$$

which is a necessary condition for an extremum.

In practice the solution of the second order differential equation (10) will yield a function $y(x)$ that involves two arbitrary constants and the unknown multiplier λ. These may be evaluated by the two boundary conditions (6) and by substituting $y(x)$ into the isoperimetric constraint (5).

Example 6.1   (Shape of a Hanging Rope)   A rope of length l with constant mass per unit length ρ hangs from two fixed points (a, y_0) and (b, y_1) in the plane. Let $y(x)$ be an arbitrary configuration of the rope with the y axis adjusted so that $y(x) > 0$. A small element of length ds at (x, y) has mass $\rho\, ds$ and potential energy $\rho gy\, ds$ relative to $y = 0$. Therefore the total potential energy of the rope hanging in the arbitrary configuration $y = y(x)$ is given by the functional

$$J(y) = \int_0^l \rho gy\, ds = \int_a^b \rho gy\sqrt{1 + y'^2}\, dx \tag{11}$$

It is known that the actual configuration minimizes the potential energy. Thus we are faced with minimizing (11) subject to the isoperimetric condition

$$W(y) = \int_a^b \sqrt{1 + y'^2}\, dx = l \tag{12}$$

By the previous prescription we form the auxiliary function

$$L^* = L + \lambda G$$
$$= \rho gy\sqrt{1 + y'^2} + \lambda\sqrt{1 + y'^2}$$

and write down the associated Euler equation (10). In this case, however, L^* does not depend explicitly on x and thus a first integral is given by

$$L^* - y'L_{y'}^* = C$$

or

$$(\rho gy + \lambda)\left(\sqrt{1 + y'^2} - \frac{y'^2}{\sqrt{1 + y'^2}}\right) = C$$

Solving for y' and separating variables yields

$$\frac{dy}{\sqrt{(\rho gy + \lambda)^2 - C^2}} = \frac{1}{C}\, dx$$

Letting $u = \rho gy + \lambda$ and using the antiderivative formula

$$\int \frac{du}{\sqrt{u^2 - C^2}} = \cosh^{-1} \frac{u}{c}$$

gives

$$\frac{1}{\rho g} \cosh^{-1} \frac{u}{C} = \frac{1}{C}x + C_1$$

Then

$$y = -\frac{\lambda}{\rho g} + \frac{C}{\rho g} \cosh\left(\frac{\rho gx}{C} + C_2\right)$$

Therefore the shape of a hanging rope is a *catenary*. The constants C, C_2, and λ may be determined from (12) and the endpoint conditions $y(a) = y_0$ and $y(b) = y_1$. In practice this calculation may be difficult.

Generalizations of the isoperimetric problem (4), (5),and (6) can take several directions. Some of these are left to the Exercises. We take up one example.

Example 6.2 Extremize the multiple integral

$$J(\psi) = \iiint_D \left(\frac{k^2}{2m}\left(\psi_x^2 + \psi_y^2 + \psi_z^2\right) + V\psi^2\right) dx\, dy\, dz$$

subject to the isoperimetric constraint

$$\iiint_D \psi^2 \, dx\, dy\, dz = 1 \tag{13}$$

Here k and m are constants, $V = V(x, y, z)$ is a given function, and D is a domain in R^3. We proceed by introducing the auxiliary Lagrangian

$$L^* = \frac{k^2}{2m}\left(\psi_x^2 + \psi_y^2 + \psi_z^2\right) + V\psi^2 + \lambda\psi^2$$

where λ is a Lagrange multiplier. The Euler equation is

$$L_\psi^* - \frac{\partial}{\partial x}L_{\psi_x}^* - \frac{\partial}{\partial y}L_{\psi_y}^* - \frac{\partial}{\partial z}L_{\psi_z}^* = 0$$

or

$$-\frac{k^2}{2m}\nabla^2\psi + V\psi = -\lambda\psi \tag{14}$$

where ∇^2 is the Laplacian $\nabla^2 \equiv \partial^2/\partial x^2 + \partial^2/\partial y^2 + \partial^2/\partial z^2$. Equation (14) is the *Schrödinger equation* in quantum mechanics for a particle of mass m under the influence of a potential V. In this context D is the whole of R^3 and (13) is the normalization condition for the wave function ψ whose square is a probability density. In general, solutions ψ of (14) will exist only for discrete values of the multiplier λ, which are identified with the possible energy levels of the particle and are called the eigenvalues. Thus the Schrödinger eigenvalue problem is equivalent to an isoperimetric problem. The constant k is identified with $h/2\pi$, where h is Planck's constant.

EXERCISES

6.1   Find extremals for the isoperimetric problem

$$J(y) = \int_0^1 (y'^2 + x^2)\, dx, \qquad y(0) = 0, \quad y(1) = 0$$

$$\int_0^1 y^2\, dx = 2$$

Solution: $y_n(x) = \pm 2 \sin n\pi x, \qquad n = 1, 2, 3, \ldots$.

6.2   Derive a necessary condition for an extremum for the isoperimetric problem.
Minimize

$$J(y_1, y_2) = \int_a^b L(x, y_1, y_2, y_1', y_2')\, dx$$

subject to

$$\int_a^b G(x, y_1, y_2, y_1', y_2')\, dx = C$$

and

$$y_1(a) = A_1, \qquad y_2(a) = A_2, \qquad y_1(b) = B_1, \qquad y_2(b) = B_2$$

where C, A_1, A_2, B_1, and B_2 are constants.

6.3   Use the result of Exercise 6.2 to maximize

$$J(x, y) = \int_{t_0}^{t_1} (x\dot{y} - y\dot{x})\, dt$$

subject to

$$\int_{t_1}^{t_2} \sqrt{\dot{x}^2 + \dot{y}^2}\, dt = l$$

Show that J represents the area enclosed by a curve with parametric equations $x = x(t)$, $y = y(t)$ and the constraint fixes the length of the curve. Thus, this is the problem of determining which curve of a specified perimeter encloses the maximum area. (The term *isoperimetric*, meaning *same perimeter*, originated in this context).
Answer: Circles $(x - c_1)^2 + (y - c_2)^2 = \lambda^2$.

6.4 Determine the equation of the shortest arc in the first quadrant that passes through $(0,0)$ and $(1,0)$ and encloses a prescribed area A with the x axis, where $0 < A \leq \pi/8$.

6.5 Write down the equations that determine the solution of the isoperimetric problem.
Minimize

$$\int_a^b \left(p(x) y'^2 + q(x) y^2 \right) dx$$

subject to

$$\int_a^b r(x) y^2 \, dx = 1$$

where p, q, and r are given functions and $y(a) = y(b) = 0$. The result is a *Sturm-Liouville problem* (see Chapter 4).

6.6 Consider the problem of minimizing the functional

$$J(y, z) = \int_a^b L(t, y, z, \dot{y}, \dot{z}) \, dt \tag{15}$$

subject to the algebraic constraint

$$G(t, y, z) = 0, \qquad G_z \neq 0 \tag{16}$$

and boundary conditions

$$y(a) = y(b) = z(a) = z(b) = 0$$

with all of the functions possessing the usual differentiability conditions. If $y(t)$ and $z(t)$ provide a local minimum, show that there exists a function $\lambda(t)$ such that

$$L_y - \frac{d}{dx} L_{\dot{y}} = \lambda(t) G_y, \qquad L_z - \frac{d}{dt} L_{\dot{z}} = \lambda(t) G_z$$

Hint: Solve (16) to get $z = g(t, y)$ and substitute into (15) to obtain a functional depending only on y. Find the Euler equation and use the condition $(\partial/\partial y) G(x, y, g(x, y)) = 0$.

6.7   Find extremals of the isoperimetric problem

$$J(y) = \int_0^\pi y'^2 \, dx, \qquad y(0) = y(\pi) = 0$$

subject to

$$\int_0^\pi y^2 \, dx = 1$$

REFERENCES

1. I. M. Gelfand and S. V. Fomin, *Calculus of Variations*, Prentice-Hall, Inc., Englewood Cliffs, N.J. (1963).
2. J. D. Logan, *Invariant Variational Principles*, Academic Press, Inc., New York (1977).
3. H. Rund, *The Hamilton-Jacobi Theory in the Calculus of Variations*, D. Van Nostrand Company, London (1966).
4. J. L. Troutman, *Variational Calculus with Elementary Convexity*, Springer-Verlag, New York (1983).

4

EQUATIONS OF
APPLIED MATHEMATICS

$$\frac{d}{dx}\left(\frac{dy}{dt}\right) = \frac{d^2y}{dt^2}\frac{dt}{dx}$$

i.e. set $y = \frac{dy}{dt}$ so $\frac{dy}{dx} = \frac{dy}{dt}\frac{dt}{dx} = \frac{d^2y}{dt^2}\frac{dt}{dx}$

The reader has encountered many physical problems that are modeled by ordinary differential equations and from elementary courses has been exposed to some of the basic solution techniques for such equations. In this chapter we expand our view and begin to examine partial differential equations and integral equations, as well as some standard methods for obtaining their solution. We also study the diffusion equation and illustrate common techniques, which apply to many other problems, on solving problems of heat conduction. We end with a short introduction to integral equations. In Chapter 5 the study of partial differential equations continues in the context of wave propagation and mathematical models for continuous systems. Therefore two of the most fundamental processes in nature, diffusion and wave phenomena, are the main subjects of this and the succeeding chapter.

4.1 PARTIAL DIFFERENTIAL EQUATIONS

Definitions

Partial differential equations is one of the fundamental areas of applied analysis and it is hard to imagine any area of scientific endeavor where the impact of this subject is not felt. The basic issues in the subject traditionally deal with (i) existence and uniqueness of solutions, (ii) stability of solutions to small perturbations, and (iii) methods for constructing solutions. Here we focus on the latter with the advisory that only rarely is it possible to solve a partial differential equation in closed form, that is, find a formula for its

151

solution. For special linear problems Fourier analysis and integral transform methods often lead to an infinite series or integral representation of the solution. But for the majority of problems approximation techniques or numerical methods are more in order. In this chapter we present some of the basic methods and results while maintaining contact with the origins in empirical science. Our aim is to present partial differential equations in a broad context, both from the application and solutions point of view.

Many physical problems lead to ordinary differential equations, and several have been encountered in the first three chapters or in previous experience. In general an ordinary differential equation of second order is an equation of the form

$$F(t, y, \dot{y}, \ddot{y}) = 0$$

where t ranges over some interval I. A solution is a twice continuously differentiable function $y(t)$ of the single independent variable t that reduces the above equation to an identity on I, that is,

$$F(t, y(t), \dot{y}(t), \ddot{y}(t)) = 0, \qquad t \in I$$

We can regard y as a state function for the system and the differential equation as an equation that governs the evolution of the state y in time.

Many physical processes, however, cannot be modeled by ordinary differential equations because the state of the system depends on more than one independent variable. For example, the state u of a given physical system may depend on time t and a location x. Hence the system evolves in both space and time. For example, the temperature u in a bar of length l depends upon the location x in the bar and the time t from when the initial conditions were applied. We found in Chapter 1 that $u = u(x, t)$ had to satisfy a partial differential equation, the so-called heat equation

$$u_t - ku_{xx} = 0, \qquad t > 0, \quad 0 < x < l$$

where the constant k is the diffusivity of the bar.

In general, a *second order partial differential equation* in two independent variables is an equation of the form

$$G(x, t, u, u_x, u_t, u_{xx}, u_{xt}, u_{tt}) = 0 \tag{1}$$

where (x, t) lies in some domain D in R^2. By a *solution* we mean a twice continuously differentiable function $u = u(x, t)$ on D, which when substituted into (1), reduces it to an identity for (x, t) in D. We assume that u is twice continuously differentiable so that it makes sense to calculate the second order derivatives and substitute them into (1). A solution of (1) may be represented graphically as a smooth surface in three dimensional xtu space lying above the domain D in the xt plane as shown in Fig. 4.1. Here we regard x as a position or spatial coordinate and t as time. The domain D in R^2 where the problem is

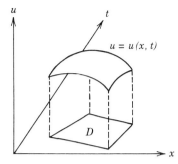

Figure 4.1. Solution surface $u = u(x, t)$ in xtu space.

defined is referred to as a space–time domain and problems that include time as an independent variable are called *evolution problems*. When two spatial coordinates, say x and y, are the independent variables we refer to the problem as an *equilibrium* or *steady-state problem*.

A partial differential equation of type (1) has infinitely many solutions. Just as the general solution of an ordinary differential equation depends on arbitrary constants, the general solution of a partial differential equation depends on arbitrary functions.

Example 1.1 Consider the simple partial differential equation

$$u_{tx} = tx$$

Integrating with respect to x gives

$$u_t = \tfrac{1}{2}tx^2 + f(t)$$

where f is an arbitrary function. Integrating with respect to t gives the general solution

$$u = \tfrac{1}{4}t^2x^2 + g(t) + h(x)$$

where h is an arbitrary function and $g(t) = \int f(t)\, dt$ is also arbitrary. Thus the general solution depends on two arbitrary functions; any choice of g and h will yield a solution.

For ordinary differential equations initial or boundary conditions fix the arbitrary constants of integration and thus often pick out a unique solution. Partial differential equations are usually accompanied by initial or boundary conditions that again select out one of the many solutions of the partial differential equation. A condition given at $t = 0$ along some segment of the x axis is called an *initial condition*. A condition given along any other curve in the xt plane is called a *boundary condition*. Initial or boundary conditions may involve specifying values of u, its derivatives, or combinations of both along the given curves in the xt plane. Partial differential equations with auxiliary conditions are called *boundary value problems*.

Example 1.2 Heat flow in a bar of length l is governed by the partial differential equation

$$u_t - ku_{xx} = 0, \qquad t > 0, \quad 0 < x < l$$

where k is a physical constant and $u = u(x, t)$ is the temperature in the bar at location x at time t. An auxiliary condition of the form

$$u(x, 0) = f(x), \qquad 0 < x < l$$

is an initial condition, since it is given at $t = 0$. We regard $f(x)$ as the initial temperature distribution in the bar. Conditions of the form

$$u(0, t) = h(t), \qquad u(l, t) = g(t), \qquad t > 0$$

are boundary conditions, and $h(t)$ and $g(t)$ represent the given temperatures held at the boundaries $x = 0$ and $x = l$ for $t > 0$, respectively. These functions are represented in Fig. 4.2. Physically, we infer that there is a unique solution to the partial differential equation subject to these initial and boundary conditions; graphically the surface representing the solution would have f, g, and h as its boundaries.

Frequently of interest is a so-called *snapshot* of the solution frozen in time, or in other words, a graph of $u(x, t_0)$ for some fixed t_0. Figure 4.2 indicates such a snapshot or time cross section of the solution surface. One way to view a solution $u(x, t)$ geometrically is to graph a sequence of snapshots $u(x, t_1), u(x, t_2), \ldots,$ for $t_1 < t_2 < \ldots$. For example, Fig. 4.3 depicts a sequence of temperature profiles in a bar of length l whose ends are held at zero degrees and whose initial temperature profile is $f(x) = x(l - x)$.

The general solution of a partial differential equation is frequently difficult to find. Thus for partial differential equations we seldom solve a boundary value problem by determining the general solution and then finding the

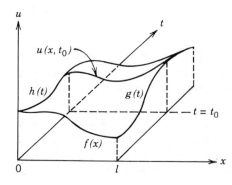

Figure 4.2. Temperature surface $u = u(x, t)$ bounded by initial temperature distribution $f(x)$ and boundary temperatures $h(t)$ and $g(t)$ at $x = 0$ and $x = l$. A time snapshot $u(x, t_0)$ is shown at time t_0.

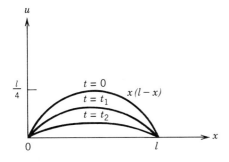

Figure 4.3. Temperature profiles or snapshots at $t = 0, t_1, t_2$.

arbitrary functions from the initial and boundary data. This is in sharp contrast to ordinary differential equations where the general solution is found and the arbitrary constants are evaluated from the initial or boundary conditions.

Generalizations of the partial differential equation (1) can be made in various directions. Higher order derivatives, several independent variables, and several unknown functions (governed by several equations) are all possibilities. As we shall see there are fundamentally three types of partial differential equations—those that govern diffusion processes, those that govern wave propagation, and those that model equilibrium processes. These types are called *parabolic*, *hyperbolic*, and *elliptic*, respectively.

Linearity versus Nonlinearity

In the next few paragraphs we introduce the ideas of linearity and nonlinearity. The separation of partial differential equations and ordinary differential equations into these two distinct classes is a significant one. Besides being generally easier to solve, linear equations have a linear algebraic structure to their solution set, that is, the sum of two solutions of a linear homogeneous equation is again a solution, as is a constant multiple of a solution. Such is not the case for nonlinear equations. The sum of two solutions or a constant multiple of a solution may not be a solution. For nonlinear equations solutions do not add or superimpose. Superposition for linear equations often allows one to construct a variety of solutions that can meet diverse boundary or initial requirements. In fact, this observation is the basis of the Fourier method or method of eigenfunction expansions for linear equations. Linear equations are also susceptible to transform methods for finding solutions. These include methods based on Laplace transforms and Fourier transforms. In summary, there is a profound difference between these two classes of problems.

To formulate the concepts more precisely we regard the partial differential equation (1) as defining a differential operator L acting on the unknown function $u(x, t)$ and we write (1) as

$$Lu(x, t) = f(x, t), \qquad (x, t) \in D$$

or suppressing the independent variables that are understood, just

$$Lu = f, \qquad (x, t) \in D \qquad (2)$$

In (2) all terms involving u are put on the left in the term Lu, and f is a known function. If $f = 0$ on D, then (2) is *homogeneous*; if f is not identically zero, then (2) is *nonhomogeneous*. The heat equation $u_t - ku_{xx} = 0$ can be written $Lu = 0$, where L is the partial differential operator $\partial/\partial t - k\partial^2/\partial x^2$, and it is clearly homogeneous. The partial differential equation $uu_t + 2txu - \sin tx = 0$ can be written $Lu = \sin tx$, where L is the differential operator defined by $Lu = uu_t + 2txu$. This equation is nonhomogeneous. The definition of linearity depends upon the operator L in (2). We say that (2) is a *linear equation* if L has the properties

(i) $L(u + w) = Lu + Lw$.
(ii) $L(cu) = cLu$.

where u and w are functions and c is a constant. If (2) is not linear, then it is *nonlinear*.

Example 1.3 The heat equation is linear since

$$L(u + w) = (u + w)_t - k(u + w)_{xx}$$
$$= u_t + w_t - ku_{xx} - kw_{xx}$$
$$= Lu + Lw$$

and

$$L(cu) = (cu)_t - k(cu)_{xx}$$
$$= cu_t - cku_{xx}$$
$$= cLu$$

Example 1.4 The differential equation $uu_t + 2txu = \sin tx$ is nonlinear since

$$L(u + w) = (u + w)(u + w)_t + 2tx(u + w)$$
$$= uu_t + wu_t + ww_t + uw_t + 2txu + 2txw$$

but

$$Lu + Lw = uu_t + 2txu + ww_t + 2txw$$

Note that the nonhomogeneous term $\sin tx$ does not affect linearity or nonlinearity.

It is clear that $Lu = f$ is linear if Lu is first degree in u and its derivatives; that is, no products involving u and its derivatives occur. Hence the most general *linear equation of second order* is of the form

$$a(x,t)u_{tt} + b(x,t)u_{xt} + c(x,t)u_{xx} + d(x,t)u_t$$
$$+ e(x,t)u_x + g(x,t)u = f(x,t), \qquad (x,y) \in D \qquad (3)$$

where the functions a, b, c, d, e, g, and f are given continuous functions on D. If any of the coefficients $a, \ldots, g$ depend on u, we say that the equation is *quasi linear*.

Referring to an earlier remark we say that (3) is *hyperbolic, parabolic,* or *elliptic* on a domain D if $b(x,t)^2 - 4a(x,t)c(x,t)$ is positive, zero, or negative, respectively, on that domain. For example, the heat equation $u_t - ku_{xx} = 0$ has $b^2 - 4ac = 0$ and is parabolic on all of R^2.

Superposition

If $Lu = 0$ is a linear homogeneous equation and u_1 and u_2 are two solutions, then it obviously follows that $u_1 + u_2$ is a solution, since $L(u_1 + u_2) = Lu_1 + Lu_2 = 0 + 0 = 0$. Also cu_1 is a solution, since $L(cu_1) = cLu_1 = c \cdot 0 = 0$. A simple induction argument shows that if $u_1, \ldots, u_n$ are solutions of $Lu = 0$ and $c_1, \ldots, c_n$ are constants, then the finite sum or linear combination $c_1u_1 + \cdots + c_nu_n$ is also a solution; this is the *superposition principle* for linear equations. If certain convergence properties hold, the superposition principle can be extended to infinite sums $c_1u_1 + c_2u_2 + \cdots$.

Another form of a superposition principle is a continuous version of the one just cited. In this case let $u(x, t, \alpha)$ be a family of solutions on D, where α is a real number ranging over some interval Γ. That is, suppose $u(x, t, \alpha)$ is a solution of $Lu = 0$ for *each* value of $\alpha \in \Gamma$. Then *formally*[†] we may superimpose these solutions by forming

$$u(x,t) = \int_\Gamma c(\alpha)u(x,t,\alpha)\,d\alpha$$

where $c(\alpha)$ is a function representing a continuum of coefficients, the analog of $c_1, c_2, \ldots, c_n$. If we can write

$$Lu = L\int_\Gamma c(\alpha)u(x,t,\alpha)\,d\alpha$$

$$= \int_\Gamma c(\alpha)Lu(x,t,\alpha)\,d\alpha$$

$$= \int_\Gamma c(\alpha) \cdot 0\,d\alpha = 0$$

[†]Generally, a *formal* calculation in mathematics is one that lacks complete rigor but that can usually be justified under certain circumstances.

then u is also a solution. This sequence of steps, and hence the superposition principle, depends upon the validity of pulling the differential operator L inside the integral sign. For such a superposition principle to hold, these formal steps would need careful analytical verification.

Example 1.5 Consider the heat equation

$$u_t - ku_{xx} = 0, \qquad t > 0, \quad x \in R^1 \tag{4}$$

It is straightforward to verify that

$$u(x, t, \alpha) = \frac{1}{\sqrt{4\pi kt}} \exp\left(-\frac{(x - \alpha)^2}{4kt}\right), \qquad t > 0, \quad x \in R^1$$

is a solution of (4) for any $\alpha \in R^1$. This solution is called the *fundamental solution*. We may formally superimpose these solutions to obtain

$$u(x, t) = \int_{-\infty}^{\infty} c(\alpha) \frac{1}{\sqrt{4\pi kt}} \exp\left(-\frac{(x - \alpha)^2}{4kt}\right) d\alpha$$

where $c(\alpha)$ is some function. It can be shown that if $c(\alpha)$ is continuous and bounded, then differentiation under the integral sign can be justified and $u(x, t)$ is therefore a solution (see John [1]).

EXERCISES

1.1 By direct integration find the general solution of the following partial differential equations in terms of arbitrary functions.

(a) $u_x = 3xt + 4$, where $u = u(x, t)$.

(b) $u_{xx} = 6xy$, where $u = u(x, y)$.

(c) $u_{xy} + (1/x)u_y = y/x^2$, where $u = u(x, y)$. (Let $v = u_y$ and solve the resulting first order equation for v by treating y as a parameter.)

(d) $u_{yx} + u_x = 1$, where $u = u(x, y)$.

1.2 By introducing polar coordinates $x = r\cos\theta$ and $y = r\sin\theta$ find the general solution of the equation

$$yu_x - xu_y = 0$$

Solution: $u = f(x^2 + y^2)$, where f is an arbitrary function.

$$\frac{du}{dx} = \frac{du}{dr}\frac{dr}{dx} + \frac{du}{d\varphi}\frac{d\varphi}{dx} \qquad \frac{dr}{dx} = \cos\varphi \qquad \frac{dr}{dy} = \sin\varphi \qquad \frac{d}{dx}\arctan(x) = \frac{1}{1+x^2}$$

$$\frac{d\varphi}{dx} = -\frac{\sin\varphi}{r} \qquad \frac{d\varphi}{dy} = \frac{\cos\varphi}{r}$$

▷ **1.3** Determine regions where the following equations are hyperbolic, elliptic, or parabolic.

(a) $tu_{tt} + u_{xx} = 0$ (c) $u_{tt} + (1 + x^2)u_x - u_t = e^t$

(b) $u_{tt} - u_{xx} = 0$ (d) $u_{xx} + u_{yy} = f(x, y)$

▷ **1.4** Determine whether the following equations are linear or nonlinear.

(a) $u_t u_{tt} + 3tu = 0$ (c) $u_{tt} - u_{xx} = 0$

(b) $\exp(t)u_{tx} - x^2 u = \cos t$ (d) $u_{tx} + u^2 = \sin x$

Which of the equations in Exercises 1.3 and 1.4 are homogeneous?

1.5   The solution to a certain partial differential equation with initial and boundary conditions on the domain $t > 0, 0 < x < \pi$, is given by

$$u(x, t) = e^{-2t}\cos \frac{x}{2}$$

Sketch snapshots of the solution at the times $t = 0, 1$, and 2.

1.6   Sketch snapshots of the fundamental solution of the heat equation

$$u(x, t, 0) = \frac{1}{\sqrt{4\pi kt}} \exp\left(-\frac{x^2}{4kt}\right)$$

for different times t. What do the temperature profiles look like as t approaches zero? Comment on the differences one would observe in the profiles for large k and for small k.

4.2   THE DIFFUSION EQUATION

Heat Conduction

In Example 3.1 of Chapter 1 heat conduction in a rod of finite length was discussed in a simplified setting with the goal of scaling the problem and reducing it to dimensionless form. We obtained from the analysis the *heat* or *diffusion equation*

$$u_t - ku_{xx} = 0, \qquad k > 0 \tag{1}$$

a second order, linear, parabolic equation. Before proceeding with a more refined derivation to include sources and variable parameters the reader may wish to review Example 3.1. As before we consider a cylindrical rod extending

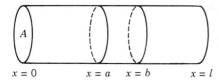

Figure 4.4. Bar of cross-sectional area A. Arbitrary cross sections at $x = a$ and $x = b$ are indicated.

$x = 0$ $x = a$ $x = b$ $x = l$

from $x = 0$ to $x = l$ with constant cross-sectional area A. We implicitly assume that any physical parameters are constant over each cross section and that heat energy flows only in the longitudinal or x direction. It is this assumption that gives the problem a one dimensional spatial character. The rod, shown in Fig. 4.4, is assumed to vary in its physical composition, which is defined by its density ρ, specific heat c_v, and thermal conductivity K. We do not assume that these quantities are constant, but rather depend explicitly on the location x as well as the temperature u. That is

$$\rho = \rho(x, u), \qquad c_v = c_v(x, u), \qquad K = K(x, u)$$

For materials with *memory* these parameters could depend on time t as well. The diffusivity k is defined by $k = K/\rho c_v$. With only a little reflection it seems reasonable that the physical parameters should vary with temperature and it is this variation that gives a problem its nonlinear character.

There are two basic physical laws that come to the forefront in deriving an equation for heat conduction. The first is an empirical law, Fourier's heat law, and the second is more fundamental, the conservation of energy law. Fourier's heat law states that the heat flux across a cross section at location x is proportional to the area and to the temperature gradient at that location; that is,

$$\text{flux at } x = -K\big(x, u(x, t)\big)u_x(x, t)A \qquad (2)$$

The thermal conductivity K is the proportionality factor. Equation (2), being an empirical result, may be inappropriate for some problems; one might imagine a diffusion problem, for example, where the flux is proportional to the cube of the temperature gradient u_x. By convention the flux is positive if the heat flow is in the positive direction, or to the right. In Section 1.3 of Chapter 1 we derived the equation of heat flow by requiring that energy conservation hold in a small box of width Δx and then taking the limit as $\Delta x \to 0$. Here we proceed differently by taking an arbitrary element of the rod of finite extent $I: a \le x \le b$ and applying the conservation law (see Fig. 4.4). Although the *infinitesimal element* method is suitable for most arguments, the *finite element* is sometimes more theoretically satisfying and its use will provide another *method* of applied mathematics.

By energy conservation or energy balance we mean

$$
\left\{ \begin{array}{l} \text{Time rate of change} \\ \text{of heat energy in } I \end{array} \right\} = \left\{ \begin{array}{l} \text{heat flux through the} \\ \text{cross section at } x = a \end{array} \right\}
$$

$$
- \left\{ \begin{array}{l} \text{heat flux through the} \\ \text{cross section at } x = b \end{array} \right\} \tag{3}
$$

$$
+ \left\{ \begin{array}{l} \text{amount of energy added in} \\ I \text{ from an external source} \end{array} \right\}
$$

The last term in (3) is called the *source term* (if negative, it is a *sink*). If the bar is radiating heat to the external medium, for example, then there is a negative source; a sleeve that injects energy into the bar would be a positive source. In general the source may depend on time, position, or temperature and is represented by a given function $f(x, t, u)$ given in dimensions of energy per unit volume per unit time. Thus the total source in I is

$$
A \int_a^b f(x, t, u(x, t)) \, dx \tag{4}
$$

given in energy per unit time. The amount of energy in I at any given time is

$$
A \int_a^b \rho(x, u(x, t)) c_v(x, u(x, t)) u(x, t) \, dx \tag{5}
$$

and the fluxes through the faces $x = a$ and $x = b$ at time t are given according to Fourier's heat law by

$$
-AK(a, u(a, t)) u_x(a, t) \tag{6}
$$

and

$$
-AK(b, u(b, t)) u_x(b, t) \tag{7}
$$

respectively. Substituting (4) through (7) into the energy balance equation (3) in the appropriate places yields

$$
\frac{d}{dt} \int_a^b \rho c_v u \, dx = -K(a, u(a, t)) u_x(a, t) + K(b, u(b, t)) u_x(b, t) + \int_a^b f \, dx \tag{8}
$$

which is an integral form of the energy balance law. By the fundamental theorem of calculus, assuming the functions have the requisite degree of smoothness,

$$\int_a^b \frac{\partial}{\partial x}\left(K(x, u(x, t))u_x(x, t)\right) dx$$

$$= K(b, u(b, t))u_x(b, t) - K(a, u(a, t))u_x(a, t)$$

Further

$$\frac{d}{dt}\int_a^b \rho c_v u \, dx = \int_a^b \frac{\partial}{\partial t}\left(\rho c_v u\right) dx$$

provided $(\rho c_v u)_t$ is continuous. Therefore (8) can be written

$$\int_a^b \left\{ \frac{\partial}{\partial t}(\rho c_v u) - \frac{\partial}{\partial x}(Ku_x) - f \right\} dx = 0 \qquad (9)$$

Since a and b are arbitrary we can write, assuming the integrand is continuous,[†]

$$\frac{\partial}{\partial t}(\rho c_v u) - \frac{\partial}{\partial x}(Ku_x) = f \qquad (10)$$

which is the general *diffusion* or *heat conduction equation* in differential form. We recognize $(\rho c_v u)_t$ as the energy rate of change term, $(Ku_x)_x$ as the flux term, and f as the source term. Because ρ, c_v, K, and perhaps f depend on u the equation is nonlinear. If ρ, c_v, and K are constants, then (10) reduces to

$$u_t - ku_{xx} = \frac{1}{\rho c_v} f \qquad (11)$$

which is the classical *linear nonhomogeneous heat equation*.

Well-Posed Problems

An auxiliary condition on the temperature u at time $t = 0$ is called an initial condition and is of the form

$$u(x, 0) = f(x), \qquad 0 < x < l \qquad (12)$$

where $f(x)$ is the given initial temperature distribution. Conditions prescribed at $x = 0$ and $x = l$ are called boundary conditions. If the temperature u at the ends of the bar are prescribed, then the boundary conditions take the form

$$u(0, t) = g(t), \qquad u(l, t) = h(t), \qquad t > 0 \qquad (13)$$

[†] If $f(x)$ is continuous and $\int_a^b f(x) \, dx = 0$ for every subinterval $[a, b]$ of $[0, l]$, then $f(x) = 0$ for $x \in [0, l]$. A proof is requested in the Exercise 2.2.

where g and h are prescribed functions. Other boundary conditions are possible. It is easy to imagine physically that one end of the bar, say at $x = 0$, is insulated so that no heat can pass through; this means by Fourier's heat law (2) that the flux at $x = 0$ is zero or

$$u_x(0, t) = 0, \qquad t > 0 \tag{14}$$

Condition (14) is called the *insulated end* boundary condition. Or one may prescribe the flux at an end as a given function of t, for example,

$$-KAu_x(0, t) = \phi(t), \qquad t > 0 \tag{15}$$

If the conductivity K depends on u, then condition (15) will be nonlinear. Quite generally some combination of the temperature and flux may be prescribed, such as

$$\alpha u(0, t) + \beta u_x(0, t) = \psi(t), \qquad t > 0$$

where α and β are constants and ψ is a given function.

If an initial condition is given and a boundary condition is prescribed at one end (say at $x = 0$) of a very long rod, then the problem can be considered on an infinite medium $x \geq 0$ on which the heat conduction equation holds. For example the boundary value problem

$$u_t - ku_{xx} = 0, \qquad\qquad t > 0, \quad 0 < x < \infty$$
$$u(x, 0) = 0, \qquad\qquad 0 < x < \infty$$
$$u(0, t) = 1 - \cos t, \qquad t > 0$$

governs heat flow in an infinite medium $x \geq 0$ of constant diffusivity k, initially at zero degrees, subject to the maintenance of the end at $x = 0$ at $1 - \cos t$ degrees. Here we expect the problem to model heat conduction in a long bar for a long enough time so that any condition at the far end would not affect the temperature distribution in the portion of the rod that we are interested in studying. As a practical example heat flow in an infinite medium arises from the study of the underground temperature variations, given the temperature changes at ground level on the earth. It seems clear that problems on the infinite interval $-\infty < x < \infty$ are also relevant in special physical situations. In this case only an initial condition can be prescribed. In order to restrict the class of solutions to those that are physically meaningful a condition is sometimes prescribed at infinity, for example u is bounded at infinity or $\lim_{x \to +\infty} u(x, t) = 0, t > 0$.

In summary, by a *boundary value problem* for the heat conduction equation we mean the problem of solving some form of the heat equation, for example (10) or (11), subject to initial or boundary conditions. It may be clear from the physical context which auxiliary conditions should be prescribed to obtain a unique solution; in some cases it is not clear. A large body of mathematical literature is devoted to proving the existence and uniqueness of solutions to various kinds of partial differential equations subject to sundry auxiliary data.

Existence–uniqueness theorems are useful even in the most applied contexts. For example, before embarking upon a large numerical calculation of the solution to a heat conduction problem it is helpful to know that a unique solution to the problem exists. An example of such an existence–uniqueness theorem is the following:

Theorem 2.1 *The boundary value problem*

$$u_t - ku_{xx} = 0, \qquad 0 < x < l, \quad 0 < t < T$$
$$u(x,0) = f(x), \qquad 0 < x < l \qquad\qquad\qquad (16)$$
$$u(0,t) = g(t), \qquad u(l,t) = h(t), \qquad 0 < t < T$$

where $f \in C[0, l]$ *and* $g, h \in C[0, T]$ *has a unique solution* $u(x, t)$ *on the rectangle* $R: 0 \le x \le l, 0 \le t \le T$, *for any* $T > 0$.

For this theorem we present the uniqueness proof to show an example of the *energy method*. By way of contradiction assume solutions are not unique and that there are two distinct solutions $u_1(x, t)$ and $u_2(x, t)$ to (16). Then their difference $w(x, t) \equiv u_1(x, t) - u_2(x, t)$ must satisfy the boundary value problem

$$w_t - kw_{xx} = 0, \qquad\qquad 0 < x < l, \quad 0 < t < T \qquad (17)$$
$$w(x,0) = 0, \qquad\qquad 0 < x < l \qquad\qquad\qquad (18)$$
$$w(0,t) = w(l,t) = 0, \qquad 0 < t < T \qquad\qquad\qquad (19)$$

If we show $w(x, t) \equiv 0$ on R, then $u_1(x, t) = u_2(x, t)$ on R, which is a contradiction. To this end define the *energy integral*

$$E(t) = \int_0^l w^2(x, t)\, dx \qquad\qquad (20)$$

Clearly $E(t) \ge 0$ and $E(0) = 0$. Furthermore

$$E'(t) = \int_0^l 2ww_t\, dx = 2k \int_0^l ww_{xx}\, dx = 2kww_x\big|_0^l - 2k \int_0^l w_x^2\, dx$$

where the last equality was obtained using integration by parts. From (19) the boundary term vanishes and

$$E'(t) = -2k \int_0^l w_x^2(x, t)\, dx \le 0$$

Thus $E(t)$ is nonincreasing, which along with the facts that $E(t) \ge 0$ and

$E(0) = 0$, implies $E(t) = 0$. Therefore the integrand in (20) must vanish identically in $0 \leq x \leq l, 0 \leq t \leq T$, since w is continuous in both its arguments. Thus $w \equiv 0$ on R and the uniqueness part of the theorem is proved. Existence will be established later by actually exhibiting a solution.

Uniqueness can also be proved using a maximum principle that also leads to comparison theorems and continuity with respect to initial or boundary data. We save this proof until the next section where it is given for three dimensional heat flow. The argument will be easily adapted to a single dimension.

In addition to existence and uniqueness questions the notion of continuous dependence of the solution on the initial or boundary data is important. This concept is called *stability*. From a physical viewpoint it is reasonable that small changes in the initial or boundary temperatures should not lead to large changes in the overall temperature distribution. The mathematical model should reflect this stability in that small changes in the auxiliary data should lead to only small changes in the solution. Stated differently, the solution should be stable under small perturbations of the initial or boundary data.

If a given boundary value problem satisfies the three conditions—(i) there is a solution, (ii) the solution is unique, and (iii) the solution is stable—then the problem is said to be *well-posed*. (We remark, parenthetically, that although stability seems desirable, many important physical processes are unstable, and so *ill-posed problems* are frequently studied in mathematics as well.)

Example 2.1 (Hadamard's example)   Consider the partial differential equation

$$u_{tt} + u_{xx} = 0, \qquad t > 0, \quad x \in R^1 \tag{21}$$

subject to the initial conditions

$$u(x,0) = 0, \qquad u_t(x,0) = 0, \qquad x \in R^1 \tag{22}$$

Note that (21) is elliptic and not parabolic like the heat equation. The solution of this problem is clearly the zero solution $u(x, t) \equiv 0$ for $t \geq 0, x \in R^1$. Now let us change (22) to

$$u(x,0) = 0, \qquad u_t(x,0) = 10^{-4}\sin 10^4 x \tag{23}$$

a tiny high-frequency wiggle

which represents a very small change in the initial data. The solution of (21) subject to (23) is

$$u(x, t) = 10^{-8}\sin(10^4 x)\sinh(10^4 t) \tag{24}$$

For large values of t the function $\sinh(10^4 t)$ behaves like $\exp(10^4 t)$. Therefore the solution (24) grows exponentially with t. Hence for the initial value problem (21)–(22) an arbitrarily small change in the initial data leads to an arbitrarily large solution, and the problem is not well-posed.

In numerical calculations stability is essential. For example, suppose a numerical scheme is devised to propagate the initial and boundary conditions. Those conditions can never be represented exactly in the computer, as small errors will exist because of roundoff or truncation of the data. If the problem itself is unstable, then these small errors in the data may be propagated in the numerical scheme in such a way that the calculation becomes meaningless. In Chapter 8 we observe that numerical schemes themselves may propagate errors and a different type of stability must be examined, namely numerical stability.

Diffusion in Higher Dimensions[†]

To derive an equation governing diffusion in three dimensions let Ω be an open region in R^3 with boundary $\partial\Omega$. The union of Ω and $\partial\Omega$ is called the *closure* of Ω and is denoted by $\overline{\Omega}$. Volume and surface integrals over Ω and $\partial\Omega$ will be denoted by

$$\iiint_{\Omega} f(x_1, x_2, x_3)\, dx_1\, dx_2\, dx_3 \equiv \int_{\Omega} f\, dx$$

$$\iint_{\partial\Omega} g(x_1, x_2, x_3)\, d\tau \equiv \int_{\partial\Omega} g\, d\tau$$

where $d\tau$ is a surface element on $\partial\Omega$. Points in R^3 will be denoted by $\mathbf{x} = (x_1, x_2, x_3)$. Two preliminary results are required for the sequel, the Divergence Theorem and Green's first identity.

Theorem 2.2   (*Divergence Theorem*)   *Let Ω be an open bounded region in R^3 with piecewise smooth boundary surface $\partial\Omega$ and let $\mathbf{f}$ be a vector field continuous on $\overline{\Omega}$ and class C^1 on Ω. Then*

$$\int_{\Omega} \operatorname{div} \mathbf{f}\, dx = \int_{\partial\Omega} \mathbf{f} \cdot \mathbf{n}\, d\tau$$

where $\mathbf{n}$ is the outward unit normal to $\partial\Omega$.

Using the vector identity

$$\operatorname{div}(u\, \mathbf{grad}\, v) = \mathbf{grad}\, u \cdot \mathbf{grad}\, v + u\nabla^2 v \tag{25}$$

[†] This section requires a knowledge of vector calculus. The reader may omit this section with no loss of continuity.

where ∇^2 is the Laplacian $\partial^2/\partial x_1^2 + \partial^2/\partial x_2^2 + \partial^2/\partial x_3^2$, it is easily shown that the following result holds.

Corollary 2.1 (*Green's Identity*) *Let u and v be scalar functions of class* $C^2(\Omega)$ *and continuous on* $\bar{\Omega}$. *Then*

$$\int_\Omega u\nabla^2 v \, dx = \int_{\partial\Omega} u \frac{\partial v}{\partial n} \, d\tau - \int_\Omega \mathbf{grad} \, u \cdot \mathbf{grad} \, v \, dx \qquad (26)$$

where $\partial v/\partial n \equiv \mathbf{n} \cdot \mathbf{grad} \, v$ *is the normal derivative.*

Rather than restrict the analysis to heat flow we broaden the perspective and derive a balance law for general diffusive processes. Let $c(\mathbf{x}, t)$ be the concentration of a substance measured in amount per unit volume that is diffusing in an arbitrary region Ω. Let $\mathbf{q}$ denote the flux vector such that $-\int_{\partial\Omega}\mathbf{q} \cdot \mathbf{n} \, d\tau$ represents the total amount of the substance flowing out of Ω per unit time. If p denotes the local production rate of the substance (measured in amount per unit volume per unit time) caused by some mechanism for creating or removing the substance, for example, a chemical reaction, then $\int_\Omega p \, dx$ is the total amount produced in Ω per unit time. The general *balance law in integral form* is therefore

$$\frac{d}{dt}\int_\Omega c \, dx = -\int_{\partial\Omega}\mathbf{q} \cdot \mathbf{n} \, d\tau + \int_\Omega p \, dx \qquad (27)$$

which in words states that the rate of change of the amount of the substance in Ω equals the amount flowing through the boundary $\partial\Omega$ plus the amount produced by sources (or sinks) inside Ω. A local version of (27) can be obtained by appealing to Theorem 2.2 and bringing the time derivative inside the integral on the left side (assuming c_t is continuous) to obtain

$$\int_\Omega c_t \, dx = -\int_\Omega \mathrm{div} \, \mathbf{q} \, dx + \int_\Omega p \, dx$$

Since Ω is arbitrary, it follows that

$$c_t = -\mathrm{div} \, \mathbf{q} + p \qquad (28)$$

provided the integrands are sufficiently smooth. Equation (28) is a partial differential equation for c and $\mathbf{q}$; and p is assumed to be known. To proceed further we must assume a *constitutive relation*, or material property, that relates $\mathbf{q}$ to c. Generally, we assume that the only mechanism causing the substance to diffuse is the concentration gradient; then the generalization of the Fourier law is

$$\mathbf{q} = -D \, \mathbf{grad} \, c$$

where D is the *diffusivity* of the substance measured in area per unit time. With this material assumption (28) becomes

$$c_t = \operatorname{div}(D \operatorname{\mathbf{grad}} c) + p \tag{29}$$

In the general case D and p could depend on c, thereby making (29) nonlinear. If D is constant, then (27) is

$$c_t - D\nabla^2 c = p \tag{30}$$

where we have used the fact that $\operatorname{div}(\operatorname{\mathbf{grad}} c) = \nabla^2 c$. Equation (30) is the *three dimensional nonhomogeneous diffusion equation*.

Example 2.2 (Heat Flow) We may specialize the previous results to heat conduction. Let $u(\mathbf{x}, t)$ denote the temperature at $(\mathbf{x}, t)$ in a region Ω characterized by specific heat c_v and density ρ. Then identifying $c \equiv \rho c_v u$ and $D \equiv K/\rho c_v$, where K is the thermal conductivity, we obtain the energy balance law

$$(\rho c_v u)_t = \operatorname{div}\left(\frac{K}{\rho c_v} \operatorname{\mathbf{grad}}(\rho c_v u)\right) + p$$

If ρ, c_v, and K are constants, then we obtain the *three dimensional heat conduction equation*

$$u_t - k\nabla^2 u = \frac{p}{\rho c_v} \tag{31}$$

where $k = K/\rho c_v$ (compare (11)). Here p represents a heat source or sink in the region Ω. We may always rescale and set $k = 1$.

Example 2.3 Motivated by their widespread occurrence in both chemical and biological settings, *reaction-diffusion equations* are of great interest in applied mathematics and modeling. The simplest example of such an equation is (30), where the production p depends, often nonlinearly, on c.

By intuition, if there are no sources the temperatures inside a heat conductor are bounded by the extreme temperatures attained either inside initially or on the boundary subsequently. The mathematical statement of this result is called the *maximum principle*.

Theorem 2.3 *Let Ω be an open bounded region in R^3 whose boundary is a piecewise smooth closed surface $\partial\Omega$ and let $u(\mathbf{x}, t) \in C(\overline{\Omega} \times [0, T])$, where*

[0, T] *is an interval of time. Denote*

$$\Omega = \overset{\circ}{\Omega}$$

$$M_0 = \max\{u(\mathbf{x}, t): \mathbf{x} \in \Omega \quad \text{and} \quad t = 0\}$$

$$M_{\partial\Omega} = \max\{u(\mathbf{x}, t): \mathbf{x} \in \partial\Omega \quad \text{and} \quad 0 \le t \le T\}$$

$$M = \max\{M_0, M_{\partial\Omega}\} \quad \text{max over bottom and sides of the "can"}$$

If $u_t - \nabla^2 u \le 0$ *in* $\Omega \times (0, T)$, *then* $u \le M$ *in* $\overline{\Omega} \times [0, T]$.

Proof Let $v(\mathbf{x}, t) = u(\mathbf{x}, t) + \varepsilon(x_1^2 + x_2^2 + x_3^2)$. Then $v_t - \nabla^2 v = -6\varepsilon \le 0$ in $\Omega \times (0, T)$. Now assume v attains its maximum value at $\mathbf{x}_0 \in \Omega$ and $0 < t_0 \le T$. Then at $(\mathbf{x}_0, t_0)$ we have $v_t \ge 0$ and $\nabla^2 v \le 0$, which is a contradiction. Thus v must assume its maximum value for $\mathbf{x}$ on $\partial\Omega$ and $0 \le t \le T$ or else for $t = 0$ and $\mathbf{x} \in \Omega$. But

$$u \le v \le \max v = M + \varepsilon r^2 \quad \text{in} \quad \overline{\Omega} \times [0, T]$$

where $r^2 = \max\{x_1^2 + x_2^2 + x_3^2\}$ over $\partial\Omega$. Since $\varepsilon > 0$ is arbitrary it follows that $u \le M$ in $\overline{\Omega} \times [0, T]$.

Corollary 2.2 *Under the same hypotheses as Theorem* 2.3 *with*

$$m_0 = \min\{u(\mathbf{x}, t): \mathbf{x} \in \Omega \quad \text{and} \quad t = 0\}$$

$$m_{\partial\Omega} = \min\{u(\mathbf{x}, t): \mathbf{x} \in \partial\Omega \quad \text{and} \quad 0 \le t \le T\}$$

$$m = \min\{m_0, m_{\partial\Omega}\}$$

if $u_t - \nabla^2 u \ge 0$ *in* $\Omega \times (0, T)$, *then* $u \ge m$ *in* $\overline{\Omega} \times [0, T]$, *and if* $u_t - \nabla^2 u = 0$ *in* $\Omega \times (0, T)$, *then* $m \le u \le M$ *in* $\overline{\Omega} \times [0, T]$.

Proof Apply Theorem 2.3 to $-u$ to obtain the first implication. The second implication is a combination of the two results.

The maximum principle and its corollary give direct access to a uniqueness theorem. Consider the boundary value problem

$$u_t - \nabla^2 u = p(\mathbf{x}, t), \qquad \mathbf{x} \in \Omega, \quad 0 < t < T$$
$$u(\mathbf{x}, 0) = f(\mathbf{x}), \qquad \mathbf{x} \in \Omega \qquad\qquad (32)$$
$$u(\mathbf{x}, t) = g(\mathbf{x}, t), \qquad \mathbf{x} \in \partial\Omega, \quad 0 \le t \le T$$

where Ω is as above and p, f, and g are given functions. The triple $\{p, f, g\}$ is called the *data* associated with (32).

Theorem 2.4   (*Uniqueness*)   *If the boundary value problem* (32) *has a continuous solution u in* $\overline{\Omega} \times [0, T]$, *then the solution is unique.*

Proof   If u_1 and u_2 are two continuous solutions of (32), then $v \equiv u_1 - u_2$ is a continuous solution to (32) with data $\{0, 0, 0\}$. Thus for v we have $m_0 = M_0 = 0$ and $m_{\partial\Omega} = M_{\partial\Omega} = 0$. Hence $m = M = 0$, and by Corollary 2.2 we have $v \equiv 0$.

The maximum principle also yields comparison results for solutions of (32) with different data.

Theorem 2.5   *Let* u_1 *and* u_2 *be solutions of* (32) *corresponding to data* $\{p_1, f_1, g_1\}$ *and* $\{p_2, f_2, g_2\}$, *respectively. If* $p_1 \leq p_2$ *in* $\Omega \times (0, T)$, $f_1 \leq f_2$ *in* Ω, *and* $g_1 \leq g_2$ *in* $\partial\Omega \times [0, T]$, *then* $u_1 \leq u_2$ *in* $\overline{\Omega} \times [0, T]$.

Proof   Let $v = u_1 - u_2$. Then $v_t - \nabla^2 v \leq 0$ on $\Omega \times (0, T)$, $M_0 \leq 0$ on Ω, and $m_{\partial\Omega} \leq 0$ on $\partial\Omega \times [0, T]$. By Theorem 2.3 $v \leq 0$ on $\overline{\Omega} \times [0, T]$, which proves the result.

In the case of steady-state diffusion, or the equilibrium case where the temperature u and production p do not depend on the time t, equation (31) reduces

$$\nabla^2 u = -\frac{p}{K}, \qquad \mathbf{x} \in \Omega \tag{33}$$

which is *Poisson's equation*. If there are no sources then

$$\nabla^2 u = 0, \qquad \mathbf{x} \in \Omega \tag{34}$$

which is *Laplace's equation*. These equations govern equilibrium processes that occur physically after a long enough time so that the transients die away. Example 2.1 shows that the initial value problem associated with Laplace's equation is not well-posed. Generally (33) or (34) is accompanied by a boundary condition of the form $u = f$ on $\partial\Omega$ (called a *Dirichlet condition*) or $\mathbf{n} \cdot \mathbf{grad}\, u = f$ on $\partial\Omega$ (called a *Neumann condition*). In other words u or its normal derivative is prescribed on $\partial\Omega$. In the Exercises a maximum principle is developed for solutions to Laplace's equation.

In some applications it is convenient to work in either cylindrical or spherical coordinates. These coordinates are defined by the equations

$$x = r\cos\theta$$
$$y = r\sin\theta \qquad \text{(cylindrical coordinates } r, \theta, z)$$
$$z = z$$

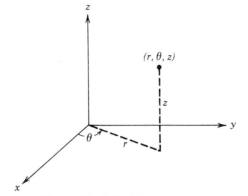

Figure 4.5. Cylindrical coordinates.

and

$$x = r \sin \phi \cos \theta$$
$$y = r \sin \phi \sin \theta \quad \text{(spherical coordinates } r, \phi, \theta)$$
$$z = r \cos \phi$$

Geometrically the coordinates are shown in Figs. 4.5 and 4.6. An application of the chain rule permits us to write the Laplacian in cylindrical and spherical coordinates as

$$\nabla^2 u = u_{rr} + \frac{1}{r} u_r + \frac{1}{r^2} u_{\theta\theta} + u_{zz} \quad \text{(cylindrical)} \tag{35}$$

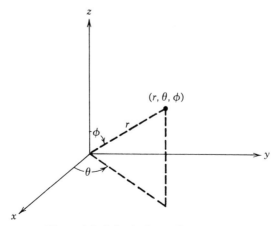

Figure 4.6. Spherical coordinates.

and

$$\nabla^2 u = \frac{1}{r^2}\frac{\partial}{\partial r}(r^2 u_r) + \frac{1}{r^2\sin\phi}\frac{\partial}{\partial\phi}(\sin\phi u_\phi) + \frac{1}{r^2\sin^2\phi}u_{\theta\theta} \qquad \text{(spherical)}$$

$$(36)$$

We leave the details of these calculations to the reader (Exercise 2.8).

EXERCISES

2.1   Using the energy method as in the uniqueness proof presented in Theorem 2.1 prove that solutions to the boundary value problem

$$u_t - ku_{xx} = 0, \qquad 0 < x < l, \quad 0 < t < T$$
$$u(x,0) = f(x), \qquad 0 < x < l$$
$$u_x(0,t) = 0, \qquad u_x(l,t) = g(t), \qquad 0 < t < T$$

are unique for any $T > 0$. Assume f and g are continuous functions.

2.2   Prove that if $f(x)$ is a continuous function and

$$\int_a^b f(x)\, dx = 0$$

for all subintervals $[a, b]$ of $[0, l]$, then $f(x)$ must vanish identically on $[0, l]$.

2.3   A homogeneous (constant ρ, c_v, and K) bar has cross-sectional area $A(x), 0 < x < l$, and there is only a small variation of $A(x)$ with x, so that the assumption of constant temperature in any cross section remains valid. There are no sources and the flux is given by $-KA(x)u_x(x, t)$. From a conservation law obtain a partial differential equation for the temperature $u(x, t)$, which reflects the area variation of the bar.

2.4   Transform the boundary value problem

$$u_t - u_{xx} = 0, \qquad 0 < x < l, \quad t > 0$$
$$u(x,0) = f(x), \qquad 0 < x < l$$
$$u(0,t) = g(t), \qquad u(l,t) = h(t), \qquad t > 0$$

to a problem with homogeneous boundary conditions.

$$v(x,t) = u(x,t) + (1-\tfrac{x}{l})(g(0)-g(t)) + \tfrac{x}{l}(h(0) - h(t)) - f(x)$$

2.5   A homogeneous region is bounded by two infinitely long concentric cylindrical surfaces of radii $r = 1$ and $r = r_0 > 1$, respectively. The inner surface is kept at zero degrees and the outer is kept at u_0 degrees, where u_0 is a constant. Find the equilibrium temperature $u(r)$ in the region $1 < r < r_0$ between the two cylinders. *Solution:* $u(r) = u_0(\ln r/\ln r_0)$.

2.6   A homogeneous region is bounded by two concentric spherical surfaces of radii $r = 1$ and $r = r_0 > 1$, respectively. The inner surface is kept at zero degrees and the outer surface is maintained at u_0 degrees, where u_0 is a constant. Find the equilibrium temperature $u(r)$ in the region $1 < r < r_0$ between the two spheres. Compare to the previous problem. *Solution:* $u(r) = (u_0 r_0)/(1 - r_0)(1/r - 1)$.

2.7   Consider the boundary value problem

$$u_t - ku_{xx} = q(x), \qquad 0 < x < l, \quad t > 0$$

$$u_x(0, t) = A, \qquad u_x(l, t) = B, \qquad t > 0$$

$$u(x,0) = f(x), \qquad 0 < x < l$$

Interpret this problem in a physical context and find a condition under which an equilibrium solution can exist. *Solution:* $A - B = (1/k)\int_0^l q(x)\, dx$.

2.8   Derive (35) and (36).

2.9   Prove (25) and (26).

2.10   Let Ω be an open bounded region in R^3 with piecewise smooth boundary $\partial\Omega$ and let $\nabla^2 u \geq 0$ in Ω and $u \in C(\bar{\Omega})$. Prove that if $u \leq M$ on $\partial\Omega$, then $u \leq M$ in Ω. State and prove a corresponding minimum principle.

2.11   Prove Theorem 2.4 using an energy argument as in the proof of Theorem 2.1 (*Hint:* Define $E(t) = \int_\Omega v^2\, dx$, where $v = u_1 - u_2$ and apply Corollary 2.1.)

2.12   Under the assumptions in Corollary 2.1 prove *Green's second identity*

$$\int_\Omega (u\nabla^2 v - v\nabla^2 u)\, dx = \int_{\partial\Omega} \left(u\frac{\partial v}{\partial n} - v\frac{\partial u}{\partial n} \right) d\tau$$

2.13   Let Ω be an open bounded region in R^3. Prove that the *Dirichlet problem*

$$\nabla^2 u = 0, \qquad \mathbf{x} \in \Omega$$
$$u = f, \qquad \mathbf{x} \in \partial\Omega$$

has at most one solution.

2.14   Let Ω be an open bounded region in R^3. Show that any solution to the *Neumann problem*

$$\nabla^2 u = p, \qquad \mathbf{x} \in \Omega$$
$$\frac{\partial u}{\partial n} = f, \qquad \mathbf{x} \in \partial\Omega$$

must necessarily satisfy the condition $\int_\Omega p \, dx = \int_{\partial\Omega} f \, d\tau$. Interpret physically.

4.3   CLASSICAL TECHNIQUES

Separation of Variables

Now we discuss a classical method for obtaining solutions to certain linear boundary value problems. The method is due to Fourier (1768–1830) and is called *separation of variables* or the *eigenfunction expansion method*. The idea is to superimpose infinitely many functions $u_1(x, t), u_2(x, t), \ldots$ that satisfy the partial differential equation and boundary conditions by writing

$$c_1 u_1(x, t) + c_2 u_2(x, t) + \cdots$$

and then choose the constants $c_1, c_2, \ldots$ so that the combination satisfies the initial conditions as well. From a theoretical viewpoint the method is valuable in that analytic solutions in the representation of an infinite series can be obtained for certain problems, thereby providing existence. In practical engineering applications, however, boundary value problems are almost always solved numerically on a computer. One example followed by some general remarks on when the method is applicable should provide ample preparation for the reader to apply the technique to other problems.

Example 3.1   The problem we consider is

$$u_t - ku_{xx} = 0, \qquad 0 < x < l, \quad t > 0 \tag{1}$$
$$u(x, 0) = f(x), \qquad 0 < x < l \tag{2}$$
$$u(0, t) = u(l, t) = 0, \qquad t > 0 \tag{3}$$

which models heat flow in a homogeneous bar with constant diffusivity k whose ends are held fixed at zero degrees and whose initial temperature distribution is $f(x)$. The first step is to assume a solution of the form

$$u(x, t) = X(x)T(t) \qquad (4)$$

that is, a product of a function of x and a function of t. When (4) is substituted into the partial differential equation (1) the result can be written as

$$\frac{T'(t)}{kT(t)} = \frac{X''(x)}{X(x)}$$

with the variables x and t separated, and hence the name *separation of variables*. As x and t vary the only way a function of t can be equal to a function of x is if both functions are equal to the same constant, which we call $-\lambda$. Hence

$$X''(x) + \lambda X(x) = 0 \qquad (5)$$

and

$$T'(t) = -\lambda k T(t) \qquad (6)$$

Calling the constant $-\lambda$ does not mean it is negative; we use the negative sign only for convenience. Therefore the assumption (4) has led to a pair of ordinary differential equations (5) and (6). Now we apply (4) to the boundary conditions (3) and obtain

$$T(t) X(0) = 0, \qquad T(t) X(l) = 0 \qquad (7)$$

Excluding the uninteresting possibility that $T(t) = 0$, we get $X(0) = 0$ and $X(l) = 0$. Therefore we are led to a boundary value problem for the function X

$$X''(x) + \lambda X(x) = 0, \qquad 0 < x < l \qquad (8a)$$
$$X(0) = 0, \qquad X(l) = 0 \qquad (8b)$$

In this last step, in order to get the boundary conditions on X it was essential that zero appear on the right sides in (7); hence, the boundary conditions in the original problem had to be homogeneous.

The plan can be described briefly as follows. We determine values of λ for which (8) has nontrivial solutions. Often there will be infinitely many such values $\lambda_1, \lambda_2, \ldots$ that accomplish this and to each will correspond a solution $X_n(x)$, $n = 1, 2, \ldots$, of (8) and a corresponding solution $T_n(t)$ of (6). Thus we will have constructed infinitely many solutions $u_n(x, t) = X_n(x)T_n(t)$, $n = 1, 2, \ldots$, to the partial differential equation that satisfy the boundary conditions. We then superimpose these solutions by defining

$$u(x, t) = \sum_{n=1}^{\infty} b_n u_n(x, t)$$

and we choose the constants b_n so that $u(x, t)$ satisfies the initial condition as well. Being a sum of solutions to a linear problem we expect u to be a solution and to satisfy the homogeneous boundary conditions.

To solve (8) we examine three cases: $\lambda < 0$, $\lambda = 0$, and $\lambda > 0$.

(i) $\lambda < 0$. If $\lambda < 0$, then the general solution of (8a) is

$$X(x) = A \cosh \sqrt{-\lambda}\, x + B \sinh \sqrt{-\lambda}\, x$$

where A and B are arbitrary constants. $X(0) = 0$ implies $A = 0$ and $X(l) = 0$ implies $B = 0$. Therefore the boundary value problem (8) has only the trivial solution if $\lambda < 0$.

(ii) $\lambda = 0$. In this case the general solution of (8a) is $X = Ax + B$. Application of the boundary conditions (8b) forces $A = B = 0$. Again no nontrivial solutions exist.

(iii) $\lambda > 0$. Here the general solution of (8a) is

$$X(x) = A \sin \sqrt{\lambda}\, x + B \cos \sqrt{\lambda}\, x$$

$X(0) = 0$ forces $B = 0$. Then $X(l) = 0$ implies

$$A \sin \sqrt{\lambda}\, l = 0$$

Now we have the possibility of selecting values of λ that will make this equation hold without choosing $A = 0$, which would again lead to a trivial solution. Therefore take

$$\sqrt{\lambda}\, l = n\pi, \qquad n = 1, 2, 3, \ldots$$

or

$$\lambda = \frac{n^2 \pi^2}{l^2}, \qquad n = 1, 2, 3, \ldots \tag{9}$$

The case $n = 0$ was the subject of Case (ii) above. Corresponding to each value of λ in (9) there is a solution of (8) given by

$$X_n(x) = \sin \frac{n\pi x}{l}, \qquad n = 1, 2, \ldots \tag{10}$$

We have chosen the constant A in front of the sine function to have value unity.

We have succeeded in solving the boundary value problem (8). The values

$$\lambda_n = \frac{n^2\pi^2}{l^2}, \qquad n = 1, 2, \ldots \tag{11}$$

for which (8) has a nontrivial solution are called the *eigenvalues* and the corresponding solutions (10) are called the *eigenfunctions*. Each choice of λ_n provides a solution to the T equation (6). Equation (6) becomes

$$T_n'(t) = -\frac{n^2\pi^2 k}{l^2} T_n(t), \qquad n = 1, 2, \ldots$$

from which we infer

$$T_n(t) = \exp\left(-\frac{n^2\pi^2 kt}{l^2}\right), \qquad n = 1, 2, \ldots$$

As indicated we attempt to find a solution to the original boundary value problem (1)–(3) of the form

$$u(x, t) = \sum_{n=1}^{\infty} b_n X_n(x) T_n(t)$$

$$= \sum_{n=1}^{\infty} b_n \sin\frac{n\pi x}{l} \exp\left(-\frac{n^2\pi^2 kt}{l^2}\right) \tag{12}$$

The constants b_n are determined from the initial condition (2), which becomes

$$u(x, 0) = f(x) = \sum_{n=1}^{\infty} b_n \sin\frac{n\pi x}{l} \tag{13}$$

We proceed formally. Let m be a fixed but arbitrary positive integer. Multiplying (13) by $\sin(m\pi x/l)$ and integrating the result from 0 to l gives

$$\int_0^l f(x)\sin\frac{m\pi x}{l}\, dx = \sum_{n=1}^{\infty} b_n \int_0^l \sin\frac{m\pi x}{l} \sin\frac{n\pi x}{l}\, dx \tag{14}$$

In obtaining (14) we have interchanged the order of integration and summation. From calculus, for positive integers m and n

$$\int_0^l \sin\frac{m\pi x}{l} \sin\frac{n\pi x}{l}\, dx = \begin{cases} 0, & n \neq m \\ \dfrac{l}{2}, & n = m \end{cases} \tag{15}$$

Consequently there is just one nonzero term in the infinite sum in (14) occurring when $n = m$; that is, $b_m(l/2)$. Hence

$$b_m = \frac{2}{l} \int_0^l f(x) \sin \frac{m\pi x}{l} \, dx$$

Since m was an arbitrary index we must in fact have

$$b_n = \frac{2}{l} \int_0^l f(x) \sin \frac{n\pi x}{l} \, dx, \qquad n = 1, 2, 3 \ldots \tag{16}$$

We repeat that the calculations here are formal, meaning that we have not rigorously justified each step, for example, the term-by-term integration of an infinite series to obtain (14). The calculation has led to an expression that is the solution to (1)–(3), however, namely the infinite sum (12) where the constants b_n are given by (16). It can be shown that this does indeed supply a solution to the problem.

For the separation of variables method to be successful the partial differential equation as well as the boundary conditions must be linear and homogeneous. Upon substitution of $u = X(x)T(t)$ they must separate into a t *part* and an x *part*. Problems that do not meet these criteria, in particular the homogeneity conditions, can frequently be transformed into problems that do. The separation of variables method is discussed in detail in texts on boundary value problems (for example, see Churchill [2]).

Fourier Series

A key to the preceding computation was the claim that the function $f(x)$ could be expanded in the series $b_1 \sin(\pi x/l) + b_2 \sin(2\pi x/l) + \cdots$ by choosing the coefficients $b_1, b_2, \ldots$ appropriately. The validity of such a claim lies in the domain of Fourier analysis that addresses the more general question of when a given function $f(x)$ can be expanded in a *trigonometric series* of the form

$$\frac{a_0}{2} + \sum_{n=1}^{\infty} \left(a_n \cos \frac{n\pi x}{l} + b_n \sin \frac{n\pi x}{l} \right) \tag{17}$$

It is common knowledge that an infinitely differentiable function f can be expanded in a Taylor series

$$f(x) = \sum_{n=0}^{\infty} a_n (x - x_0)^n \tag{18}$$

in powers of $x - x_0$ with coefficients given by $a_n = f^{(n)}(x_0)/n!$. But it is not immediately clear that an arbitrary function can be written in terms of periodic functions like sines and cosines. The basic questions are: (i) For what functions f is the representation (17) possible? (ii) How does the series (17) converge to f? (iii) How are the coefficients a_n and b_n calculated? In the sequel we present a theorem that will answer these questions.

First we address question (iii) in a special case. Suppose the series (17) converges uniformly to a function f defined on $[-l, l]$. Then

$$f(x) = \frac{a_0}{2} + \sum_{n=1}^{\infty}\left(a_n\cos\frac{n\pi x}{l} + b_n\sin\frac{n\pi x}{l}\right) \tag{19}$$

and the coefficients a_n and b_n can be calculated as follows. Multiply (19) by $\sin(m\pi x/l)$, where $m > 0$, and integrate from $x = -l$ to $x = l$ to get

$$\int_{-l}^{l} f(x)\sin\frac{m\pi x}{l}\,dx = \frac{a_0}{2}\int_{-l}^{l}\sin\frac{m\pi x}{l}\,dx$$
$$+ \sum_{n=1}^{\infty}\left(a_n\int_{-l}^{l}\cos\frac{n\pi x}{l}\sin\frac{m\pi x}{l}\,dx\right.$$
$$\left. + b_n\int_{-l}^{l}\sin\frac{n\pi x}{l}\sin\frac{m\pi x}{l}\,dx\right) \tag{20}$$

Here the order of summation and integration has been interchanged, a valid step because of the uniform convergence. The first integral on the right side of (20) vanishes and the remaining integrals can be calculated using the calculus formulas

$$\int_0^{l}\left\{\begin{matrix}\sin\\\cos\end{matrix}\right\}\frac{n\pi x}{l}\left\{\begin{matrix}\sin\\\cos\end{matrix}\right\}\frac{m\pi x}{l}\,dx = \begin{cases}l/2, & \text{if } m = n\\0, & \text{if } m \neq n\end{cases}$$
$$\int_{-l}^{l}\sin\frac{n\pi x}{l}\cos\frac{m\pi x}{l}\,dx = 0 \tag{21}$$

which determines the coefficients $b_1, b_2, \ldots$ in (19). By nearly the same argument the coefficients $a_0, a_1, a_2, \ldots$ can be found. Multiply (19) by $\cos(m\pi x/l)(m > 0)$ and integrate from $-l$ to l to obtain, after using (21),

$$\int_{-l}^{l} f(x)\sin\frac{m\pi x}{l}\,dx = b_m l$$

which determines the coefficients $b_1, b_2, \ldots$ in (19). By nearly the same argument the coefficients $a_0, a_1, a_2, \ldots$ can be found. Multiply (19) by $\cos(m\pi x/l)(m > 0)$ and integrate from $-l$ to l to obtain, after using (21),

$$\int_{-l}^{l} f(x)\cos\frac{m\pi x}{l}\,dx = a_m l$$

To determine a_0 integrate (19) from $-l$ to l to obtain

$$\int_{-l}^{l} f(x)\,dx = \frac{a_0}{2}(2l) \tag{21}$$

We have shown that if the series (17) converges uniformly to $f(x)$ on $[-l, l]$, then the coefficients are given by the formulas

$$a_n = \frac{1}{l} \int_{-l}^{l} f(x) \cos \frac{n\pi x}{l} \, dx, \qquad n = 0, 1, 2, \ldots \qquad (22)$$

$$b_n = \frac{1}{l} \int_{-l}^{l} f(x) \sin \frac{n\pi x}{l} \, dx, \qquad n = 1, 2, \ldots \qquad (23)$$

The integrals necessarily exist since f, being the uniform sum of continuous functions, is itself continuous.

Whether or not the series (17) converges uniformly and the representation (19) is valid, it is still possible to formally compute the coefficients a_n and b_n by the formulas (22) and (23), provided the function f is chosen such that the integrals exist. In this case the formal series (17) is called the *Fourier series* associated with f and the coefficients a_n and b_n given by (22) and (23) are called the *Fourier* coefficients. Question (ii) may now be rephrased by asking what conditions on f guarantee the convergence of its Fourier series and does the series actually converge to f in one sense or the other (pointwise, uniform, etc.)?

One large class of functions for which the theory can be developed is the piecewise smooth functions, that is, functions f for which both f and f' are continuous except at possibly finitely many points.

Definition 3.1 *A function f is piecewise continuous on $[a, b]$ if it has at most finitely many discontinuities on $[a, b]$, and at those points f has finite one-sided limits. A function f is piecewise smooth on $[a, b]$ if both f and f' are piecewise continuous on $[a, b]$. The one-sided limits are denoted by*

$$f(x_0^+) = \lim_{x \to x_0^+} f(x), \qquad f(x_0^-) = \lim_{x \to x_0^-} f(x)$$

One can prove the following Fourier-type theorem that ensures pointwise convergence. Its proof can be found in [2].

Theorem 3.1 *If $f(x)$ is piecewise smooth on the interval $(-l, l)$, then its Fourier series converges at each x in $(-l, l)$ to*

$$\tfrac{1}{2}(f(x^-) + f(x^+))$$

At the endpoints $x = l$ and $x = -l$ the series converges to

$$\tfrac{1}{2}(f(l^-) + f(-l^+))$$

Thus the Fourier series (18) converges at each x to the average value of the left and right limits $f(x^-)$ and $f(x^+)$ at x. If f is continuous at x, then this means that the series converges to $f(x)$ at that point.

If $f(x)$ is an odd function on $(-l, l)$, that is, $f(-x) = -f(x)$, then the coefficients a_n are all zero and b_n are given by

$$b_n = \frac{2}{l} \int_0^l f(x) \sin \frac{n\pi x}{l} \, dx, \qquad n = 1, 2, \ldots \tag{24}$$

and the Fourier series (18) becomes

$$\sum_{n=1}^{\infty} b_n \sin \frac{n\pi x}{l} \tag{25}$$

If f is any piecewise smooth function on $(0, l)$, then we may form the Fourier series (25) where the b_n are given by (24). In this context (25) is called the *Fourier sine series* of f. By Theorem 3.1 the series will converge to the average value $\frac{1}{2}(f(x^+) + f(x^-))$. To apply the theorem extend f to the entire interval $(-l, l)$ to make it odd.

Similarly, if $f(x)$ is an even function on $(-l, l)$, that is, $f(-x) = f(x)$, then it has a Fourier expansion

$$\frac{a_0}{2} + \sum_{n=1}^{\infty} a_n \cos \frac{n\pi x}{l} \tag{26}$$

where the a_n are given by

$$a_n = \frac{2}{l} \int_0^l f(x) \cos \frac{n\pi x}{l} \, dx, \qquad n = 0, 1, 2, \ldots \tag{27}$$

If f is any piecewise smooth function on the interval $(0, l)$, then we may form its *Fourier cosine series* (26) with the coefficients given by (27). By extending f evenly to the symmetric interval $(-l, l)$ we can apply the Fourier theorem and conclude that (26) converges to $\frac{1}{2}(f(x^+) + f(x^-))$.

Example 3.2   Find the Fourier series for $f(x) = |x|$ on the interval $-\pi < x < \pi$. Using (22) and (23)

$$a_0 = \frac{1}{\pi} \int_{-\pi}^{\pi} |x| \, dx = \frac{1}{\pi} \int_{-\pi}^{0} -x \, dx + \frac{1}{\pi} \int_0^{\pi} x \, dx = \pi$$

$$a_n = \frac{1}{\pi} \int_{-\pi}^{0} (-x) \cos nx \, dx + \frac{1}{\pi} \int_0^{\pi} x \cos nx \, dx$$

$$= \frac{1}{\pi} \left[-\frac{x \sin nx}{n} - \frac{\cos nx}{n^2} \right]_{-\pi}^{0} + \frac{1}{\pi} \left[\frac{x \sin nx}{n} + \frac{\cos nx}{n^2} \right]_0^{\pi}$$

$$= 2 \frac{(-1)^n - 1}{\pi n^2}, \qquad n \geq 1$$

$$b_n = \frac{1}{\pi} \int_{-\pi}^{0} (-x) \sin nx \, dx + \frac{1}{\pi} \int_0^{\pi} x \sin nx \, dx = 0, \qquad n \geq 1$$

The Fourier series associated with $|x|$ on $(-\pi, \pi)$ is therefore

$$\frac{\pi}{2} + \sum_{n=1}^{\infty} 2\frac{(-1)^n - 1}{\pi n^2} \cos nx$$

or

$$\frac{\pi}{2} - \frac{4}{\pi} \cos x - \frac{4}{9\pi} \cos 3x - \cdots$$

Since $|x|$ is continuous the series will converge to $|x|$ on $-\pi < x < \pi$.

Proving uniform convergence requires more severe restrictions on f. We state another Fourier-type theorem.

Theorem 3.2   *Assume f is continuous on $[-l, l]$, $f(-l) = f(l)$, and f' is piecewise continuous on $[-l, l]$. Then the Fourier series associated with f converges uniformly and absolutely to $f(x)$ on the interval $[-l, l]$.*

Fourier-type theorems can also be obtained for less restrictive classes of functions. For the class of *square integrable functions* on $[-l, l]$, that is, the set of all functions f on $[-l, l]$ for which the integral $\int_{-l}^{l} |f(x)|^2\, dx$ exists, the following theorem holds.

Theorem 3.3   *If f is square integrable on $[-l, l]$, then the Fourier series associated with f converges in the mean to f, that is,*

$$\lim_{N \to \infty} \int_{-l}^{l} \left[f(x) - \left(\frac{a_0}{2} + \sum_{n=1}^{N} a_n \cos \frac{n\pi x}{l} + b_n \sin \frac{n\pi x}{l} \right) \right]^2 dx = 0$$

Sturm-Liouville Problems

In the separation of variables method we are often led to a boundary value problem called a *Sturm-Liouville problem*, which is of the form

$$(p(x)y')' + (-q(x) + \lambda r(x))y = 0, \qquad 0 < x < l$$
$$c_1 y(0) + c_2 y'(0) = 0 \qquad\qquad (28)$$
$$c_3 y(l) + c_4 y'(l) = 0$$

where p, q, and r are real functions of x and the c_i are constants. If for a certain value of λ this problem has a nontrivial solution $y(x)$, then λ is called an *eigenvalue* and $y(x)$ the corresponding *eigenfunction*. If p, p', q, and r are continuous on $[0, l]$ and p and r are strictly positive on $[0, l]$, then the Sturm-Liouville problem is *regular* and the following theorem can be proved.

Theorem 3.4   *For a regular Sturm-Liouville problem*

(i) *The eigenvalues are real and to each eigenvalue there corresponds a single eigenfunction unique up to a constant multiple.*

(ii) *The eigenvalues form an infinite sequence $\lambda_1, \lambda_2, \ldots$ and can be ordered according to*

$$0 \le \lambda_1 < \lambda_2 < \lambda_3 < \cdots$$

with

$$\lim_{n \to \infty} \lambda_n = \infty$$

(iii) *If $y_1(x)$ and $y_2(x)$ are two eigenfunctions corresponding to distinct eigenvalues λ_1 and λ_2, then*

$$\int_0^l y_1(x) y_2(x) r(x)\, dx = 0$$

We introduce some useful notation and terminology in the following discussion. If for any two functions f and g we have

$$\int_0^l f(x) g(x) r(x)\, dx = 0 \tag{29}$$

then we say f and g are *orthogonal* on $[0, l]$ with respect to the *weight function* $r(x) > 0$. Thus Part (iii) of Theorem 3.4 states that eigenfunctions corresponding to distinct eigenvalues are orthogonal. The integral on the left side of (29) is called the *inner product* of f and g and is denoted by $\langle f, g \rangle$; thus

$$\langle f, g \rangle \equiv \int_0^l f(x) g(x) r(x)\, dx$$

The number $\|f\|$ defined by

$$\|f\| = \left(\int_0^l f(x)^2 r(x)\, dx \right)^{1/2}$$

is called the *norm* of f. Clearly $\|f\| = \langle f, f \rangle^{1/2}$. The norm measures the *size* of a function and generalizes the common notion of length of a vector; the inner product generalizes the idea of the scalar or dot product for vectors. A set of functions $f_1(x), f_2(x), \ldots$ is *orthogonal* if $\langle f_m, f_n \rangle = 0$ for the $m \ne n$ and *orthonormal* if in addition each function in the set has norm one, that is, $\langle f_n, f_n \rangle = 1$ for $n = 1, 2, \ldots$. It is clear that an orthogonal set can be made into an orthonormal set by dividing each function by its norm.

Example 3.3   In Example 3.1 the Sturm-Liouville problem

$$y'' + \lambda y = 0, \qquad 0 < x < l$$
$$y(0) = 0, \qquad y(l) = 0$$

appeared. In this case $r(x) = 1$, $p(x) = 1$, and $q(x) = 0$. The eigenvalues are

$$\frac{\pi^2}{l^2} < \frac{4\pi^2}{l^2} < \frac{9\pi^2}{l^2} < \cdots$$

and the corresponding eigenfunctions are

$$y_n(x) = \sin \frac{n\pi x}{l}, \qquad n = 1, 2, \ldots$$

The y_n are orthogonal on $[0, l]$ with respect to weight function $r(x) = 1$ as expressed by (15). The norm of y_n is

$$\|y_n\| = \left(\int_0^l \sin^2 \frac{n\pi x}{l} \, dx \right)^{1/2} = \sqrt{\frac{l}{2}}$$

Therefore the orthonormal set of eigenfunctions is

$$\sqrt{\frac{2}{l}} \sin \frac{n\pi x}{l}, \qquad n = 1, 2 \ldots$$

In Example 3.1 we observed that a given piecewise smooth function $f(x)$ could be expanded in a series of the eigenfunctions. More generally we may ask if a function can be expanded in terms of the orthonormalized eigenfunctions $y_1(x)$, $y_2(x), \ldots$ of a regular Sturm-Liouville problem (28). That is, in one sense or another, does

$$f(x) = \sum_{n=1}^{\infty} c_n y_n(x) \tag{30}$$

for some choice of the coefficients c_n? As before this question can be answered affirmatively by a formal calculation. The argument depends heavily upon the orthogonality of the eigenfunctions. Assuming (30) holds multiply by $r(x)y_m(x)$ and integrate from 0 to l to get

$$\langle f, y_m \rangle = \left\langle \sum_{n=1}^{\infty} c_n y_n, y_m \right\rangle = \sum_{n=1}^{\infty} c_n \langle y_n, y_m \rangle = c_m \|y_m\|^2$$

Since the y_n are orthonormal and m is an arbitrary index

$$c_n = \langle f, y_n \rangle, \qquad n = 1, 2, 3 \dots \tag{31}$$

In the previous calculation the order of summation and integration was interchanged and we used the orthogonality of the y_n to collapse the infinite sum to a single term. The interchange depends on the series converging uniformly. The series (30) with coefficients defined by (31) is called the *generalized Fourier series* for f. The c_n are called the *generalized Fourier coefficients*.

Theorem 3.5 *Let $y_1, y_2, \dots$ be an orthonormal set of eigenfunctions for the regular Sturm-Liouville problem (28) and let f be a piecewise smooth function on the interval $(0, l)$. Then the series (30) with coefficients given by (31) converges at each x in $(0, l)$ to the value $\frac{1}{2}[f(x^-) + f(x^+)]$.*

The interval $(0, l)$ in the previous discussion can be replaced by any finite interval (a, b) and in some cases the results can be extended to infinite intervals. If p or r vanishes at one or both of the endpoints of the interval then the Sturm-Liouville problem is *singular*. Under suitable conditions Theorems 3.4 and 3.5 can be generalized to singular problems.

Example 3.4 A rod extending from $x = 1$ to $x = e$ has end temperatures given by

$$u(1, t) = u(e, t) = 0, \qquad t > 0$$

and an initial temperature distribution given by

$$u(x, 0) = f(x), \qquad 1 < x < e$$

No sources are present and the rod has constant density ρ and specific heat c_v, but its thermal conductivity K varies according to $K(x) = x^2$. From (10) in Section 4.2 the equation governing the temperature $u(x, t)$ is

$$c_v \rho u_t = \frac{\partial}{\partial x}(x^2 u_x), \qquad 1 < x < e, \quad t > 0 \tag{32}$$

To apply the Fourier method to find the solution let $u = X(x)T(t)$. Substituting into (32) and separating variables gives

$$c_v \rho \frac{T'}{T} = \frac{1}{X}\frac{d}{dx}(x^2 X') = -\lambda$$

where $-\lambda$ is constant and

$$X(1) = X(e) = 0 \tag{33}$$

Thus T satisfies the equation

$$T' = -\frac{\lambda}{c_v \rho} T \tag{34}$$

and X satisfies

$$\frac{d}{dx}(x^2 X') + \lambda X = 0, \qquad 1 < x < e \tag{35}$$

The ordinary differential equation (35) and boundary conditions (33) define a regular Sturm-Liouville problem on $[1, e]$. To determine the eigenvalues and eigenfunctions we rewrite (35) as

$$x^2 X'' + 2x X' + \lambda X = 0$$

and recognize it as a Cauchy-Euler equation with auxiliary equation $m(m - 1) + 2m + \lambda = 0$. The roots are

$$m = -\tfrac{1}{2} \pm \sqrt{\tfrac{1}{4} - \lambda} \tag{36}$$

If $\lambda = \tfrac{1}{4}$, then the roots are $m = -\tfrac{1}{2}, -\tfrac{1}{2}$, and (35) has general solution $X = (A + B \ln x) x^{-1/2}$. Applying the boundary conditions (33) gives $A = B = 0$, and therefore there are only trivial solutions. If $\lambda < \tfrac{1}{4}$, then the roots (36) are real and the general solution of (35) is

$$X = A x^{-\frac{1}{2} + \sqrt{\frac{1}{4} - \lambda}} + B x^{-\frac{1}{2} - \sqrt{\frac{1}{4} - \lambda}}$$

Applying the boundary conditions again gives $A = B = 0$. In the case $\lambda > \tfrac{1}{4}$ the roots are complex and (35) has general solution

$$X = \frac{A}{\sqrt{x}} \sin\left(\sqrt{\lambda - \tfrac{1}{4}} \ln x\right) + \frac{B}{\sqrt{x}} \cos\left(\sqrt{\lambda - \tfrac{1}{4}} \ln x\right)$$

The boundary condition $X(1) = 0$ forces $B = 0$. Then $X(e) = 0$ becomes

$$\frac{A}{\sqrt{e}} \sin\left(\sqrt{\lambda - \tfrac{1}{4}}\right) = 0$$

which forces

$$\sqrt{\lambda - \tfrac{1}{4}} = n\pi, \qquad n = 1, 2, \dots$$

($n = 0$ is the case $\lambda = \tfrac{1}{4}$, which was discussed above). Therefore the eigenvalues of the Sturm-Liouville problem (35) and (33)

$$\lambda_n = n^2 \pi^2 + \tfrac{1}{4}, \qquad n = 1, 2, \dots$$

and the corresponding eigenfunctions are

$$X_n(x) = \frac{1}{\sqrt{x}} \sin(n\pi \ln x), \qquad n = 1, 2, \ldots$$

The normalized eigenfunctions are

$$\phi_n(x) = \frac{(1/\sqrt{x})\sin(n\pi \ln x)}{\|X_n\|} = \sqrt{\frac{2}{x}} \sin(n\pi \ln x)$$

since

$$\|X_n\| = \left(\int_1^e \frac{1}{x} \sin^2(n\pi \ln x) \, dx \right)^{1/2} = \frac{1}{\sqrt{2}}$$

For each n the T equation (34) becomes

$$T_n' = -\frac{\lambda_n}{c_v \rho} T_n$$

with solution

$$T_n(t) = \exp\left(-\frac{\lambda_n}{c_v \rho} t \right), \qquad n = 1, 2, \ldots$$

Now we superimpose the solutions $T_n(t)X_n(x)$ to form

$$u(x, t) = \sum_{n=1}^{\infty} a_n \exp\left(-\frac{\lambda_n t}{c_v \rho} \right) \frac{1}{\sqrt{x}} \sin(n\pi \ln x) \tag{37}$$

and choose the coefficients a_n so that u satisfies the initial condition. We have

$$u(x, 0) = f(x) = \sum_{n=1}^{\infty} a_n \frac{1}{\sqrt{x}} \sin(n\pi \ln x) = \sum_{n=1}^{\infty} (a_n \|X_n\|) \phi_n(x)$$

The right side is the generalized Fourier series of $f(x)$ in terms of the normalized eigenfunctions $\phi_n(x)$. By Theorem 3.5 the coefficients are given by the generalized Fourier coefficients. Therefore

$$a_n \|X_n\| = \int_1^e f(x) \phi_n(x) \, dx, \qquad n = 1, 2, \ldots$$

or

$$a_n = 2 \int_1^e \frac{f(x)}{\sqrt{x}} \sin(n\pi \ln x) \, dx, \qquad n = 1, 2, \ldots \tag{38}$$

Consequently the solution of (32) subject to the given initial and boundary data is given by (37), where the a_n are defined by (38). For a given $f(x)$ one must apply numerical integration techniques to actually determine the a_n.

A treatment of the mathematical basis of expansions in eigenfunctions in an abstract setting can be found in Stakgold [3].

Integral Transforms

Another important class of techniques for solving linear partial differential equations, especially on infinite domains, are the methods based on integral transforms. An *integral transform* is a relation of the form

$$F(s) = \int_I K(s, t) f(t) \, dt \tag{39}$$

whereby a given function f is transformed to another function F, called the *transform* of f. The known function K is called the *kernel* of the transform and I is a given interval of integration. The basic idea is to use (39) to change a problem for f into a simpler problem for F, solve the resulting problem for F, and then recover the solution f of the original problem via an *inversion formula*

$$f(t) = \int_{I'} G(s, t) F(s) \, ds$$

where G and I' are given. In this context f is called the *inverse transform* of F. Many different transforms are used in applied analysis, each specialized to a particular class of problems. By definition the integral transform (39) is *linear*. That is, if the transform is denoted by $F(s) = \mathcal{T}[f(t)]$, then it easily follows that

$$\mathcal{T}\left[c_1 f_1(t) + c_2 f_2(t)\right] = c_1 \mathcal{T}\left[f_1(t)\right] + c_2 \mathcal{T}\left[f_2(t)\right]$$

In elementary courses one learns that an ordinary differential equation can be transformed via Laplace transforms to an equivalent algebraic equation. The algebraic equation can be solved and the solution of the original ordinary differential equation can be obtained by an inversion formula or by looking up the inverse transform in a table. In a similar manner it is possible to reduce certain partial differential equations to ordinary differential equations.

If $f(t)$ is a function with domain $t \geq 0$, then the *Laplace transform* of f is given by

$$F(s) = \int_0^\infty e^{-st} f(t) \, dt \tag{40}$$

where $F(s)$ is defined for those values of s for which the improper integral converges. One class of functions for which the transform (40) exists is described in the following theorem. A proof is requested in Exercise 3.18.

Theorem 3.6   *Let f be a piecewise continuous function on the interval* $[0, t_1]$ *for any* $t_1 > 0$ *and let* $|f(t)| \le M \exp(at)$ *on* $t > t_1 > 0$ *for some constants* a, t_1, *and* $M > 0$. *Then the Laplace transform defined by* (40) *exists for all* $s > a$.

If $F(s)$ is the Laplace transform of $f(t)$, then $f(t)$ can be recovered from $F(s)$ via the *inversion formula*

$$f(t) = \frac{1}{2\pi i} \int_{a-i\infty}^{a+i\infty} F(s) e^{st} \, ds \tag{41}$$

Therefore to compute the inverse transform it is required to compute a contour integral over the infinite straight line path in the complex plane from $a - i\infty$ to $a + i\infty$, where a is chosen to be any real number exceeding the real part of s_0, where s_0 is a value where the integral (40) converges. Fortunately many inverse transforms have already been computed and are listed in available tables. The derivation of (41) is given in more advanced treatments. For our purposes several *transform-inverse transform* pairs are listed in Table 4.1. The unfamiliar entries will be defined as they occur in examples or exercises.

It is convenient to use the operator notation $\mathcal{L}$ and $\mathcal{L}^{-1}$ for the Laplace transform and its inverse given by (40) and (41). Thus $\mathcal{L}[f] = F(s)$ and $\mathcal{L}^{-1}[F] = f(t)$. Although $\mathcal{L}$ is linear and hence additive, it is not multiplicative; that is, $\mathcal{L}[fg]$ is not $\mathcal{L}[f] \cdot \mathcal{L}[g]$. The following theorem, called the *convolution theorem*, is useful in the sequel and gives the correct result.

Theorem 3.7   *We have*

$$\mathcal{L}[f * g] = \mathcal{L}[f]\mathcal{L}[g]$$

where $f * g$ *is the convolution of f and g defined by*

$$(f * g)(t) = \int_0^t f(\tau) g(t - \tau) \, d\tau \tag{42}$$

Stated differently, Theorem 3.7 requires that the inverse transform of $F(s)G(s)$ be the convolution integral defined by the right side of (42).

For partial differential equations functions of two variables are transformed by holding one of the variables fixed and transforming on the other. For example, we define

$$U(x, s) = \int_0^\infty u(x, t) e^{-st} \, dt$$

TABLE 4.1. Laplace Transforms

$f(t)$	$F(s)$
1	$s^{-1}, \quad s > 0$
$\exp(at)$	$\dfrac{1}{s - a}, \quad s > a$
$t^n, \quad n$ a positive integer	$\dfrac{n!}{s^{n+1}}, \quad s > 0$
$\sin at$ and $\cos at$	$\dfrac{a}{s^2 + a^2}$ and $\dfrac{s}{s^2 + a^2}, \quad s > 0$
$\sinh at$ and $\cosh at$	$\dfrac{a}{s^2 - a^2}$ and $\dfrac{s}{s^2 - a^2}, \quad s > |a|$
$e^{at}\sin bt$	$\dfrac{b}{(s - a)^2 + b^2}, \quad s > a$
$e^{at}\cos bt$	$\dfrac{s - a}{(s - a)^2 + b^2}, \quad s > a$
$t^n\exp(at)$	$\dfrac{n!}{(s - a)^{n+1}}, \quad s > a$
$H(t - a)$	$s^{-1}\exp(-as), \quad s > 0$
$\delta(t - a)$	$\exp(-as)$
$H(t - a)f(t - a)$	$F(s)\exp(-as)$
$\operatorname{erf}\sqrt{t}$	$s^{-1}(1 + s)^{-1/2}, \quad s > 0$
$\dfrac{1}{\sqrt{t}}\exp\dfrac{-a^2}{4t}$	$\sqrt{\pi/s}\,\exp(-a\sqrt{s}), \quad (s > 0)$
$\operatorname{erfc}\dfrac{a}{2\sqrt{t}}$	$s^{-1}\exp(-a\sqrt{s}), \quad s > 0$
$\dfrac{a}{2t^{3/2}}\exp\dfrac{-a^2}{4t}$	$\sqrt{\pi}\,\exp(-a\sqrt{s}), \quad s > 0$
$f^{(n)}(t)$	$s^n F(s) - s^{n-1}f(0) - s^{n-2}f'(0) - \cdots - f^{(n-1)}(0)$
$\int_0^t f(\tau)g(t - \tau)\,d\tau$	$F(s)G(s)$

In this formula x acts as a parameter and t is the variable on which the transform is taken. As in the one dimensional case time derivatives are transformed via (see Table 4.1)

$$\int_0^\infty u_t(x, t)e^{-st}\,dt = sU(x, s) - u(x, 0)$$

$$\int_0^\infty u_{tt}(x, t)e^{-st}\,dt = s^2 U(x, s) - su(x, 0) - u_t(x, 0)$$

These equations are easily obtained using integration by parts. However,

$$\int_0^\infty u_x(x, t)e^{-st}\,dt = \frac{\partial}{\partial x}\int_0^\infty u(x, t)e^{-st}\,dt = \frac{\partial}{\partial x}U(x, s)$$

so that differentiation with respect to the parameter x just amounts to differentiation under the integral sign. We implicitly assume that these operations are valid. The following example illustrates the Laplace transform method on a heat conduction problem.

Example 3.5   Consider the boundary value problem

$$u_t - ku_{xx} = 0, \qquad t > 0, \quad x > 0$$
$$u(x, 0) = 0, \qquad x > 0$$
$$u(0, t) = 1, \qquad t > 0$$
$$u \quad \text{bounded}$$

Taking the Laplace transform of both sides of the partial differential equation, that is multiplying by e^{-st} and integrating with respect to t from 0 to ∞, we obtain

$$-u(x, 0) + sU(x, s) - kU_{xx}(x, s) = 0$$

or

$$U_{xx}(x, s) - (s/k)U(x, s) = 0$$

Solving this ordinary differential equation with s as a parameter we obtain

$$U(x, s) = Ae^{-\sqrt{s/k}\,x} + Be^{\sqrt{s/k}\,x}$$

where A and B are functions of s. Since bounded solutions are sought we set $B = 0$ to discard the growing exponential term and we obtain

$$U(x, s) = Ae^{-\sqrt{s/k}\,x}$$

But from the boundary condition

$$U(0, s) = \int_0^\infty u(0, t)e^{-st}\,dt = \frac{1}{s}$$

Therefore

$$U(x, s) = \frac{1}{s}e^{-\sqrt{s/k}\,x}$$

Consulting Table 4.1 we find that the solution is

$$u(x, t) = \text{erfc}\left(\frac{x}{2\sqrt{kt}}\right)$$

where erfc is the *complementary error function* defined by

$$\text{erfc}(t) = 1 - \text{erf}(t)$$

and erf is the *error function*

$$\text{erf}(t) = \frac{2}{\sqrt{\pi}} \int_0^t \exp(-\xi^2) \, d\xi$$

Several additional examples are contained in the exercises.

Another transform technique particularly well-suited to problems on an infinite interval $-\infty < x < \infty$ is the Fourier transform method. If $f(x)$ is an absolutely integrable function on R^1, that is, $\int_{-\infty}^{\infty} |f(x)| \, dx < \infty$, then the *Fourier transform* of f is defined by

$$F(s) = \frac{1}{\sqrt{2\pi}} \int_{-\infty}^{\infty} f(x) e^{-isx} \, dx \tag{43}$$

Thus to each absolutely integrable function $f(t)$ there is associated a unique transform $F(s)$ in the transform domain s. One reason for the usefulness of the Fourier transform is the similar form of the inversion formula

$$f(x) = \frac{1}{\sqrt{2\pi}} \int_{-\infty}^{\infty} F(s) e^{isx} \, ds \tag{44}$$

which permits the recovery of the function $f(x)$ if its transform $F(s)$ is known. As in the case of the Laplace transform the application of Fourier transforms to partial differential equations is carried out by transforming the dependent variable and one of the independent variables via (43) while holding the other independent variable fixed as a parameter. When the partial differential equation is transformed in this way there results an ordinary differential equation in the transformed variables that may be solved, and finally a return to the original variables is accomplished by the inversion formula (44). We illustrate this method by solving the *pure initial value problem* for the diffusion equation.

Example 3.6 Consider the problem

$$u_t - ku_{xx} = 0, \qquad x \in R^1, \quad t > 0 \tag{45}$$

$$u(x, 0) = f(x), \qquad x \in R^1 \tag{46}$$

We assume there is a solution u having the property that u, u_t, u_x, and u_{xx} are continuously differentiable and absolutely integrable on R^1 and u and u_x

tend to zero as $|x| \to \infty$. We transform on x with t acting as a parameter. If $U(s, t)$ denotes the transform of $u(x, t)$, then

$$\frac{1}{\sqrt{2\pi}} \int_{-\infty}^{\infty} u_t(x, t) e^{-isx} \, dx = U_t(s, t)$$

and

$$\frac{1}{\sqrt{2\pi}} \int_{-\infty}^{\infty} u_{xx}(x, t) e^{-isx} \, dx = -s^2 U(s, t)$$

where the last equation was obtained by integrating by parts twice and using the conditions that u and u_x tend to zero as $|x| \to \infty$. Therefore taking the Fourier transform of equation (45), that is, multiplying by $(2\pi)^{-1/2} \exp(-isx)$ and integrating from $-\infty$ to ∞, gives

$$U_t(s, t) + ks^2 U(s, t) = 0$$

which has general solution

$$U(s, t) = C(s) e^{-ks^2 t}$$

To determine $C(s)$ we take the transform of (46) to get

$$U(s, 0) = \frac{1}{\sqrt{2\pi}} \int_{-\infty}^{\infty} u(x, 0) e^{-isx} \, dx = \frac{1}{\sqrt{2\pi}} \int_{-\infty}^{\infty} f(x) e^{-isx} \, dx = F(s)$$

Hence $C(s) = F(s)$ and

$$U(s, t) = F(s) e^{-ks^2 t}$$

Applying the inversion formula (44) gives the solution

$$
\begin{aligned}
u(x, t) &= \frac{1}{\sqrt{2\pi}} \int_{-\infty}^{\infty} F(s) e^{-ks^2 t} e^{isx} \, ds \\
&= \frac{1}{2\pi} \int_{-\infty}^{\infty} \left(\int_{-\infty}^{\infty} f(\xi) e^{-is\xi} \, d\xi \right) e^{-ks^2 t} e^{isx} \, ds \\
&= \frac{1}{2\pi} \int_{-\infty}^{\infty} f(\xi) \left(\int_{-\infty}^{\infty} e^{is(x-\xi) - ks^2 t} \, ds \right) d\xi
\end{aligned}
\tag{47}
$$

It is possible to recast the last integral into a more familiar form using Euler's formula $\exp(i\theta) = \cos\theta + i \sin\theta$. Then

$$\int_{-\infty}^{\infty} e^{is(x-\xi) - ks^2 t} \, ds = 2 \int_{0}^{\infty} e^{-ks^2 t} \cos s(x - \xi) \, ds = \left(\frac{\pi}{kt} \right)^{1/2} e^{-(x-\xi)^2 / 4kt}$$

where in the last step we used the formula

$$\int_0^\infty e^{-z^2} \cos az \, dz = \frac{\sqrt{\pi}}{2} e^{-a^2/4}$$

and the substitution $z = s\sqrt{kt}$. Consequently (47) takes the form

$$u(x, t) = \frac{1}{\sqrt{4\pi kt}} \int_{-\infty}^{\infty} f(\xi) e^{-(x-\xi)^2/4kt} \, d\xi \qquad (48)$$

As mentioned in Example 1.5 the function

$$u(x, t, \xi) = \frac{1}{\sqrt{4\pi kt}} e^{-(x-\xi)^2/4kt} \qquad (49)$$

is the *fundamental solution* of the heat equation on the domain R^1. It is instructive to graph time snapshots of (49) to understand the nature of (48). To this end we fix ξ and note that

$$\lim_{t\to 0^+} u(x, t, \xi) = 0, \qquad x \neq \xi$$

$$\lim_{t\to\infty} u(x, t, \xi) = 0, \qquad x \in R^1$$

$$\lim_{t\to 0^+} u(x, t, \xi) = +\infty, \qquad x = \xi$$

Consequently, for large times the fundamental solution goes to zero for all $x \in R^1$. As $t \to 0^+$, however, the solution develops an infinite spike at $x = \xi$ while tending to zero for $x \neq \xi$. Therefore we may interpret the fundamental solution as the solution to the heat equation in an infinite bar initially at zero degrees with an instantaneous unit heat source applied at $x = \xi$, $t = 0$. It is a unit source since

$$\int_{-\infty}^{\infty} u(x, t, \xi) \, dx = 1, \qquad \text{for all} \quad t > 0$$

that is, the area (which is proportional to the energy) under each profile in Fig. 4.7 is unity. Stated differently, the function $u(x, t, \xi)$ is the temperature effect at (x, t) caused by a unit heat source applied initially at $x = \xi$. Functions that represent effects due to point sources are called *influence functions* or *Green's functions*, and knowledge of the Green's function for a given problem permits the solution to be written down in the general case when there is a distribution of sources. In the present heat flow problem if $u(x, t, \xi)$ gives the effect due to a unit heat source at ξ, then $f(\xi)u(x, t, \xi)$ is the effect at (x, t) due to a

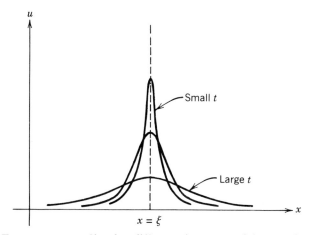

Figure 4.7. Temperature profiles for different times caused by a unit point energy source at $x = \xi$, $t = 0$.

source of magnitude $f(\xi)$ applied at $x = \xi$. If these effects are superimposed for all $\xi \in R^1$ then (48) results. Therefore the integral solution (48) of the initial value problem (45) and (46) can be thought of as the superposition of the effect of infinitely many point sources.

For a detailed discussion of transform methods see Zauderer [4].

EXERCISES

3.1 Find eigenvalues and eigenfunctions for the following problems.

▷**(a)** $(x^3 y')' + \lambda x y = 0$, $1 < x < e$ Euler Equation:
$\qquad y(1) = y(e) = 0$
$$A x^2 y'' + B x y' + C y = 0$$
$$x = e^s \implies A \frac{d^2 y}{ds^2} + (B-A) \frac{dy}{ds} + C y = 0$$

(b) $y'' + \lambda y = 0$, $0 < x < 1$
$\qquad y(0) = 0$, $y(1) - y'(1) = 0$

▷**(c)** $y'' + 2\beta y' + \lambda y = 0$, $0 < x < l$
$\qquad y(0) = y(l) = 0$

(d) $y'' + \lambda y = 0$, $0 < x < l$
$\qquad y'(0) = y(l) = 0$

3.2 Let $f(x) = x^2$, $-\pi \le x \le \pi$, and $f(x + 2\pi) = f(x)$ for all x. Show that

$$f(x) = \frac{\pi^2}{3} + 4 \sum_{1}^{\infty} (-1)^n \frac{\cos nx}{n^2}$$

3.3   Prove that

$$\frac{\pi^2}{12} = 1 - \frac{1}{4} + \frac{1}{9} - \frac{1}{16} + \cdots$$

(Use Exercise 3.2.)

3.4   Find the Fourier sine series of $f(x) = 1$ on $0 \leq x \leq \pi$. *Solution:*

$$\frac{4}{\pi}\left(\sin x + \frac{\sin 3x}{3} + \frac{\sin 5x}{5} + \cdots\right)$$

3.5   Find the Fourier cosine series of $f(x) = \sin x$ on $0 \leq x \leq \pi$.

3.6   Find the Fourier series for the function

$$f(x) = \begin{cases} 0, & \text{for } -\pi < x < 0 \\ 1 & \text{for } 0 \leq x < \pi \end{cases}$$

To what value does the series converge at $x = 0$?

3.7   For the boundary value problem

$$u_t - ku_{xx} = 0, \qquad 0 < x < l, \quad t > 0$$
$$u_x(0, t) = u_x(l, t) = 0, \qquad t > 0$$
$$u(x, 0) = f(x), \qquad 0 < x < l$$

show that the asymptotic or long time solution as $t \to +\infty$ is a constant that is the average value of the initial temperature distribution $f(x)$.

3.8   Use the method of separation of variables to find a series representation of the solution to the problem

$$u_{tt} - c^2 u_{xx} + hu = 0, \qquad t > 0, \quad 0 < x < l$$
$$u(x, 0) = f(x), \qquad u_t(x, 0) = 0, \quad 0 < x < l$$
$$u(0, t) = u(l, t) = 0, \qquad t > 0$$

Solution:

$$u(x, t) = \sum_1^\infty b_n \sin\frac{n\pi x}{l} \cos\sqrt{\frac{n^2\pi^2 c^2}{l^2} + ht}$$

where

$$b_n = \frac{2}{l} \int_0^l f(x)\sin \frac{n\pi x}{l} \, dx$$

3.9 Use the method of separation of variables to find a series representation of the solution to the problem

$$u_{xx} + u_{yy} = 0, \qquad 0 < x < a, \quad 0 < y < b$$
$$u(x,0) = 0, \qquad u(x,b) = f(x), \quad 0 < x < a$$
$$u(0,y) = u(a,y) = 0, \qquad 0 < y < b$$

Solution:

$$u(x,y) = \sum_1^\infty b_n \frac{\sinh \dfrac{n\pi y}{a}}{\sinh \dfrac{n\pi b}{a}} \sin \frac{n\pi x}{a}$$

where

$$b_n = \frac{2}{a} \int_0^a f(x)\sin \frac{n\pi x}{a} \, dx$$

3.10 Solve by separation of variables:

$$u_{tt} - x^2 u_{xx} - xu_x = 0, \qquad 1 < x < e, \quad t > 0$$
$$u(1,t) = u_x(e,t) = 0, \qquad t > 0$$
$$u(x,0) = 0, \qquad u_t(x,0) = 1, \quad 1 < x < e$$

3.11 Use separation of variables to find a bounded solution of

$$u_{xx} + u_{yy} = 0, \qquad 0 < x < \pi, \quad y > 0$$
$$u(0,y) = u(\pi,y) = 0, \qquad y > 0$$
$$u(x,0) = 1, \qquad 0 < x < \pi$$

Solution:

$$u(x,y) = \frac{4}{\pi} \sum_{k=1}^\infty \frac{1}{2k-1} \exp((1-2k)y)\sin(2k-1)x$$

3.12   Use Laplace transforms to solve

$$u_t = u_{xx}, \qquad 0 < x < 1, \quad t > 0$$
$$u(0, t) = u(1, t) = 1, \qquad t > 0$$
$$u(x, 0) = 1 + \sin \pi x, \qquad 0 < x < 1$$

Solution: $u = 1 + e^{-\pi^2 t}\sin \pi x$

3.13   Show that the solution of the initial value problem

$$u_t - k u_{xx} = 0, \qquad x \in R^1, \quad t > 0$$
$$u(x, 0) = \begin{cases} u_0, & |x| < l \\ 0, & |x| \geq l \end{cases}$$

can be represented as

$$u(x, t) = -\frac{u_0}{2}\left[\mathrm{erf}\left(\frac{x - l}{\sqrt{4kt}}\right) - \mathrm{erf}\left(\frac{x + l}{\sqrt{4kt}}\right)\right]$$

Show that for large t

$$u(x, t) \cong \frac{u_0 l}{\sqrt{\pi kt}}$$

Hint: Show that $\mathrm{erf}(x) \cong (2/\sqrt{\pi})x$ for small x with an error of not more than

$$\frac{2}{\sqrt{\pi}}\frac{|x|^3}{3}$$

3.14   Solve

$$u_{tt} - u_{xx} = 0, \qquad x \in R^1, \quad t > 0$$
$$u(x, 0) = \sin x, \qquad u_t(x, 0) = 1, \qquad x \in R^1$$

3.15   Solve

$$u_{tt} - u_{xx} = 0, \qquad 0 < x < \pi, \quad t > 0$$
$$u(0, t) = u(\pi, t) = 0, \qquad t > 0$$
$$u(x, 0) = 0, \qquad u_t(x, 0) = 4\sin x, \qquad 0 < x < \pi$$

3.16 Solve the problem

$$u_t - u_{xx} = 0, \qquad x > 0, \quad t > 0$$
$$u(x,0) = u_0, \qquad x > 0$$
$$u(0,t) = u_1, \qquad t > 0$$
$$\lim_{x \to \infty} u(x,t) = u_0$$

where u_0 and u_1 are positive constants. *Solution*:

$$u = u_0 + (u_1 - u_0)\left(1 - \frac{2}{\sqrt{\pi}} \int_0^{x/2\sqrt{t}} e^{-z^2}\, dz\right)$$

3.17 Obtain the solution to the boundary value problem

$$u_t - u_{xx} = 0, \qquad x > 0, \quad t > 0$$
$$u(x,0) = 0, \qquad x > 0$$
$$u(0,t) = f(t), \qquad t > 0$$
$$\lim_{x \to \infty} u(x,t) = 0$$

in the form

$$u(x,t) = \int_0^t f(t - \xi) \frac{x}{2(\pi\xi^3)^{1/2}} \exp\left(\frac{-x^2}{4\xi}\right) d\xi$$

3.18 Prove Theorem 3.6. *Hint*: Write

$$\int_0^\infty e^{-st} f(t)\, dt = \int_0^{t_1} e^{-st} f(t)\, dt + \int_{t_1}^\infty e^{-st} f(t)\, dt$$

and show that each integral on the right side converges.

4.4 INTEGRAL EQUATIONS

Classification and Origins

Another type of fundamental equation that occurs frequently in applied mathematics is an integral equation. An integral equation is an equation where the unknown function occurs under an integral sign. For example

$$x^2 y(x) = \int_0^1 e^{x\xi} y(\xi)\, d\xi \tag{1}$$

$$y(x) = \cos x - \int_0^x (x^2 + \xi) y(\xi)\, d\xi \tag{2}$$

and

$$\sin x = \int_0^\infty e^{y(\xi)x}\, d\xi \tag{3}$$

are all integral equations since the unknown function y appears under an integral. In this section we mainly discuss two types of linear integral equations, the *Fredholm equation*

$$\alpha(x)y(x) = f(x) + \int_a^b k(x, \xi)y(\xi)\, d\xi, \qquad a \le x \le b \tag{4}$$

and the *Volterra equation*

$$\alpha(x)y(x) = f(x) + \int_a^x k(x, \xi)y(\xi)\, d\xi, \qquad a \le x \le b \tag{5}$$

Here α, f, and k are given functions, a and b are constants, and y is the function to be determined. The function $k(x, \xi)$ is called the *kernel* of the integral equation. The difference between Fredholm and Volterra equations is the variable upper limit of integration in the Volterra equation. By a *solution* of (4) or (5) we mean a continuous function $y = y(x)$ on the interval $a \le x \le b$, which when substituted into the equation reduces it to an identity on the interval. If $f(x) \equiv 0$, the equation is *homogeneous* and it is *nonhomogeneous* if $f(x) \ne 0$. If $\alpha(x) \equiv 0$, the unknown appears only under the integral sign and the equation is said to be an integral equation of the *first kind*. When $\alpha(x) =$ constant it is of the *second kind*, and when $\alpha(x)$ is not a constant it is of the *third kind*. We note that (1) is a homogeneous Fredholm integral equation of the third kind and (2) is a nonhomogeneous Volterra equation of the second kind. Equation (3) is neither of Volterra nor Fredholm type, but rather a nonlinear integral equation, since a nonlinear function of the unknown y appears. If $k(x, \xi) = k(\xi, x)$, the kernel is *symmetric*, and if k is unbounded on the interval of integration, then k is *singular*. Although Fredholm and Volterra equations appear quite similar they require very different methods of solution.

The reader is familiar with many problems in science and engineering that lead in their formulation to a differential equation as a mathematical model. Less familiar are physical situations leading directly to an integral equation. We present some examples for which the formulation of an integral equation arises naturally. The importance of integral equations, however, not only lies in the fact that they provide a description of how nature behaves but they are also important in reformulating and solving differential equations. It is common practice to recast a differential equation to integral equation form both as a solution technique and as a basis for inventing stable numerical algorithms to solve the differential equation.

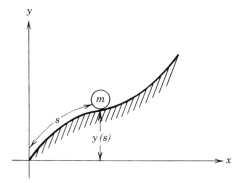

Figure 4.8. Hill of height $y(s)$ where s is arclength along the contour.

Example 4.1 Consider a monotone increasing hill of an unknown shape. By rolling a ball of mass m up the hill and measuring the round-trip transit time for different initial energies given the ball we determine the shape of the hill. Suppose the profile of the hill (see Fig. 4.8) is given by parametric equations $x = x(s)$ $y = y(s)$, where s is arclength along the contour of the hill measured from the bottom, that is, from the point $(0,0)$. Let $V = V(s)$ be the potential energy of the ball at distance s with $V(0) = 0$. Then $V(s) = mgy(s)$ and if we can find $V(s)$ we will have solved the problem. Let E be the initial kinetic energy given the ball and $T(E)$ be the time required for the ball to go up the hill and return. The assumption is that $T(E)$ is known for $E > 0$. By conservation of energy

$$E = \frac{1}{2}m\left(\frac{ds}{dt}\right)^2 + V(s)$$

$$\underbrace{\qquad\qquad\qquad}_{KE = \frac{1}{2}mv^2}$$

or

$$\frac{ds}{dt} = \pm\sqrt{\frac{2}{m}}\left(E - V(s)\right)^{1/2}$$

where the positive sign is to be taken for the outward trip and the negative sign for the return. Since $T(E)$ is twice the time taken to go up the hill,

$$T(E) = 2\int_0^{s_1(E)} \frac{dt}{ds}\,ds$$

where $s_1(E)$ is the distance that the ball travels in going up. Thus

$$T(E) = \sqrt{2m}\int_0^{s_1(E)}\left(E - V(s)\right)^{-1/2}\,ds \qquad (6)$$

202 EQUATIONS OF APPLIED MATHEMATICS

We observe that $V(s_1(E)) = E$. Now we change variables in (6) via

$$\tilde{V} = V(s) \tag{7}$$

Then $d\tilde{V} = V'(s)\,ds$ and

$$T(E) = \sqrt{2m} \int_0^E (E - \tilde{V})^{-1/2} \frac{1}{V'(s(\tilde{V}))}\, d\tilde{V} \tag{8}$$

But from the derivative formula for inverse functions,

$$s'(\tilde{V}) = \frac{1}{V'(s(\tilde{V}))}$$

where $s = s(\tilde{V})$ is the inverse of (7). Hence (8) becomes

$$T(E) = \sqrt{2m} \int_0^E (E - \tilde{V})^{-1/2} s'(\tilde{V})\, d\tilde{V} \tag{9}$$

which is a Volterra integral equation of the first kind for the function $s'(\tilde{V})$. Solving this integral equation for $s'(\tilde{V})$ will give $s(\tilde{V})$ by integration and thus $V(s)$, and the resulting solution to the problem can be determined.

Example 4.2 In R^3 the potential V at a point (x, y, z) due to a mass distribution ρ is given by

$$V(x, y, z) = -G \int_{R^3} \frac{\rho(\xi, \eta, \zeta)}{r}\, d\xi\, d\eta\, d\zeta$$

where G is a constant and $r^2 = (x - \xi)^2 + (y - \eta)^2 + (z - \zeta)^2$. Therefore knowledge of ρ gives the potential by direct integration. The inverse problem of determining ρ from a given potential V is an integral equation. One can show that ρ and $\bar{V}$ are related by

$$\nabla^2 V = 4\pi G\rho, \qquad (x, y, z) \in R^3$$

which is Poisson's differential equation.

Example 4.3 (Inventory Control Problem) A shop manager determines that a percentage $k(t)$ of goods remains unsold at time t after he has purchased the goods. At what rate should he purchase goods so that the stock remains constant? Here we are assuming all processes are continuous. At time $t = 0$ suppose that the manager stocks an amount A of goods and he buys at the rate (goods/time) $y(t)$ for $t > 0$. In the time interval $[\tau, \tau + \Delta\tau]$ he buys an amount $y(\tau)\Delta\tau$ and at time t the portion remaining unsold is

$k(t - \tau)y(\tau)\,\Delta\tau$. Therefore the amount of goods in the shop unsold at time t is

$$Ak(t) + \int_0^t k(t - \tau)y(\tau)\,d\tau \qquad \text{\tiny looks like convolution}$$

The first term represents the amount of the initial purchase left unsold at time t. Hence this is the total amount of goods in the shop, and so the manager requires that

$$A = Ak(t) + \int_0^t k(t - \tau)y(\tau)\,d\tau$$

The solution $y(t)$ to the manager's problem satisfies a nonhomogeneous Volterra integral equation of the first kind.

Example 4.4 (Inversion of Thermodynamic Data) One of the major problems of statistical mechanics is to determine the equation of state of a gas in terms of the potential $V(r)$ between the molecules of the gas. If the number density $\tilde{\rho}$ in moles per unit volume of the gas is small, then the equation of state is given by the *virial expansion*

$$\frac{p}{\tilde{\rho}RT} = 1 - 2\pi B(kT)\tilde{\rho} + O(\tilde{\rho}^2) \tag{10}$$

where p is the pressure, T is the temperature, R is the gas constant, k is Boltzman's constant, and

$$B(kT) = \int_0^\infty (1 - e^{-V(r)/kT})r^2\,dr \tag{11}$$

which is known as the second virial coefficient. Here $O(\tilde{\rho}^2)$ denotes higher order terms in $\tilde{\rho}$ that can be neglected if $\tilde{\rho}$ is small. Given a potential $V(r)$ equation (10) gives the equation of state where B is given by (11). This is the direct problem. The inverse problem of determining the intermolecular potential $V(r)$ from the equation of state requires solving the nonlinear integral equation (11) for V given B.

Many problems in science and engineering are the same as Examples 4.2 and 4.4, namely a *direct problem* and an *inverse problem*. Inverse problems often lead to integral equations.

Example 4.5 (Population Modeling) Let p_0 be the number of individuals in a human population at time $t = 0$ and let $p_s(t)$ be the surviving population at time $t > 0$. Define the survival function $f(t)$ to be the fraction of individuals surviving to age t. Thus

$$p_s(t) = p_0 f(t)$$

If children are born at the rate $r(t)$, then $r(\tau_i)\,\Delta_i\,\tau$ individuals will be added to the population in the interval $\Delta_i\,\tau \equiv \tau_{i+1} - \tau_i$. At time t these children would be of age $t - \tau_i$, but only $f(t - \tau_i)r(\tau_i)\,\Delta_i\,\tau$ of them survive. Repeating this argument for the m subintervals $[\tau_0, \tau_1], \ldots, [\tau_{m-1}, \tau_m]$ of $[0, t]$, we find that the total number of individuals added through new births is

$$B_m(t) = \sum_{i=0}^{m-1} f(t - \tau_i)r(\tau_i)\,\Delta_i\,\tau$$

Passing to the limit as $m \to \infty$ and as the length of each subinterval goes to zero we find that

$$B(t) = \int_0^t f(t - \tau)r(\tau)\,d\tau$$

where $B(t)$ the total number of births in $[0, t]$ that have survived to time t. Adding to $p_s(t)$ gives

$$p(t) = p_0 f(t) + \int_0^t f(t - \tau)r(\tau)\,d\tau$$

which is the total population at time t. If the rate r is proportional to the population p, that is, $r(t) = kp(t)$, then

$$p(t) = p_0 f(t) + \int_0^t kf(t - \tau)p(\tau)\,d\tau$$

is a Volterra integral equation of the second kind for $p(t)$. The survival function is commonly found in insurance tables and is assumed to be known.

Relationship to Differential Equations

In differential equations the initial value problem

$$y' = f(x, y), \qquad y(x_0) = y_0 \tag{12}$$

is studied in detail. This equation can be easily formulated as an integral equation by direct integration. For, let $y(x)$ be the solution; then

$$y'(x) = f(x, y(x))$$

Replacing x by ξ and integrating from x_0 to x gives

$$\int_{x_0}^x y'(\xi)\,d\xi = \int_{x_0}^x f(\xi, y(\xi))\,d\xi$$

$y(x) - y(x_0) = \int \sim$

$\frac{d}{dx}\int_{x_0}^x y'(\xi)\,d\xi = \frac{d}{dx}(y(x) - y(x_0)) = y'(x)$

or

$$y(x) = y_0 + \int_{x_0}^{x} f(\xi, y(\xi))\, d\xi \tag{13}$$

where in the last step the fundamental theorem of calculus was applied. Equation (13) is an integral equation for y, and it is equivalent to the differential equation and the initial condition (12). That is, y is a solution of (12) if, and only if, y is a solution to (13). To recover (12) from (13) we differentiate (13) with respect to x to get

$$y'(x) = \frac{d}{dx} \int_{x_0}^{x} f(\xi, y(\xi))\, d\xi = f(x, y(x))$$

where the last equality follows from the rule for differentiating an integral. This example points out an important fact, namely that the initial condition in (12) is automatically included in the integral equation (13).

In general it is possible to reformulate certain initial and boundary value problems as integral equations, and vice versa. In practice, to carry out this procedure the following lemma, which is a concise formula for a repeated integral, is helpful.

Lemma 4.1 *Let $f(x)$ be a continuous function for $x \geq a$. Then*

$$\int_{x_2=a}^{x_2=x} \int_{x_1=a}^{x_1=x_2} f(x_1)\, dx_1\, dx_2 = \int_{a}^{x} (x - \xi) f(\xi)\, d\xi \tag{14}$$

Proof The double iterated integral on the left is over the triangular region R shown in Fig. 4.9. By interchanging the order of integration

$$\int_{x_2=a}^{x_2=x} \int_{x_1=a}^{x_1=x_2} f(x_1)\, dx_1\, dx_2 = \int_{x_1=a}^{x_1=x} \int_{x_2=x_1}^{x_2=x} f(x_1)\, dx_2\, dx_1$$

$$= \int_{x_1=a}^{x_1=x} f(x_1)(x - x_1)\, dx_1$$

$$= \int_{a}^{x} f(\xi)(x - \xi)\, d\xi$$

Formula (14) can be generalized to n repeated integrals to obtain

$$\int_{a}^{x} \int_{a}^{x_n} \cdots \int_{a}^{x_2} f(x_1)\, dx_1\, dx_2 \cdots dx_n = \frac{1}{(n-1)!} \int_{a}^{x} (x - \xi)^{n-1} f(\xi)\, d\xi \tag{15}$$

The repeated integral on the left side of (15) is often symbolically represented

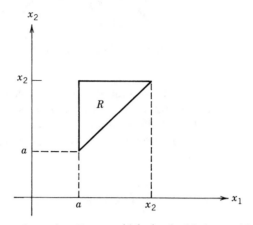

Figure 4.9. Triangular region R over which the double integral in (14) is taken.

by

$$\underbrace{\int_a^x \int_a^x \cdots \int_a^x f(x) \underbrace{dx \ldots dx}_{n \text{ times}}}_{n \text{ times}}$$

but care must be taken to interpret this expression as in (15).

Example 4.6   Consider the second order initial value problem

$$y''(x) + A(x)y'(x) + B(x)y(x) = f(x), \qquad x > a \qquad (16)$$

$$y(a) = y_0, \qquad y'(a) = y_0' \qquad (17)$$

Solving for $y''(x)$ in (16) and integrating from a to x we obtain

$$y'(x) - y_0' = -\int_a^x A(\xi)y'(\xi)\,d\xi - \int_a^x B(\xi)y(\xi)\,d\xi + \int_a^x f(\xi)\,d\xi$$

If the first integral is integrated by parts the result is

$$y'(x) = -A(x)y(x) - \int_a^x [B(\xi) - A'(\xi)]\,y(\xi)\,d\xi$$

$$+ \int_a^x f(\xi)\,d\xi + A(a)y_0 + y_0'$$

Integrating again

$$\int^x \text{ is a function of } x \text{— the}$$
$$\text{upper limit of integration}$$

$$y(x) - y_0 = -\int_0^x A(\xi)y(\xi)\,d\xi - \int_a^x \int_a^{x_1} [B(\xi) - A'(\xi)]\,y(\xi)\,d\xi\,dx_1$$

$$0? \qquad + \int_a^x \int_a^{x_1} f(\xi)\,d\xi\,dx_1 + [A(a)y_0 + y_0'](x - a)$$

Using Lemma 4.1

$$y(x) = -\int_a^x \{ A(\xi) + (x - \xi)[B(\xi) - A'(\xi)] \} y(\xi)\, d\xi$$
$$+ \int_a^x (x - \xi) f(\xi)\, d\xi + [A(a)y_0 + y_0'](x - a) + y_0$$

The last equation is of the form

$$y(x) = \int_a^x k(x, \xi) y(\xi)\, d\xi + F(x) \tag{18}$$

which is a Volterra equation of second kind. By differentiating (18) twice with respect to x using Leibniz' rule (see Exercise 4.1) we can recover (16). We note that the integral equation automatically contains the initial data (17) and therefore (18) is equivalent to (16) and (17).

Boundary value problems can be recast as integral equations in a similar fashion.

Example 4.7 Consider

$$y''(x) + \lambda y(x) = 0, \quad 0 < x < l$$
$$y(0) = y(l) = 0 \tag{19}$$

where λ is a constant. Integrating from 0 to x gives

$$y'(x) - y'(0) = -\lambda \int_0^x y(\xi)\, d\xi$$

Integrating again and using Lemma 4.1 then shows

$$y(x) = y'(0)x - \lambda \int_0^x (x - \xi) y(\xi)\, d\xi$$

The number $y'(0)$ is not known since we are dealing with a boundary value problem. But evaluating the last equation at $x = l$ gives

$$y'(0) = \frac{\lambda}{l} \int_0^l (l - \xi) y(\xi)\, d\xi$$

Hence

$$y(x) = \frac{\lambda x}{l} \int_0^l (l - \xi) y(\xi)\, d\xi - \lambda \int_0^x (x - \xi) y(\xi)\, d\xi$$
$$= \int_0^x \frac{\lambda x}{l} (l - \xi) y(\xi)\, d\xi + \int_x^l \frac{\lambda x}{l} (l - \xi) y(\xi)\, d\xi$$
$$- \lambda \int_0^x (x - \xi) y(\xi)\, d\xi$$
$$= \lambda \int_0^x \frac{\xi(l - x)}{l} y(\xi)\, d\xi + \lambda \int_x^l \frac{x(l - \xi)}{l} y(\xi)\, d\xi$$

If we define the kernel by

$$k(x, \xi) = \begin{cases} \dfrac{\xi(l - x)}{l}, & \xi < x \\[2mm] \dfrac{x(l - \xi)}{l}, & x < \xi \end{cases}$$

then

$$y(x) = \lambda \int_0^l k(x, \xi) y(\xi) \, d\xi \tag{20}$$

which is a homogeneous Fredholm equation for y of the second kind. By differentiating (20) twice one can recover the differential equation (see Exercise 4.2).

Fredholm Equations

Consider the Fredholm equation of the second kind

$$y(x) = f(x) + \lambda \int_a^b k(x, \xi) y(\xi) \, d\xi \tag{21}$$

where f is a given continuous function, λ is a constant, and the kernel k is continuous and has the form

$$k(x, \xi) = \sum_{i=1}^n \alpha_i(x)\beta_i(\xi) \tag{22}$$

Such a kernel is called *separable* or *degenerate* and integral equations of the form (21) with separable kernels can be reduced to solving a linear system of algebraic equations. To accomplish this we substitute (22) into (21) to obtain

$$y(x) = f(x) + \lambda \sum_{i=1}^n \left\{ \int_a^b \beta_i(\xi) y(\xi) \, d\xi \right\} \alpha_i(x)$$

The quantities in braces denoted by

$$c_i \equiv \int_a^b \beta_i(\xi) y(\xi) \, d\xi$$

are unknown constants. Once they are determined the solution is given by

$$y(x) = f(x) + \lambda \sum_{i=1}^n c_i \alpha_i(x) \tag{23}$$

If (23) is multiplied by $\beta_j(x)$ and the result is integrated from a to b, we obtain

$$c_j = f_j + \lambda \sum_{i=1}^{n} c_i a_{ji}, \qquad j = 1, \ldots, n \qquad (24)$$

where

$$f_j = \int_a^b f(x)\beta_j(x)\,dx, \qquad a_{ij} = \int_a^b \alpha_j(x)\beta_i(x)\,dx$$

In matrix form (24) is

$$(I - \lambda A)\mathbf{c} = \mathbf{f} \qquad (25)$$

where I is the identity matrix, $A = (a_{ij}), \mathbf{c} = (c_1 \ldots c_n)^{\mathrm{T}}, \mathbf{f} = (f_1 \ldots f_n)^{\mathrm{T}}$, where T denotes transpose. Therefore (25) represents a system of n linear algebraic equations for $\mathbf{c}$. From matrix theory we recall the following basic results for a linear system

$$B\mathbf{x} = \mathbf{f} \qquad (26)$$

where B is an n by n matrix, $\mathbf{f}$ is a given n vector, and $\mathbf{x}$ is the unknown vector.

1. If $\mathbf{f} = \mathbf{0}$ and $\det B \neq 0$, then (26) has only the trivial solution $\mathbf{x} = \mathbf{0}$. If $\mathbf{f} = \mathbf{0}$ and $\det B = 0$, then (26) has infinitely many solutions.
2. If $\mathbf{f} \neq \mathbf{0}$ and $\det B \neq 0$, then (26) has a unique solution. If $\mathbf{f} \neq \mathbf{0}$ and $\det B = 0$, then (26) has either no solution or infinitely many solutions.

Therefore we have the following *Fredholm alternative theorem.*

Theorem 4.1   *Consider the Fredholm integral equation* (21) *with a separable kernel k.*

(i)   *If*

$$\int_a^b f(x)\beta_j(x)\,dx, \qquad j = 1, \ldots, n$$

are not all zero and if $\det(I - \lambda A) \neq 0$, *then there exists a unique solution to* (21) *given by* (23), *where* $\mathbf{c} = (c_1 \ldots c_n)^T$ *is the unique solution of* (25). *If* $\det(I - \lambda A) = 0$, *then either no solution exists or infinitely many solutions exist.*

(ii)   *If*

$$\int_a^b f(x)\beta_j(x)\,dx = 0, \qquad \text{for} \quad j = 1, \ldots, n$$

and if $\det(I - \lambda A) \neq 0$, *then* (21) *has the solution* $y = f(x)$. *If* $\det(I - \lambda A) = 0$, *then* (21) *has infinitely many solutions.*

In the case of the homogeneous equation ($f \equiv 0$) the values of λ for which

$$y(x) = \lambda \int_a^b k(x, \xi) y(\xi) \, d\xi \tag{27}$$

has a nontrivial solution are called the *eigenvalues* and the corresponding solutions are called the *eigenfunctions*. From Theorem 4.1 (ii) the eigenvalues λ are found from the equation

$$\det(I - \lambda A) = 0$$

and thus there are at most n of them. If λ_0 is an eigenvalue of (27), then the corresponding eigenfunctions are given by (23) or

$$y(x) = \lambda_0 \sum_{i=1}^{n} c_i \alpha_i(x) \tag{28}$$

where the c_i are found from (25) with $\mathbf{f} = \mathbf{0}$.

Example 4.8   Find the eigenvalues and eigenfunctions of the integral equation

$$y(x) = \lambda \int_0^1 (1 - 3x\xi) y(\xi) \, d\xi$$

Here

$$\alpha_1(x) = 1, \qquad \beta_1(\xi) = 1, \qquad \alpha_2(x) = -3x, \qquad \beta_2(\xi) = \xi$$

Then

$$A = \begin{pmatrix} \int_0^1 \beta_1 \alpha_1 \, dx & \int_0^1 \beta_1 \alpha_2 \, dx \\ \int_0^1 \beta_2 \alpha_1 \, dx & \int_0^1 \beta_2 \alpha_2 \, dx \end{pmatrix} = \begin{pmatrix} 1 & -\dfrac{3}{2} \\ \dfrac{1}{2} & -1 \end{pmatrix}$$

and

$$\det(I - \lambda A) = \det \begin{vmatrix} 1 - \lambda & \dfrac{3}{2}\lambda \\ -\dfrac{\lambda}{2} & 1 + \lambda \end{vmatrix} = 1 - \frac{\lambda^2}{4}$$

Setting $\det(I - \lambda A) = 0$ gives eigenvalues

$$\lambda = +2, -2$$

Thus the given integral equation has nontrivial solutions whenever $\lambda = 2$ or

$\lambda = -2$. In the case $\lambda = 2$ the system (25) becomes

$$-c_1 + 3c_2 = 0$$
$$-c_1 + 3c_2 = 0$$

or $c_2 = A$ and $c_1 = 3A$, where A is an arbitrary constant. From (28) the corresponding eigenfunctions are

$$y(x) = 2(3A + A(-3x))$$
$$= B(1-x), \qquad B \text{ constant}$$

By a similar argument $\lambda = -2$ gives eigenfunctions $y(x) = B(1 - 3x)$.

Example 4.9   Consider the nonhomogeneous equation

$$y(x) = f(x) + \lambda \int_0^1 (1 - 3x\xi) y(\xi) \, d\xi$$

If

$$\int_0^1 f(x) \, dx \neq 0 \quad \text{or} \quad \int_0^1 xf(x) \, dx \neq 0$$

then the equation has a unique solution provided $\lambda \neq \pm 2$. If $\lambda = 2$, then the system (25) becomes

$$-c_1 + 3c_2 = \int_0^1 f(x) \, dx$$

$$-c_1 + 3c_2 = \int_0^1 xf(x) \, dx$$

If $\int_0^1 f(x) \, dx \neq \int_0^1 xf(x) \, dx$, then there is no solution. If $\int_0^1 f(x) \, dx = \int_0^1 xf(x) \, dx$, then there are infinitely many solutions to the linear system, namely $c_2 = A$, $c_1 = 3A - \int_0^1 f(x) \, dx$, where A is an arbitrary constant. Hence for $\lambda = 2$, solutions to the integral equation are given by

$$y(x) = f(x) + 2\left\{ \left(3A - \int_0^1 f(x) \, dx \right) - 3Ax \right\}$$

$$= f(x) - 2\int_0^1 f(x) \, dx + 6A(1 - x)$$

where A is an arbitrary constant. Similar calculations can be made when $\lambda = -2$.

Symmetric Kernels

In many problems having origins in the physical sciences Fredholm equations with symmetric kernels arise naturally. Therefore we consider the integral equation

$$y(x) = \lambda \int_a^b k(x, \xi) y(\xi) \, d\xi \qquad (29)$$

where

$$k(x, \xi) = k(\xi, x)$$

and k is real and continuous on the square $a \leq x \leq b, a \leq \xi \leq b$. The following theorem expresses the basic facts about the eigenvalues and eigenfunctions of (29).

Theorem 4.2   *For the integral equation* (29)

   (i)   *If $y_m(x)$ and $y_n(x)$ are eigenfunctions corresponding to distinct eigenvalues λ_m and λ_n, then y_m and y_n are orthogonal over the interval $[a, b]$, that is*

$$\int_a^b y_m(x) y_n(x) \, dx = 0$$

   (ii)   *The eigenvalues are real.*

   (iii)   *If the kernel is not separable, then there are infinitely many eigenvalues $\lambda_1, \lambda_2, \ldots$ that can be ordered in such a way that*

$$0 < |\lambda_1| \leq |\lambda_2| \leq |\lambda_3| \leq \cdots$$

   and

$$\lim_{n \to \infty} |\lambda_n| = \infty$$

   (iv)   *Corresponding to each eigenvalue λ there are at most finitely many independent eigenfunctions; thus every eigenvalue has finite multiplicity.*

Proof   If $\lambda_m, y_m(x)$ and $\lambda_n, y_n(x)$ are two eigenvalue–eigenfunction pairs, then

$$y_m(x) = \lambda_m \int_a^b k(x, \xi) y_m(\xi) \, d\xi$$

$$y_n(x) = \lambda_n \int_a^b k(x, \xi) y_n(\xi) \, d\xi$$

Multiplying the first equation by $y_n(x)$, integrating from a to b, and then interchanging the order of integration gives the following sequence of steps.

$$\int_a^b y_m(x) y_n(x)\, dx = \lambda_m \int_a^b y_n(x) \left(\int_a^b k(x, \xi) y_m(\xi)\, d\xi \right) dx$$

$$= \lambda_m \int_a^b y_m(\xi) \left(\int_a^b k(x, \xi) y_n(x)\, dx \right) d\xi$$

$$= \lambda_m \int_a^b y_m(\xi) \left(\int_a^b k(\xi, x) y_n(x)\, dx \right) d\xi$$

$$= \frac{\lambda_m}{\lambda_n} \int_a^b y_m(\xi) y_n(\xi)\, d\xi$$

Thus $(\lambda_m - \lambda_n) \int_a^b y_m(x) y_n(x)\, dx = 0$, and since $\lambda_n \neq \lambda_m$ Part (i) is proved. To prove Part (ii) let $\lambda = \alpha + i\beta$ be an eigenvalue of (29). Then

$$\bar{y}(x) = \bar{\lambda} \int_a^b k(x, \xi) \bar{y}(\xi)\, d\xi$$

where *bar* denotes complex conjugation. By the argument in Part (i)

$$0 = (\lambda - \bar{\lambda}) \int_a^b y(x) \bar{y}(x)\, dx = 2i\beta \int_a^b |y(x)|^2\, dx$$

and therefore $\beta = 0$, showing λ is real.

The proof of Part (iii) is outside the scope of our treatment and we refer the reader to Hochstadt [5], Courant and Hilbert [6], or Stakgold [3]. We emphasize that Theorem 4.2 holds only for real, symmetric, continuous kernels. For example, a Fredholm equation with a nonsymmetric kernel may possess complex eigenvalues. Other examples may be found in the references.

Example 4.10   Consider the integral equation

$$y(x) = \lambda \int_0^1 k(x, \xi) y(\xi)\, d\xi \tag{30}$$

where

$$k(x, \xi) = \begin{cases} x(1 - \xi), & 0 \leq x \leq \xi \\ \xi(1 - x), & \xi \leq x \leq 1 \end{cases}$$

The kernel k is continuous and symmetric on the unit square. Differentiating (30) twice shows that $y(x)$ satisfies the boundary value problem

$$y'' + \lambda y = 0, \qquad 0 < x < 1$$
$$y(0) = y(1) = 0$$

This problem can have a nontrivial solution only when λ is positive, in which case $y(x) = c_1\cos\sqrt{\lambda}\,x + c_2\sin\sqrt{\lambda}\,x$. The condition $y(0) = 0$ implies $c_1 = 0$ and the condition $y(1) = 0$ implies $\sin\sqrt{\lambda} = 0$. Therefore the eigenvalues are

$$\lambda = \lambda_n = n^2\pi^2, \qquad n = 1, 2, 3\ldots$$

and the corresponding eigenfunctions are

$$y_n(x) = \sin n\pi x, \qquad n = 1, 2, 3, \ldots$$

For Sturm-Liouville problems (see Section 4.3) we asked which functions could be expanded in a series of eigenfunctions of the given problem. For integral equations we have the following expansion theorem that we state without proof.

Theorem 4.3   (*Hilbert-Schmidt*)   *Suppose there is a continuous function g for which*

$$F(x) = \int_a^b k(x, \xi) g(\xi)\, d\xi$$

$\left(k \text{ is continuous, real } \& \text{ symmetric}\right)$

Then F(x) can be expanded as

$$F(x) = \sum_{n=1}^\infty c_n y_n(x) \tag{31}$$

where the $y_n(x)$ are the normalized eigenfunctions of (29).

Because of the orthogonality of the $y_n(x)$ the coefficients c_n are the generalized Fourier coefficients (see Section 4.3) given by the standard formula

$$c_n = \int_a^b F(x) y_n(x)\, dx \tag{32}$$

We note that if k is separable then the infinite series in (31) collapses to a finite sum. In general the convergence of (31) is uniform and absolute on $[a, b]$.

With the aid of Theorem 4.3 we can write a formula for the solution of the nonhomogeneous integral equation

$$y(x) = f(x) + \lambda\int_a^b k(x, \xi) y(\xi)\, d\xi \tag{33}$$

in terms of the eigenvalues and eigenfunctions of (29). As before k in (33) is continuous, real, and symmetric. If $y(x)$ is a solution to (33), then

$$y(x) - f(x) = \int_a^b k(x, \xi)(\lambda y(\xi))\, d\xi$$

and hence $y - f$ is generated by the continuous function λy. By Theorem 4.3 the function $y - f$ can be expanded as

$$y(x) - f(x) = \sum_{n=1}^{\infty} c_n y_n(x)$$

where the y_n are the orthonormal eigenfunctions of (29). The c_n are given by

$$c_n = \int_a^b (y(x) - f(x)) y_n(x) \, dx = \int_a^b y(x) y_n(x) \, dx - f_n$$

where $f_n = \int_a^b f(x) y_n(x) \, dx$. Therefore

$$\int_a^b y(x) y_n(x) \, dx = f_n + \lambda \int_a^b \left(\int_a^b k(x, \xi) y(\xi) \, d\xi \right) y_n(x) \, dx$$

$$= f_n + \lambda \int_a^b \left(\int_a^b k(x, \xi) y_n(\xi) d\xi \right) y(x) \, dx$$

where the symmetry of k was used along with a change in the order of integration. Hence, if λ_n is the eigenvalue corresponding to y_n, then

$$\int_a^b y(x) y_n(x) \, dx = f_n + \frac{\lambda}{\lambda_n} \int_a^b y_n(x) y(x) \, dx$$

and

$$c_n = \frac{\lambda f_n}{\lambda_n - \lambda} \tag{34}$$

Thus we have the following theorem.

Theorem 4.4   *Let* $y(x)$ *be the solution to* (33) *where* λ *is not an eigenvalue of* (29). *Then*

$$y(x) = f(x) + \lambda \sum_{n=1}^{\infty} \frac{f_n}{\lambda_n - \lambda} y_n(x) \tag{35}$$

where

$$f_n = \int_a^b f(x) y_n(x) \, dx$$

and the λ_n *and* y_n *are the eigenvalues and eigenfunctions of* (29).

Corollary 4.1   *Equation* (35) *can be written*

$$y(x) = f(x) - \lambda \int_a^b \Gamma(x, \xi, \lambda) f(\xi) \, d\xi$$

where Γ is the resolvent kernel defined by

$$\Gamma(x, \xi, \lambda) = \sum_{n=1}^{\infty} \frac{y_n(x) y_n(\xi)}{\lambda - \lambda_n}, \qquad \lambda \neq \lambda_n$$

Proof   Exercise 4.20.

Example 4.11   Solve

$$y(x) = x + \lambda \int_0^1 k(x, \xi) y(\xi) \, d\xi \qquad k(x,\xi) = \begin{cases} x(1-\xi) & 0 \leq x \leq \xi \\ \xi(1-x) & \xi \leq x \leq 1 \end{cases}$$

where k is given in Example 4.10. In this case the normalized eigenfunctions are $y_n(x) = \sqrt{2} \sin n\pi x$ and the eigenvalues are $\lambda_n = n^2 \pi^2$. Then

$$f_n = \int_0^1 x(\sqrt{2} \sin n\pi x) \, dx = \frac{(-1)^{n+1} \sqrt{2}}{n\pi}$$

and (35) gives

$$y(x) = x + \frac{\sqrt{2}\lambda}{\pi} \sum_{n=1}^{\infty} \frac{(-1)^{n+1} \sin n\pi x}{n(n^2\pi^2 - \lambda)}, \qquad \lambda \neq n^2\pi^2$$

If $\lambda = \lambda_k$ for some k, then the solution will not exist unless f_k is zero, that is, unless f is orthogonal to y_k. In this case Equation (34) with $n = k$ is satisfied identically for any choice of c_k and the solution to (33) may be written

$$y(x) = f(x) + c_k y_k(x) + \lambda \sum_{\substack{n=1 \\ n \neq k}}^{\infty} \frac{f_n}{\lambda_n - \lambda} y_n(x)$$

where c_k is arbitrary. Therefore infinitely many solutions will exist.

Volterra Equations

It is common practice in applied mathematics to apply iterative methods to determine approximate solutions to various kinds of equations.

Example 4.12   Suppose it is desired to find the root of the algebraic equation $x = \cos x$. Beginning with an initial guess x_0 we can generate a sequence of approximations by the iterative scheme

$$x_{n+1} = \cos x_n, \qquad n = 0, 1, 2, \ldots$$

Graphically we observe that the root must occur between $x = 0.5$ and $x = 1.0$. Taking $x_0 = 0.75$ we find

$$x_1 = 0.7316$$
$$x_2 = 0.7440$$
$$x_3 = 0.7357$$
$$\vdots$$
$$x_9 = 0.7387$$
$$x_{10} = 0.7392,\ldots$$

This sequence converges to the root of $x = \cos x$, which to four decimal places is 0.7390. Generally the iterative scheme $x_{n+1} = F(x_n)$, $n = 0, 1, 2, \ldots$, will converge to a root $\bar{x}$ of $x = F(x)$ provided the initial guess x_0 is sufficiently close to $\bar{x}$ and $|F'(\bar{x})| < 1$.

Example 4.13 Consider the problem of determining a solution to the initial value problem

$$y' = f(x, y), \qquad y(x_0) = y_0$$

where f is a given continuously differentiable function. By (13) this problem is equivalent to the integral equation

$$y(x) = y_0 + \int_{x_0}^{x} f(\xi, y(\xi))\, d\xi \tag{36}$$

Beginning with an initial approximation $\phi_0(x)$ to the solution of (36) we can generate a sequence of successive approximations $\phi_1(x), \phi_2(x), \ldots$ via

$$\phi_{n+1}(x) = y_0 + \int_{x_0}^{x} f(\xi, \phi_n(\xi))\, d\xi, \qquad n = 0, 1, 2, \ldots$$

This method is called *Picard iteration* and it can be shown that the sequence $\phi_{n+1}(x)$ converges to the unique solution of (36) as n tends to infinity. For example the initial value problem

$$y' = 2x(1 + y), \qquad y(0) = 0 \tag{37}$$

is equivalent to

$$y(x) = \int_{0}^{x} 2\xi(1 + y(\xi))\, d\xi$$

The Picard scheme is

$$\phi_{n+1}(x) = \int_{0}^{x} 2\xi(1 + \phi_n(\xi))\, d\xi, \qquad n = 0, 1, 2, \ldots$$

Taking $\phi_0(x) = 0$ we obtain

$$\phi_1(x) = \int_0^x 2\xi(1 + 0) \, d\xi = x^2$$

$$\phi_2(x) = \int_0^x 2\xi(1 + \xi^2) \, d\xi = x^2 + \frac{1}{2}x^4$$

$$\phi_3(x) = \int_0^x 2\xi\left(1 + \xi^2 + \frac{1}{2}\xi^4\right) d\xi = x^2 + \frac{x^4}{2} + \frac{x^6}{2 \cdot 3}$$

and so on. Generally one can find

$$\phi_{n+1}(x) = x^2 + \frac{x^4}{2!} + \frac{x^6}{3!} + \cdots + \frac{x^{2n+2}}{(n+1)!}$$

In this case one can identify the limit of $\phi_{n+1}(x)$ as $\exp(x^2) - 1$, which is the solution of (37).

The general Volterra equation

$$y(x) = f(x) + \lambda \int_a^x k(x, \xi) y(\xi) d\xi, \qquad a \le x \le b \qquad (38)$$

where f and k are continuous can be attacked in a similar manner. We define the sequence of successive approximations by

$$y_{n+1}(x) = f(x) + \lambda \int_a^x k(x, \xi) y_n(\xi) \, d\xi, \qquad n = 0, 1, 2, \ldots \qquad (39)$$

where $y_0(x)$ is a given initial approximation. To study this iterative process further we introduce the *linear integral operator* K defined by

$$K(y) \equiv \int_a^x k(x, \xi) y(\xi) \, d\xi$$

and write (39) as

$$y_{n+1} = f + \lambda K(y_n), \qquad n = 0, 1, 2, \ldots$$

Then the first two iterates are given by

$$y_1 = f + \lambda K(y_0)$$
$$y_2 = f + \lambda K(y_1) = f + \lambda K(f + \lambda K(y_0)) = f + \lambda K(f) + \lambda^2 K^2(y_0)$$

where $K^2(y_0)$ means $K(K(y_0))$. Continuing in this fashion

$$y_{n+1} = f + \sum_{i=1}^{n} \lambda^i K^i(f) + \lambda^{n+1} K^{n+1}(y_0) \qquad (40)$$

where $K^m(f)$ denotes $K(K(K \cdots (K(f)) \cdots)$, m times. To examine the convergence of (40) we calculate the iterated transformation K^{n+1}. We have

$$|K(y_0)| = \left| \int_a^x k(x, \xi) y_0(\xi) \, d\xi \right| \le \int_a^x |k(x, \xi)||y_0(\xi)| \, d\xi \le (x - a)MC$$

where M is the maximum of $|k|$ on the square $a \le x, \xi \le b$, and C is the maximum of $|y_0(x)|$ on $a \le x \le b$. Next

$$|K^2(y_0)| = \left| \int_a^x k(x, \xi) K(y_0)(\xi) \, d\xi \right| \le \int_a^x |k(x, \xi)||K(y_0)(\xi)| \, d\xi$$

$$\le \int_a^x M(\xi - a) MC \, d\xi = \frac{(x - a)^2}{2} CM^2$$

By an induction argument

$$|K^{n+1}(y_0)| \le \frac{(x - a)^{n+1}}{(n + 1)!} CM^{n+1} \le \frac{(b - a)^{n+1}}{(n + 1)!} CM^{n+1}$$

Therefore

$$|\lambda|^{n+1}|K^{n+1}(y_0)| \to 0, \qquad \text{as } n \to \infty$$

uniformly on $a \le x \le b$ for any choice of the initial iterate $y_0(x)$. This leads to the following result, which we state without proof.

Theorem 4.5 *If f and k are continuous, then the iterative scheme* (39) *converges uniformly to the unique solution y(x) of* (38) *and*

$$y(x) = f(x) + \sum_{i=1}^{\infty} \lambda^i K^k(f) \tag{41}$$

The representation (41) *is called the Neumann series for y(x).*

Example 4.14 Consider

$$y(x) = x + \lambda \int_0^x (x - \xi) y(\xi) \, d\xi$$

Then

$$K^1(x) = \int_0^x (x - \xi) \xi \, d\xi = \frac{x^3}{3!}$$

$$K^2(x) = \int_0^x (x - \xi) \frac{\xi^3}{6} \, d\xi = \frac{x^5}{5!}$$

and so on. The Neumann series is

$$y(x) = x + \frac{\lambda x^3}{3!} + \frac{\lambda^2 x^5}{5!} + \cdots$$

We have an immediate corollary.

Corollary 4.2 *For any value of* λ *the Volterra equation*

$$y(x) = \lambda \int_0^x k(x, \xi) y(\xi) \, d\xi \qquad (42)$$

has only the trivial solution $y(x) \equiv 0$. *Hence* (42) *has no eigenvalues.*

There is an alternate representation of (41) in terms of iterated kernels. We note that

$$(Kf)(x) = \int_a^x k(x, \xi) f(\xi) \, d\xi$$

$$(K^2 f)(x) = K(Kf)(x)$$

$$= \int_a^x k(x, \xi) \int_a^\xi k(\xi, \xi_1) f(\xi_1) \, d\xi_1 \, d\xi$$

$$= \int_a^x \left(\int_{\xi_1}^x k(x, \xi) k(\xi, \xi_1) \, d\xi \right) f(\xi_1) \, d\xi_1$$

We define

$$k_2(x, \xi_1) \equiv \int_{\xi_1}^x k(x, \xi) k(\xi, \xi_1) \, d\xi$$

so that

$$(K^2 f)(x) = \int_a^x k_2(x, \xi_1) f(\xi_1) \, d\xi_1$$

Following the same steps we find

$$(K^3 f)(x) = \int_a^x k_3(x, \xi_1) f(\xi_1) \, d\xi_1$$

where

$$k_3(x, \xi_1) = \int_{\xi_1}^x k(x, \xi) k_2(\xi, \xi_1) \, d\xi$$

and in general

$$(K^n f)(x) = \int_a^x k_n(x, \xi_1) f(\xi_1) \, d\xi_1$$

where

$$k_n(x, \xi_1) = \int_{\xi_1}^x k(x, \xi) k_{n-1}(\xi, \xi_1) \, d\xi$$

The kernels $k_1 = k, k_2, k_3, \ldots$ are called the *iterated kernels* and the Neumann series (41) can be written

$$
\begin{aligned}
y(x) &= f(x) + \lambda \sum_{i=1}^{\infty} \lambda^{i-1} \int_a^x k_i(x, \xi) f(\xi) \, d\xi \\
&= f(x) + \lambda \int_a^x \left(\sum_{i=1}^{\infty} \lambda^{i-1} k_i(x, \xi) \right) f(\xi) \, d\xi \\
&= f(x) + \lambda \int_a^x \Gamma(x, \xi, \lambda) f(\xi) \, d\xi
\end{aligned}
\tag{43}
$$

where

$$\Gamma(x, \xi, \lambda) \equiv \sum_{i=1}^{\infty} \lambda^{i-1} k_i(x, \xi)$$

is the *resolvent kernel*.

Some Volterra equations have kernels of a special type that allow a direct solution. For example a Volterra equation of the form

$$y(x) = f(x) + \lambda \int_0^x k(x - \xi) y(\xi) \, d\xi \tag{44}$$

is said to be of *convolution type*. Such equations can be solved by Laplace transforms. We recognize the integral in (44) as the convolution $k * y$. Therefore taking Laplace transforms of (44) gives

$$\mathscr{L}[y] = \mathscr{L}[f] + \lambda \mathscr{L}[k] \mathscr{L}[y]$$

Thus

$$\mathscr{L}[y] = \frac{\mathscr{L}[f]}{1 - \lambda \mathscr{L}[k]}$$

and the solution is

$$y(x) = \mathscr{L}^{-1} \left\{ \frac{\mathscr{L}[f]}{1 - \lambda \mathscr{L}[k]} \right\}, \qquad \lambda \mathscr{L}[k] \neq 1$$

Example 4.15 Solve

$$y(x) = x - \int_0^x (x - \xi) y(\xi) \, d\xi$$

by Laplace transform methods. Here $f(x) = x$ and $k(x) = x$ and $\mathscr{L}[x] = 1/s^2$. Taking the Laplace transform of the equation gives

$$\mathscr{L}[y] = \frac{1}{s^2} - \frac{1}{s^2} \mathscr{L}[y]$$

or

$$\mathscr{L}[y] = \frac{1}{1 + s^2}$$

Thus

$$y(x) = \mathscr{L}^{-1}\left\{\frac{1}{1 + s^2}\right\} = \sin x$$

An integral equation is termed *singular* if its kernel is singular or if the interval of integration is infinite. We examine a few integral equations to illustrate the variety of phenomena that can occur.

Example 4.16 Consider the equation

$$y(x) = \lambda \int_0^\infty \sin(x\xi) y(\xi) \, d\xi \tag{45}$$

If $\lambda = \sqrt{2/\pi}$, then

$$y_\alpha(x) = \sqrt{\frac{2}{\pi}} e^{-\alpha x} + \frac{x}{\alpha^2 + x^2}, \qquad x > 0 \tag{46}$$

is a solution for any $\alpha > 0$. Therefore $\sqrt{2/\pi}$ is an eigenvalue with infinite multiplicity (compare Theorem 4.2). We recognize the integral in (45) as the Fourier sine transform of y.

Example 4.17 The eigenvalues of the integral equation

$$y(x) = \lambda \int_0^\infty \exp(-x\xi) y(\xi) \, d\xi$$

are not discrete but rather form a continuum of values. They are given by

$$\lambda = \sqrt{\frac{\sin \alpha \pi}{\pi}} \, , \qquad 0 < \alpha < 1$$

and so $0 < \lambda \le 1/\sqrt{\pi}$. The corresponding eigenfunctions are

$$y_\alpha(x) = \sqrt{\Gamma(1 - \alpha)}\, x^{\alpha - 1} + \sqrt{\Gamma(\alpha)}\, x^{-\alpha}$$

where $\Gamma(x)$ is the gamma function (see F. Hildebrand, *Methods of Applied Mathematics*, 2nd ed., Prentice-Hall, Englewood Cliffs, N.J. (1965), page 273).

Example 4.18   In Example 4.1 we encountered the equation

$$f(x) = \int_0^x \frac{y(\xi)}{\sqrt{x - \xi}}\, d\xi \tag{47}$$

which is known as *Abel's equation*. It arises in many applications and since it is convolution type it can be solved by Laplace transform methods. We shall, however, illustrate its solution by another elementary technique. Multiplying (47) by $(s - x)^{-1/2}$ and integrating from 0 to s gives

$$\int_0^s \frac{f(x)}{\sqrt{s - x}}\, dx = \int_0^s \int_0^x \frac{y(\xi)}{\sqrt{s - x}\sqrt{x - \xi}}\, d\xi\, dx$$

$$= \int_0^s \int_\xi^s \frac{y(\xi)}{\sqrt{s - x}\sqrt{x - \xi}}\, dx\, d\xi$$

where in the last step the order of integration was interchanged. We leave as an exercise to show

$$\int_\xi^s \frac{dx}{\sqrt{s - x}\sqrt{x - \xi}} = \pi \tag{48}$$

Thus

$$\int_0^s \frac{f(x)}{\sqrt{s - x}}\, dx = \pi \int_0^s y(\xi)\, d\xi$$

Taking the derivative with respect to s yields

$$y(s) = \frac{1}{\pi} \frac{d}{ds} \int_0^s \frac{f(x)}{\sqrt{s - x}}\, dx$$

which is the solution to (47).

EXERCISES

4.1   (Leibniz' Formula)   Prove

$$\frac{d}{dt} \int_{a(t)}^{b(t)} u(x, t)\, dx = \int_{a(t)}^{b(t)} u_t(x, t)\, dx$$
$$+ u(b(t), t) b'(t) - u(a(t), t) a'(t)$$

Hint: Use the chain rule to differentiate

$$G(t, a, b) = \int_{a}^{b} u(x, t)\, dx$$

Assume u, a, and b are class C^1.

4.2   Differentiate (20) to obtain (19).

4.3   Verify that $y(x) = 1/(\pi\sqrt{x})$ is a solution of the integral equation

$$1 = \int_{0}^{x} \frac{y(\xi)}{\sqrt{x - \xi}}\, d\xi$$

4.4   Show that the Fredholm equation (4) of the third kind can be trans-formed to one of the second kind provided $\alpha(x) > 0$ for $a \le x \le b$. *Hint*: Divide by $\alpha(x)$.

4.5   Formulate the initial value problem

$$y'' - \lambda y = f(x), \qquad x > 0, \quad \lambda \text{ a constant}$$
$$y(0) = 1, \qquad y'(0) = 0$$

as a Volterra integral equation.

4.6   **(a)** Show that the Volterra equation

$$f(x) = \lambda \int_{0}^{x} K(x, \xi) y(\xi)\, d\xi, \qquad K(x, x) \ne 0$$

of the first kind and can be transformed to a Volterra equation of the second kind. *Hint*: Differentiate.

(b) Apply Part (a) to the equation

$$\sin x = \int_{0}^{x} \exp(x - \xi)\, y(\xi)\, d\xi$$

▷**4.7**   Find the eigenvalues for the following equations.

(a) $y(x) = \lambda \int_0^1 y(\xi)\, d\xi$ (c) $y(x) = \lambda \int_0^1 (x^2 + \xi^2) y(\xi)\, d\xi$

(b) $y(x) = \lambda \int_0^1 x\xi y(\xi)\, d\xi$ (d) $y(x) = \lambda \int_0^1 e^{x+\xi} y(\xi)\, d\xi$

4.8   Show that the equation

$$y(x) = \lambda \int_0^\pi \sin x \sin 2\xi\, y(\xi)\, d\xi$$

has no eigenvalues.

4.9   In the integral equation

$$y(x) = x^2 + \int_0^1 \sin(x\xi) y(\xi)\, d\xi$$

replace $\sin(x\xi)$ by the first two terms in its power series expansion and therefore obtain an approximate solution.

▷ **4.10**   Solve the integral equation

$$y(x) = 1 + \lambda \int_{-\pi}^\pi e^{i\omega(x-\xi)} y(\xi)\, d\xi$$

considering all cases.

4.11   Solve the equation

$$y(x) = f(x) + \lambda \int_0^{2\pi} \sin(x + \xi) y(\xi)\, d\xi$$

and find a condition on $f(x)$ so that the equation has a solution when λ is an eigenvalue. Obtain the solution if $f(x) = \sin x$, considering all cases.

4.12   Solve

$$y(x) = x + \int_0^1 (x\xi^2 + \xi x^2) y(\xi)\, d\xi$$

4.13   Does the equation

$$y(x) = \sin x + 3 \int_0^\pi (x + \xi) y(\xi)\, d\xi$$

have a solution?

4.14 Show that the solution of the integral equation

$$y(x) = f(x) + \lambda \int_0^1 x e^\xi y(\xi) \, d\xi, \qquad \lambda \neq 1$$

is

$$y(x) = f(x) + \frac{\lambda}{1 - \lambda} \int_0^1 x e^\xi f(\xi) \, d\xi$$

Comment on the case $\lambda = 1$.

4.15 Solve the following nonlinear integral equations and discuss solutions for different values of λ.

(a) $y(x) = 1 + \lambda \int_0^1 y(\xi)^2 \, d\xi$

(b) $y(x) = \lambda \int_0^1 \xi y(\xi)^2 \, d\xi$

4.16 Find the iterated kernels $k_1, k_2, k_3, \ldots$ for the Volterra equation

$$y(x) = f(x) + \lambda \int_0^x e^{x+\xi} y(\xi) \, d\xi$$

Determine the resolvent kernel in closed form and write the solution in the form (43).

▷ **4.17** Use Laplace transforms to solve

(a) $y(x) = f(x) + \lambda \int_0^x e^{x-\xi} y(\xi) \, d\xi$

(b) $x = \int_0^x e^{x-\xi} y(\xi) \, d\xi$, (*Solution:* $y = 1 - x$)

4.18 Find the resolvent kernel and solve

$$y(x) = f(x) + \lambda \int_0^x y(\xi) \, d\xi$$

▷ **4.19** Solve

$$x = \int_0^x \frac{y(\xi) \, d\xi}{\sqrt{x - \xi}}$$

Solution:

$$y = \frac{2\sqrt{x}}{\pi}$$

4.20 Prove Corollary 4.1

4.21 Consider the Fredholm equation of the first kind

$$f(x) = \int_a^b k(x, \xi) y(\xi) \, d\xi$$

where the kernel is continuous and of the form

$$k(x, \xi) = \sum_{i=1}^n \alpha_i(x) \beta_i(\xi)$$

(a) Show that no solution can exist unless f is of the form

$$f(x) = \sum_{i=1}^n a_i \alpha_i(x), \qquad a_i \text{ constants}$$

(b) If $y(x)$ is a solution show that $y(x) + \phi(x)$ is also a solution for any continuous function $\phi(x)$ that is orthogonal to each $\beta_i(x)$, $i = 1, \ldots, n$.

(c) If the consistency condition in (a) holds, show that a solution of the form

$$y(x) = \sum_{j=1}^n b_j \beta_j(x)$$

exists for some choice of constants b_j provided that

$$\det \int_a^b \beta_i(x) \beta_j(x) \, dx \neq 0$$

4.22 If possible, solve the integral equations

(a) $e^x = \int_0^\pi \sin(x + \xi) y(\xi) \, d\xi$

(b) $3x^2 + 4x = \int_{-1}^1 (6x^2\xi + 4x\xi^2) y(\xi) \, d\xi$

4.23 **(a)** Consider the Fredholm equation

$$y(x) = f(x) + \lambda \int_a^b k(x, \xi) y(\xi) \, d\xi$$

where f and k are continuous. Letting K be the integral operator defined by

$$Kf(x) \equiv \int_a^b k(x, \xi) f(\xi) \, d\xi$$

show that the solution may be represented by the convergent series

$$y(x) = f(x) + \sum_{n=1}^{\infty} \lambda^n K^n f(x)$$

provided $|\lambda| < M^{-1}(b-a)^{-1}$, where $|k(x, \xi)| \le M$.

(b) Under the conditions of Part (a) show that

$$y(x) = f(x) + \lambda \int_a^b \left(\sum_{n=0}^{\infty} \lambda^n k_{n+1}(x, \xi) \right) f(\xi)\, d\xi$$

where

$$k_n(x, \xi) = \int_a^b k(x, \xi_1) k_{n-1}(\xi_1, \xi)\, d\xi_1, \qquad n = 2, 3, \ldots$$

and $k_1(x, \xi) = k(x, \xi)$. The quantity

$$\Gamma(x, \xi, \lambda) \equiv \sum_{n=0}^{\infty} \lambda^n k_{n+1}(x, \xi)$$

is called the resolvent kernel.

(c) Show that

$$\Gamma(x, t, \lambda) = k(x, t) + \lambda \int_a^b k(x, \xi) \Gamma(\xi, t, \lambda)\, d\xi$$

(d) Find the resolvent kernel for $k(x, \xi) = 1 - 3x\xi$ and find the solution of

$$y(x) = 1 + \lambda \int_0^1 (1 - 3x\xi) y(\xi)\, d\xi$$

where $|\lambda| < 2$. Investigate the cases $\lambda = \pm 2$ and $|\lambda| > 2$.

4.24 Let A be a real, symmetric $n \times n$ matrix with eigenvalues λ_j and eigenvectors e_j, $j = 1, \ldots, n$. Show that the solution x to the system $Ax - \lambda x = f$, $\lambda \ne \lambda_j$, can be written as

$$x = \sum_{j=1}^{n} \frac{f_j}{\lambda_j - \lambda} e_j$$

where the f_j are the components of f in the basis e_j. Compare to Theorem 4.4.

4.25 Consider the boundary value problem

$$y'' + \lambda y = f(x), \qquad 0 < x < \pi$$
$$y(0) = y(\pi) = 0$$

where $\lambda \neq n\pi$, $n = 1, 2, \ldots$, and where $f \in C^1[0, \pi]$, $f(0) = f(\pi) = 0$. By expanding y and f in terms of the eigenfunctions and eigenvalues of the problem

$$u'' + \lambda u = 0, \qquad u(0) = u(\pi) = 0, \quad 0 < x < \pi$$

show that

$$y(x) = \int_0^\pi G(x, \xi, \lambda) f(\xi) \, d\xi$$

for some function G, and find G (G is called the Green's function for the problem). Compare to Theorem 4.4 and Exercise 4.23.

REFERENCES

1. F. John, *Partial Differential Equations*, 4th ed., Springer-Verlag, New York (1982).
2. R. V. Churchill, *Fourier Series and Boundary Value Problems*, 2nd ed., McGraw-Hill Book Publishing Company, New York (1969).
3. I. Stakgold, *Green's Functions and Boundary Value Problems*, Wiley-Interscience, New York (1979).
4. E. Zauderer, *Partial Differential Equations of Applied Mathematics*, Wiley-Interscience, New York (1983).
5. H. Hochstadt, *Integral Equations*, Wiley-Interscience, New York (1973).
6. R. Courant and D. Hilbert, *Methods of Mathematical Physics*, Vol. 1, Wiley-Interscience, New York (1953).

5

WAVE PHENOMENA IN CONTINUOUS SYSTEMS

Two of the fundamental processes in nature are diffusion and wave propagation. In the last chapter we studied the equation of heat conduction, a parabolic partial differential equation that is the prototype of equations governing linear diffusion processes. In the present chapter we investigate wave phenomena and obtain equations that govern the propagation of waves first in simple model settings and then in continuum mechanics. The evolution equations governing such phenomena are hyperbolic and are fundamentally different from their parabolic counterparts in diffusion and from elliptic equations that govern equilibrium states.

5.1 WAVE PROPAGATION

Waves

By a *wave* is meant an identifiable signal or disturbance in a medium that is propagated in time, carrying energy with it. A few familiar examples are electromagnetic waves, waves on the surface of water, sound waves, and stress waves in solids, as occur in earthquakes. Material or matter is not necessarily convected with the wave; it is the disturbance, that carries energy, that is propagated. In this section we investigate several model equations that occur in wave phenomena and point out basic features that are encountered in the study of the propagation of waves.

A simple mathematical model of a wave is the function

$$u(x, t) = f(x - ct) \tag{1}$$

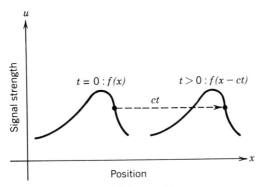

Figure 5.1. Right traveling wave.

which represents an undistorted right traveling wave moving at constant velocity c. The coordinate x represents position, t time, and u the strength of the disturbance. At $t = 0$ the wave profile is $u = f(x)$ and at $t > 0$ units of time the disturbance has moved to the right ct units of length (see Fig. 5.1). Of particular importance is that the wave profile described by (1) moves without distortion. Not all waves have this property; it is characteristic of linear waves, or wave profiles that are solutions to linear partial differential equations. On the other hand, waves that distort and break are characteristic of nonlinear processes. To find a partial differential equation that governs a process described by (1) we compute u_t and u_x to get

$$u_t = -cf'(x - ct), \qquad u_x = f'(x - ct)$$

Hence

$$u_t + cu_x = 0 \tag{2}$$

Equation (2) is a first order linear partial differential equation that, in the sense just described, is the simplest wave equation. It is called the *advection equation* and its general solution is (1), where f is an arbitrary function. The name comes from the fact that (2) describes what would happen if dye were squirted into a stream moving by at velocity c; the color would be advected downstream without distortion. Similarly, a traveling wave of the form $u = f(x + ct)$ is a left moving wave and is a solution of the partial differential equation $u_t - cu_x = 0$.

Other waves of interest in many calculations are periodic, or sinusoidal waves. These traveling waves are represented by expressions of the form

$$u = A\cos(kx - \omega t) \tag{3}$$

(see Fig. 5.2). The positive number A is the *amplitude*, k is the *wave number* (the number of oscillations in 2π units of space observed at a fixed time), and

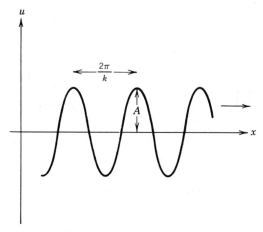

Figure 5.2. Periodic wave.

ω the *angular frequency* (the number of oscillations in 2π units of time observed at a fixed location x). The number $\lambda = 2\pi/k$ is the *wavelength* and $P = 2\pi/\omega$ is the *time period*. The wavelength measures the distance between successive crests and the time period is the time for an observer located at a fixed position x to see a repeat pattern. If we write (3) as

$$u = A \cos k\left(x - \frac{\omega}{k}t \right)$$

then we note that (3) represents a traveling wave moving to the right with velocity $c = \omega/k$. This number is called the *phase velocity* and it is the speed one would have to move to remain at the same point on the traveling wave. For calculations the complex exponential form

$$u = \exp(i(kx - \omega t)) \tag{4}$$

is often used rather than (3). Computations involving differentiations are easier with (4) and afterwards, making use of Euler's formula $\exp(i\theta) = \cos\theta + i\sin\theta$, the real or imaginary part may be taken to recover real solutions. Again, waves of the type (3) (or (4)) are characteristic of linear processes and linear equations.

Not all waves propagate in such a way that their profile remains unchanged or undistorted; surface waves on the ocean are obvious examples. Less familiar perhaps but just as common, are stress or pressure waves that propagate in solids or gases. To fix the idea and to indicate how nonlinearity affects the shape of a wave let us consider a stress wave propagating in a metallic bar, caused, for example, by a force on one end of the bar. The distortion of a wave profile results from the property of most materials to transmit signals at a

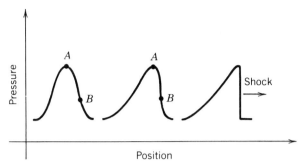

Figure 5.3. Distortion of a wave into a shock.

speed that increases with increasing pressure. Therefore a stress wave that is propagating in a medium will gradually distort and steepen until it propagates as a discontinuous disturbance, or shock wave. Figure 5.3 shows various snapshots of a stress wave propagating into a material. The wave steepens as time increases because signals or disturbances travel faster when the pressure is higher. Thus the point A moves to the right faster than the point B. The shock that forms in pressure is accompanied by discontinuous jumps in the other flow parameters, such as density, particle velocity, temperature, energy, and entropy. Here we are considering finite amplitude waves rather than small amplitude waves. The latter can propagate as linear waves.

Physically, a shock wave is not strictly a discontinuity but rather an extremely thin region where the change in the state is steep. The width of the shock is small, usually of the order of a few mean free paths, or average distance to a collision, of the molecules. In a shock there are two competing effects that cause this thinness, the nonlinearity of the material that is causing the shock to form and the dissipative effects (e.g., viscosity) that are tending to smear the wave out. Usually these two effects just cancel and the front assumes a shape that does not change in time.

The same mechanism that causes pressure waves to steepen into shocks, that is, an increase in signal transmission speed at higher pressures, causes

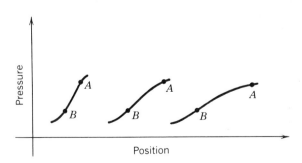

Figure 5.4. Snapshots of a rarefaction wave.

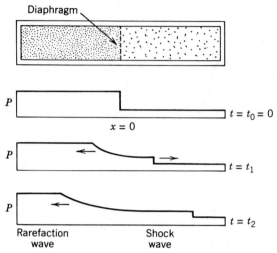

Figure 5.5. Pressure profiles in a shock tube.

release waves or *rarefaction waves* to form that lower the pressure. Figure 5.4 shows a spreading rarefaction. Again, point *A* at higher pressure moves to the right faster than point *B*, thereby flattening the wave.

A simple example of a situation in which both shock and rarefaction waves form is in a shock tube where a long cylinder contains a gas at high pressure and low pressure separated by a thin diaphragm. Figure 5.5 shows the pressure profiles at times after the diaphragm is punctured and indicates the presence of both a rarefaction and a shock wave.

So far we have mentioned two types of waves, those that propagate undistorted at constant velocity and those that distort because the speed of propagation depends on the amplitude of the wave. There is a third phenomenon that is relevant in some problems, namely that of *dispersion*. In this case the speed of propagation depends on the wavelength of the particular wave. So, for example, longer waves can travel faster than shorter ones. Thus an observer of a wave at fixed location x_0 may see a different temporal wave pattern from another observer at fixed location x_1. Dispersive wave propagation arises from both linear and nonlinear equations.

Linear Waves

In the preceding paragraphs we introduced the simplest wave equation, the advection equation,

$$u_t + cu_x = 0, \qquad x \in R^1, \quad t > 0 \tag{5}$$

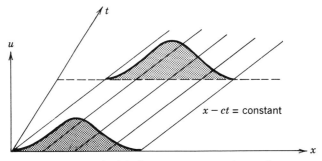

Figure 5.6. Solution to $u_t + cu_x = 0$, $c > 0$.

whose general solution is a right traveling wave

$$u = f(x - ct) \tag{6}$$

propagating at constant velocity c, where f is an arbitrary function. If we impose the initial condition

$$u(x,0) = \phi(x), \qquad x \in R^1 \tag{7}$$

then it follows that $f(x) = \phi(x)$ and so the solution to the initial value problem (5) and (7) is

$$u(x, t) = \phi(x - ct) \tag{8}$$

The straight lines $x - ct =$ constant play a special role in this problem; along those lines the initial values are propagated with constant value. We may interpret them as lines in space–time along which the signals are carried (see Fig. 5.6). Furthermore, along these lines the partial differential equation (5) reduces to the ordinary differential equation $du/dt = 0$. That is to say, if $\mathscr{C}$ is the curve $x = ct + k$ for some constant k, then the directional derivative of u along that curve is

$$\frac{du}{dt}(x(t), t) = u_x(x(t), t)\frac{dx}{dt} + u_t(x(t), t)$$

$$= u_x(x(t), t)c + u_t(x(t), t)$$

which is the left side of (5) evaluated along $\mathscr{C}$. The family of straight lines $x - ct = k$ (k constant) is called the family of *characteristic curves* for this problem. Note that their speed c is the reciprocal of their slope in the xt coordinate system.

Now let us complicate the partial differential equation (5) by replacing the constant c by a function of the independent variables t and x and consider the initial value problem

$$u_t + c(x, t)u_x = 0, \qquad\qquad x \in R^1, \quad t > 0 \qquad\qquad (9)$$
$$u(x, 0) = \phi(x), \qquad\qquad x \in R^1$$

where $c(x, t)$ is a given function. Let $\mathscr{C}$ be the family of curves defined by the differential equation

$$\frac{dx}{dt} = c(x, t) \qquad\qquad (10)$$

Then along a member of $\mathscr{C}$

$$\frac{du}{dt} = u_x\frac{dx}{dt} + u_t = u_xc(x, t) + u_t = 0$$

Hence u is constant on each member of $\mathscr{C}$. The curves $\mathscr{C}$ defined by (10) are the *characteristic curves*.

Example 1.1   Consider the initial value problem

$$u_t + 2tu_x = 0, \qquad\qquad x \in R^1, \quad t > 0$$
$$u(x, 0) = \exp(-x^2), \qquad\qquad x \in R^1$$

The characteristic curves are defined by the differential equation $dx/dt = 2t$ that yields the family of parabolas

$$x = t^2 + k, \qquad k \text{ constant}$$

Knowing that u is constant on these characteristic curves allows us to find a solution to the initial value problem. Let (x, t) be an arbitrary point with $t > 0$. The characteristic curve through (x, t) passes through $(\xi, 0)$ and has equation $x = t^2 + \xi$ (see Fig. 5.7). Since u is constant on this curve

$$u(x, t) = \exp(-\xi^2) = \exp\left(-(x - t^2)^2\right)$$

which is the unique solution to the initial value problem. The speed of the signal at (x, t) is $2t$ which is dependent on t. In general equation (9) propagates signals at the speed $c(x, t)$. In the present example the wave speeds up as time increases but it retains its initial shape.

By following the same reasoning as in Example 1.1 we can write the solution to the initial value problem (9) as $u(x, t) = \phi(k)$, where $a(x, t) = k$ define the characteristic curves given by (10). The partial differential equation (9) itself has general solution $u(x, t) = f(a(x, t))$, where f is an arbitrary function (see Exercise 1.1).

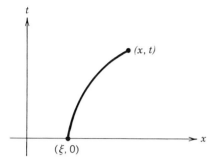

Figure 5.7. Characteristic $x = t^2 + k$.

Nonlinear Waves

In the last section we examined the two simple model wave equations $u_t + cu_x = 0$ and $u_t + c(x, t)u_x = 0$ that are both first order and linear. Now we study the same type of equation when a nonlinearity is introduced. In particular we consider

$$u_t + c(u)u_x = 0, \qquad x \in R^1, \quad t > 0 \tag{11}$$

where $c'(u) > 0$, with initial condition

$$u(x, 0) = \phi(x), \qquad x \in R^1 \tag{12}$$

Using the guidance of the earlier examples we define the *characteristic curves* by the differential equation

$$\frac{dx}{dt} = c(u) \tag{13}$$

Then along a particular such curve $x = x(t)$ we have

$$\frac{d}{dt}u(x(t),t) = \frac{du}{dt}(x(t), t) = u_x(x(t), t)c(u(x(t), t)) + u_t(x(t), t) = 0$$

$$dx/dt$$

Therefore u is constant along the characteristics, and the characteristics are straight lines since

$$\frac{d^2x}{dt^2} = \frac{d}{dt}\left(\frac{dx}{dt}\right) = \frac{d}{dt}c(u(x(t), t)) = c'(u)\frac{du}{dt} = 0$$

this thing is zero from previous line

In the nonlinear case, however, the speed of the characteristics as defined by (13) depends on the value u of the solution at a given point. To find the equation of the characteristic $\mathscr{C}$ through (x, t) we note that its speed is

$$\frac{dx}{dt} = c(u(\xi, 0)) = c(\phi(\xi))$$

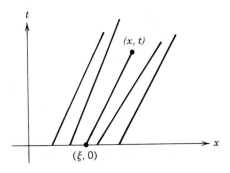

Figure 5.8. Characteristics of $u_t + c(u)u_x = 0$.

(see Fig. 5.8). This results from applying (13) at $(\xi, 0)$. Hence, after integrating,

The integration step produces a constant of integration, which must be chosen to be ξ so that $x(t) = \xi$ when $t = 0$.

$$x = c(\phi(\xi))t + \xi \tag{14}$$

gives the equation of the desired characteristic $\mathscr{C}$. Equation (14) defines $\xi = \xi(x, t)$ implicitly as a function of x and t, and the solution $u(x, t)$ of the initial value problem (11) and (12) is given by

$$u(x, t) = \phi(\xi) \tag{15}$$

where ξ is defined by (14).

Example 1.2 Consider the initial value problem

$$u_t + uu_x = 0, \qquad x \in R^1, \quad t > 0$$

$$\phi(x) \equiv u(x, 0) = \begin{cases} 2, & x < 0 \\ 2 - x, & x \in [0, 1] \\ 1, & x > 1 \end{cases}$$

The initial curve is sketched in Fig. 5.9. Since $c(u) = u$ the characteristics are straight lines emanating from $(\xi, 0)$ with speed $c(\phi(\xi)) = \phi(\xi)$. These are

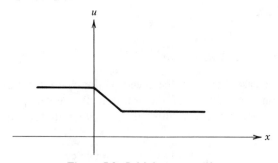

Figure 5.9. Initial wave profile.

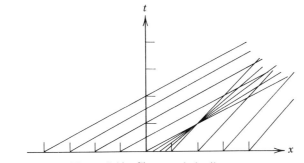

Figure 5.10. Characteristic diagram.

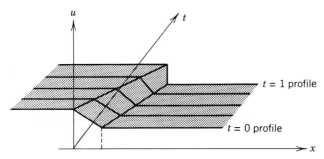

Figure 5.11. Solution surface.

plotted in Fig. 5.10. For $x < 0$ the lines have speed 2; for $x > 1$ the lines have speed 1; for $0 \leq x \leq 1$ the lines have speed $2 - x$ and these all intersect at $(2, 1)$. Immediately one observes that a solution cannot exist for $t > 1$, since the characteristics cross beyond that time and they carry different constant values of u. Figure 5.11 shows several wave profiles that indicate the steepening that is occurring. At $t = 1$ *breaking* of the wave occurs, which is the first instant when the solution becomes multiple valued. To find the solution for $t < 1$ we first note that $u(x, t) = 2$ for $x < 2t$ and $u(x, t) = 1$ for $x > t + 1$. For $2t < x < t + 1$ Equation (14) becomes

$$x = (2 - \xi)t + \xi$$

which gives

solve for $\frac{\xi}{2}$

$$\xi = \frac{x - 2t}{1 - t}$$

Equation (15) then yields $u(x,t) = \varphi(\xi) = \varphi\left(\frac{x-2t}{1-t}\right) = 2 - \frac{x-2t}{1-t}$ $x \in [0,1]$

$$u(x, t) = \frac{2 - x}{1 - t}, \qquad 2t < x < t + 1, \quad t < 1$$

$u_t = \frac{2-x}{(1-t)^2}$ $u_x = -\frac{1}{1-t}$

This explicit form of the solution also indicates the difficulty at the breaking time $t = 1$.

In general the initial value problem (11) and (12) may have a solution only up to a finite time t_b, which is called the *breaking time*. Let us assume in addition to $c'(u) > 0$ that the initial wave profile satisfies the conditions

$$\phi(x) \geq 0, \qquad \phi'(x) < 0$$

At the time when breaking occurs the gradient u_x will become infinite. To compute u_x we differentiate (14) implicitly with respect to x to obtain

$$\xi_x = \frac{1}{1 + c'(\phi(\xi))\phi'(\xi)t}$$

Then from (15)

$$u_x = \frac{\phi'(\xi)}{1 + c'(\phi(\xi))\phi'(\xi)t}$$

The gradient catastrophe will occur at the minimum value of t, which makes the denominator zero. Hence

$$t_b = \min_{\xi} \frac{-1}{\phi'(\xi)c'(\phi(\xi))}, \qquad t_b \geq 0$$

In Example 1.2 we have $c(u) = u$ and $\phi(\xi) = 2 - \xi$. Hence $\phi'(\xi)c'(\phi(\xi)) = (-1)(1) = -1$ and $t_b = 1$ is the time when breaking occurs.

In summary we have observed that the nonlinear partial differential equation

$$u_t + c(u)u_x = 0, \qquad c'(u) > 0$$

propagates the initial wave profile at a speed $c(u)$, which depends on the value of the solution u at a given point. Since $c'(u) > 0$ large values of u are propagated faster than small values and distortion of the wave profile occurs. This is consistent with our earlier remarks of a physical nature, namely that wave distortion and shocks develop in materials because of the property of the medium to transmit signals more rapidly at higher levels of stress or pressure. Mathematically, distortion and the development of shocks or discontinuous solutions are distinctively nonlinear phenomena caused by the term $c(u)u_x$ in the last equation.

Heretofore all of the wave equations we have examined were considered on an infinite domain $-\infty < x < \infty$. Now we investigate problems on a finite spatial domain. To illustrate the basic points it is sufficient to study the simple

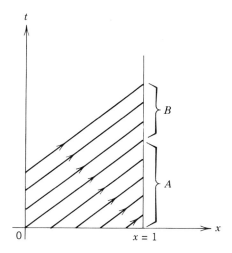

Figure 5.12. Characteristic diagram.

advection equation

$$u_t + cu_x = 0, \qquad 0 < x < 1, \quad t > 0, \quad c > 0$$

with an initial condition

$$u(x,0) = f(x), \qquad 0 < x < 1$$

On the interval $(-\infty, \infty)$ the solution to the problem was a unidirectional right traveling wave $u = f(x - ct)$. On a bounded interval what kinds of boundary conditions can be imposed at $x = 0$ and $x = 1$? We make reference to Fig. 5.12. Since u is given by the initial condition $f(x)$ along the initial line $t = 0$, $0 < x < 1$, data cannot be prescribed arbitrarily on the segment A along the boundary $x = 1$. This is because the characteristics $x - ct = $ constant carry the initial data to the segment A. Boundary data can be imposed along the line $x = 0$, since those data would be carried along the forward-going characteristics to the segment B along $x = 1$. Then boundary conditions along B cannot be prescribed arbitrarily. Thus it is clear that the problem

$$u_t + cu_x = 0, \qquad 0 < x < 1, \quad t > 0, \quad c > 0$$
$$u(x,0) = f(x), \qquad 0 < x < 1$$
$$u(0, t) = g(t), \qquad t > 0$$

is properly posed in the sense that there is a unique solution. There are no backward-going characteristics in this problem so there are no left traveling waves. Thus waves are not reflected from the boundary $x = 1$. In summary, care must be taken to properly formulate boundary value problems for unidirectional wave equations like the ones studied thus far. We shall see that the situation is much different for second order hyperbolic partial differential equations like $u_{tt} - u_{xx} = 0$. In this case both forward- and backward-going characteristics exist, and so left traveling waves are also possible as well as a mechanism for reflections from boundaries.

Burgers' Equation

One approach to the study of partial differential equations is to examine certain model equations to gain a sense of the role played by various terms in the equation and how those terms relate to fundamental physical processes. In the preceding paragraphs we observed that the advection equation

$$u_t + cu_x = 0 \tag{16}$$

propagates an initial disturbance or signal at velocity c while maintaining the precise form of the signal. On the other hand, the nonlinear equation

$$u_t + uu_x = 0 \tag{17}$$

propagates signals in such a way that distortion occurs in the signal profile; that is, the nonlinear convection term uu_x causes either a shocking up or rarefaction effect. In Chapter 4 we also examined the heat equation

$$u_t - \nu u_{xx} = 0, \quad \nu > 0 \tag{18}$$

which contains a diffusion term νu_{xx}. Frequently, insight can be gained into the nature of the various terms in an evolution equation by attempting to find either traveling wave solutions

$$u = f(x - ct) \tag{19}$$

or else complex exponential solutions of the form

$$u = A \exp i(kx - \omega t) \tag{20}$$

where A is the amplitude, k the wave number, and ω the frequency.

Example 1.3   Consider the partial differential equation

$$u_t + cu_x - \nu u_{xx} = 0, \quad (c, \nu > 0) \tag{21}$$

which contains a linear convection term cu_x and a diffusion term νu_{xx}. If (20) is substituted into (21), then after simplification we find that

$$\omega = ck - i\nu k^2$$

Hence, for any k Equation (21) admits solutions of the form

$$u = A \exp(-\nu k^2 t)\exp(ik(x - ct))$$

The factor $\exp(ik(x - ct))$ represents a harmonic right traveling wave with wave number k, and the factor $A \exp(-\nu k^2 t)$ represents a decaying amplitude. Qualitatively two conclusions can be drawn.

(i) For waves of constant wavelength (k constant) the attenuation of the wave increases with increasing ν; hence ν is a measurement of diffusion.

(ii) For constant ν the attenuation increases as k increases; hence smaller wavelengths attenuate faster than longer wavelengths.

Example 1.4 Consider the nonlinear equation

$$u_t + uu_x - \nu u_{xx} = 0, \qquad \nu > 0 \tag{22}$$

which is known as *Burgers' equation*. The term uu_x will have a shocking up effect that will cause waves to break and the term νu_{xx} is a diffusion term like the one occurring in the heat equation (18) or in Equation (21). We attempt to find a traveling wave solution of (22) of the form

$$u = f(x - ct) \tag{23}$$

where f and c are to be determined. Substituting (23) into (22) gives

$$-cf'(s) + f(s)f'(s) - \nu f''(s) = 0$$

where

$$s = x - ct$$

Noting that $ff' = \frac{1}{2}(d/ds)f^2$ and performing an integration gives

$$-cf + \frac{1}{2}f^2 - \nu f' = B$$

where B is a constant of integration. Hence

$$\frac{df}{ds} = \frac{1}{2\nu}(f - f_1)(f - f_2) \tag{24}$$

where

$$f_1 = c - \sqrt{c^2 + 2B}, \qquad f_2 = c + \sqrt{c^2 + 2B}$$

We assume $c^2 > -2B$, and so $f_2 > f_1$. Separating variables in (24) and integrating again gives

$$\frac{s}{2\nu} = \int \frac{df}{(f - f_1)(f - f_2)} = \frac{1}{f_2 - f_1} \ln \frac{f_2 - f}{f - f_1}, \qquad f_1 < f < f_2$$

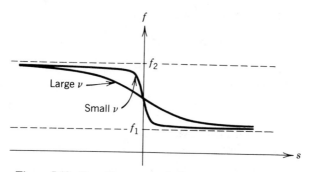

Figure 5.13. Traveling wave solutions to Burgers' equation.

Solving for f yields

$$f(s) = \frac{f_2 + f_1 e^{Ks}}{1 + e^{Ks}} \tag{25}$$

where

$$K = \frac{1}{2\nu}(f_2 - f_1) > 0$$

For large positive s we have $f(s) \sim f_1$ and for large negative s we have $f(s) \sim f_2$. It is easy to see that $f'(s) < 0$ for all s and that $f(0) = \frac{1}{2}(f_1 + f_2)$. The graph of f is shown in Fig. 5.13 for different values of ν. The weaker the diffusive effect (small ν) the sharper the gradient in f. The traveling wave solution to (22) is

$$u(x, t) = \frac{f_2 + f_1 e^{K(x - ct)}}{1 + e^{K(x - ct)}}$$

where the speed is determined from the definition of f_1 and f_2 to be

$$c = \frac{1}{2}(f_1 + f_2)$$

Graphically the traveling wave solution is the profile $f(s)$ in Fig. 5.13 moving to the right at speed c. The solution of Burgers' equation, because it resembles the actual profile of a shock wave, is called the *shock structure* solution; it joins the asymptotic states f_1 and f_2. Without the νu_{xx} term the solutions of (22) would shock up or tend to break. The presence of the diffusion term prevents this breaking effect by countering the nonlinearity. The result is competition and balance between the nonlinear term uu_x and the diffusion term $-\nu u_{xx}$ much the same as occurs in a real shock wave in the narrow region where the gradient is steep. In this shock wave context the $-\nu u_{xx}$ term could be interpreted as a viscosity term.

The Korteweg-deVries Equation

Up to now we have seen how the effects of convection, diffusion, and nonlinear distortion can be modeled by simple partial differential equations. There is another effect called *dispersion*, which plays an important role in many physical processes. Dispersive systems are characterized by partial differential equations that admit solutions of the form (20) where the frequency ω is a definite, *real* function of the wave number k, that is,

$$\omega = \omega(k)$$

This relation is called the *dispersion relation*. We have already obtained one such relation in (21) where

$$\omega = ck - i\nu k^2$$

In this example ω is a *complex* function of k. In general when $\omega(k)$ is complex we call the wave (20) *diffusive*.

Example 1.5 Consider the partial differential equation

$$u_{tt} + \gamma u_{xxxx} = 0, \qquad \gamma > 0 \tag{26}$$

which arises in studying the transverse vibrations of a beam. Substituting (20) into (26) gives the dispersion relation

$$\omega = \pm \sqrt{\gamma}\, k^2$$

Hence (26) admits dispersive wave solutions.

In the case of a dispersive wave the phase velocity is given by

$$\frac{\omega(k)}{k}$$

This is the velocity with which one must move to remain on the crest of a wavelet. For dispersive waves we impose the further condition that $\omega(k)/k$ is not constant but rather dependent on k. This means that waves of different wavelengths or wave numbers will propagate at different speeds; that is, they will disperse. The general definition is that (20) represents a *dispersive wave* if $\omega(k)$ is real and $\omega''(k) \neq 0$. The most familiar example of dispersive waves is ocean waves or water waves. Other common dispersive systems are vibrating beams and the propagation of electromagnetic waves in dielectrics.

Example 1.6 A perturbation analysis of the equations that govern long waves in shallow water leads to another fundamental equation of applied mathematics, the *Korteweg-deVries equation* (*KdV equation* for short)

$$u_t + uu_x + ku_{xxx} = 0, \qquad k > 0 \tag{27}$$

This equation is a model equation for a nonlinear dispersive process. The uu_x term causes a shocking up effect. In Burgers' equation this effect was balanced by a diffusion term $-\nu u_{xx}$ creating a shock structure solution. Now we replace this term by ku_{xxx}, which can be interpreted as a dispersion term. The KdV equation and slight modifications of it arise in many physical contexts (see C. S. Gardner *et al.*, *SIAM Review*, Vol. 18, p. 412, 1976). We follow the same reasoning as in the determination of a traveling wave solution of Burgers' equation. That is, we assume a solution of the form

$$u = f(s), \qquad s = x - ct \tag{28}$$

where the waveform f and the wave speed c are to be determined. Substituting (28) into (27) gives

$$-cf' + \frac{1}{2}\frac{d}{ds}f^2 + kf''' = 0$$

which when integrated gives

$$-cf + \tfrac{1}{2}f^2 + kf'' = A, \qquad A \text{ constant}$$

Multiplying by f' and integrating again yields

$$-\frac{c}{2}f^2 + \frac{1}{6}f^3 + \frac{1}{2}kf'^2 = Af + B, \qquad B \text{ constant}$$

Hence one can write

$$\frac{df}{ds} = \pm\sqrt{\frac{1}{3k}}\,\phi(f)^{1/2}, \qquad \phi(f) = -f^3 + 3cf^2 + 6Af + 6B \tag{29}$$

Since $\phi(f)$ is a cubic polynomial we must consider five possibilities.

  (i) ϕ has one real root α.
  (ii) ϕ has three distinct real roots $\gamma < \beta < \alpha$.
  (iii) ϕ has three real roots satisfying $\gamma = \beta < \alpha$.
  (iv) ϕ has three real roots satisfying $\gamma < \alpha = \beta$.
  (v) ϕ has a triple root γ.

Clearly, if α is a real root of ϕ, then $f = \alpha$ is a constant solution. We seek real, nonconstant, bounded solutions that will exist only when $\phi(f) \geq 0$. Upon examining the direction field it is easy to see that only unbounded solutions of (29) exist for Cases (i) and (iv). We leave Case (v) and Case (ii) to the reader. Case (ii) leads to cnoidal waves.

We examine Case (iii) in which a class of solutions known as *solitons* arise. In this case (29) becomes

$$\frac{ds}{\sqrt{3k}} = \frac{df}{(f - \gamma)\sqrt{\alpha - f}} \tag{30}$$

Letting $f = \gamma + (\alpha - \gamma)\operatorname{sech}^2 w$ and noting $df = -2(\alpha - \gamma)\operatorname{sech}^2 w \tanh w \, dw$, it follows that (30) can be written

$$\frac{ds}{\sqrt{3k}} = \frac{-2(\alpha - \gamma)\operatorname{sech}^2 w \tanh w \, dw}{(\alpha - \gamma)\operatorname{sech}^2 w\sqrt{(\alpha - \gamma) - (\alpha - \gamma)\operatorname{sech}^2 w}} = \frac{-2 \, dw}{\sqrt{\alpha - \gamma}}$$

Integrating gives

$$w = -\sqrt{\frac{\alpha - \gamma}{12k}}\, s$$

and therefore a solution to (30) is given by

$$f(s) = \gamma + (\alpha - \gamma)\operatorname{sech}^2\left(\sqrt{\frac{\alpha - \gamma}{12k}}\, s\right) \tag{31}$$

Clearly $f(s) \to \gamma$ as $s \to \pm\infty$. A graph of the waveform f is shown in Fig. 5.14. It is instructive to write the roots α and γ in terms of the original

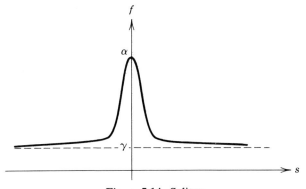

Figure 5.14. Soliton.

parameters. To this end

$$\begin{aligned}\phi(f) &= -f^3 + 3cf^2 + 6Af + 6B\\ &= (f - \gamma)^2(\alpha - f)\\ &= -f^3 + (\alpha + 2\gamma)f^2 + (-2\alpha\gamma - \gamma^2)f + \gamma^2\alpha\end{aligned}$$

Thus the wave speed c is given by

$$c = \frac{\alpha + 2\gamma}{3}$$

and we can write the solution as

$$u(x, t) = \gamma + a\operatorname{sech}^2\left(\sqrt{\frac{a}{12k}}\left(x - \left[\gamma + \frac{a}{3}\right]t\right)\right), \qquad a \equiv \alpha - \gamma$$

We notice several features of this traveling wave solution. The velocity relative to the asymptotic state at $\pm\infty$ is proportional to the amplitude a. The width of the wave, defined by $\sqrt{12k/a}$, increases as k increases; that is, the wave disperses. Finally, the amplitude a is independent of the asymptotic state at $\pm\infty$. Such a waveform is known as a soliton and many of the important equations of mathematical physics have solitonlike solutions (e.g., the Boussinesq equation, the Sine-Gordon equation, the Born-Infeld equation, and nonlinear Schrödinger-type equations). In applications the value of such solutions is that if a pulse or signal travels as a soliton, then the information contained in the pulse can be carried over long distances with no distortion or loss of intensity. Solitons, or solitary waves, have been observed in various canals and waterways.

Nonlinear wave phenomena are discussed in detail in Whitham [1].

Conservation Laws

Consider the first order partial differential equation

$$u_t + F(u)_x = 0, \quad x \in R^1, \quad t > 0 \tag{32}$$

where F is a given continuously differentiable function. An equation of this form is called a *conservation law* for the following reasons. If we integrate (32) from $x = a$ to $x = b$ we obtain

$$\frac{d}{dt}\int_a^b u(x, t)\, dx + \int_a^b F(u)_x\, dx = 0 \tag{33}$$

By the fundamental theorem of calculus

$$\int_a^b F(u)_x\, dx = F(u(b, t)) - F(u(a, t))$$

and so (33) may be written

$$\frac{d}{dt} \int_a^b u(x, t)\, dx = F(u(a, t)) - F(u(b, t)) \tag{34}$$

If u is the amount of a quantity per unit length, then the left side of (34) is the time rate of change of the total amount of the quantity inside the interval $[a, b]$. If $F(u(x, t))$ is the flux through x, that is, the amount of the quantity per unit time positively flowing across x, then (34) states that the rate of change of the quantity in $[a, b]$ equals the flux in at $x = a$ minus the flux out through $x = b$. Thus (34) is a statement of conservation of u. If u is sufficiently smooth, for example of class C^1, then the steps from (32) to (34) may be reversed and (32) may be regarded as a conservation law as well since it is equivalent to (34). The important fact to notice, however, is that (34) holds even if u is not smooth. We call (34) the *integral form* of the conservation law and (32) the *differential form*. A smooth solution $u(x, t)$ of (32) is called a *genuine solution*.

Example 1.7 The model nonlinear equation

$$u_t + uu_x = 0$$

can be written in conservation form as

$$u_t + \left(\tfrac{1}{2}u^2\right)_x = 0$$

where the flux is given by $F(u) = \tfrac{1}{2}u^2$.

In general we can write (32) in the form

$$u_t + c(u)u_x = 0 \tag{35}$$

where

$$c(u) = F'(u)$$

We noted earlier that the initial value problem associated with (32) does not always have a solution for all $t > 0$. We showed that u is constant on the characteristic curves given by $dx/dt = c(u)$ and the initial condition

$$u(x, 0) = f(x) \tag{36}$$

is propagated along straight lines with speed $c(u)$. If $c'(u) > 0$ and $f'(x) < 0$, then the characteristics issuing from two points ξ_1 and ξ_2 on the x axis with $\xi_1 < \xi_2$ have speeds $c(f(\xi_1))$ and $c(f(\xi_2))$, respectively. It follows that $c(f(\xi_1)) > c(f(\xi_2))$ and therefore the characteristics must intersect (see Fig. 5.15). Since u is constant on characteristics the solution is meaningless at the

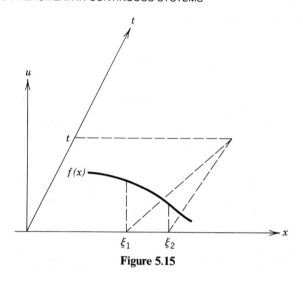

Figure 5.15

value of t where the intersection occurs. The first such t, denoted earlier by t_b, is termed the breaking time. A smooth solution exists only up to time t_b; beyond t_b something else must be tried. The key to continuing the solution for $t > t_b$ is to take a hint from the propagation of stress waves in a real physical continuum. There, under certain conditions, a smooth wave profile evolves into a shock wave. Hence it is reasonable to expect that a discontinuous solution may exist and propagate for $t > t_b$. This discontinuous solution cannot satisfy the differential form (35) of the conservation law, but the integral form (34) remains valid. Therefore let us turn to a study of the simplest kinds of such solutions in an effort to obtain a condition that must hold at the discontinuity.

Let F be continuously differentiable everywhere and let $x = s(t)$ be a smooth curve across which u is discontinuous. Suppose u is smooth on each side of the curve (see Fig. 5.16). Choose a and b such that the curve $x = s(t)$ intersects $a \leq x \leq b$ at time t and let u_0 and u_1 denote the right and left

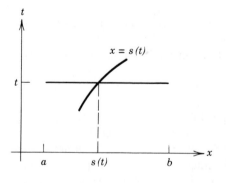

Figure 5.16

limits of u at $s(t)$, respectively. That is

$$u_0 = \lim_{x \to s(t)^+} u(x, t), \qquad u_1 = \lim_{x \to s(t)^-} u(x, t)$$

From the conservation law (34) and Leibniz's rule for differentiating integrals

$$F(u(a, t)) - F(u(b, t))$$

$$= \frac{d}{dt} \int_a^{s(t)} u(x, t) \, dt + \frac{d}{dt} \int_{s(t)}^b u(x, t) \, dt$$

$$= \int_a^{s(t)} u_t(x, t) \, dt + u_1 \frac{ds}{dt} + \int_{s(t)}^b u_t(x, t) \, dx - u_0 \frac{ds}{dt}$$

Since u_t is bounded in each of the intervals $[a, s(t)]$ and $[s(t), b]$ separately, both integrals tend to zero in the limit as $a \to s(t)^-$ and $b \to s(t)^+$. Therefore

$$F(u_1) - F(u_0) = (u_1 - u_0) \frac{ds}{dt} \tag{37}$$

Equation (37) is a *jump condition* relating the values of u and the flux in front of and behind the discontinuity to the speed ds/dt of the discontinuity. Thus the integral form of the conservation law provides a restriction on possible jumps across a simple discontinuity. A conventional notation is to rewrite (37) as

$$[F(u)] = [u] \frac{ds}{dt}$$

where $[\cdot]$ indicates the jump in a quantity across the discontinuity.

Example 1.8 Let

$$u_t + \left(\tfrac{1}{2} u^2 \right)_x = 0, \qquad x \in R^1, \quad t > 0$$

$$u(x, 0) = f(x) = \begin{cases} 1, & \text{for } x \le 0 \\ 1 - x, & \text{for } 0 < x < 1 \\ 0, & \text{for } x \ge 1 \end{cases}$$

The characteristic diagram is shown in Fig. 5.17. The lines emanating from the x axis have speed $c(f(x)) = f(x)$. It is clear from the geometry that $t_b = 1$ and that a single valued solution exists for $t < 1$. For $t > 1$ we fit a shock or discontinuity beginning at $(1, 1)$, separating the state $u_1 = 1$ on the left from the state $u_0 = 0$ on the right. Thus $[u] = 1 - 0 = 1$ and $[F] = \tfrac{1}{2} u_1^2 - \tfrac{1}{2} u_0^2 = \tfrac{1}{2}$. Therefore the discontinuity must have speed $s'(t) = [F]/[u] = \tfrac{1}{2}$. The resulting characteristic diagram and solution are shown in Fig. 5.18. We have obtained a solution for all $t > 0$ satisfying the conservation law.

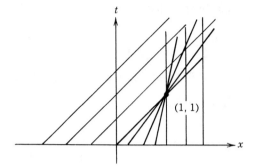

Figure 5.17. Characteristic diagram for Example 1.7.

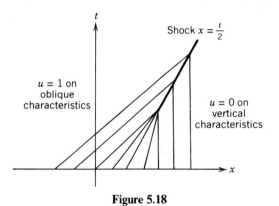

Figure 5.18

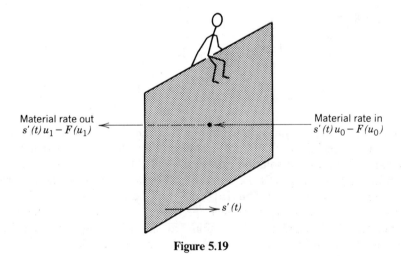

Figure 5.19

Another popular method for deriving the jump condition (37) is to imagine an observer riding on the discontinuity (see Fig. 5.19). At each instant the amount of material the observer sees entering must equal the amount leaving. The amount entering is $s'(t)u_0$ (because of the observer's motion) plus the flux of material $-F(u_0)$ (the negative sign occurs because flux is positive if material motion is to the right). Similarly, the amount leaving is $s'(t)u_1 - F(u_0)$. Hence

$$s'(t)u_0 - F(u_0) = s'(t)u_1 - F(u_1)$$

from which (37) follows.

Quasi-Linear Equations

Up to now we have examined special first order partial differential equations subject to initial data. In the next few paragraphs we consider the general quasi-linear wave equation

$$a(x, t, u)u_x + b(x, t, u)u_t = c(x, t, u) \tag{38}$$

and discuss the initial value problem as well as the concept of a general solution. We assume that a, b, $c \in C^1(D)$, where D is some open connected set in R^3 and $a^2 + b^2 + c^2 \neq 0$. First we focus on the *initial value problem*. For simplicity let us consider the problem

$$a(x, t, u)u_x + u_t = c(x, t, u), \qquad t > 0, \quad x \in R^1 \tag{39}$$

$$u(x, 0) = f(x), \qquad x \in R^1 \tag{40}$$

As in earlier examples we search for curves that can play the role of characteristics. In this case, along the family of curves defined by

$$\frac{dx}{dt} = a(x, t, u) \tag{41}$$

the partial differential equation reduces to an ordinary differential equation

$$\frac{du}{dt} = c(x, t, u) \tag{42}$$

In contrast to Example 1.2 u is not constant on the characteristics defined by (41). The initial data (40) can be written

$$x = \xi, \qquad u = f(\xi), \qquad \text{at} \quad t = 0 \tag{43}$$

Equations (41) and (42) represent a nonautonomous system of two differential equations for x and u with initial values given in (43). The general solution of (41) and (42) involves two arbitrary constants and is of the form

$$x = F(t; c_1, c_2), \qquad u = G(t; c_1, c_2)$$

The constants can be evaluated from the initial data (43) to obtain $c_1 = c_1(\xi)$ and $c_2 = c_2(\xi)$. Then the solution of (39) and (40) is given implicitly by

$$x = F(t; c_1(\xi), c_2(\xi)) \tag{44a}$$

$$u = G(t; c_1(\xi), c_2(\xi)) \tag{44b}$$

In principle (44a) can be solved for $\xi = \xi(x, t)$, and then ξ can be substituted into (44b) to obtain an explicit representation for u in terms of x and t.

Example 1.9   Consider the initial value problem

$$u_t + uu_x + u = 0, \qquad x \in R^1, \quad t > 0$$

$$u(x, 0) = -\frac{x}{2}, \qquad x \in R^1 \tag{45}$$

This problem is equivalent to

$$\frac{du}{dt} = -u, \qquad \frac{dx}{dt} = u \tag{46}$$

with

$$x = \xi, \qquad u = -\frac{\xi}{2}, \qquad \text{at} \quad t = 0$$

The general solution of (46) is

$$x = -c_1 \exp(-t) + c_2, \qquad u = c_1 \exp(-t)$$

Applying the initial data gives

$$x = \frac{\xi}{2} + \frac{\xi}{2}\exp(-t), \qquad u = -\frac{\xi}{2}\exp(-t)$$

Hence $\xi = 2x/(1 + \exp(-t))$ and

$$u(x, t) = -\frac{x \exp(-t)}{1 + \exp(-t)}$$

is a solution to (45) for all $t > 0$, $x \in R^1$.

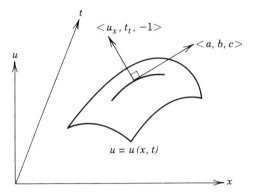

Figure 5.20. Integral surface.

In some contexts the general solution of the quasi-linear equation (38) is required. Intuitively we suspect the expression for the general solution should contain one arbitrary function. To obtain the general solution we proceed geometrically by noting that the partial differential equation (38) can be written as the scalar product of two vectors

$$\langle a, b, c \rangle \cdot \langle u_x, u_t, -1 \rangle = 0$$

If $u = u(x, t)$ is an *integral surface*, or *solution* of (38), then the normal vector $\langle u_x, u_t, -1 \rangle$ to the surface at a given point is perpendicular to the vector $\langle a, b, c \rangle$ at that point. It follows that the vector $\langle a, b, c \rangle$ must be tangent to the surface and therefore integral surfaces must be formed from the integral curves of the vector field $\langle a, b, c \rangle$. The *integral curves* are given as solutions to the system of ordinary differential equations

$$\frac{dx}{d\xi} = a(x, t, u), \qquad \frac{dt}{d\xi} = b(x, t, u), \qquad \frac{du}{d\xi} = c(x, t, u) \qquad (47)$$

where ξ is a parameter along the curves (see Fig. 5.20). The parameter ξ in (47) can be suppressed and (47) can be written as

$$\frac{dx}{a(x, t, u)} = \frac{dt}{b(x, t, u)} = \frac{du}{c(x, t, u)} \qquad (48)$$

The system (47) or (48) is called the *characteristic system* associated with (38).

A relation of the form $\phi(x, t, u) = $ constant is called a *first integral* of the system (48) if $\phi(x, t, u)$ has a constant value along any integral curve of (48). We show in the sense of the following lemma that the general solution of (48) consists of any two independent first integrals.

Lemma 1.1   *Let ϕ, $\psi \in C^1(D)$ and suppose that*

$$\phi(x, t, u) = c_1, \qquad \psi(x, t, u) = c_2 \tag{49}$$

are two first integrals of (48) satisfying the independence condition

$$\text{grad } \phi \times \text{grad } \psi \neq 0 \quad \text{in} \quad D$$

If C is any integral curve of (48), then there exist values of the constants c_1 and c_2 such that the intersection of the two surfaces (49) is a curve that coincides with C.

Proof   Let P_0 be a point on C and take $c_1 = \phi(P_0)$ and $c_2 = \psi(P_0)$. With these values of c_1 and c_2, let C' be the intersection of the two surfaces in (49). Since $\langle a, b, c \rangle$ is tangent to it at each of its points, C' is an integral curve of (48). Therefore both C and C' are integral curves that contain a common point P_0, and therefore by the uniqueness theorem for solutions of ordinary differential equations the curves C and C' must coincide. (The independence condition guarantees that $\text{grad } \phi$ and $\text{grad } \psi$ are not parallel, and thus the surfaces defined in (49) intersect in a curve.)

It follows from Lemma 1.1 that the integral curves of (48) form a two parameter family of curves that are intersections of $\phi = c_1$, $\psi = c_2$, where ϕ and ψ are independent first integrals of (48). Next we observe that any first integral of (48) generates a solution of the given partial differential equation (38).

Lemma 1.2   *If $\phi(x, t, u) \in C^1(D)$ is a first integral of (48), then*

$$a\phi_x + b\phi_t + c\phi_u = 0 \tag{50}$$

Moreover, if $\phi_u \neq 0$, then the function $u = u(x, t)$ found from $\phi(x, t, u) = 0$ is a solution of (38).

Proof   Along an integral curve $(x(\xi), t(\xi), u(\xi))$ we have

$$\phi(x(\xi), t(\xi), u(\xi)) = k$$

where k is a constant. Differentiating with respect to ξ gives

$$\frac{d\phi}{d\xi} = a\phi_x + b\phi_t + c\phi_u = 0$$

which is (50). Since $\phi_u \neq 0$ the equation $\phi(x, t, u) = 0$ may be solved locally for u to obtain $u = u(x, t)$. By the chain rule

$$\phi_x + \phi_u u_x = 0, \qquad \phi_t + \phi_u u_t = 0$$

and so (50) shows that u is a solution of (38).

We now state the main result that shows that there is a connection between independent first integrals of (48) and the general solution of (38). That there should be such a connection is geometrically clear. Any two one parameter families of independent first integrals of (48) intersect in a two parameter family of curves that coincide with the integral curves of (48). Any integral surface must be composed of integral curves; that is, it must pass through a family of these integral curves.

Theorem 1.1 *Let* $\phi(x, t, u)$, $\psi(t, x, u) \in C^1(D)$ *be two independent first integrals of the characteristic system* (48), *with* $\phi_u^2 + \psi_u^2 \neq 0$ *in* D. *Then the general solution of* (38) *is given by*

$$H(\phi(x, t, u), \psi(x, t, u)) = 0 \tag{51}$$

where H is an arbitrary function.

Proof By Lemma 1.2

$$a\phi_x + b\phi_t + c\phi_u = 0$$

and

$$a\psi_x + b\psi_t + c\psi_u = 0$$

Now let $f(x, t, u) = 0$ be any integral surface of (38). Then

$$af_x + bf_t + cf_u = 0$$

Since $a^2 + b^2 + c^2 \neq 0$, it follows that

$$\det \begin{bmatrix} \phi_x & \phi_t & \phi_u \\ \psi_x & \psi_t & \psi_u \\ f_x & f_t & f_u \end{bmatrix} = 0$$

Therefore ϕ, ψ, and f are not independent functions and $f = H(\phi, \psi)$ for some function H.

This theorem shows that the integral surface (51) is generated by a one parameter subfamily of integral curves obtained by restricting the values of c_1 and c_2 by the relation

$$H(c_1, c_2) = 0 \tag{52}$$

where $\phi = c_1$ and $\psi = c_2$. With this information it is easy to see in principle how to solve the initial value problem

$$a(x, t, u)u_x + b(x, t, u)u_t = c(x, t, u), \qquad t > 0, \quad x \in R^1 \qquad (53)$$
$$u(x,0) = f(x), \qquad\qquad x \in R^1$$

The problem is equivalent to solving

$$\frac{dx}{a} = \frac{dt}{b} = \frac{du}{c}$$

subject to the initial data

$$x = \xi, \qquad t = 0, \qquad u = f(\xi), \qquad \xi \in R^1$$

If $\phi = c_1$ and $\psi = c_2$ are independent first integrals, then the parameters c_1 and c_2 are given by

$$c_1 = \phi(\xi, 0, f(\xi)), \qquad c_2 = \psi(\xi, 0, f(\xi))$$

If ξ is eliminated from this pair of equations, we obtain a relation of the form (52), which determines H. The solution of the initial value problem is then given implicitly by (51).

Example 1.10   Consider the initial value problem

$$(t + 2xu)u_x - (x + 2tu)u_t = \tfrac{1}{2}(x^2 - t^2), \qquad x \in R^1, \quad t > 0$$
$$u(x,0) = x, \qquad\qquad x \in R^1$$

The characteristic system is

$$\frac{dx}{t + 2xu} = \frac{dt}{-(x + 2tu)} = \frac{du}{\tfrac{1}{2}(x^2 - t^2)}$$

which has first integrals

$$\phi \equiv x^2 + t^2 - 4u^2 = c_1, \qquad \psi \equiv xt + 2u = c_2$$

The initial data can be written parametrically as

$$x = \xi, \qquad t = 0, \qquad u = \xi, \qquad \xi \in R^1$$

Then

$$c_1 = -3\xi^2, \qquad c_2 = 2\xi$$

giving

$$H(c_1, c_2) = \tfrac{3}{4}c_2^2 + c_1$$

Thus the solution is given implicitly by

$$\tfrac{3}{4}(xt + 2u)^2 + x^2 + t^2 - 4u^2 = 0$$

Solving for u gives

$$u = \frac{3xt}{2} + \sqrt{3x^2t^2 + x^2 + t^2}$$

The solution is uniquely determined for all $t > 0$, $x \in R^1$.

We have not proved that there always exists a solution to the initial value problem (53). It seems clear that the steps outlined in the discussion following (53) will not always have validity or will not be able to be carried out. Without proof we state the following existence–uniqueness theorem.

Theorem 1.2 *Consider the initial value problem* (53) *where a, b, and c are continuously differentiable in R^3 and f is continuously differentiable on R^1. If $b(x_0, 0, f(x_0)) \neq 0$, then there exists a unique solution $u = u(x, t)$ of* (53) *in a neighborhood of* $(x_0, 0)$.

From this theorem it follows immediately that the initial value problem (39)–(40) has a unique solution in a neighborhood of the entire x axis provided a, c, and f are class C^1. Geometrically it is easy to understand the condition $b(x_0, 0, f(x_0)) \neq 0$. It guarantees that the vector field $\langle a, b, c \rangle$ will have a component in the t direction at P_0: $(x_0, 0, f(x_0))$. Hence the integral surface through P_0, which must contain the integral curve, is defined for $t > 0$. Otherwise, if $b(x_0, 0, f(x_0)) = 0$, the integral curve through P_0 would not

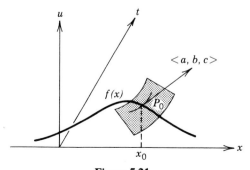

Figure 5.21

need to be defined for $t > 0$, since its tangent at P_0 would have a vanishing t component (see Fig. 5.21). We have already observed that even though a solution may exist locally for a short period of time, it need not exist for all $t > 0$.

for 11-19   1.34, 6, 9, 12

EXERCISES

1.1 Verify that the equation $u_t + c(x, t)u_x = 0$ has a solution $u(x, t) = f(a(x, t))$ for an arbitrary differentiable function f, where $a(x, t) = $ *constant* are the integral curves of $dx/dt = c(x, t)$.

1.2 Solve the following initial value problems on $t > 0$, $x \in R^1$. Sketch the characteristic curves in each case.

(a) $u_t + 3u_x = 0$, $u(x, 0) = \sin x$

(b) $u_t + xu_x = 0$, $u(x, 0) = \exp(-x)$

(c) $u_t - x^2 t u_x = 0$, $u(x, 0) = x + 1$

1.3 Find the solution and sketch a characteristic diagram

$$u_t + cu_x = 0, \quad x \in R^1, \quad t > 0, \quad c > 0$$

$$u(x, 0) = \begin{cases} 1 - x^2, & |x| \le 1 \\ 0, & |x| > 1 \end{cases}$$

1.4 Consider the initial value problem

$$u_t + uu_x = 0, \quad x \in R^1, \quad t > 0$$

$$u(x, 0) = \begin{cases} 1 - x^2, & |x| \le 1 \\ 0, & |x| > 1 \end{cases}$$

Sketch the characteristic diagram. At what time t_b does the wave break? Find a formula for the solution.

1.5 Consider the initial value problem

$$u_t + uu_x = 0, \quad x \in R^1, \quad t > 0$$

$$u(x, 0) = \exp(-x^2), \quad x \in R^1$$

Sketch the characteristic diagram and find the point (x_b, t_b) in space–time where the wave breaks.

1.6   Consider the *signaling problem*

$$u_t + c(u)u_x = 0, \qquad t > 0, \quad x > 0$$
$$u(x,0) = u_0, \qquad x > 0$$
$$u(0,t) = g(t), \qquad t > 0$$

where c and g are given functions and u_0 is a positive constant. If $c'(u) > 0$, under what conditions on the signal g will no shocks form? Determine the solution in this case in the domain $x > 0$, $t > 0$.

1.7   Consider

$$u_t + uu_x + u = 0, \qquad x \in R, \quad t > 0$$
$$u(x,0) = k \exp(-x^2), \qquad x \in R, \quad k > 0$$

For what values of k does breaking occur? If $k = 2$, prove that the shock cannot progress beyond a certain fixed line $x = $ constant.

1.8   Consider the *damped* nonlinear equation

$$u_t + uu_x + au = 0, \qquad x \in R^1, \quad t > 0$$

where $a > 0$ and u is subject to the initial condition

$$u(x,0) = f(x), \qquad x \in R^1$$

Show that this problem is equivalent to the system

$$\frac{dx}{dt} = u, \qquad \frac{du}{dt} = -au$$

with $x = \xi$, $u = f(\xi)$ holding at $t = 0$. Show that the solution is given implicitly by

$$u = f(\xi)e^{-at}, \qquad x = \frac{f(\xi)}{a}(1 - e^{-at}) + \xi$$

Determine a condition on f that will guarantee that no shocks will form. If a shock does form, show that the breaking time t_b is greater than that for the undamped equation when $a = 0$.

1.9   For the initial value problem

$$u_t + c(u)u_x = 0, \qquad x \in R^1, \quad t > 0$$
$$u(x,0) = f(x), \qquad x \in R^1$$

show that if the functions $c(u)$ and $f(x)$ are both nonincreasing or both nondecreasing, then no shocks develop for $t \geq 0$.

1.10 Consider the problem

$$u_t + u^2 u_x = 0, \qquad x \in R^1, \quad t > 0$$
$$u(x,0) = x, \qquad x \in R^1$$

Derive the solution

$$u(x, t) = \begin{cases} x, & \text{if } t = 0 \\ \dfrac{\sqrt{1 + 4xt} - 1}{2t}, & \text{if } t \neq 0 \quad \text{and} \quad 1 + 4tx > 0 \end{cases}$$

When do shocks develop? Verify that $\lim_{t \to 0^+} u(x, t) = x$.

1.11 In the traveling wave solution obtained to Burgers' equation in Example 1.4 the shock thickness is defined by $(f_2 - f_1)/\max|f'(s)|$. Give a geometric interpretation and show that the shock thickness is given by $8\nu/(f_2 - f_1)$.

▷ **1.12** Show that Burgers' equation $u_t + u u_x = \nu u_{xx}$ can be reduced to the diffusion equation via the transformation $u = -2\nu \phi_x/\phi$ (this is the Cole-Hopf transformation). If the initial condition $u(x,0) = f(x)$ is imposed, derive the solution

$$u(x, t) = \int_{-\infty}^{\infty} \frac{x - \xi}{t} \exp\left(\frac{-G}{2\nu}\right) d\xi \bigg/ \int_{-\infty}^{\infty} \exp\left(\frac{-G}{2\nu}\right) d\xi$$

where

$$G = G(\xi, x, t) = \int_0^\xi f(\eta) \, d\eta + \frac{(x - \xi)^2}{2t}$$

1.13 Consider the initial value problem

$$u_t + u u_x + u = 0, \qquad x \in R^1, \quad t > 0$$
$$u(x,0) = -2x, \qquad x \in R^1$$

Find the time t_b that a gradient catastrophe occurs and beyond which there is no continuously differentiable solution.

1.14 Consider the initial value problem

$$u_t + uu_x + \alpha u^2 = 0, \qquad x \in R^1, \quad t > 0, \quad \alpha > 0$$
$$u(x,0) = 1, \qquad x \in R^1$$

Derive the solution $u(x, t) = (1 + \alpha t)^{-1}$

1.15 Consider the equation $u_t + c(u)u_x = 0$, $x \in R^1$, $t > 0$, with initial data $u(x,0) = u_1$ for $x < 0$ and $u(x,0) = u_0$ for $x \geq 0$, with $c'(u) > 0$.

(a) Find a nonsmooth solution in the case $u_0 > u_1$

(b) Find a discontinuous solution in the case $u_0 < u_1$

(c) Discuss (a) and (b) in the context of shock waves and rarefaction waves.

1.16 Consider the problem

$$2u_t + u_x = 0, \qquad x > 0, \quad t > 0$$
$$u(0, t) = 0, \qquad t > 0$$
$$u(x,0) = 1, \qquad x \geq 0$$

Let $H(\tau)$ be the unit step function defined by $H(\tau) = 0$ for $\tau < 0$ and $H(\tau) = 1$ for $\tau \geq 0$. In what sense are $u_1(x, t) = 1 - H(t - 2x)$ and $u_2(x, t) = 1 - H(t - x)$ both solutions to the problem? Which one is the appropriate solution?

1.17 Find the general solution of the following quasilinear equations.

(a) $(t + u)u_x + tu_{x_t} = x - t$ (e) $uu_x + tu_t = x$

(b) $x^2 u_x + t^2 u_t = 2xt$ (f) $uu_x - uu_t = t - x$

(c) $uu_t = -t$ (g) $2xuu_x + 2tuu_t = u^2 - x^2 - t^2$

(d) $xu_x + tu_t = xt(u^2 + 1)$

1.18 In Example 1.6 verify that bounded solutions of (29) exist only on Cases (ii) and (v).

1.19 For the following equations find the dispersion relation and phase velocity.

(a) $u_t + cu_x + ku_{xxx} = 0$

(b) $u_{tt} - c^2 u_{xx} = 0$

5.2 MATHEMATICAL MODELS OF CONTINUA

Continuum mechanics (fluid dynamics, elasticity, gasdynamics, etc.) is the study of motion, kinematics and dynamics, of continuous systems of particles such as fluids, solids, and gases. Out of this subject evolve some of the most important partial differential equations of mathematical physics and some of the most important techniques for solving applied problems. For example, the origins of singular perturbation theory lie in the study of the flow of air around a wing or airfoil, and many of the problems of bifurcation theory have their beginnings in fluid mechanics. More than any other area of engineering or physics continuum mechanics remains a paradigm for the development of techniques and examples in applied mathematics.

The field equations, or governing equations, of continuum mechanics are a set nonlinear partial differential equations for unknown quantities such as density, pressure, displacement, particle velocity, and so on, which describe the state of the continuum at any instant. They arise as in the case of the heat conduction equation from conservation laws such as conservation of mass, momentum, and energy, and assumptions regarding the makeup of the continuum called constitutive relations or equations of state. Under certain cases, say when the amplitude of waves is small, the equations reduce to some of the simpler equations of applied mathematics, like the one dimensional wave equation.

The equations are developed from the concept that the material under investigation is a continuum, that is, it exhibits no structure however finely it is divided. This development gives rise to a model in which the fluid parameters are defined at all points as continuous functions of space and time. The molecular aggregation of matter is completely extraneous to the continuum model. The model may be expected to be invalid when the size of the continuum region of interest is the same order as the characteristic dimension of the molecular structure. For gases this dimension is on the order of 10^{-7} meters, which is the mean free path, and for liquids it is on the order of 10^{-10} meters, which is a few intermolecular spacings. Thus the continuum model is violated only in extreme cases.

Kinematics

The underlying physical model to which we apply the general ideas is a one dimensional continuum that can be thought of as a cylindrical pipe through which a fluid is flowing. The fluid may be a liquid or gas and the lateral wall of the pipe is assumed to have no effect on the flow parameters. Throughout the motion we assume that each cross section remains planar and moves longitudinally down the cylinder and that there is no variation of any of the flow parameters in any cross section. It is this assumption, that the only variation is longitudinal, that gives the description a one dimensional character (see Fig. 5.22). The situation is similar to the one dimensional heat conduction model studied in Chapter 4.

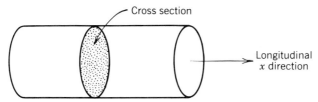

Figure 5.22. One dimensional continuum.

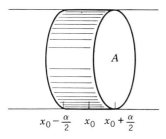

Figure 5.23. Cross section of a fluid disk D centered at x_0.

To understand how physical quantities are defined in the one dimensional continuum model we consider the following experiment to observe how the density of a fluid is related to its molecular structure. At time t we consider a disk D of fluid of width α centered at x_0 and having cross-sectional area A (see Fig. 5.23). The average density of the fluid in D is $\rho_\alpha = M_\alpha/\alpha A$, where M_α is the mass of fluid in D. To define the density $\rho(x_0, t)$ we examine what happens as α approaches zero. The graph in Fig. 5.24 records the results of the experiment. In Region II, since there are many particles (molecules) inside D, one would expect that the average density ρ_α would vary very little. If, however, α were on the order of molecular distances, say $\alpha = 10^{-9}$ meters,

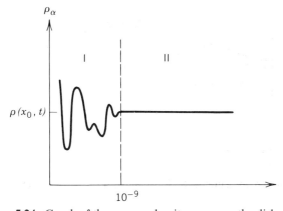

Figure 5.24. Graph of the average density ρ_α versus the disk width α.

there may be only a few molecules in D and we might expect large fluctuations in ρ_α even for small changes in α. Such rapid fluctuations are depicted in Region I in Fig. 5.24. It seems unreasonable therefore to define $\rho(x_0, t)$ as the limiting value of ρ_α as $\alpha \to 0$. Rather $\rho(x_0, t)$ should be defined as

$$\rho(x_0, t) = \lim_{\alpha \downarrow \alpha^*} \rho_\alpha$$

where α^* is the value of α where density nonuniformities begin to occur; here, for example, $\alpha^* = 10^{-9}$ meters. In a similar fashion other physical quantities can be considered as point functions in continuously distributed matter without regard to its molecular or atomistic structure. This continuum assumption is often phrased as follows—the fluid is composed of small regions or *fluid elements* that can be idealized as points at which the flow variables become continuous functions of position and time, but they are not so small that discernable fluctuations of these quantities exist within the element.

We now discuss the kinematics of fluid motion. In a one dimensional moving fluid there are two basic coordinate systems used to keep track of the motion. For example, suppose water is flowing in a stream and one wishes to measure the temperature. One could stand on the bank at a fixed location x measured from some reference position $x = 0$ and insert a thermometer, thereby measuring the temperature $\theta(x, t)$ as a function of time and the fixed position x (see Fig. 5.25). The coordinate x is called an *Eulerian* or *laboratory* coordinate. On the other hand, one could measure the temperature from a boat riding with the flow. In this case at time $t = 0$ each particle is labeled

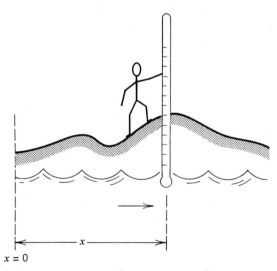

Figure 5.25. Eulerian measurement: The temperature $\theta(x, t)$ is measured at a fixed location x.

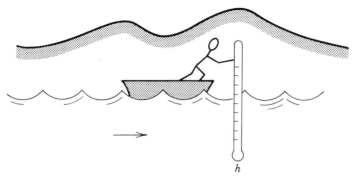

Figure 5.26. Lagrangian measurement: The temperature $\Theta(h, t)$ is measured with respect to a moving particle labeled h.

with a particle label h and each particle always retains its label as it moves downstream (see Fig. 5.26). The result of the measurement would be the temperature $\Theta(h, t)$ as a function of time t and the *Lagrangian* or *material coordinate h*. The variable x is a fixed spatial coordinate and the representation of the field functions or physical variables (temperature, density, pressure, etc.) as functions of t and x gives an *Eulerian* description of the flow. A representation in terms of t and the material variable h gives a *Lagrangian* description of the flow. For the physical variables we shall reserve lowercase letters for Eulerian quantities and capital letters for Lagrangian quantities. Thus $f(x, t)$ denotes the measurement of a physical quantity in Eulerian coordinates and $F(h, t)$ denotes the corresponding description in terms of Lagrangian coordinates.

Now let I be an interval representing the one dimensional region along the axis of the cylinder where the fluid is in motion. At time $t = 0$ we label all of the particles (in a one dimensional continuum the word *particle* is synonomous with *cross section* and we shall use them interchangeably) with a Lagrangian coordinate h such that

$$x = h \quad \text{at} \quad t = 0$$

where x is a fixed Eulerian coordinate in I. By a *fluid motion* or *flow* we mean a twice continuously differentiable mapping $\phi: I \times [0, t_1] \rightarrow I$ defined by

$$x = \phi(h, t) \tag{1}$$

which for each $t \in [0, t_1]$ is invertible on I. Rather than write ϕ in (1) we use the symbol x and write (1) as

$$x = x(h, t) \tag{2}$$

The convention of using x to denote both a spatial coordinate and a function

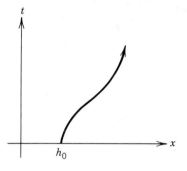

Figure 5.27. Eulerian particle path.

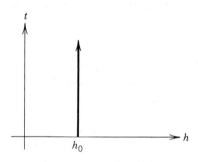

Figure 5.28. Lagrangian particle path.

is a common one and often preferred over (1) unless confusion arises. In words, Equation (2) gives the position x at time t of the particle or cross section labeled h. For a fixed $h = h_0$ the curve

$$x = x(h_0, t), \quad 0 \le t \le t_1$$

defines the *particle path* of the particle labeled h_0. Particle paths are often sketched on space–time diagrams as shown in Figs. 5.27 and 5.28.

The invertibility assumptions on ϕ guarantee that (2) can be solved for h to obtain

$$h = h(x, t) \tag{3}$$

This equation determines the particle or cross section at position x at time t. Since (2) and (3) are inverses,

$$x = x(h(x, t), t) \tag{4}$$

and

$$h = h(x(h, t), t) \tag{5}$$

Consequently, if $f(x, t)$ is a physical quantity represented in Eulerian form, then

$$F(h, t) = f(x(h, t), t) \tag{6}$$

gives the Lagrangian description. Conversely, if $F(h, t)$ is a Lagrangian quantity, then

$$f(x, t) = F(h(x, t), t) \tag{7}$$

gives its description in Eulerian form. The duality expressed by (6) and (7) is physically meaningful; for example, (7) states that the Eulerian measurement f made by an individual at x at time t coincides with the measurement F made by one moving on the particle h at the instant t when h is at x. Using the chain rule it is easy to show that the derivatives are related by

$$F_h(h, t) = f_x(x(h, t), t) x_h(h, t) \tag{8}$$

and

$$F_t(h, t) = f_t(x(h, t), t) + f_x(x(h, t), t) x_t(h, t) \tag{9}$$

where $f_x(x(h, t), t)$ means $f_x(x, t)$ evaluated at $x = x(h, t)$, for example. Good notation in fluid dynamics is essential. Here we have indicated explicitly the points at which the derivatives are evaluated. Shortcuts can sometimes lead to confusing and little understood expressions and equations, and writing out the general formulas in detail at least once can dismiss much of the confusion.

As in classical mechanics we define the *velocity* of a particle or cross section h as the time rate of change of its position or

$$V(h, t) = x_t(h, t) \quad = \quad \tfrac{\partial}{\partial t} \left(\text{position of particle } h \right)$$

Therefore the Eulerian form of the velocity is defined by

$$v(x, t) = V(h(x, t), t)$$

which gives the velocity of the cross section now at location x. The *acceleration* of the cross section labeled h is

$$A(h, t) = V_t(h, t) = x_{tt}(h, t)$$

Therefore the acceleration of the cross section now at location x is given in Eulerian form by

$$a(x, t) = V_t(h(x, t), t) = V_t(h, t)|_{h(x, t)} \tag{10}$$

From (9) it follows that

$$a(x, t) = v_t(x, t) + v(x, t) v_x(x, t)$$

The time rate of change of a physical quantity following a cross section or particle h that is now located at x, such as occurs on the right side of (10), is called the material derivative of that quantity. Precisely, the *material* or *convective derivative* of $f(x, t)$ is defined by

$$\frac{Df}{Dt}(x, t) \equiv F_t(h(x, t), t) = F_t(h, t)\big|_{h = h(x, t)} \tag{11}$$

where on the right we have explicitly noted that the derivative is taken before the evaluation at $h(x, t)$ is made. The quantity Df/Dt may be interpreted as the time derivative of F following the cross section h frozen at the instant h is located at x. From (9) it easily follows that

$$\frac{Df}{Dt} = f_t + vf_x \tag{12}$$

Therefore in Eulerian coordinates the acceleration can be written

$$a = \frac{Dv}{Dt}$$

One additional kinematical relation will play an important role in the derivation of the equations that govern the flow. It is an equation that gives a relation for the time rate of change of the Jacobian of the mapping ϕ defining the flow. The *Jacobian* is defined by

$$J(h, t) \equiv x_h(h, t)$$

Hence

$$J_t(h, t) = \frac{\partial}{\partial h} x_t(h, t) = V_h(h, t) = v_x(x(h, t), t) x_h(h, t)$$

or

$$J_t(h, t) = v_x(x(h, t), t) J(h, t) \tag{13}$$

Equation (13) is the *Euler expansion formula* in Lagrangian form. To obtain the Eulerian form we let $h = h(x, t)$ to obtain

$$\frac{Dj}{Dt}(x, t) = v_x(x, t) j(x, t)$$

where $j(x, t) = J(h(x, t), t)$.

Mass Conservation

We now derive the *field equations* that govern the motion of a one dimensional continuous medium. These equations express conservation of mass, momentum, and energy and are universal in that they are valid for any medium. The

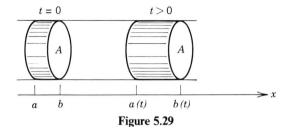

Figure 5.29

derivation of the field equations is based on a Lagrangian or material approach. First we want to find the analytic implications of the fact that the mass in an arbitrary material portion of the cylinder at $t = 0$ does not change as that portion of material moves in time. At time $t = 0$ we consider an arbitrary portion of fluid between $x = a$ and $x = b$ (see Fig. 5.29), and after time t we suppose that this portion of fluid has moved to the region between $x = a(t) \equiv x(a, t)$ and $x = b(t) \equiv x(b, t)$. If $\rho(x, t)$ denotes the Eulerian density of the fluid, then the amount of fluid between $a(t)$ and $b(t)$ is

$$\int_{a(t)}^{b(t)} \rho(x, t) A \, dx$$

The assumption of mass conservation then requires that

$$\frac{d}{dt} \int_{a(t)}^{b(t)} \rho(x, t) \, dx = 0 \tag{14}$$

Using Leibniz's formula (see Exercise 4.1 in Chapter 4) for differentiating integrals with variable limits we could proceed directly and compute the left side of (14). We prefer, however, a method that easily generalizes to higher dimensions and is more in the spirit of techniques in fluid dynamics. We change variables in (14) according to $x = x(h, t)$. Then $dx = J(h, t) \, dh$ and (14) becomes

$$\frac{d}{dt} \int_a^b \Delta(h, t) J(h, t) \, dh = 0 \tag{15}$$

where $\Delta(h, t)$ is the Lagrangian density defined by $\Delta(h, t) = \rho(x(h, t), t)$. We note that (14) has been changed to an integral with constant limits so that the derivative may be brought directly under the integral since Δ and J are continuously differentiable. Consequently

$$\int_a^b (\Delta J_t + J\Delta_t) \, dh = \int_a^b [\Delta(h, t) v_x(x(h, t), t) + \Delta_t(h, t)] J \, dh = 0$$

where we have used the Euler expansion formula (13). Since a and b are arbitrary and J is nonzero, we conclude that

$$\Delta_t(h, t) + \Delta(h, t)v_x(x(h, t), t) = 0 \tag{16}$$

from which we get

$$\Delta_t + \frac{\Delta}{J}V_h = 0 \tag{17}$$

which is the *conservation of mass law* in Lagrangian form. The Eulerian form can be obtained directly from (16) by substituting $h = h(x, t)$. We get

$$\frac{D\rho}{Dt} + \rho v_x = 0 \tag{18}$$

or

$$\rho_t + v\rho_x + \rho v_x = 0 \tag{19}$$

Equation (18) or (19) is known as the *continuity equation* and it is one of the fundamental equations of one dimensional flow. It is a first order nonlinear partial differential equation in terms of the density ρ and velocity v. The fluid motion is said to be *incompressible* if

$$\frac{D\rho}{Dt} = 0 \quad \text{or} \quad \Delta_t = 0$$

and *steady* if ρ and v are independent of t.

Momentum Conservation

In classical mechanics a particle of mass m having velocity **v** has linear momentum $m\mathbf{v}$. Newton's second law states that the time rate of change of momentum of the particle is equal to the net external force acting upon it. To generalize this law to one dimensional continuous media we *assume* the *balance of linear momentum principle*, which states that the time rate of change of linear momentum of any portion of the fluid equals the sum of the external forces acting upon it. The *linear momentum* at time t possessed by material of density $\rho(x, t)$ in the material region $a(t) \le x \le b(t)$ is defined by

$$\mathbf{i}A\int_{a(t)}^{b(t)}\rho(x, t)v(x, t)\, dx$$

where $\mathbf{i}$ is the unit vector of the positive x direction, $v(x, t)$ is the velocity, and A is the cross-sectional area.

The precise characterization of the forces acting on a material region in a continuous medium was the result of ideas evolving from works of Newton,

Euler, and Cauchy. Basically there are two types of forces that act on the material region, body forces and surface forces. *Body forces* are forces such as gravity, or an electric or magnetic field. Such a force is assumed to act on each cross section of the region and it is represented by

$$\mathbf{f}(x, t) = f(x, t)\mathbf{i}$$

The units of f are force per unit mass, and the total body force acting upon the region $a(t) \le x \le b(t)$ is therefore

$$\mathbf{i}A \int_{a(t)}^{b(t)} \rho(x, t) f(x, t)\, dx$$

Surface forces are forces like pressure that act across sections in the fluid medium. More specifically consider a cross section at time t located at x. By $\boldsymbol{\sigma}(x, t, \mathbf{i})$ we denote the force per unit area on the material *on* the negative (left) side of the cross section *due to* the material on the positive (right) side of the section. Similarly $\boldsymbol{\sigma}(x, t, -\mathbf{i})$ denotes the force per unit area exerted on the material *on* the right side of the section *by* the material on the left side of the section. By convention the third argument in $\boldsymbol{\sigma}$, here either $\mathbf{i}$ or $-\mathbf{i}$, is a vector that points outward and normal from the surface on which the force is acting. The vectors $\boldsymbol{\sigma}(x, t, \mathbf{i})$ and $\boldsymbol{\sigma}(x, t, -\mathbf{i})$ are called *surface tractions* or *stress vectors*. Figure 5.30 shows a geometrical description of these notions. Before a cross section is indicated, no stress is defined; however, once a cross section x is indicated the forces $A\boldsymbol{\sigma}(x, t, \mathbf{i})$ and $A\boldsymbol{\sigma}(x, t, -\mathbf{i})$, which are exerted on the shaded and unshaded portions, respectively, are defined. In anticipation of what we shall prove later we have drawn these forces opposite and equal in Fig. 5.30. At the present stage it is not known in which direction these point. And we emphasize that the third argument in $\boldsymbol{\sigma}$, either $\mathbf{i}$ or $-\mathbf{i}$, does not define the direction of the stress but serves only to indicate the orientation or normal direction of the surface of the section.

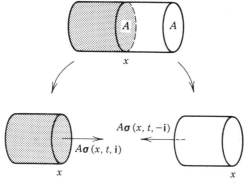

Figure 5.30

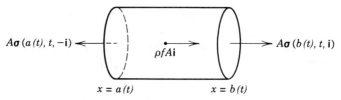

$A\boldsymbol{\sigma}\,(a\,(t),\,t,\,-\mathbf{i})$ ← — — $\rho fA\mathbf{i}$ — → $A\boldsymbol{\sigma}\,(b\,(t),\,t,\,\mathbf{i})$

$x = a\,(t)$ $x = b\,(t)$

Figure 5.31. Forces on a fluid element.

We can now write a quantitative description of the balance of the linear momentum principle for a material region $a(t) \le x \le b(t)$

$$\mathbf{i}\frac{d}{dt}A\int_{a(t)}^{b(t)}\rho(x,t)v(x,t)\,dx = \mathbf{i}A\int_{a(t)}^{b(t)}\rho(x,t)f(x,t)\,dx$$
$$+A\boldsymbol{\sigma}(b(t),t,\mathbf{i}) + A\boldsymbol{\sigma}(a(t),t,-\mathbf{i}) \quad (20)$$

In words this states that the time rate of change of momentum of the material region $[a(t), b(t)]$ equals the total body force plus the surface forces on $[a(t), b(t)]$. A schematic of the forces is shown in Fig. 5.31. To put (20) in a more workable form we calculate the left side in a similar manner as was done for the integral form of the conservation of mass equation in the preceding section. Changing variables to the Lagrangian coordinate h and using the Euler expansion formula gives

$$\frac{d}{dt}\int_{a(t)}^{b(t)}\rho v\,dx = \frac{d}{dt}\int_{a}^{b}\Delta VJ\,dh = \int_{a}^{b}\left(J\frac{\partial}{\partial t}(\Delta V) + \Delta VJ_t\right)dh$$
$$= \int_{a}^{b}\left[\frac{\partial}{\partial t}(\Delta V) + \Delta Vv_x\right]J\,dh = \int_{a(t)}^{b(t)}\left[\frac{D}{Dt}(\rho v) + \rho vv_x\right]dx$$

From the mass conservation equation (18) we have

$$\frac{D}{Dt}(\rho v) = \rho\frac{Dv}{Dt} + v\frac{D\rho}{Dt} = \rho\frac{Dv}{Dt} - v\rho v_x$$

Hence

$$\frac{d}{dt}\int_{a(t)}^{b(t)}\rho v\,dx = \int_{a(t)}^{b(t)}\rho\frac{Dv}{Dt}\,dx \quad (21)$$

and (20) can be written

$$\mathbf{i}\int_{a(t)}^{b(t)}\left(\rho\frac{Dv}{Dt} - \rho f\right)dx = \boldsymbol{\sigma}(b(t),t,\mathbf{i}) + \boldsymbol{\sigma}(a(t),t,-\mathbf{i}) \quad (22)$$

This equation can be further simplified using the following lemma, which is the continuum mechanical version of the action–reaction principle expressed in Newton's third law.

Lemma 2.1   *The balance of linear momentum principle expressed by* (22) *implies*

$$\sigma(x, t, -\mathbf{i}) = -\sigma(x, t, \mathbf{i}) \tag{23}$$

for any cross section x.

Proof   Let x_0 be arbitrary with $a(t) < x_0 < b(t)$. Applying (22) to the region between $a(t)$ and x_0 and to the region between x_0 and $b(t)$ we get

$$\mathbf{i} \int_{a(t)}^{x_0} \left[\rho \frac{Dv}{Dt} - \rho f \right] dx = \sigma(x_0, t, \mathbf{i}) + \sigma(a(t), t, -\mathbf{i})$$

and

$$\mathbf{i} \int_{x_0}^{b(t)} \left[\rho \frac{Dv}{Dt} - \rho f \right] dx = \sigma(b(t), t, \mathbf{i}) + \sigma(x_0, t, -\mathbf{i})$$

Adding these two equations and then subtracting (22) from the result gives

$$\sigma(x_0, t, \mathbf{i}) + \sigma(x_0, t, -\mathbf{i}) = 0$$

which because of the arbitrariness of x_0 proves (23).

This result permits us to define the *stress component* $\sigma(x, t)$ by the equation

$$\sigma(x, t, \mathbf{i}) = \sigma(x, t)\mathbf{i}$$

Therefore (22) becomes

$$\int_{a(t)}^{b(t)} \left(\rho \frac{Dv}{Dt} - \rho f \right) dx = \sigma(b(t), t) - \sigma(a(t), t)$$

$$= \int_{a(t)}^{b(t)} \sigma_x(x, t) \, dx$$

Because the interval $[a(t), b(t)]$ is arbitrary we have

$$\rho \frac{Dv}{Dt} = \rho f + \sigma_x \tag{24}$$

which is the Eulerian form of the partial differential equation expressing balance of linear momentum. A familiar form of the equation can be written if we define the *pressure p* by

$$p(x, t) = -\sigma(x, t)$$

Then (24) becomes

$$v_t + vv_x + \frac{1}{\rho}p_x = f \tag{25}$$

Thermodynamics and Energy Conservation

In summary, the Eulerian equations of motion for a one dimensional continuum with no body forces are

$$\rho_t + v\rho_x + \rho v_x = 0 \tag{26}$$

$$v_t + vv_x + \frac{1}{\rho}p_x = 0 \tag{27}$$

These are a pair of first order nonlinear partial differential equations for the three unknowns ρ, v, and p. Intuition tells us that a third equation is needed. Further reflection dictates that particular physical properties of the medium must play a role, and such properties are not included in (26) and (27), which are completely general and hold for any continuum. Equations that specify properties of the medium are known as *equations of state* or *constitutive relations*. Such equations give relations between observable effects and the internal constitution of the material and they are generally expressible in terms of thermodynamic variables such as density, pressure, energy, entropy, temperature, and so on.

Example 2.1 A simple equation of state is one in which the pressure depends only on the density, that is,

$$p = F(\rho) \tag{28}$$

where F is a given differentiable function and $F'(\rho) > 0$. Equation (28) is known as a *barotropic* equation of state and it allows the pressure to be eliminated from (27) to obtain

$$v_t + vv_x + \frac{F'(\rho)}{\rho}p_x = 0 \tag{29}$$

Therefore equations (26) and (29) govern *barotropic flow* and represent a complete set of equations for the two unknowns ρ and v. It is not evident at this point, but implicit in (28) are assumptions regarding the other thermodynamic variables that may not be valid for a general fluid motion.

In contrast to (28) it is more often the case that the equation of state introduces yet another unknown variable in the problem, for example the temperature, internal energy, or entropy. Therefore still another equation is required and it comes from the consideration of energy conservation. For general fluid motions a complete set of field equations usually consists of conservation equations for mass, momentum, and energy, as well as one or more constitutive relations.

A discussion of energy conservation naturally entails the development of some basic concepts in equilibrium thermodynamics. Classical thermodynamics deals with relations between equilibrium states of uniform matter and the laws that govern changes in those states. It is implicitly assumed that the changes take place slowly so that at each instant there are no spatial gradients in any of the thermodynamic quantities. That is, changes occur through a sequence of uniform equilibrium states. Although fluid motions do not appear to be slow, it is nevertheless found that the usual results of classical thermodynamics can be directly applied provided we consider the instantaneous *local thermodynamic state*. Hence each of the thermodynamic quantities, such as temperature, pressure, and so on, are assumed to be functions of position and time and thermodynamical relations are assumed to hold locally. To illustrate this principle consider a hypothetical gas in a container and let P, Δ, and Θ denote its pressure, density, and temperature, respectively. It is observed experimentally that the ratio of pressure to density is proportional to the temperature, or

$$\frac{P}{\Delta} = R\Theta \tag{30}$$

where R is the constant of proportionality characteristic of the gas. Equation (30) is a Lagrangian statement, since it holds for a material volume of gas. The assumption of local thermodynamic equilibrium allows us to assume that

$$\frac{p}{\rho} = R\theta$$

where p, ρ, and the temperature θ are the local Eulerian quantities that are functions of x and t. (It is possible to imagine processes that take place so rapidly that local fluid elements are incapable of establishing instantaneous equilibrium. Such processes belong to the study of nonequilibrium fluid dynamics.)

Example 2.2 (The Ideal Gas)   Under normal conditions most gases obey the *ideal gas law*

$$p = R\rho\theta \tag{31}$$

$$e = c_v\theta + \text{constant} \tag{32}$$

where p is pressure, ρ is density, θ is temperature, and e is the internal energy per unit mass. The constant c_v is the specific heat at constant volume and R is the *gas constant* for the particular gas. A gas satisfying (31) and (32) is called an ideal gas and many compressible fluids of practical interest can be approximately treated as ideal gases. Hence (31) and (32) describe a wide range of phenomena in fluid dynamics. These equations coupled with (26) and (27) give only four equations for the five unknowns p, v, ρ, θ, and e, so it is again evident that another equation is required.

A common statement of the first law of thermodynamics is that energy is conserved if heat is taken into account. Thus the first law of thermodynamics provides a concise statement of energy conservation. It arises in considering the consequences of adding a small amount of heat to a unit mass of a material substance such that equilibrium conditions are established at each step. Some of the energy will go into the work $pd(1/\rho)$ done in expanding the specific volume $1/\rho$ by $d(1/\rho)$ and the remainder will go into increasing the internal energy e by de. The precise relationship is

$$q = de + pd\rho^{-1} \tag{33}$$

where q is the heat added. For an equilibrium process Equation (33) is the *first law of thermodynamics*. In general the differential form q is not exact, that is, there does not exist a state function Q for $q = dQ$. If $q = 0$, then the process is called *adiabatic*.

Example 2.3   For an ideal gas described by Equations (31) and (32) we have

$$q = de + pd\rho^{-1} = c_v\, d\theta + R\theta\, d\left(\ln \rho^{-1}\right) \tag{34}$$

From (34) we notice that θ^{-1} is an integrating factor for the differential form q. Therefore

$$\theta^{-1}q = d\left(c_v \ln \theta + R \ln \rho^{-1}\right) = ds$$

where

$$s \equiv c_v \ln \theta + R \ln \rho^{-1} + \text{constant} \tag{35}$$

is the *entropy*. Consequently, for an ideal gas the first law (33) takes the form

$$de = \theta\, ds - pd\rho^{-1} \tag{36}$$

Equation (35) may also be written in terms of the pressure as

$$p = k\rho^\gamma \exp \frac{s}{c_v} \tag{37}$$

where

$$\gamma \equiv 1 + \frac{R}{c_v}$$

and k is a constant. Combining (31) and (32) gives another form of the equation of state, namely

$$e = \frac{p}{(\gamma - 1)\rho} + \text{constant} \tag{38}$$

Finally, in some contexts it is useful to introduce the enthalpy h defined by $h = e + p/\rho$. For an ideal gas $h = c_p\theta$, where $c_p = R + c_v$ is the specific heat at constant pressure. Hence $\gamma = c_p/c_v$ is the ratio of specific heats. For air $\gamma = 1.4$ and for a monatomic gas $\gamma = \frac{5}{3}$. Generally, $\gamma > 1$.

Other equations of state have been developed to include various effects. The *Abel* or *Clausius equation of state*

$$p\left(\frac{1}{\rho} - \alpha\right) = R\theta, \qquad \alpha \text{ constant}$$

introduces a constant covolume α to account for the size of the molecules. The *van der Waals equation of state*

$$\left(p + \beta\rho^2\right)\left(\frac{1}{\rho} - \alpha\right) = R\theta, \qquad \alpha, \beta \text{ constants}$$

contains the term $\beta\rho^2$ to further account for intermolecular forces. The *Tait equation*

$$\frac{p + B}{p} = \left(\frac{\rho}{\rho_0}\right)^{\bar{\gamma}}$$

where $\bar{\gamma}$ and B are constants has been used to model the behavior of liquids at high pressures. Thompson [2] can be consulted for a more detailed discussion of the material properties of matter.

We now take up the formulation of a partial differential equation governing the flow of energy in a system. We first follow an approach consistent with our earlier development of balance laws in integral form. As before let $[a(t), b(t)]$ be a one dimensional material region with cross-sectional area A. We define the kinetic energy of the fluid in the region by $A \int_{a(t)}^{b(t)} \frac{1}{2} \rho v^2 \, dx$ and the internal energy in the region by $A \int_{a(t)}^{b(t)} \rho e \, dx$. By the general conservation of energy principle the time rate of change of the total energy equals the rate that the forces do work on the region plus the rate that heat flows into the region. There are two forces, the body force $f(x, t)$ acting at each cross section of the region and the stress $\sigma(x, t)$ acting at the two ends. Since force times velocity equals the rate work is done, the total rate that work is done on the region is $A \int_{a(t)}^{b(t)} \rho f v \, dx - A\sigma(a(t), t)v(a(t), t) + A\sigma(b(t), t)v(b(t), t)$. The rate that heat flows into the region is $A\phi(a(t), t) - A\phi(b(t), t)$, where $\phi(x, t)$ is the heat flux in energy units per unit area per time. Therefore we *postulate* the balance of energy law

$$\frac{d}{dt} \int_{a(t)}^{b(t)} \left(\frac{1}{2} \rho v^2 + \rho e \right) dx = \int_{a(t)}^{b(t)} \rho f v \, dx - \sigma(a(t), t)v(a(t), t)$$
$$+ \sigma(b(t), t)v(b(t), t) + \phi(a(t), t) - \phi(b(t), t)$$

or

$$\frac{d}{dt} \int_{a(t)}^{b(t)} \left(\frac{1}{2} \rho v^2 + \rho e \right) dx = \int_{a(t)}^{b(t)} \left(\rho f v + (\sigma v)_x - \phi_x \right) dx$$

Assuming sufficient smoothness of the state variables and using the arbitrariness of the interval $[a(t), b(t)]$ we have the following Eulerian differential form of the conservation of energy law.

$$\frac{1}{2} \rho \frac{Dv^2}{Dt} + \rho \frac{De}{Dt} = \rho f v + (\sigma v)_x - \phi_x \tag{39}$$

Equation (39) can be cast into various forms. First, multiplication of the momentum balance equation (24) by v gives

$$\frac{1}{2} \rho \frac{Dv^2}{Dt} = \rho f v + v \sigma_x$$

Subtracting this from (39) gives an equation for the rate of change of internal energy

$$\rho \frac{De}{Dt} = \sigma v_x - \phi_x \tag{40}$$

An alternate approach to energy conservation comes from an examination of the first law of thermodynamics (33). The second law of thermodynamics states that the differential form q in general has integrating factor θ^{-1} and

$\theta^{-1}q = ds$, where s is the entropy and θ is the absolute temperature. Thus

$$\theta \, ds = de + p d\rho^{-1}$$

This combined form of the first and second laws of thermodynamics can be reformulated as a partial differential equation. Since it refers to a given material region, we *postulate*

$$\theta \frac{Ds}{Dt} = \frac{De}{Dt} + p \frac{D(1/\rho)}{Dt} \tag{41}$$

which is another local form of the conservation of energy principle. Noting that $D\rho^{-1}/Dt = -(1/\rho^2)\, D\rho/Dt = -(1/\rho)v_x$ and putting $\sigma = -p$, equation (41) may be subtracted from (40) to obtain

$$\theta \frac{Ds}{Dt} = -\frac{1}{\rho}\phi_x$$

which relates the entropy change to the heat flux. If we assume the *constitutive relation* $\phi = -K\theta_x$, which is Fourier's law, then the energy equation (41) can be expressed as

$$\frac{De}{Dt} + p \frac{D(1/\rho)}{Dt} = \frac{1}{\rho}(K\theta_x)_x \tag{42}$$

Example 2.4 (Adiabatic Flow)   In adiabatic flow $\theta \, Ds/Dt = 0$ and the energy equation (41) becomes

$$\frac{De}{Dt} + p \frac{D(1/\rho)}{Dt} = 0 \tag{43}$$

Equations (26), (27), and (43) along with a *thermal equation of state* $p = p(\rho, \theta)$ and a *caloric equation of state* $e = e(\rho, \theta)$ give a set of five equations for the unknowns ρ, v, p, e, and θ.

Example 2.5 (Ideal Gas)   For the adiabatic flow of an ideal gas the expression (38) for the energy can be substituted into (43) to obtain

$$\frac{Dp}{Dt} - \frac{\gamma p}{\rho} \frac{D\rho}{Dt} = 0 \tag{44}$$

Equations (26), (27), and (44) give three equations for ρ, v, and p. Since

$Ds/Dt = 0$ the entropy is constant for a given fluid particle (cross section); in Lagrangian form $S = S(h)$. In this case the equations governing adiabatic flow are often written

$$\frac{D\rho}{Dt} + \rho v_x = 0, \qquad \rho\frac{Dv}{Dt} + p_x = 0, \qquad \frac{Ds}{Dt} = 0$$

along with the equation of state (37) giving $p = p(s, \rho)$. If the entropy is constant initially, that is, $S(h, 0) = s_0$ for all h, then $s(x, t) = s_0$ for all x and t and the resulting adiabatic flow is called *isentropic*. This case reduces to barotropic flow as discussed in Example 2.1.

In general, for nonadiabatic flow involving heat conduction equations (26), (27), and (42) along with thermal and caloric equations of state give five equations for ρ, v, p, e, and θ. If there is no motion ($v = 0$) and $s = c_v \ln \theta +$ constant, then (42) reduces to the classical diffusion equation

$$\rho c_v \theta_t = (K\theta_x)_x$$

The Acoustic Approximation

The equations governing the adiabatic flow of a gas are nonlinear and cannot be solved in general. Rather, only special cases can be resolved. The simplest case, that of acoustics, arises when the derivations from a constant equilibrium state $v = 0$, $\rho = \rho_0$, and $p = p_0$ are assumed to be small. Therefore we consider the isentropic equations

$$\rho_t + v\rho_x + \rho v_x = 0 \tag{45}$$
$$\rho v_t + \rho v v_x + p_x = 0 \tag{46}$$

with the barotropic equation of state

$$p = F(\rho)$$

It is common in acoustics to introduce the *condensation* δ defined by

$$\delta \equiv \frac{\rho - \rho_0}{\rho_0}$$

which is the relative change in density from the constant state ρ_0. We assume that v, δ, and all their derivatives are small compared to unity. Then the equation of state can be expanded in a Taylor series as

$$p = F(\rho_0) + F'(\rho_0)(\rho - \rho_0) + \tfrac{1}{2}F''(\rho_0)(\rho - \rho_0)^2 + O((\rho - \rho_0)^3)$$
$$= F(\rho_0) + \rho_0 F'(\rho_0)\delta + \tfrac{1}{2}\rho_0^2 F''(\rho_0)\delta^2 + O(\delta^3)$$

Whence

$$p_x = \rho_0 F'(\rho_0)\delta_x + O(\delta^2)$$

and therefore (45) and (46) become

$$\delta_t + v\delta_x + (1 + \delta)v_x = 0$$
$$(1 + \delta)v_t + (1 + \delta)vv_x + F'(\rho_0)\delta_x + O(\delta^2) = 0$$

Discarding products of small terms gives a set of linearized equations for the small deviations δ and v, namely

$$\delta_t + v_x = 0 \tag{47}$$
$$v_t + F'(\rho_0)\delta_x = 0 \tag{48}$$

These equations are known as the *acoustic approximation*. Eliminating v from (47) and (48) gives

$$\delta_{tt} - c_0^2 \delta_{xx} = 0 \tag{49}$$

where

$$c_0^2 = F'(\rho_0) \tag{50}$$

Equation (49) is a second order linear partial differential equation known as the *wave equation*, and it is the subject of Section 5.3. The quantity c_0 in (50) is the propagation speed of acoustic waves. For an ideal gas $p = F(\rho) = k\rho^\gamma$ and thus

$$c_0 = \sqrt{\frac{\gamma p_0}{\rho_0}}$$

is the speed that small amplitude signals are propagated in an ideal gas. By eliminating δ from (47) and (48) we note that the small velocity amplitude v satisfies the wave equation as well.

The science of acoustics is the study of fluid motions associated with the propagation of sound. The fundamental equation is the wave equation that was obtained as an approximation for small amplitude signals. At this point it is worthwhile to test the acoustic approximation to determine its range of validity. Let U and Γ denote scales for the velocity v and condensation δ, respectively. From Chapter 1 we know that such scales can be defined as the maximum value of the quantity, for example, $U = \max|v|$ and $\Gamma = \max|\delta|$. If l and τ denote length and time scales for the problem, then the balance of the two terms in (47) and the two terms in (48) require

$$\Gamma\tau^{-1} \approx Ul^{-1}, \qquad U\tau^{-1} \approx c_0^2\Gamma l^{-1}$$

Thus

$$c_0 \approx l\tau^{-1}$$

In the approximation we assumed that the convective term vv_x was small compared to v_t. Thus for acoustics

$$U^2 l^{-1} \ll U\tau^{-1}$$

or

$$U \ll c_0 \tag{51}$$

That is, the acoustic approximation can be justified when the maximum particle velocity is small compared to the speed c_0 that sound waves travel in the medium. The ratio U/c_0 is called the *Mach number M* and therefore acoustics is valid when $M \ll 1$. For phenomena involving ordinary audible sounds the approximation is highly satisfactory, but for flows resulting from the presence of high-speed aircraft the full nonlinear equations (45) and (46) would be required for an accurate description. In the theory of *nonlinear acoustics* a few, but not all, of the nonlinear terms are retained in the perturbation equations preceding (47) and (48). An introductory treatment of physical acoustics can be found in Thompson [2].

Stress Waves in Solids

An alternate direction of application of the conservation laws for a continuous medium is solid mechanics. In such media we write the governing equations as (see (19) and (24))

$$\rho_t + v\rho_x + \rho v_x = 0 \tag{52}$$

$$v_t + vv_x = f + \frac{1}{\rho}\sigma_x \tag{53}$$

and our physical model is a cylindrical bar in which we seek to describe the longitudinal vibrations, that is, the motion of the planar cross sections of the bar. In this case the equation of state or constitutive relation is generally a relation between the distortion (compression or elongation) the bar undergoes subject to an applied stress. To define the distortion we consider at time $t = 0$ a small portion of the bar between h and $h + \Delta h$. At time $t > 0$ this material is located between $x(h, t)$ and $x(h + \Delta h, t)$. The *distortion* is the fractional change given by

$$\text{Distortion} = \frac{\text{new length} - \text{original length}}{\text{original length}}$$

$$= \frac{x(h + \Delta h, t) - x(h, t) - \Delta h}{\Delta h}$$

$$= \frac{\partial x}{\partial h}(h, t) - 1 + O(\Delta h)$$

We define the *strain E* to be the lowest order approximation of the distortion, that is,

$$E \equiv \frac{\partial x}{\partial h}(h, t) - 1$$

If $U(h, t)$ denotes the *displacement* of a cross section h at time t, then $x(h, t) = h + U(h, t)$ and

$$E = \frac{\partial U}{\partial h}(h, t) \tag{54}$$

The constitutive relationship is an equation that gives the Lagrangian stress $\Sigma(h, t)$ as a definite function of the strain $E(h, t)$. Quite generally the graph may look like the solid curve shown in Fig. 5.32. If, however, only small strains are of interest, then we may approximate the curve by a straight line (dotted in Fig. 5.11) that has equation

$$\Sigma(h, t) = Y(h)E(h, t) \tag{55}$$

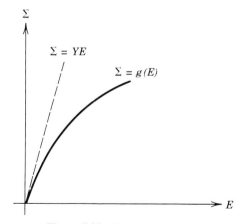

Figure 5.32. Stress versus strain.

The proportionality factor $Y(h)$ is called *Young's modulus* or the *stiffness* and the linear stress–strain relation (55) is called *Hooke's law*.[†] In Eulerian coordinates (55) becomes

$$\sigma(x, t) = y(x, t)\varepsilon(x, t) \tag{56}$$

where $\sigma(x, t) = \Sigma(h(x, t), t)$, $y(x, t) = Y(h(x, t))$, and $\varepsilon(x, t) = E(h(x, t), t)$. Using the relation

$$\left.\frac{\partial U}{\partial h}\right|_{h=h(x,t)} = \frac{u_x}{1 - u_x}$$

where u is the Eulerian displacement, the constitutive relation (56) can be written

$$\sigma(x, t) = y(x, t)\frac{u_x(x, t)}{1 - u_x(x, t)} \tag{57}$$

Then σ can be eliminated from the momentum equation (53) to obtain

$$\rho v_t + \rho v v_x = \rho f + \frac{\partial}{\partial x}\left(\frac{y u_x}{1 - u_x}\right) \tag{58}$$

Thus with the assumption of a constitutive relation (55) we have obtained two equations, (52) and (58), for the three unknowns v, u, and ρ. The third equation is the defining relationship between u and v,

$$v = \frac{Du}{Dt} = u_t + v u_x \tag{59}$$

As they stand, Equations (52), (58), and (59) are nonlinear and cannot be resolved by analytical means. As in the case of the acoustic approximation, however, we can obtain a linearized theory for small displacements u. Therefore consider a bar with initial density $\rho_0(x)$ and stiffness $y_0(x)$. For a small displacement theory assume u and its derivatives are small compared to unity. From (59)

$$v = u_t(1 - u_x)^{-1} = u_t(1 + u_x + u_x^2 - \cdots) = u_t + O(u^2)$$

and

$$v_t = u_{tt} + O(u^2)$$

[†] We assume explicitly that Y does not depend on time t. Materials for which Y is time dependent are said to have *memory*.

Let $\bar{\rho}(x, t)$ denote the deviation from the initial density $\rho_0(x)$, that is,

$$\rho(x, t) = \rho_0(x) + \bar{\rho}(x, t)$$

Then by introducing the Lagrangian density Δ and using the mean value theorem

$$\begin{aligned}|\bar{\rho}(x, t)| &= |\rho(x, t) - \rho_0(x)| \\ &= |\Delta(x - u, t) - \Delta(x, 0)| \\ &\leq |\Delta_h(\tilde{x}, \tilde{t})u| + |\Delta_t(\tilde{x}, \tilde{t})|t\end{aligned}$$

where $0 < \tilde{t} < t$ and $x < \tilde{x} < x - u$. Hence the deviation $\bar{\rho}$ is small provided that u *and* t are small and that the derivatives Δ_h and Δ_t are bounded. A similar calculation shows that if $y(x, t) = y_0(x) + \bar{y}(x, t)$, then the deviation $\bar{y}$ satisfies the equality $|\bar{y}(x, t)| = |Y_h(\tilde{x})u|$, and therefore $\bar{y}$ is small if u is small and Y_h is bounded.

With all of this information the momentum equation (58) can be written (taking $f = 0$)

$$\begin{aligned}(\rho_0 + \bar{\rho})(u_{tt} &+ O(u^2)) + (\rho_0 + \bar{\rho})(vv_x) \\ &= \frac{\partial}{\partial x}\left[(y_0 + \bar{y})(u_x)(1 + u_x + O(u^2))\right]\end{aligned}$$

Retaining only lowest order terms we get the linearized small displacement equation

$$\rho_0(x)u_{tt} = \frac{\partial}{\partial x}(y_0(x)u_x) \tag{60}$$

Subject to the assumptions in the last paragraph we may expect (60) to govern the small longitudinal vibrations of a bar of density $\rho_0(x)$ and stiffness $y_0(x)$.

We notice that the unknown density dropped out of (58) and could be replaced by the initial density $\rho_0(x)$. Therefore (60) is one equation in the single unknown u. Exercise 2.13 shows that the conservation of mass equation (52) is a consistency relation. If $\rho_0(x) = \rho_0$ and $y_0(x) = y_0$, where ρ_0 and y_0 are constants, then (60) reduces to the wave equation

$$u_{tt} - c_0^2 u_{xx} = 0, \qquad c_0^2 = \frac{y_0}{\rho_0}$$

Generally a rod of finite extent will be subject to boundary conditions at its ends, say at $x = 0$ and at $x = l$. For definiteness we focus on the endpoint $x = l$. Clearly if the end is held fixed then

$$u(l, t) = 0, \qquad t > 0 \qquad \text{(fixed end)}$$

If the end is free, or no force acts on the face at $x = l$, then $\Sigma(l, t) = 0$ or $Y(l)E(l, t) = 0$. Hence

$$Y(l)\frac{\partial U}{\partial h}(l, t) = 0, \qquad t > 0 \qquad \text{(free end)}$$

or in Eulerian variables

$$y(l, t)\frac{u_x(l, t)}{1 - u_x(l, t)} = 0 \qquad \text{(free end)}$$

A properly posed problem for determining the small displacements $u(x, t)$ of a bar of length l with stiffness $y_0(x)$ and density $\rho_0(x)$ consists of the partial differential equation (60) along with boundary conditions at $x = 0$ and $x = l$, and with initial conditions of the form

$$u(x, 0) = u_0(x), \qquad u_t(x, 0) = v_0(x), \qquad 0 < x < l$$

where $u_0(x)$ and $v_0(x)$ are the given initial displacement and velocity, respectively.

EXERCISES

2.1   Verify (8) and (9).

2.2   The reciprocal W of the Lagrangian density Δ is called the *specific volume*. If at time $t = 0$ the density is a constant ρ_0, prove that conservation of mass and momentum can be expressed as

$$W_t - \rho_0^{-1}V_h = 0$$
$$V_t + \rho_0^{-1}P_h = 0$$

2.3   Show that the Jacobian J is given by

$$J(h, t) = \frac{\Delta(h, 0)}{\Delta(h, t)}$$

and therefore measures the ratio of the initial density to the density at time t.

2.4   Prove that

$$\frac{d}{dt}\int_{a(t)}^{b(t)}\rho(x, t)g(x, t)\, dx = \int_{a(t)}^{b(t)}\rho(x, t)\frac{Dg}{Dt}(x, t)\, dx$$

for any sufficiently smooth function g, where $a(t) \le x \le b(t)$ is a material region.

2.5 Write the equation $F_t + F^2 F_h = 0$, where $F = F(h, t)$ in Eulerian form.

2.6 A one dimensional flow is defined by

$$x = (h - t)(1 + ht)^{-1}, \qquad 0 < t < t_1$$

(a) Sketch the particle paths on xt and ht diagrams.

(b) Find $V(h, t)$ and $v(x, t)$.

(c) Verify the Eulerian expansion formula.

(d) Verify that $Dv/Dt = v_t + vv_x$.

(e) If the density is $\Delta(h, t) = h^2$, find the density that an observer would measure at $x = 1$.

(f) Repeat (a) through (e) for the flow defined by $x = \frac{h}{2}(1 + \exp t)$

2.7 Derive the conservation of mass and momentum equations for one dimensional flow in a cylinder of variable cross-sectional area $A(h, t)$. Here $A(h, t)$ denotes the cross-sectional area of the section with Lagrangian label h at time t. Assume $A(h, t) = A(h, 0)$ for all $t > 0$. *Solution*:

$$\frac{D(a\rho)}{Dt} + a\rho v_x = 0, \qquad \rho a \frac{Dv}{Dt} = \rho a f + \frac{\partial}{\partial x}(a\sigma)$$

where $a(x, t) = A(h(x, t), 0)$.

2.8 A point source produces a spherically symmetric outward motion of the surrounding material in such a way that all the physical parameters are functions of radius r and time t. By considering the time evolution of the material between two concentric spherical shells derive the mass conservation equation in the form

$$\frac{\partial}{\partial t}\left(\Delta(a, t)r^2(a, t)\frac{\partial r}{\partial a}(a, t)\right) = 0$$

where Δ is the Lagrangian density and a is a radial Lagrangian coordinate. Conclude that

$$\Delta(a, t)r^2(a, t)\frac{\partial r}{\partial a}(a, t) = \rho_0 a^2$$

and derive the Eulerian form

$$\frac{D\rho}{Dt} + \rho\frac{\partial v}{\partial r} + \frac{2\rho v}{r} = 0$$

and the Lagrangian form

$$\frac{1}{\Delta}\Delta_t + \frac{r^2(a,t)\Delta}{a^2\rho_0}\frac{\partial V}{\partial a} + \frac{2V}{r(a,t)} = 0$$

where ρ_0 is the constant initial density of the medium.

2.9 In a one dimensional flow at time $t = 0$ a shock wave with velocity $D(t) > 0$ moves to the right from $x = 0$. Transform the mass and momentum equations (26) and (27) with $f = 0$ to *shock fixed coordinates* t and z, where t is time and z is the distance measured positively from the shock front backward into the flow.
Solution:

$$\rho_t - \frac{\partial}{\partial z}(\rho\bar{v}) = 0, \qquad \bar{v}_t + D_t - \bar{v}\bar{v}_z - \frac{1}{\rho}p_z = 0$$

where $\bar{v} \equiv v - D$ is fluid velocity relative to the shock front.

2.10 By introducing the enthalpy $h = e + p/\rho$ into the energy equation $De/Dt + pD(\rho^{-1})/Dt = 0$ show that for steady flow

$$h_x - \frac{1}{\rho}p_x = 0$$

Prove that

$$h + \frac{1}{2}v^2 = \text{constant}$$

2.11 A metallic bar with constant density ρ_0 and stiffness y_0 has length l and cross sectional area A, and it undergoes small longitudinal displacements $u(x, t)$. The left end at $x = 0$ is fixed, but the right end at $x = l$ is attached to a spring that exerts a force proportional to its elongation with stiffness k (see Fig. 5.33). What is the boundary condition on u at $x = l$?
Answer: $Ay_0u_x(l, t) + ku(l, t) = 0$.

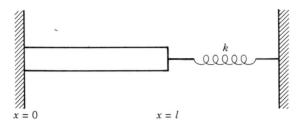

Figure 5.33. Exercise 2.11.

2.12 Time harmonic standing wave solutions of the form

$$u(x, t) = w(x)\exp(i\omega t)$$

often exist for the wave equation

$$\rho_0(x)u_{tt} = \frac{\partial}{\partial x}(y_0(x)u_x), \qquad 0 < x < l$$

subject to boundary conditions at $x = 0$ and $x = l$. The set of values ω and the corresponding spatial distributions $w(x)$ are called the *fundamental frequencies* and *normal modes of vibration*, respectively. Determine the frequencies and normal modes for the following problems.

(a) $u_{tt} - c_0^2 u_{xx} = 0, \qquad 0 < x < \pi$

$u(0, t) = 0, \qquad u_x(\pi, t) = 0$

(b) $u_{tt} - c_0^2 u_{xx} = 0, \qquad 0 < x < \dfrac{\pi}{2}$

$u(0, t) = 0, \qquad u_x\left(\dfrac{\pi}{2}, t\right) + \dfrac{3}{2}u\left(\dfrac{\pi}{2}, t\right) = 0$

(c) $u_{tt} = \dfrac{\partial}{\partial x}(x^2 u_x), \qquad 1 < x < e$

$u(1, t) = u(e, t) = 0$

(d) $xu_{tt} = \dfrac{\partial}{\partial x}(xu_x), \qquad 0 < x < 2$

$\lim_{x \to 0^+} xu_x(x, t) = 0, \qquad u(2, t) = 0$

2.13 In the linearized theory for small longitudinal displacements of a bar following Equation (59) in the text, show that the conservation of mass equation (52) becomes

$$\bar{\rho}_t = -\frac{\partial}{\partial x}(\rho_0(x)v)$$

and hence provides a consistency check for the smallness of $\bar{\rho}$.

2.14 Derive the continuity equation (19) by directly calculating

$$\frac{d}{dt} \int_{a(t)}^{b(t)} \rho(x, t)\, dx$$

using Leibniz's formula (Exercise 4.1 in Chapter 4).

2.15 Derive (37) and (41).

2.16 Show that the equations governing a one dimensional time dependent flow of a compressible fluid under constant pressure are

$$v_t + vv_x = 0$$
$$\rho_t + (\rho v)_x = 0$$
$$e_t + (ev)_x + pv_x = 0$$

Subject to initial conditions $v(x,0) = f(x)$, $\rho(x,0) = g(x)$, and $e(x,0) = h(x)$, where f, g, and h are given $C^1(R^1)$ functions with nonnegative derivatives, derive the solution $v = f(x - vt)$ and

$$\rho = \frac{g(x - vt)}{1 + tf'(x - vt)}, \qquad e = \frac{h(x - vt) + p}{1 + tf'(x - vt)} - p$$

Here e is measured in energy per unit volume.

5.3 THE WAVE EQUATION

D'Alembert's Solution

We have observed that the one dimensional wave equation

$$u_{tt} - c^2 u_{xx} = 0 \tag{1}$$

arises naturally in acoustics and in the small vibrations of a bar. There are many other physical situations that lead to this important equation, electromagnetism being a significant one. In this section we examine various aspects of this equation and its solutions. From earlier remarks we note that (1) is hyperbolic. It is easy to see (Exercise 3.2) that the general solution of the wave equation is given by

$$u(x, t) = f(x + ct) + g(x - ct) \tag{2}$$

where f and g are arbitrary functions; hence solutions are in general the superposition of right and left traveling waves moving at speed c. The functions f and g are determined by the initial and boundary data, although it is difficult to do this in specific cases.

The pure initial value problem for the wave equation is

$$u_{tt} - c^2 u_{xx} = 0, \qquad x \in R^1, \quad t > 0$$
$$u(x,0) = F(x), \qquad u_t(x,0) = G(x), \qquad x \in R^1 \tag{3}$$

where F and G are given functions, and it can be solved exactly by D'Alembert's solution.

Theorem 3.1   *In (3) let $F \in C^2(R)$ and $G \in C^1(R)$. Then the solution of (3) is given by*

$$u(x,t) = \frac{1}{2}(F(x + ct) + F(x - ct)) + \frac{1}{2c} \int_{x-ct}^{x+ct} G(y) \, dy \tag{4}$$

Proof   The derivation of (4) follows directly by applying the initial conditions to determine the arbitrary functions f and g in (2). We have

$$u(x,0) = f(x) + g(x) = F(x) \tag{5}$$
$$u_t(x,0) = cf'(x) - cg'(x) = G(x) \tag{6}$$

Dividing (6) by c then integrating gives

$$f(x) - g(x) = \frac{1}{c} \int_0^x G(y) \, dy + A \tag{7}$$

where A is a constant of integration. Adding and subtracting (5) and (7) yield the two equations

$$f(x) = \frac{1}{2} F(x) + \frac{1}{2c} \int_0^x G(y) \, dy + \frac{1}{2} A$$

$$g(x) = \frac{1}{2} F(x) - \frac{1}{2c} \int_0^x G(y) \, dy - \frac{1}{2} A$$

from which (4) follows. Since $F \in C^2$ and $G \in C$ it follows that $u \in C^2$ and satisfies the wave equation.

Much insight can be gained by examining a simple problem. Therefore let us consider the initial value problem

$$u_{tt} - c^2 u_{xx} = 0, \qquad x \in R^1, \quad t > 0$$
$$u(x,0) = F(x), \qquad u_t(x,0) = 0, \qquad x \in R^1$$

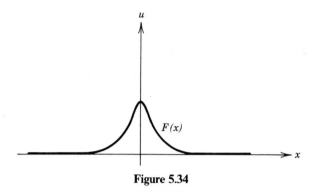

Figure 5.34

By (4) the solution is

$$u(x, t) = \tfrac{1}{2}(F(x - ct) + F(x + ct))$$

To fix the idea suppose $F(x)$ is the initial signal shown in Fig. 5.34. At time $t > 0$ we note that u is the average of $F(x - ct)$ and $F(x + ct)$, which is just the average of the two signals resulting from shifting $F(x)$ to the right ct units and to the left ct units. A sequence of time snapshots at $t_1 < t_2 < t_3$ in Fig. 5.35 shows schematically how the two signals $F(x - ct)$ and $F(x + ct)$ are averaged to produce u. Thus the initial signal splits into two signals moving in opposite directions each at speed c. The left moving signal travels along the characteristics $x + ct = $ constant and the right moving signal travels along the characteristics $x - ct = $ constant. Therefore there are two families of characteristic lines along which disturbances propagate.

If the initial data F and G in (3) are *supported* in an interval I on the x axis (i.e., F and G are zero outside I), then the region R of the xt plane that is affected by the disturbances within I is called the *region of influence* of I (see Fig. 5.36). This region has as its lateral boundaries the two characteristics $x + ct = $ constant and $x - ct = $ constant emanating from the left and right endpoints of I, respectively. A signal in I can never affect the solution outside R; or stated differently, if the initial data are supported in I, then the solution is supported in R. These statements follow directly from (4).

We may also pose the question of determining which initial values affect the value of u at a give point (x_0, t_0). From D'Alembert's solution

$$u(x_0, t_0) = \frac{1}{2}(F(x_0 - ct_0) + F(x_0 + ct_0)) + \frac{1}{2c}\int_{x_0 - ct_0}^{x_0 + ct_0} G(y)\, dy$$

Because of the integrated term u will depend only on the values in the interval J: $[x_0 - ct_0, x_0 + ct_0]$. This interval, called the *interval of dependence*, is found by following the characteristics $x - ct = x_0 - ct_0$ and $x + ct = x_0 + ct_0$ back to the initial $t = 0$ line (Fig. 5.37).

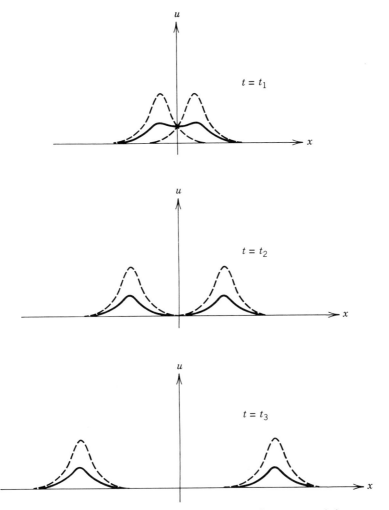

Figure 5.35. Time snapshots showing $u(x, t)$ (solid) as the average of the two profiles $F(x + ct)$ and $F(x - ct)$ (dashed).

When boundary conditions are present the situation is more complicated, since waves are reflected from the boundaries. For example, consider the mixed initial-boundary-value problem

$$u_{tt} - c^2 u_{xx} = 0, \qquad 0 < x < l, \quad t > 0$$
$$u(x,0) = F(x), \qquad u_t(x,0) = G(x), \qquad 0 \le x \le l$$
$$u(0, t) = a(t), \qquad u(l, t) = b(t), \qquad t > 0 \tag{8}$$

where $F \in C^2[0, l]$, $G \in C^1[0, l]$, and $a, b \in C^2[0, \infty)$. If $u(x, t)$ is a class C^2

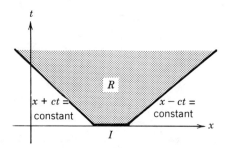

Figure 5.36. Region of influence R of an interval I.

solution on $t \geq 0$ and $0 \leq x \leq l$, then necessarily the compatibility conditions $F(0) = a(0)$, $F(l) = b(0)$, $G(0) = a'(0)$, $g(l) = b(0)$, $a''(0) = c^2 F''(0)$, $b''(0) = c^2 F(l)$ must hold at the corners $x = 0$, $x = l$, and $t = 0$. This problem can be transformed into one with homogeneous boundary conditions that can be attacked analytically by the Fourier method discussed in Section 4.2 (see Exercise 3.8). A geometrical construction of the solution based on the following lemma gives insight into the nature of the solution.

Lemma 3.1   *Let $u(x, t)$ be of class C^3 for $t > 0$ and $x \in R^1$. Then u satisfies the wave equation (1) if and only if u satisfies the difference equation*

$$u(x - ck, t - h) + u(x + ck, t + h)$$
$$= u(x - ch, t - k) + u(x + ch, t + k) \tag{9}$$

for all $h, k > 0$.

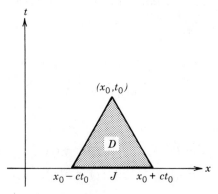

Figure 5.37. Domain D and interval J of dependence of a point (x_0, t_0).

Proof Necessity follows easily from the fact that the general solution of the wave equation is given by (2) and both the forward and backward going waves $f(x + ct)$ and $g(x - ct)$ satisfy the difference equation. Conversely, if u satisfies the difference equation, then set $h = 0$, add $-2u(x, t)$ to both sides and then divide by $c^2 k^2$ to get

$$\frac{u(x - ck, t) + u(x + ck, t) - 2u(x, t)}{c^2 k^2}$$
$$= \frac{u(x, t - k) + u(x, t + k) - 2u(x, t)}{c^2 k^2} \tag{10}$$

By Taylor's theorem

$$u(x \pm ck, t) = u(x, t) \pm u_x(x, t)ck + \tfrac{1}{2}u_{xx}(x, t)c^2 k^2 + o(k^2)$$
$$u(x, t \pm k) = u(x, t) \pm u_t(x, t)k + \tfrac{1}{2}u_{tt}(x, t)k^2 + o(k^2)$$

Substituting these quantities into (10) gives

$$u_{xx}(x, t) = \frac{1}{c^2}u_{tt}(x, t) + \frac{o(k^2)}{k^2}$$

Taking $k \to 0$ shows that u satisfies (1) and completes the proof.

Geometrically, the points A: $(x - ck, t - h)$, B: $(x + ch, t + k)$, C: $(x + ck, t + h)$, and D: $(x - ch, t - k)$ are the vertices of a *characteristic parallelogram ABCD* formed by two pairs of characteristics, one forward going pair and one backward going pair. Then (9) may be written

$$u(A) + u(C) = u(B) + u(D)$$

This equation may be used to geometrically construct a solution to (8) as follows. We divide the region $0 < x < l$, $t > 0$ into Regions I, II, III,... as shown in Fig. 5.38 by drawing in the characteristics emanating from the lower corners $(0, 0)$ and $l, 0)$ and continually reflecting them from the boundaries. The solution in Region I is completely determined by D'Alembert's solution (4). To find the solution at a point R in Region II we sketch the characteristic parallelogram $PQRS$ and use Lemma 3.1 to find $u(R) = u(S) + u(Q) - u(P)$. The quantity $u(S)$ is determined by the boundary condition at $x = 0$ and $u(P)$ and $u(Q)$ are known from the solution in Region I. A similar argument can be made for points in Region III. To determine the solution at a point M in Region IV we complete the characteristic parallelogram $KLMN$ as shown in Fig. 5.38 and use Lemma 3.1 to conclude $u(M) = u(L) + u(N) - u(K)$. Again the three quantities on the right are known and we can proceed to the next step and obtain u in the entire region $t > 0$, $0 < x < l$.

Further properties of the wave equation are discussed in Smoller [3].

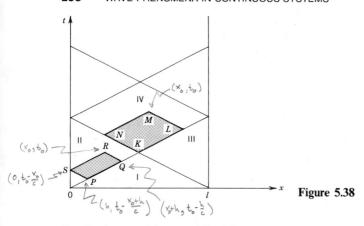

Figure 5.38

Scattering and Inverse Problems

In this section we consider an application of the mathematical theory of wave propagation. The exercises will contain further extensions of the theory and additional applications. By a scattering problem we mean the following. Let us consider a situation where an object, called a *scatterer*, lies in some medium. An incoming *incident wave* impinges upon the scatterer and produces a *reflected wave* and a *transmitted wave* (Fig. 5.39). The *direct scattering problem* is to determine the reflected and transmitted waves (amplitude, wave number, frequency, etc.) if the properties of the incident wave and the scatterer are known. The *inverse scattering problem* is to determine the properties of the scatterer, given the incident, reflected, and transmitted waves. Inverse scattering problems arise in a number of physical situations. For example, in radar or sonar theory a known incident wave and observed reflected wave are used to detect the properties or presence of aircraft or submarine objects. In tomography X-rays and sound waves are used to determine the presence or properties of tumors by detecting density variations, for example; in this case the incident, reflected, and transmitted waves are all known. In geological explora-

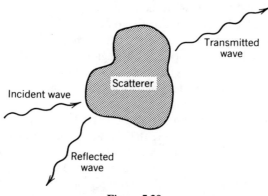

Figure 5.39

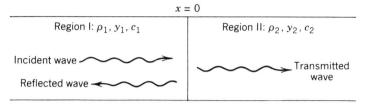

Figure 5.40. Interface between two materials as a scatterer.

tion explosions may be set off on the earth's surface producing waves that, when reflected from underground layers, may indicate the presence of oil or something else of geological interest. Inverse scattering theory is currently an active area of applied mathematics research.

In the succeeding paragraphs we set up and solve a simple inverse problem involving the wave equation. We assume that an infinite bar $-\infty < x < \infty$ is composed of two homogeneous materials separated by an interface at $x = 0$. One material in Region I ($x < 0$) has constant density ρ_1 and stiffness (Young's modulus) y_1 and the other in Region II ($x > 0$) has constant density ρ_2 and stiffness y_2. The sound velocities in Region I and Region II are c_1 and c_2, respectively. Recall that $c^2 = y/\rho$. The inverse scattering problem can be stated as follows: Suppose an observer at $x = -\infty$ sends out an incident right traveling wave $\exp(i(x - c_1 t))$ of unit amplitude and unit wave number. When the wave impinges upon the interface a wave is reflected back into Region I and another wave is transmitted into Region II (see Fig. 5.40). If the material properties ρ_1 and y_1 of Region I are known, it is possible to determine the properties ρ_2 and y_2 of Region II by measuring the amplitude and wave number of the reflected wave?

We answer this question in the context of small displacement theory where the displacements u_1 and u_2 in Region I and Region II, respectively, satisfy the one dimensional wave equation

$$u_{tt} - c^2 u_{xx} = 0 \tag{11}$$

At the interface $x = 0$ the displacements must be continuous, since we assume that the regions do not separate. Therefore

$$u_1(0^-, t) = u_2(0^+, t), \qquad t > 0 \tag{12}$$

Here $u_1(0^-, t)$ denotes the limit of $u_1(x, t)$ as $x \to 0^-$ and $u_2(0^+, t)$ denotes the limit of $u_2(x, t)$ as $x \to 0^+$. Further we require that the force across the interface be continuous or

$$y_1 \frac{\partial u_1}{\partial x}(0^-, t) = y_2 \frac{\partial u_2}{\partial x}(0^+, t), \qquad t > 0 \tag{13}$$

Equations (12) and (13) are called the *interface conditions*. We seek u_1 and u_2 satisfying (11) subject to (12) and (13). A reasonable attempt at a solution is to take u_1 to be the incident wave superimposed with the reflected wave and u_2 to be the transmitted wave. Hence we try a solution of the form

$$u_1 = \exp(i(x - c_1 t)) + A \exp(i\alpha(x + c_1 t))$$
$$u_2 = B \exp(i\beta(x - c_2 t))$$

The quantities A and α denote the amplitude and wave number of the reflected wave moving at speed c_1 in Region I to the left, and B and β denote the amplitude and wave number of the transmitted wave moving at speed c_2 in Region II to the right. Clearly u_1 and u_2 satisfy the wave equation with $c_1^2 = y_1/\rho_1$ and $c_2^2 = y_2/\rho_2$, respectively. The inverse scattering problem can now be stated in analytic terms as follows: Given A, α, ρ_1, and y_1, can ρ_2 and y_2 be determined?

The interface conditions (12) and (13) will yield relations among the constants. Condition (12) implies

$$\exp(-ic_1 t) + A \exp(i\alpha c_1 t) = B \exp(-i\beta c_2 t)$$

which in turn gives

$$\alpha = -1, \qquad \beta = \frac{c_1}{c_2}, \qquad B = 1 + A \tag{14}$$

Therefore

$$u_1 = \exp\left[i(x - c_1 t)\right] + A \exp\left[-i(x + c_1 t)\right]$$
$$u_2 = (1 + A)\exp\left[i\frac{c_1}{c_2}(x - c_2 t)\right]$$

An application of (13) forces

$$y_1(1 - A) = y_2(1 + A)\frac{c_1}{c_2} \tag{15}$$

A close examination of (14) and (15) yields the information we seek. Knowledge of A and c_1 permits the calculation of the ratio y_2/c_2 or $\sqrt{y_2\rho_2}$. Consequently both y_2 and ρ_2 cannot be calculated, only their product. If, however, there is means to measure the speed c_2 of the transmitted wave or determine its wave number β, then the material properties of Region II can be found.

The direct scattering problem has a simple solution. If all of the material parameters ρ_1, y_1, ρ_2, and y_2 are known, relations (14) and (15) uniquely determine the reflected and transmitted waves.

EXERCISES

3.1   Maxwell's equations for a nonconducting medium with permeability μ and permitivity ε are

$$\frac{\partial \mathbf{B}}{\partial t} + \text{curl } \mathbf{E} = 0, \qquad \varepsilon \frac{\partial \mathbf{E}}{\partial t} = \frac{1}{\mu} \text{ curl } \mathbf{B}$$

$$\text{div } \mathbf{B} = 0, \qquad \text{div } \mathbf{E} = 0$$

where $\mathbf{B}$ is the magnetic induction and $\mathbf{E}$ is the electric field. Show that the components of $\mathbf{B}$ and $\mathbf{E}$ satisfy the three dimensional wave equation $u_{tt} - c^2(u_{xx} + u_{yy} + u_{zz}) = 0$ with propagation speed $c = (\varepsilon\mu)^{-1/2}$, where $u = u(t, x, y, z)$.

3.2   Show that the transformation

$$\xi = x - ct, \qquad \eta = x + ct$$

transforms the wave equation (1) into the partial differential equation

$$\Phi_{\xi\eta} = 0, \qquad \Phi(\xi, \eta) = u(x, y)$$

Thus show that the general solution to (1) is given by (2).

3.3   Let u_1 and u_2 be solutions to the pure initial value problems

$$u_{tt} - c^2 u_{xx} = 0, \qquad x \in R^1, \quad 0 < t < T_0$$
$$u(x,0) = F_1(x), \qquad u_t(x,0) = G_1(x), \qquad x \in R^1$$

and

$$u_{tt} - c^2 u_{xx} = 0, \qquad x \in R^1, \quad 0 < t < T_0$$
$$u(x,0) = F_2(x), \qquad u_t(x,0) = G_2(x), \qquad x \in R^1$$

respectively. For any $\varepsilon > 0$ prove there exists a $\delta > 0$ (depending on ε) such that

$$|F_1 - F_2| \le \delta \quad \text{and} \quad |G_1 - G_2| \le \delta, \qquad x \in R^1$$

implies

$$|u_1 - u_2| \le \varepsilon, \qquad x \in R^1, \quad 0 < t < T_1$$

Thus the solution depends continuously upon the initial data.

3.4 Let $v \in C^2$ satisfy the boundary value problem

$$v_{tt} - \frac{y_0}{\rho_0} v_{xx} = 0, \qquad 0 < x < l, \quad t > 0$$

$$v(x,0) = v_t(x,0) = 0, \qquad 0 < x < l$$

$$v(0,t) = v(l,t) = 0, \qquad t > 0$$

(a) Prove that $v = 0$ by introducing the *energy integral*

$$E(t) = \int_0^l \left(\tfrac{1}{2}\rho_0 v_t^2 + \tfrac{1}{2} y_0 v_x^2 \right) dx$$

and showing that $E(t) \equiv 0$ for all $t \geq 0$.

(b) Prove that solutions to the boundary value problem

$$u_{tt} - c_0^2 u_{xx} = 0, \qquad 0 < x < l, \quad t > 0, \quad c_0^2 = \frac{y_0}{\rho_0}$$

$$u(x,0) = F(x), \qquad u_t(x,0) = G(x), \qquad 0 < x < l$$

$$u(0,t) = H(t), \qquad u(l,t) = K(t), \qquad t > 0$$

are unique.

3.5 Let $u = u(x,t)$ be a solution to the boundary value problem in the unit square $\Omega = \{(x,y): 0 < x, y < 1\}$:

$$u_{xy} = 0 \text{ in } \Omega, \qquad u = f \text{ on } \partial\Omega$$

Show that f must satisfy the relation $f(O) + f(P) = f(Q) + f(R)$, where $O = (0,0)$, $P = (1,1)$, $Q = (1,0)$, and $R = (0,1)$. Discuss well-posedness for this hyperbolic equation.

$-p$ 196

3.6 Determine the solution in Exercise 3.8 of Chapter 4 when $l = \pi$, $h = 1$, $c = 1$ and

$$f(x) = \begin{cases} x, & 0 \leq x \leq \dfrac{\pi}{2} \\[2mm] \pi - x, & \dfrac{\pi}{2} \leq x \leq \pi \end{cases}$$

Solution:

$$u = \frac{4}{\pi} \sum_{k=1}^{\infty} \frac{(-1)^{k+1}}{(2k-1)^2} \sin((2k-1)x)\cos\left(\sqrt{(2k-1)^2 + 1}\, t\right)$$

3.7 Use Fourier transforms to find the solution to

$$u_{tt} - c^2 u_{xx} = 0, \qquad x \in R^1, \quad t > 0$$
$$u(x,0) = 0, \qquad u_t(x,0) = G(x), \qquad x \in R^1$$

where G vanishes outside some closed, bounded interval.
Solution:

$$u(x,t) = \frac{1}{2\pi} \int_{-\infty}^{\infty} \int_{-\infty}^{\infty} G(\xi) \frac{\sin cst}{cs} e^{is(x-\xi)} \, d\xi \, ds$$

3.8 Consider the boundary value problem

$$u_{tt} - c^2 u_{xx} = 0, \qquad t > 0, \quad 0 < x < 1$$
$$u(x,0) = F(x), \qquad u_t(x,0) = G(x), \qquad 0 < x < 1$$
$$u(0,t) = a(t), \qquad u(1,t) = b(t), \qquad t > 0$$

Transform this problem to one with a nonhomogeneous partial differential equation and homogeneous boundary conditions.

3.9 Consider the nonhomogeneous problem

$$u_{tt} - c^2 u_{xx} = \phi(x,t), \qquad x \in R^1, \quad t > 0$$
$$u(x,0) = F(x), \qquad u_t(x,0) = G(x), \qquad x \in R^1$$

If D is the characteristic triangle shown in Fig. 5.37, show that

$$u(x_0, t_0) = \frac{1}{2} [F(x_0 + ct_0) + F(x_0 - ct_0)]$$

$$+ \frac{1}{2c} \int_{x_0 - ct_0}^{x_0 + ct_0} G(s) \, ds + \frac{1}{2c} \iint_D \phi(x,t) \, dx \, dt$$

[*Hint*: Integrate the partial differential equation over D and apply Green's Theorem.]

3.10 Consider a bar with constant stiffness y_0 and density ρ_0 occupying the region $x \leq 0$. An incident small displacement wave $\exp(i(x - c_0 t))$ impinges upon $x = 0$ from $x = -\infty$.

(a) Find the reflected wave if the end at $x = 0$ is fixed.

(b) Find the reflected wave if the end at $x = 0$ is free.

(c) Find the reflected wave if a mass M is attached to the end at $x = 0$
(*Hint*: Show that the boundary condition is $Mu_{tt} + y_0 u_x = 0$.)

(d) Find the reflected wave if the end at $x = 0$ is constrained by a light spring with spring constant $k > 0$ satisfying Hooke's law (see Exercise 2.11).

3.11 Find the solution to the *outgoing signaling problem*

$$\phi_{tt} - c^2\phi_{xx} = 0, \qquad x > 0, \quad -\infty < t < \infty$$

$$\phi_x(0, t) = Q(t), \quad -\infty < t < \infty$$

3.12 A bar of constant cross-sectional area is composed of two materials. For $-\infty < x < 0$ and $1 < x < \infty$ the material parameters are y_0 and ρ_0 and for $0 < x < 1$ the parameters are y_1 and ρ_1. An incident displacement wave $\exp(i(x - c_0t))$ impinges upon the system from $-\infty$. Assuming the linearized theory and the fact that displacements are continuous across the discontinuities what are the frequency and wave number of the transmitted wave in the region $x > 1$?

3.13 In three dimensions the wave equation is

$$u_{tt} - c^2\nabla^2 u = 0$$

where ∇^2 is the Laplacian. For waves with spherical symmetry $u = u(r, t)$ and $\nabla^2 u = u_{rr} + (2/r)u_r$. By introducing the variable $U = ru$, show that the general solution for the spherically symmetric wave equation is

$$u = \frac{1}{r}f(r - ct) + \frac{1}{r}g(r + ct)$$

3.14 **(a)** In the linearized theory for a bar of constant stiffness y_0 and variable density $\rho(x)$ show that the existence of time harmonic solutions of the form $u(x, t) = u(x)\exp(i\omega t)$ requires that u satisfy the one-dimensional *Helmholtz equation*

$$u''(x) + k^2(x)u(x) = 0, \qquad k^2(x) = \frac{\omega^2\rho(x)}{y_0}$$

(b) If $\rho(x) = \rho_0(1 + \bar{\rho}(x))$, where $\bar{\rho} \ll 1$, show that $k^2(x) = k_0^2(1 + \bar{\rho}(x))$, where $k_0^2 = \omega^2\rho_0/y_0$.

(c) Assuming $u(x) = v(x)\exp(ik_0x)$ with $v''(x) \ll 1$ show that $2v'(x) - ik_0\bar{\rho}(x)v(x) = 0$ (This is a one dimensional version of the Leontovich-Fock parabolic approximation.)

(d) Obtain the WKB (Wentzel-Kramers-Brillouin) approximation

$$u(x) = v_0 \exp\left\{ ik_0\left(x + \tfrac{1}{2}\int_{x_0}^{x} \bar{\rho}(\xi)\, d\xi \right)\right\}, \qquad v_0 \quad \text{constant}$$

5.4 GASDYNAMICS

Conservation Laws

In Section 5.2 the governing equations for the dynamics of fluid flow were obtained using a Lagrangian approach. That is, the conservation laws were written as they applied to the motion of an actual material region. This approach is in the same spirit as in the classical mechanics of particles and systems of particles where the governing law, Newton's second law, refers to the actual masses themselves. In the next few paragraphs an alternate derivation of the conservation laws is presented based on an Eulerian approach. Not only is this a common method for obtaining the equations, but its presentation is consistent with the general philosophy of extending our knowledge of the equations and the modeling processes available in applied mathematics. The Eulerian approach, based upon viewing the system from a fixed laboratory window, has no counterpart in classical mechanics. But the underlying concept is easily understood, for it is based on the simple observation that the time rate of change of a quantity inside a fixed spatial region must equal the rate at which the substance flows into the region minus the rate at which the substance flows out (see Section 5.1, Conservation Laws).

As before we imagine a fluid flowing in a tube of constant cross-sectional area A in such a way that the physical parameters are constant in any cross section. We fix two cross sections $x = a$ and $x = b$ (see Fig. 5.41). The first principle, that of mass conservation, states simply that mass is neither created nor destroyed, or the rate of change of mass inside the interval $a \le x \le b$ equals the mass flux in at $x = a$ less the mass flux out at $x = b$. In symbols

$$\frac{d}{dt}\int_{a}^{b} A\rho(x,t)\, dx = A\rho(a,t)v(a,t) - A\rho(b,t)v(b,t) \qquad (1)$$

where $\rho(x,t)$ is the density and $v(x,t)$ is the velocity. We observe that

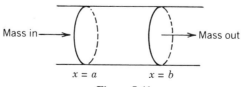

Mass in →

→ Mass out

$x = a$ $x = b$

Figure 5.41

$A\rho(x, t)v(x, t)$ gives the amount of fluid that crosses the location x per unit time and is hence the mass flux. Equation (1) can be written

$$\int_a^b \rho_t(x, t) \, dx = -\int_a^b (\rho(x, t)v(x, t))_x \, dx$$

Since the interval $a \leq x \leq b$ is arbitrary it follows that

$$\rho_t + (\rho v)_x = 0 \tag{2}$$

which is the continuity equation, or the partial differential equation expressing conservation of mass.

The momentum balance law states that the rate of change of momentum inside $[a, b]$ must equal the transport of momentum carried into and out of the region at $x = a$ and $x = b$ plus the momentum created inside $[a, b]$ by the stress (pressure) p acting at a and b. In mathematical terms

$$\frac{d}{dt} \int_a^b A\rho(x, t)v(x, t) \, dx = A\rho(a, t)v^2(a, t)$$

$$-A\rho(b, t)v^2(b, t) + Ap(a, t) - Ap(b, t) \tag{3}$$

where v is the fluid velocity. Here we have neglected body forces. The term $A\rho(x, t)v^2(x, t)$ represents the momentum times the velocity at x and thus is called the momentum flux at x. Using the fundamental theorem of calculus (3) implies

$$\int_a^b (\rho(x, t)v(x, t))_t \, dx = -\int_a^b (\rho(x, t)v^2(x, t) + p(x, t))_x \, dx$$

The arbitrariness of the interval $a \leq x \leq b$ forces

$$(\rho v)_t + (\rho v^2 + p)_x = 0 \tag{4}$$

which is a partial differential equation expressing conservation of momentum. Both (2) and (4) are of the form

$$\frac{\partial}{\partial t}[\cdots] + \frac{\partial}{\partial x}[\cdots] = 0$$

and are therefore said to be in *conservation form*.

In the simplest gasdynamical calculations equations (2) and (4) may be supplemented by the barotropic equation of state

$$p = f(\rho), \qquad f'(\rho) > 0 \tag{5}$$

which gives pressure as a function of density only. An example is provided by the isentropic flow of an ideal gas (see Examples 2.1 and 2.5). Motivated by acoustics (see Section 5.2) it is convenient to introduce the *local sound speed c* defined by

$$c^2 = f'(\rho) \tag{6}$$

Then (2) and (4) may be written

$$p_t + vp_x + c^2\rho v_x = 0 \tag{7}$$

$$\rho v_t + \rho vv_x + p_x = 0 \tag{8}$$

where we have used the facts that $p_x = c^2\rho_x$ and $p_t = c^2\rho_t$ to rewrite (2), and we have used (2) to replace the quantity ρ_t in (4), thereby giving (8). We note that (8) coincides with the momentum equation obtained directly in Section 5.2 using a Lagrangian approach. Equations (7) and (8) along with (5) provide a starting point for the study of isentropic flow of a gas.

Riemann's Method

Heretofore there has been little discussion of how fluid motion begins. It is common in gasdynamics, or compressible fluid dynamics, to imagine the flow being induced by the motion of a piston inside the cylindrical pipe. Such a device is not as unrealistic as it may first appear. The piston may represent the fluid on one side of the valve after it is opened or it may represent a detonator in an explosion process; while in aerodynamics it may represent a blunt object moving into a gas.

Accordingly we set up and solve a simple problem whose method of solution, called Riemann's method, is typical of more general problems in gasdynamics. We consider a gas in a tube initially in the constant state

$$v = 0, \qquad \rho = \rho_0, \qquad p = p_0, \qquad c = c_0 \tag{9}$$

with equation of state

$$p = k\rho^\gamma, \qquad k, \gamma > 0 \tag{10}$$

To initiate the motion a piston located initially at $x = 0$ is withdrawn slowly according to

$$x = X(t), \qquad X'(t) < 0, \qquad X''(t) < 0 \tag{11}$$

where X is given function (see Fig. 5.42). The problem is to determine v, p, c, and ρ for all $t > 0$ and $X(t) < x < \infty$. We may regard this problem as a boundary value problem where initial conditions are given along the positive x axis and v is given by the piston velocity along the piston path (see Fig. 5.43).

The method of solution is motivated by the study of the simple nonlinear model equation $u_t + c(u)u_x = 0$ from Section 5.1. There we were able to

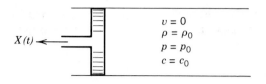

$$v = 0$$
$$\rho = \rho_0$$
$$p = p_0$$
$$c = c_0$$

$X(t)$

Figure 5.42

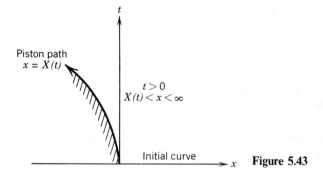

t

Piston path
$x = X(t)$

$t > 0$
$X(t) < x < \infty$

Initial curve

x

Figure 5.43

define a family of curves called characteristics along which signals propagated and along which the partial differential equation reduced to an ordinary differential equation. In the case of this model equation $du/dt = 0$ along $dx/dt = c(u)$. We follow a similar strategy for (7) and (8) and attempt to find characteristic curves along which the partial differential equations reduce to simpler equations. To this end we multiply (8) by c and then add and subtract it from (7) to obtain

$$p_t + (v + c)p_x + \rho c(v_t + (v + c)v_x) = 0 \qquad (12)$$

$$p_t + (v - c)p_x - \rho c(v_t + (v - c)v_x) = 0 \qquad (13)$$

Hence along the families of curves C^+ and C^- defined by

$$C^+: \frac{dx}{dt} = v + c \qquad (14)$$

$$C^-: \frac{dx}{dt} = v - c \qquad (15)$$

we have

$$\frac{dp}{dt} + \rho c \frac{dv}{dt} = 0, \quad \text{on} \quad C^+ \qquad (16)$$

$$\frac{dp}{dt} - \rho c \frac{dv}{dt} = 0, \quad \text{on} \quad C^- \qquad (17)$$

Equations (16) and (17) may be rewritten as

$$\frac{c}{\rho}\frac{d\rho}{dt} \pm \frac{dv}{dt} = 0, \quad \text{on} \quad C^+, C^-$$

Integrating gives

$$\int \frac{c(\rho)}{\rho} \, d\rho + v = \text{constant} \quad \text{on} \quad C^+ \tag{18}$$

$$\int \frac{c(\rho)}{\rho} \, d\rho - v = \text{constant} \quad \text{on} \quad C^- \tag{19}$$

The left sides of (18) and (19) are called the *Riemann invariants*; they are quantities that are constant along the characteristic curves C^+ and C^- defined by (14) and (15). If the equation of state is defined by (10), then

$$c^2 = k\gamma\rho^{\gamma-1}$$

and

$$\int \frac{c(\rho)}{\rho} \, d\rho = \frac{2c}{\gamma - 1}$$

Therefore the Riemann invariants are

$$r_+ \equiv \frac{2c}{\gamma - 1} + v = \text{constant} \quad \text{on} \quad C^+ \tag{20}$$

$$r_- \equiv \frac{2c}{\gamma - 1} - v = \text{constant} \quad \text{on} \quad C^- \tag{21}$$

We have enough information from the Riemann invariants to determine the solution of the piston withdrawal problem. First consider the C^- characteristics. Since $c = c_0$ when $v = 0$ the characteristics that begin on the x axis leave with speed $-c_0$ and the constant in (21) has valued $2c_0/(\gamma - 1)$. Hence on the C^- characteristics

$$\frac{2c}{\gamma - 1} - v = \frac{2c_0}{\gamma - 1} \tag{22}$$

Since the last equation must hold along every C^- characteristic it must hold everywhere and therefore r_- is constant in the entire region $t > 0$, $X(t) < x < \infty$. The C^- characteristics must end on the piston path, since their speed $v_p - c_p$ at the piston is more negative than the speed v_p of the piston (see Fig. 5.44).

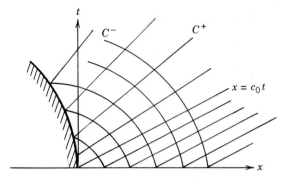

Figure 5.44. C^+ and C^- characteristics.

It is easy to see that the C^+ characteristics are straight lines. Adding and subtracting (22) and (20) shows

$$c = \text{constant} \quad \text{on a} \quad C^+ \quad \text{characteristic}$$
$$v = \text{constant} \quad \text{on a} \quad C^+ \quad \text{characteristic}$$

We have used the fact that (22) holds everywhere. The speed $v + c$ of a C^+ characteristic is therefore constant and thus the characteristic is a straight line. The C^+ characteristics emanating from the x axis have speed c_0 (since $v = 0$ on the x axis) and carry the constant state $v = 0$, $c = c_0$, $p = p_0$, $\rho = \rho_0$ into the region $x > c_0 t$. This is just the uniform state ahead of the signal $x = c_0 t$ beginning at the origin traveling at speed c_0 into the constant state. This signal indicates the initial motion of the piston.

A C^+ characteristic beginning on the piston at $(X(\tau), \tau)$ and passing through (x, t) has equation

$$x - X(\tau) = (c + v)(t - \tau) \tag{23}$$

The speed $v + c$ can be calculated as follows. Clearly $v = X'(t)$, which is the velocity of the piston. From (22)

$$\frac{2c}{\gamma - 1} - X'(\tau) = \frac{2c_0}{\gamma - 1}$$

and hence

$$c = c_0 + \frac{\gamma - 1}{2} X'(\tau)$$

Thus

$$x - X(\tau) = \left(c_0 + \frac{\gamma + 1}{2} X'(\tau) \right)(t - \tau) \tag{24}$$

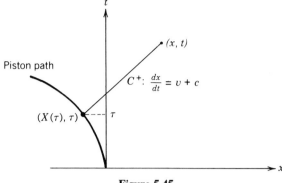

Figure 5.45

is the equation of the desired C^+ characteristic (see Fig. 5.45). The fact that v, c, p, and ρ are constant on the C^+ characteristics gives the solution

$$v(x, t) = X'(\tau)$$

where τ is given implicitly by (24). Obviously

$$c(x, t) = c_0 + \frac{\gamma - 1}{2} v(x, t)$$

The quantities p and ρ may be calculated from the equation of state and the definition of c^2.

Qualitatively we think of C^- characteristics as carrying the values $r_- = \frac{2c}{\gamma - 1} - v$ from the constant state back into the flow. The C^+ characteristics carry information from the piston forward into the flow. Whenever one of the Riemann invariants is constant throughout the flow we say that the solution in the nonuniform region is a *simple wave*. It can be shown that a simple wave solution always exists adjacent to a uniform state provided that the solution remains smooth. A complete analysis is given in Courant and Friedrichs [4].

In the piston withdrawal problem previously discussed we assumed that the piston path $x = X(t)$ satisfied the conditions $X'(t) < 0$ and $X''(t) < 0$, which means the piston is always accelerating backwards. If there is ever an instant of time when the piston slows down, that is, $X''(t) > 0$, then it is not difficult to see that a smooth solution cannot exist for all $t > 0$, for two distinct C^+ characteristics emanating from the piston will cross. Suppose, for example, that τ_1 and τ_2 are two values of t in a time interval where $X''(t) > 0$. The speed of characteristics leaving the piston is

$$v + c = c_0 + \frac{\gamma + 1}{2} X'(\tau)$$

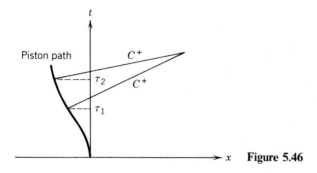

Figure 5.46

If $\tau_1 < \tau_2$, then $X'(\tau_1) < X'(\tau_2)$ and it follows that $v + c$ at τ_1 is smaller than $v + c$ at τ_2. Therefore the characteristic emanating from $(X(\tau_2), \tau_2)$ is faster than the characteristic emanating from $(X(\tau_1), \tau_1)$. Thus the two characteristics must cross (see Fig. 5.46).

Example 4.1 Beginning at time $t = 0$ a piston located at $x = 0$ moves forward into a gas under uniform conditions with equation of state given by (10). Its path is given by $X(t) = at^2$, $a > 0$. We determine the first instant of time that two characteristics cross and the wave breaks. For this problem the preceding analysis remains valid and the C^+ characteristics emanating from the piston have Equation (24) or

$$x - a\tau^2 = (c_0 + (\gamma + 1)a\tau)(t - \tau)$$

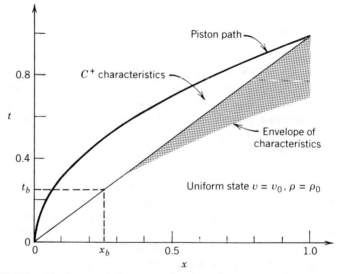

Figure 5.47. The C^+ characteristics emanating from the piston path and intersecting in the shaded region. The region is bounded above by the curve $x = c_0 t$ and below by the envelop of the characteristics. The shock begins at (x_b, t_b).

or

$$F(x, t, \tau) \equiv \gamma a \tau^2 + (c_0 - (\gamma + 1)at)\tau + x - c_0 t = 0 \qquad (25)$$

Along such a characteristic v is constant and hence

$$v(x, t) = X'(\tau) = 2a\tau$$

where $\tau = \tau(x, t)$ is to be determined from (25). To solve (25) for τ requires by the implicit function theorem that $F_\tau(x, t, \tau) \neq 0$. Since $F_\tau = 2\gamma a\tau + (c_0 - (\gamma + 1)at)$, the first instant of time t that (25) cannot be solved for τ is

$$t_b = \min_{\tau \geq 0} \frac{c_0 + 2\gamma a\tau}{(\gamma + 1)a} = \frac{c_0}{(\gamma + 1)a}$$

That is, the breaking time t_b will occur along the first characteristic (indexed by τ) where $F_\tau = 0$. The characteristic diagram is shown in Fig. 5.47 in the special case $a = c_0 = 1$ and $\gamma = 3$.

The Rankine-Hugoniot Conditions

As we observed earlier smooth solutions break down when characteristics intersect, since constant values are carried along the characteristics. The solution that develops is a discontinuous one and it propagates as a shock wave. We now determine what conditions hold across such a discontinuity. We proceed as in Section 5.1 where conservation laws are discussed.

The integral form of the conservation of mass law, namely

$$\frac{d}{dt} \int_a^b \rho(x, t) \, dx = -\rho(x, t)v(x, t)\Big|_a^b \qquad (26)$$

holds in all cases, even when the functions ρ and v are not smooth. Let $x = s(t)$ be a smooth curve in space–time, which intersects the interval $a \leq x \leq b$ at time t (see Fig. 5.48), and let v, ρ, and p suffer a simple discontinuity along the curve. Otherwise v, ρ, and p are assumed to be C^1 functions with finite limits on each side of $x = s(t)$. Then by Leibniz's rule

$$\frac{d}{dt} \int_a^b \rho(x, t) \, dx = \frac{d}{dt} \int_a^{s(t)} \rho(x, t) \, dx + \frac{d}{dt} \int_{s(t)}^b \rho(x, t) \, dx$$

$$= \int_a^{s(t)} \rho_t(x, t) \, dx + \rho(s(t)^-, t)s'(t)$$

$$+ \int_{s(t)}^b \rho_t(x, t) \, dx - \rho(s(t)^+, t)s'(t)$$

Both integrals on the right side approach zero as $a \to s(t)^-$ and $b \to s(t)^+$

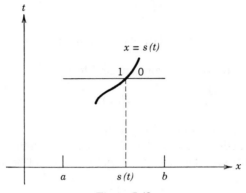

Figure 5.48

and therefore from (26) it follows that

$$\left(\rho\left(s(t)^-,t\right) - \rho\left(s(t)^+,t\right)\right)s'(t) = -\rho(x,t)v(x,t)\big|_{s(t)^-}^{s(t)^+} \qquad (27)$$

If the values (one-sided limits) of ρ and v on the right and left sides of $x = s(t)$ are denoted by the subscripts zero and one, respectively, for example,

$$\rho_0 = \lim_{x \to s(t)^+} \rho(x,t), \qquad \rho_1 = \lim_{x \to s(t)^-} \rho(x,t)$$

then (27) can be written

$$(\rho_1 - \rho_0)s'(t) = \rho_1 v_1 - \rho_0 v_0 \qquad (28)$$

In a similar fashion (see Exercise 4.1) the condition

$$(\rho_1 v_1 - \rho_0 v_0)s'(t) = \rho_1 v_1^2 + p_1 - \rho_0 v_0^2 - p_0 \qquad (29)$$

can be obtained from the integral form (3) of the conservation of momentum law. The two conditions (28) and (29) are known as the *Rankine-Hugoniot jump conditions*; they relate the values ρ_0, v_0, p_0 ahead of the discontinuity to the speed s' of the discontinuity and to the values ρ_1, v_1, and p_1 behind it. If the state ahead is at rest, that is, $v_0 = 0$, then the conditions become

$$(\rho_1 - \rho_0)s' = \rho_1 v_1 \qquad (30)$$

$$\rho_1 v_1 s' = \rho_1 v_1^2 + p_1 - p_0 \qquad (31)$$

Supplemented with the equation of state

$$p_1 = f(\rho_1) \qquad (32)$$

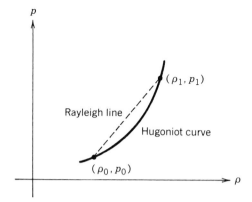

Figure 5.49. A ρp Hugoniot diagram.

the equations can be regarded as three equations in the four unknowns ρ_1, v_1, p_1, and s'. If any one of these quantities is known, the remaining three may be determined.

It is helpful to picture the information contained in (30)–(32) on a ρp diagram as shown in Fig. 5.49. The solid curve is the graph of the equation of state $p = f(\rho)$. It is known as the *Hugoniot curve*; and all states, both initial and final, must lie on this curve. The straight line connecting the state ahead (ρ_0, p_0) to the state behind (ρ_1, p_1) is the *Rayleigh line*. If (31) is rewritten using Equation (30) as

$$p_1 - p_0 = \rho_0 v_1 s'$$

then it is clear that the slope of the Rayleigh line is given by

$$\frac{p_1 - p_0}{\rho_1 - \rho_0} = \frac{\rho_0}{\rho_1} s'^2$$

These ideas extend beyond this simple case to more general problems in gasdynamics. See Courant and Friedrichs [4].

EXERCISES

4.1 Derive the jump condition (29).

4.2 Discuss the piston withdrawal problem when the piston path is given by $x = -V_0 t$ where V_0 is a positive constant. Assume an equation of state $p = k\rho^\gamma$ with uniform conditions $v = 0$, $p = p_0$, $\rho = \rho_0$, $c = c_0$ ahead of the piston at $t = 0$.

4.3 Write the following partial differential equations in conservation form, i.e. in the form $[\,\cdots\,]_t + [\,\cdots\,]_x = 0$.

(a) $u_t - 6uu_x + u_{xxx} = 0$

(b) $u_t + 3u^2 u_x + uu_{xx} + u_x^2 = 0$

4.4 At $t = 0$ the gas in a tube $x \geq 0$ is at rest. For $t > 0$ a piston initially located at $x = 0$ moves according to the law $x = X(t)$ where $X(t)$ is small. Show that in the acoustic approximation the motion induced in the gas is

$$v = X'(t - x/c_0)$$

and find the corresponding density variation.

4.5 A sphere of initial radius r_0 pulsates according to $r(t) = r_0 + A \sin \omega t$. In the linearized theory find the motion of the gas outside the sphere.

5.5 FLUID MOTIONS IN R^3

Kinematics

In the early sections of this chapter the equations governing fluid motion were developed in one spatial dimension. This limitation is severe, however, because most interesting fluid phenomena occur in higher dimensions. In this section we derive the field equations for fluid dynamics in three dimensions, with the two dimensional case being an obvious corollary.

Let Ω_0 be an arbitrary closed bounded set in R^3. We imagine that Ω_0 is a region occupied by a fluid at time $t = 0$. By a *fluid motion* we mean a mapping $\phi_t \colon \Omega_0 \to \Omega_t$ defined for all t in an interval I containing the origin, which maps the region Ω_0 into $\Omega_t = \phi_t(\Omega_0)$, the latter being the region occupied by the same fluid at time t (see Fig. 5.50). We assume that ϕ_t is represented by the formula

$$\mathbf{x} = \mathbf{x}(\mathbf{h}, t) \tag{1}$$

where $\mathbf{h} = (h_1, h_2, h_3)$ is in Ω_0 and $\mathbf{x} = (x_1, x_2, x_3)$ with

$$\mathbf{x}(\mathbf{h}, 0) = \mathbf{h}$$

Thus $\mathbf{h}$ is a Lagrangian coordinate or particle label given to each fluid particle at $t = 0$ and $\mathbf{x}$ is the Eulerian coordinate representing the laboratory position of the particle $\mathbf{h}$ at time t. We further assume that the function $\mathbf{x}(\mathbf{h}, t)$ is twice continuously differentiable on its domain $\Omega_0 \times I$ and that for each $t \in I$ the mapping ϕ_t has a unique inverse defined by

$$\mathbf{h} = \mathbf{h}(\mathbf{x}, t)$$

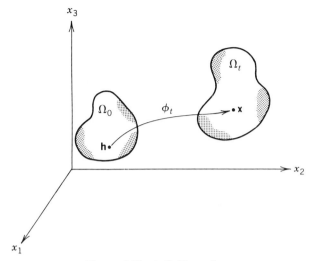

Figure 5.50. A fluid motion.

Hence

$$x(h(x, t), t) = x, \qquad h(x(h, t), t) = h$$

By convention, functions of h and t that represent the results of Lagrangian measurements will be denoted by capital letters, and functions of x and t that represent the results of a fixed laboratory measurement will be denoted by lowercase letters. The measurements are connected by the formulas

$$f(x, t) = F(h(x, t), t), \qquad F(h, t) = f(x(h, t), t)$$

In the sequel we often use the notation $F|_x$ and $f|_h$ to denote the right sides of these equations, respectively.

For a given fluid motion (1) the *velocity* is defined by

$$V(h, t) = \frac{\partial x}{\partial t}(h, t)$$

Then

$$v(x, t) = V(h(x, t), t)$$

$V(h, t) = \langle V_1(h, t), V_2(h, t), V_3(h, t) \rangle$ is the actual velocity vector of the fluid particle h at time t, whereas $v(x, t) = \langle v_1(x, t), v_2(x, t), v_3(x, t) \rangle$ is the velocity vector measured by a fixed observer at location x. If h_0 is a fixed point in Ω_0, then the curve $x = x(h_0, t)$, $t \in I$, is called the *particle path* of the particle h_0. The following theorem states that knowledge of the Eulerian velocity field $v(x, t)$ is equivalent to knowledge of all the particle paths.

Theorem 5.1   *If the velocity vector field* $\mathbf{v}(\mathbf{x}, t)$ *is known, then all of the particle paths can be determined, and conversely.*

Proof   First we prove the converse. If each particle path is given, then the function $\mathbf{x} = \mathbf{x}(\mathbf{h}, t)$ is known for all $h \in \Omega_0$. Consequently $\mathbf{h} = \mathbf{h}(\mathbf{x}, t)$ can be determined and v can be calculated from

$$\mathbf{v}(\mathbf{x}, t) = \mathbf{V}(\mathbf{h}(\mathbf{x}, t), t) = \mathbf{x}_t(\mathbf{h}(\mathbf{x}, t), t)$$

Now assume $\mathbf{v}(\mathbf{x}, t)$ is given. Then

$$\frac{\partial \mathbf{x}}{\partial t}(\mathbf{h}, t) = \mathbf{v}(\mathbf{x}(\mathbf{h}, t), t)$$

or

$$\frac{\partial \mathbf{x}}{\partial t} = \mathbf{v}(\mathbf{x}, t) \tag{2}$$

where explicit dependence on $\mathbf{h}$ has been dropped. Equation (2) is a first order system of differential equations for $\mathbf{x}$. Its solution subject to the initial condition

$$\mathbf{x} = \mathbf{h} \quad \text{at} \quad t = 0$$

gives the particle paths $\mathbf{x} = \mathbf{x}(\mathbf{h}, t)$.

It is interesting to contrast the particle paths with the so-called *streamlines* of the flow. The latter are integral curves of the vector field $\mathbf{v}(\mathbf{x}, t_0)$ frozen at some fixed but arbitrary instant t_0 of time. Thus the streamlines are found by solving the system

$$\frac{d\mathbf{x}}{ds} = \mathbf{v}(\mathbf{x}, t_0) \tag{3}$$

where $\mathbf{x} = \mathbf{x}(s)$ and s is a parameter along the curves. If $\mathbf{v}$ is independent of t, then the flow is called *steady* and the streamlines coincide with the particle paths. In a time-dependent flow they need not coincide.

Example 5.1   Consider the fluid motion

$$x_1 = t + h_1, \qquad x_2 = h_2 \exp t, \qquad x_3 = th_1 + h_3, \qquad t > 0$$

Then

$$\mathbf{V} = \left\langle \frac{\partial x_1}{\partial t}, \frac{\partial x_2}{\partial t}, \frac{\partial x_3}{\partial t} \right\rangle = \langle 1, h_2 \exp t, h_1 \rangle$$

Inverting the motion

$$h_1 = x_1 - t, \qquad h_2 = x_2 \exp(-t), \qquad h_3 = x_3 - tx_1 + t^2$$

Therefore

$$\mathbf{v} = \langle 1, x_2, x_1 - t \rangle$$

and hence the motion is not steady. The streamlines at time t_0 are given by solutions to (3), or

$$\frac{dx_1}{ds} = 1, \qquad \frac{dx_2}{ds} = x_2, \qquad \frac{dx_3}{ds} = x_1 - t_0$$

Thus

$$x_1 = s + c_1, \qquad x_2 = c_2 \exp s, \qquad x_3 = \tfrac{1}{2}s^2 + (c_1 - t_0)s + c_3$$

where c_1, c_2 and c_3 are constants.

As in the one dimensional case we define the *convective* or *material derivative* by

$$\frac{Df}{Dt}(\mathbf{x}, t) \equiv F_t(\mathbf{h}, t)|_{\mathbf{h} = \mathbf{h}(\mathbf{x}, t)}$$

where f and F are the Eulerian and Lagrangian representations of a given measurement, respectively. A straightforward calculation using the chain rule shows that

$$F_t(\mathbf{h}, t) = \frac{\partial}{\partial t} f(\mathbf{x}(\mathbf{h}, t), t)$$

$$= f_t(\mathbf{x}(\mathbf{h}, t), t) + \sum_{j=1}^{3} f_{x_j}(\mathbf{x}(\mathbf{h}, t), t) V_j(\mathbf{h}, t)$$

$$= f_t(\mathbf{x}(\mathbf{h}, t), t) + \mathbf{V} \cdot \mathbf{grad}\, f|_{\mathbf{h}}$$

Evaluating at $\mathbf{h} = \mathbf{h}(x, t)$ gives

$$\frac{Df}{Dt}(\mathbf{x}, t) = f_t(\mathbf{x}, t) + \mathbf{v}(\mathbf{x}, t) \cdot \mathbf{grad}\, f(\mathbf{x}, t) \qquad (4)$$

If $\mathbf{f} = \langle f_1, f_2, f_3 \rangle$ is a vector function, then componentwise

$$\frac{Df_i}{Dt} = \frac{\partial f_i}{\partial t} + \sum_{j=1}^{3} v_j \frac{\partial f_i}{\partial x_j}, \qquad i = 1, 2, 3$$

and we write

$$\frac{D\mathbf{f}}{Dt} = \mathbf{f}_t + \sum_{j=1}^{3} v_j \frac{\partial \mathbf{f}}{\partial x_j} \equiv \mathbf{f}_t + (\mathbf{v} \cdot \mathbf{grad})\mathbf{f}$$

Here $\mathbf{v} \cdot \mathbf{grad}$ is a special notation for the operator $v_1 \partial/\partial x_1 + v_2 \partial/\partial x_2 + v_3 \partial/\partial x_3$. The convective derivative $D\mathbf{f}/Dt$ at $(\mathbf{x}, t)$ is a measure of the time rate of change of the quantity F moving with the particle $\mathbf{h}$ frozen at the instant the particle is at $\mathbf{x}$.

Another kinematical result is the three dimensional analog of the Euler expansion theorem (see Section 5.2). It gives the time rate of change of the Jacobian

$$J(\mathbf{h}, t) = \det\left(\frac{\partial x_i}{\partial h_j} \right)$$

of the transformation ϕ_t. We record this as the next theorem.

Theorem 5.2   *In Lagrangian and Eulerian form, respectively,*

$$J_t(\mathbf{h}, t) = J(\mathbf{h}, t)\, \mathrm{div}\, \mathbf{v}(\mathbf{x}, t)\big|_{\mathbf{h}}$$
$$\frac{Dj}{Dt}(\mathbf{x}, t) = j(\mathbf{x}, t)\, \mathrm{div}\, v(\mathbf{x}, t)$$

where $j(\mathbf{x}, t) = J(\mathbf{h}(\mathbf{x}, t), t)$.

The proof of this theorem is left as an exercise (see Exercise 5.1).

In the one dimensional case the derivation of many of the results depended upon the calculation of the time derivative of an integral over the moving material region. For three dimensions we state this as a fundamental theorem.

Theorem 5.3   (*The Convection Theorem*)   *Let* $g = g(\mathbf{x}, t)$ *be a continuously differentiable function. Then*

$$\frac{d}{dt} \int_{\Omega_t} g\, dx = \int_{\Omega_t} \left(\frac{Dg}{Dt} + g\, \mathrm{div}\, \mathbf{v} \right) dx \tag{5}$$

Before proceeding with the proof we comment on some notational conventions and basic results from the calculus of several variables. In (5) the volume element is $dx = dx_1\, dx_2\, dx_3$ in the region Ω_t. The volumes we consider will always be *nice* with a smooth, or occasionally piecewise smooth, boundary $\partial \Omega_t$, so that the outward unit normal is well-defined and there is no doubt about what is meant by the bounding surface, the material inside or outside the boundary, etc. An integral over the surface $\partial \Omega_t$ will be denoted by

$$\int_{\partial \Omega_t} (\cdots) \, d\tau$$

where $d\tau$ is a surface element. A fundamental result is the *divergence theorem*, which states that for a continuously differentiable vector field $\mathbf{f}(\mathbf{x})$ on a volume Ω,

$$\int_\Omega \operatorname{div} \mathbf{f} \, dx = \int_{\partial \Omega} \mathbf{f} \cdot \mathbf{n} \, d\tau \tag{6}$$

where $\mathbf{n}$ is the outward unit normal to the surface $\partial \Omega$.

Proof of the Convection Theorem   The region of integration Ω_t of the integral on the left side of (5) depends on t, and therefore the time derivative cannot be brought under the integral sign directly. A change of variables in the integral from Eulerian coordinates $\mathbf{x}$ to Lagrangian coordinates $\mathbf{h}$ will transform the integral to one over the fixed, time-independent volume Ω_0. The derivative may then be brought under the integral sign and the calculation can proceed. Letting $\mathbf{x} = \mathbf{x}(\mathbf{h}, t)$, we have

$$\int_{\Omega_t} g(\mathbf{x}, t) \, dx = \int_{\Omega_0} G(\mathbf{h}, t) J(\mathbf{h}, t) \, dh$$

Here we have used the familiar formula for changing variables in a multiple integral where the Jacobian enters as a factor in the new volume element. Then also using Theorem 5.2

$$\frac{d}{dt} \int_{\Omega_t} g \, dx = \int_{\Omega_0} \frac{\partial}{\partial t} (GJ) \, dh$$

$$= \int_{\Omega_0} \left(G_t + G \operatorname{div} \mathbf{v} \big|_{\mathbf{x} = \mathbf{x}(\mathbf{h}, t)} \right) J \, dh$$

$$= \int_{\Omega_t} \left(\frac{Dg}{Dt} + g \operatorname{div} \mathbf{v} \right) dx$$

where in the last step we transformed back to Eulerian coordinates and used (4).

The *Reynold's transport theorem* is an immediate corollary.

Corollary 5.1 *We have*

$$\frac{d}{dt} \int_{\Omega_t} g \, dx = \int_{\Omega_t} g_t \, dx + \int_{\partial\Omega_t} g\mathbf{v} \cdot \mathbf{n} \, d\tau \tag{7}$$

where $\mathbf{n}$ *is the outer unit normal to the surface* $\partial\Omega_t$.

Proof The integrand on the right side of (5) is

$$\frac{Dg}{Dt} + g \operatorname{div} \mathbf{v} = g_t + \operatorname{div} g\mathbf{v}$$

Thus

$$\int_{\Omega_t} \left(\frac{Dg}{Dt} + g \operatorname{div} \mathbf{v} \right) dx = \int_{\Omega_t} g_t \, dx + \int_{\Omega_t} \operatorname{div} g\mathbf{v} \, dx$$

An application of the divergence theorem (6) yields

$$\int_{\Omega_t} \operatorname{div} g\mathbf{v} \, dx = \int_{\partial\Omega_t} g\mathbf{v} \cdot \mathbf{n} \, d\tau$$

and the corollary follows.

From vector analysis we recall that the integral

$$\int_{\partial\Omega_t} g\mathbf{v} \cdot \mathbf{n} \, d\tau$$

represents the flux of the vector field $g\mathbf{v}$ through the surface $\partial\Omega_t$. Hence (7) states that the time rate of change of the quantity $\int_{\Omega_t} g$, where both the integrand and the region of integration depend on t, equals the integral, frozen in time, of the change in g plus the flux or convection of g through the boundary $\partial\Omega_t$. Equation (7) is the three dimensional version of Leibniz's formula.

With the machinery developed in the preceding paragraphs it is a simple matter to obtain a mathematical expression of mass conservation. We take it as a physical axiom that a given material volume of fluid has the same mass as it evolves in time. Symbolically

$$\int_{\Omega_t} \rho(\mathbf{x}, t) \, dx = \int_{\Omega_0} \rho(\mathbf{x}, 0) \, dx$$

where $\rho(\mathbf{x}, t)$ is the Eulerian density. Thus

$$\frac{d}{dt} \int_{\Omega_t} \rho(\mathbf{x}, t) \, dx = 0$$

and Theorem 5.3 gives (taking $g = \rho$)

$$\frac{D\rho}{Dt} + \rho \operatorname{div} \mathbf{v} = 0 \qquad (8)$$

since the volume Ω_t is arbitrary. The first order nonlinear partial differential equation (8) is called the *continuity equation* and it is a mathematical expression for conservation of mass. An important class of fluid motions are those in which a material region maintains the same volume. Thus we say a fluid motion $\phi_t \colon \Omega_0 \to \Omega_t$ is *incompressible* if

$$\frac{d}{dt} \int_{\Omega_t} dx = 0, \qquad \text{for all} \quad \Omega_0$$

Theorem 5.4 *The following are equivalent*

    (i) ϕ_t is incompressible
    (ii) $\operatorname{div} \mathbf{v} = 0$
    (iii) $\Delta_t = 0$
    (iv) $\dfrac{D\rho}{Dt} = 0$

where $\Delta(\mathbf{h}, t) = \rho(\mathbf{x}(\mathbf{h}, t), t)$ is the Lagrangian density.

Proof By definition, (iii) and (iv) are equivalent, while (ii) and (iv) are equivalent by conservation of mass (8). Finally (i) and (ii) are equivalent by setting $g = 1$ in the convection theorem.

Dynamics

The basic motivation for the nature of the forces upon a fluid element was presented in Section 5.2 for one dimensional flows. The idea in three dimensions is again to generalize Newton's second law, which states for a particle of mass m that $d\mathbf{p}/dt = \mathbf{F}$, or the time rate of change of momentum is the total force. For a material region Ω_t of fluid we define the momentum by

$$\int_{\Omega_t} \rho \mathbf{v} \, dx$$

The forces on the fluid region are of two types, body forces, which act at each point of the region, and surface forces or tractions, which act upon the boundary of the region. Body forces will be denoted by $\mathbf{f}(\mathbf{x}, t)$, which represents the force per unit mass acting at the point $\mathbf{x}$ at time t. Thus the total

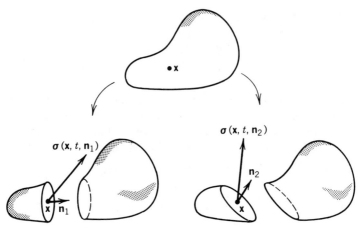

Figure 5.51. Diagram showing the dependence of the stress vector σ on the unit normal $\mathbf{n}$.

body force on the region Ω_t is

$$\int_{\Omega_t} \rho \mathbf{f} \, dx$$

At each point $\mathbf{x}$ on the boundary $\partial \Omega_t$ of the given region Ω_t we assume that there exists a vector $\sigma(\mathbf{x}, t; \mathbf{n})$ called the *stress vector*, which represents the force per unit area acting at $\mathbf{x}$ at time t on the surface $\partial \Omega_t$ by the material exterior to Ω_t. We note that the stress at $\mathbf{x}$ depends upon the orientation of the surface through $\mathbf{x}$. Simply put, different surfaces will have different stresses (see Fig. 5.51). The dependence of σ upon the orientation of the surface is denoted by the argument $\mathbf{n}$ in $\sigma(\mathbf{x}, t; \mathbf{n})$, where $\mathbf{n}$ is the outward unit normal to the surface. We think of $\mathbf{n}$ as pointing toward the material that is causing the stress. Therefore the total surface force or traction on the region Ω_t is given by

$$\int_{\partial \Omega_t} \sigma(x, t; \mathbf{n}) \, d\tau$$

We now postulate the *balance of momentum principle*, or *Cauchy's stress principle*, which states that for all material regions Ω_t

$$\frac{d}{dt} \int_{\Omega_t} \rho \mathbf{v} \, dx = \int_{\Omega_t} \rho \mathbf{f} \, dx + \int_{\partial \Omega_t} \sigma(\mathbf{x}, t; \mathbf{n}) \, d\tau \tag{9}$$

In words, the time rate of change of momentum equals the total force. Since (Exercise 5.4)

$$\frac{d}{dt} \int_{\Omega_t} \rho \mathbf{v} \, dx = \int_{\Omega_t} \rho \frac{D\mathbf{v}}{Dt} \, dx$$

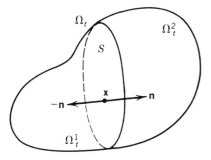

Figure 5.52

we may write (9) as

$$\int_{\Omega_t} \rho \left(\frac{D\mathbf{v}}{Dt} - \mathbf{f} \right) dx = \int_{\partial \Omega_t} \mathbf{\sigma}(\mathbf{x}, t; \mathbf{n}) \, d\tau \tag{10}$$

At a given point $\mathbf{x}$ in a fluid on a surface defined by the orientation $\mathbf{n}$ the stress on the fluid exterior to Ω_t caused by the fluid interior to Ω_t is equal and opposite to the stress on the interior caused by the fluid exterior to Ω_t. More precisely, see the following theorem.

Theorem 5.5 (*Action–Reaction*)

$$\mathbf{\sigma}(\mathbf{x}, t; -\mathbf{n}) = -\mathbf{\sigma}(\mathbf{x}, t; \mathbf{n}) \tag{11}$$

Proof Let Ω_t be a material region and divide it into two regions Ω_t^1 and Ω_t^2 by a surface S that passes through an arbitrary point $\mathbf{x}$ in Ω_t (see Fig. 5.52).

Applying (10) to Ω_t^1 and Ω_t^2 we obtain

$$\int_{\Omega_t^1} \rho \left(\frac{D\mathbf{v}}{Dt} - \mathbf{f} \right) dx = \int_{\partial \Omega_t^1} \mathbf{\sigma} \, d\tau \tag{12}$$

$$\int_{\Omega_t^2} \rho \left(\frac{D\mathbf{v}}{Dt} - \mathbf{f} \right) dx = \int_{\partial \Omega_t^2} \mathbf{\sigma} \, d\tau \tag{13}$$

Subtracting (10) from the sum of (12) and (13) gives

$$\mathbf{0} = \int_S \left(\mathbf{\sigma}(\mathbf{x}, t; \mathbf{n}) + \mathbf{\sigma}(\mathbf{x}, t; -\mathbf{n}) \right) d\tau$$

where $\mathbf{n}$ is the outward unit normal to Ω_t^1. The integral mean value theorem implies

$$\left[\mathbf{\sigma}(\mathbf{z}, t; \mathbf{n}_1) + \mathbf{\sigma}(\mathbf{z}, t; -\mathbf{n}_1) \right] \cdot \text{area}(S) = \mathbf{0} \tag{14}$$

where $\mathbf{z}$ is some point on S and $\mathbf{n}_1$ is the outward unit normal to Ω_t^1 at $\mathbf{z}$. Taking the limit as the volume of Ω_t goes to zero in such a way that the area of S goes to zero and $\mathbf{x}$ remains on S we get $\mathbf{z} \rightarrow \mathbf{x}$ and $\mathbf{n}_1 \rightarrow \mathbf{n}$, and so (14) implies the result.

As it turns out the stress vector $\sigma(\mathbf{x}, t; \mathbf{n})$ depends in a very special way upon the unit normal $\mathbf{n}$. The next theorem is one of the fundamental results in fluid mechanics.

Theorem 5.6   *(Cauchy)*   *Let* $\sigma_i(\mathbf{x}, t; \mathbf{n})$, $i = 1, 2, 3$, *denote the three components of the stress vector* $\sigma(\mathbf{x}, t; \mathbf{n})$, *and let* $\mathbf{n} = (n_1, n_2, n_3)$. *Then there exists a matrix function* $\sigma_{ji}(\mathbf{x}, t)$ *such that*

$$\sigma_i = \sum_{j=1}^{3} \sigma_{ji} n_j \qquad (15)$$

In fact

$$\sigma_{ji}(\mathbf{x}, t) = \sigma_i(\mathbf{x}, t; \mathbf{e}_j) \qquad (16)$$

where $\mathbf{e}_j$, $j = 1, 2, 3$, *denote the unit vectors in the direction of the coordinate axes* x_1, x_2, *and* x_3, *respectively.*

Proof   Let t be a fixed instant of time and let Ω_t be a tetrahedron at $\mathbf{x}$ as shown in Fig. 5.53 with the three faces S_1, S_2, and S_3 parallel to the coordinate planes and the face S is the oblique face. Using the integral mean value theorem

$$\int_{\partial \Omega_t} \sigma \, d\tau = \sum_{j=1}^{3} \int_{S_j} \sigma \, d\tau + \int_S \sigma \, d\tau$$

$$= A(S_1)\sigma(\mathbf{z}_1, t; -\mathbf{e}_1) + A(S_2)\sigma(\mathbf{z}_2, t; -\mathbf{e}_2)$$

$$+ A(S_3)\sigma(\mathbf{z}_3, t; -\mathbf{e}_3) + A(S)\sigma(\mathbf{z}, t; \mathbf{n})$$

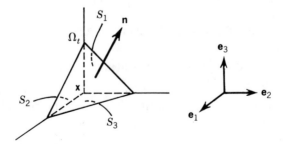

Figure 5.53

where z_i is a point on S_i and $\mathbf{z}$ is a point on S. Here A denotes the area function and $\mathbf{n}$ is the outer normal to S. Since

$$A(S_j) = A(S)n_j \equiv l^2 n_j$$

(we have defined l to be $\sqrt{A(S)}$), it follows that

$$\frac{1}{l^2} \int_{\partial\Omega_t} \boldsymbol{\sigma}\, d\tau = -\sum_{j=1}^{3} \boldsymbol{\sigma}(\mathbf{z}_j, t; \mathbf{e}_j) n_j + \boldsymbol{\sigma}(\mathbf{z}, t; \mathbf{n})$$

where Theorem 5.5 has been applied. Taking the limit as $l \to 0$ gives

$$\lim_{l\to 0} \frac{1}{l^2} \int_{\partial\Omega_t} \boldsymbol{\sigma}\, d\tau = -\sum_{j=1}^{3} \boldsymbol{\sigma}(\mathbf{x}, t; \mathbf{e}_j) n_j + \boldsymbol{\sigma}(\mathbf{x}, t; \mathbf{n})$$

We now show that

$$\lim_{l\to 0} \frac{1}{l^2} \int_{\partial\Omega_t} \boldsymbol{\sigma}(\mathbf{x}, t; \mathbf{n})\, d\tau = \mathbf{0} \tag{17}$$

To this end, it follows from (10) that

$$\left| \int_{\partial\Omega_t} \boldsymbol{\sigma}(\mathbf{x}, t; \mathbf{n})\, d\tau \right| \le \int_{\Omega_t} \left| \rho\left(\frac{D\mathbf{v}}{Dt} - \mathbf{f} \right) \right| dx \le M l^3$$

where the last inequality results from the boundedness of the integrand. Dividing by l^2 and taking the limit as $l \to 0$ proves (17). Consequently

$$\boldsymbol{\sigma}(\mathbf{x}, t; \mathbf{n}) = \sum_{j=1}^{3} \boldsymbol{\sigma}(\mathbf{x}, t; \mathbf{e}_j) n_j \tag{18}$$

or in component form

$$\sigma_i(\mathbf{x}, t; \mathbf{n}) = \sum_{j=1}^{3} \sigma_i(\mathbf{x}, t; \mathbf{e}_j) n_j, \qquad i = 1, 2, 3$$

Defining the σ_{ji} by

$$\sigma_{ji}(\mathbf{x}, t) \equiv \sigma_i(\mathbf{x}, t; \mathbf{e}_j)$$

gives the final result.

The nine quantities σ_{ji} are the components of the *stress tensor*. By definition σ_{ji} is the ith component of the stress vector on the face whose normal is $\mathbf{e}_j$.

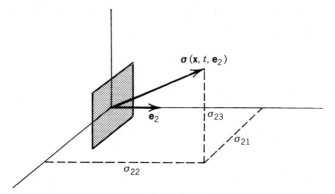

Figure 5.54

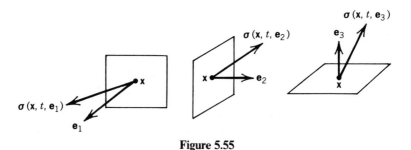

Figure 5.55

For example, σ_{21}, σ_{22}, and σ_{23} are shown in Fig. 5.54. Equation (18) shows that $\sigma(\mathbf{x}, t; \mathbf{n})$ can be resolved into a linear combination of $\sigma(\mathbf{x}, t; \mathbf{e}_1)$, $\sigma(\mathbf{x}, t; \mathbf{e}_2)$, and $\sigma(\mathbf{x}, t, \mathbf{e}_3)$, that is, into three stresses that are the stresses on the coordinate planes at $\mathbf{x}$. These stresses are shown in Fig. 5.55. The components of the three vectors shown are the σ_{ji}.

With the aid of Theorem 5.6 a vector partial differential equation can be formulated that expresses the momentum balance law (10). This calculation is facilitated by adopting a notational convention that saves in writing summation signs in complicated expressions. This practice is called the *summation convention* and it assumes that a sum is taken over any repeated index in a given term. Thus

$$a_i b_{ij} \quad \text{means} \quad \sum_i a_i b_{ij}$$

and

$$a_i b_{ij} c_j \quad \text{means} \quad \sum_i \sum_j a_i b_{ij} c_j$$

The range of the summation index (or indices) is determined from context. For example (15) may be written $\sigma_i = \sigma_{ji} n_j$ with a sum over $j = 1$ to $j = 3$ assumed on the right side. Any index not summed in a given expression is called a *free index*. Free indices vary over their appropriate ranges; for example, in (15) the index i is free and ranges over $i = 1, 2, 3$. In (16) both i and j are free with $i, j = 1, 2, 3$. In terms of this convention the divergence theorem (6) may be expressed

$$\int_\Omega \frac{\partial}{\partial x_i} f_i \, dx = \int_{\partial\Omega} f_i n_i \, d\tau \qquad (19)$$

where $\mathbf{f} = (f_1, f_2, f_3)$ and $\mathbf{n} = (n_1, n_2, n_3)$. On both sides of the equation i is assumed to be summed from $i = 1$ to $i = 3$.

In component form (10) is

$$\int_{\Omega_t} \rho \left(\frac{Dv_i}{Dt} - f_i \right) dx = \int_{\partial\Omega_t} \sigma_i \, d\tau \qquad (20)$$

But the right side is

$$\int_{\partial\Omega_t} \sigma_i \, d\tau = \int_{\partial\Omega_t} \sigma_{ji} n_j \, d\tau = \int_{\Omega_t} \frac{\partial}{\partial x_j} \sigma_{ji} \, dx$$

where we have used (15) and (19). Then (20) becomes

$$\int_{\Omega_t} \left(\rho \frac{Dv_i}{Dt} - \rho f_i - \frac{\partial}{\partial x_j} \sigma_{ji} \right) dx = 0, \qquad i = 1, 2, 3$$

Because of the arbitrariness of the region of integration

$$\rho \frac{Dv_i}{Dt} = \rho f_i + \frac{\partial}{\partial x_j} \sigma_{ji}, \qquad i = 1, 2, 3 \qquad (21)$$

Equations (21) are the *Cauchy equations* or *equations of motion*, and they represent balance of linear momentum. Along with the continuity equation (8) we have four equations for ρ, the three components v_i of the velocity, and the nine components σ_{ji} of the stress tensor.

There is a common way to write the Cauchy stress principle in terms of a fixed laboratory volume Ω_1. It is expressed by the following theorem.

Theorem 5.7 *Cauchy's stress principle* (9) *is equivalent to*

$$\frac{d}{dt} \int_{\Omega_1} \rho v_i \, dx + \int_{\partial\Omega_1} \rho v_i v_j n_j \, d\tau = \int_{\Omega_1} \rho f_i \, dx + \int_{\partial\Omega_1} \sigma_i \, d\tau \qquad (22)$$

where Ω_1 is any fixed region in R^3.

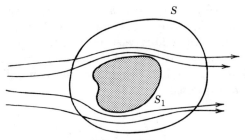

Figure 5.56

Proof   Using Corollary 5.1 with $g \equiv \rho v_i$ gives

$$\frac{d}{dt} \int_{\Omega_t} \rho v_i \, dx = \int_{\Omega_t} (\rho v_i)_t \, dx + \int_{\partial \Omega_t} \rho v_i v_j n_j \, d\tau$$

Therefore

$$\int_{\Omega_t} \rho \frac{D v_i}{Dt} \, dx = \int_{\Omega_t} (\rho v_i)_t \, dx + \int_{\partial \Omega_t} \rho v_i v_j n_j \, d\tau$$

Now let Ω_1 be the volume that coincides with Ω_t at $t = t_1$. Then

$$\int_{\Omega_1} \rho \frac{D v_i}{Dt} \, dx = \frac{d}{dt} \int_{\Omega_1} \rho v_i \, dx + \int_{\partial \Omega_1} \rho v_i v_j n_j \, d\tau$$

Applying (10) gives (22).

Example 5.2   As an application of (22) we compute the force on a stationary object in a steady fluid flow. Suppose S_1 is the boundary of the object and let S be an imaginary surface that contains the object (see Fig. 5.56). Assume the flow is steady and that there is no body force ($\mathbf{f} = \mathbf{0}$). Letting Ω_1 be the region in between S and S_1 and applying the vector form of (22) gives

$$\int_{\partial \Omega_1} (\rho \mathbf{v}(\mathbf{v} \cdot \mathbf{n}) - \boldsymbol{\sigma}) \, d\tau = \mathbf{0}, \qquad \partial \Omega_1 = S_1 \cup S$$

or

$$\int_S (\rho \mathbf{v}(\mathbf{v} \cdot \mathbf{n}) - \boldsymbol{\sigma}) \, d\tau = \int_{S_1} (\boldsymbol{\sigma} - \rho \mathbf{v}(\mathbf{v} \cdot \mathbf{n})) \, d\tau = \int_{S_1} \boldsymbol{\sigma} \, d\tau$$

where the fact that $\mathbf{v} \cdot \mathbf{n} = 0$ on S_1 has been used. But the force $\mathbf{F}$ on the obstacle is

$$\mathbf{F} = -\int_{S_1} \boldsymbol{\sigma}\, d\tau$$

$$= \int_S (\boldsymbol{\sigma} - \rho\mathbf{v}(\mathbf{v} \cdot \mathbf{n}))\, d\tau$$

This is a useful result since it permits the calculation of $\mathbf{F}$ at a control surface at a distance from the obstacle; it may be virtually impossible to obtain the stresses on the surface of the obstacle itself.

Energy

The four equations exhibited in (8) and (21) clearly do not represent a complete determined system for all of the unknowns. At this point in the one dimensional case we introduced an equation of state or constitutive relation. In the present case specification of the stress tensor σ_{ij} is required before further progress can be made.

For a fluid (gas or liquid) in a motionless state it is clear that the stress vector $\boldsymbol{\sigma}$ at a point on a surface in the fluid is always normal to the surface, that is, it is in the direction of the unit normal $\mathbf{n}$ (or $-\mathbf{n}$). Such is not true for solids, since they can sustain complicated shear or tangential stresses and not undergo motion. This normality property extends to fluids in uniform motion where the velocity field is constant over some time interval. One of the simplest classes of general nonuniform fluid motions is that class of motions termed *inviscid* where the same normality assumption holds, namely

$$\boldsymbol{\sigma}(\mathbf{x}, t; \mathbf{n}) = -p(\mathbf{x}, t)\mathbf{n} \tag{23}$$

Hence, for inviscid flow the stress across a surface is proportional to the normal vector where the proportionality factor p is the *pressure*. The concept of an inviscid fluid is highly useful in technology, since many real fluids are actually modeled by (23) and the calculations are far simpler using (23) than for more complicated constitutive relations. An *ideal* fluid is an inviscid fluid undergoing an incompressible motion, and much of the literature in fluid mechanics deals with ideal fluids.

From (23) it is easy to calculate the components σ_{ij} of the stress tensor. We have (Exercise 5.5)

$$\sigma_{ij} = -p(\mathbf{x}, t)\delta_{ij} \tag{24}$$

where δ_{ij} is the Kronecker delta symbol defined by $\delta_{ij} = 1$ if $i = j$ and $\delta_{ij} = 0$ if $i \neq j$. Since

$$\frac{\partial}{\partial x_j}\sigma_{ji} = -\frac{\partial}{\partial x_j}\left(p(\mathbf{x}, t)\delta_{ij}\right) = -\frac{\partial}{\partial x_i}p(\mathbf{x}, t)$$

we may write the Cauchy equations (21) as

$$\rho \frac{Dv_i}{Dt} = \rho f_i - \frac{\partial p}{\partial x_i}, \qquad i = 1, 2, 3$$

or in vector form as

$$\rho \frac{D\mathbf{v}}{Dt} = \rho \mathbf{f} - \mathbf{grad}\ p \tag{25}$$

Equation (25) is called the *Euler equation*. Equations (8) and (25) represent four equations for ρ, $\mathbf{v}$, and p. Supplemented by equations of state and an energy conservation equation, a complete set of equations can be found. For inviscid fluids the discussion now follows the one dimensional analysis presented in Sections 5.2.

Equation (24) suggests that in the most general case the stress tensor takes the form

$$\sigma_{ij} = -p\,\delta_{ij} + \tau_{ij} \tag{26}$$

where p is identified with the thermodynamic pressure and τ_{ij} define the components of the viscous stress tensor. These components represent effects caused by viscosity, for example, where adjacent fluid elements undergo shear stresses across surfaces separating them.

As in the one dimensional case we postulate a *balance of energy law* of the form

$$\frac{d}{dt} \int_{\Omega_t} \left(\frac{1}{2}\rho \mathbf{v} \cdot \mathbf{v} + \rho e \right) dx = \int_{\Omega_t} \rho \mathbf{f} \cdot \mathbf{v}\, dx + \int_{\partial\Omega_t} \sigma \cdot \mathbf{v}\, d\tau - \int_{\partial\Omega_t} \mathbf{q} \cdot \mathbf{n}\, d\tau \tag{27}$$

where Ω_t is an arbitrary material region with exterior unit normal $\mathbf{n}$ and $\mathbf{q}$ represents the heat flux density. In words, the time rate of change of the kinetic plus internal energy in Ω_t equals the rate work is done by the body forces $\mathbf{f}$, plus the rate work is done by the surface stresses σ, plus the rate heat flows into the region. A differential form of the energy balance law can be easily obtained by appealing to the arbitrariness of Ω_t after an application of the divergence theorem and Cauchy's theorem. We record as follows.

Theorem 5.8   *If the functions are sufficiently smooth, Equation (27) implies*

$$\rho \frac{D}{Dt}\left(\frac{1}{2}\mathbf{v} \cdot \mathbf{v} + e \right) = \rho \mathbf{f} \cdot \mathbf{v} + \frac{\partial}{\partial x_j}(\sigma_{ji}v_i) - \mathrm{div}\,\mathbf{q} \tag{28}$$

The proof is left as an exercise. We also have the following theorem.

Theorem 5.9 *The change in internal energy is given by*

$$\rho \frac{De}{Dt} = -\operatorname{div} \mathbf{q} + \sigma_{ji} \frac{\partial v_i}{\partial x_j} \tag{29}$$

Proof Multiplying the momentum balance law (21) by v_i and summing over $i = 1, 2, 3$ gives

$$\frac{1}{2} \rho \frac{D}{Dt} (\mathbf{v} \cdot \mathbf{v}) = \rho f_i v_i + \frac{\partial}{\partial x_j} (\sigma_{ji} v_i) - \sigma_{ji} \frac{\partial v_i}{\partial x_j}$$

Subtracting from (28) yields (29) and hence the result.

From the combined form of the first and second laws of thermodynamics

$$\theta \frac{Ds}{Dt} = \frac{De}{Dt} + p \frac{D(1/\rho)}{Dt} \tag{30}$$

we can obtain an expression for how the entropy of a fluid particle changes. Combining (29) and (30) and using $D\rho^{-1}/Dt = -\rho^{-2} D\rho/Dt = \rho^{-1} \operatorname{div} \mathbf{v}$ we get the following theorem.

Theorem 5.10

$$\rho \theta \frac{Ds}{Dt} = p \operatorname{div} \mathbf{v} - \operatorname{div} \mathbf{q} + \sigma_{ji} \frac{\partial v_i}{\partial x_j} \tag{31}$$

The term $-\operatorname{div} \mathbf{q}$ represents the heat flow and the sum $\Psi \equiv p \operatorname{div} \mathbf{v} + \sigma_{ji} \partial v_i/\partial x_j = \tau_{ji} \partial v_i/\partial x_j$ represents the heat generation due to deformation. Thus we may write

$$\rho \frac{Ds}{Dt} = -\frac{1}{\theta} \operatorname{div} \mathbf{q} + \frac{1}{\theta} \Psi \tag{32}$$

The function Ψ is called the *dissipation function*, and it is the rate per unit volume that mechanical energy is dissipated into heat.

Thermodynamics also requires that the entropy increase equal or exceed the heat added divided by the absolute temperature. Thus we *postulate* the inequality

$$\frac{d}{dt} \int_{\Omega_t} \rho s \, dx \geq -\int_{\partial\Omega_t} \frac{1}{\theta} \mathbf{q} \cdot \mathbf{n} \, d\tau$$

The differential form of this axiom is called the *Clausius-Duhem inequality* and is given in the following theorem. The proof is left as an exercise.

Theorem 5.11

$$\rho \frac{Ds}{Dt} \geq -\operatorname{div}(\theta^{-1}\mathbf{q}) \tag{33}$$

Equations (32) and (33) may be combined to obtain

$$\theta^{-1}\Psi - \theta^{-2}\mathbf{q} \cdot \operatorname{grad} \theta \geq 0 \tag{34}$$

Sufficient conditions for (34) are

$$\Psi \geq 0, \qquad -\mathbf{q} \cdot \operatorname{grad} \theta \geq 0$$

That is, deformation does not convert heat into mechanical energy and heat flows against the temperature gradient.

To proceed further, some assumption is required regarding the form of the viscous stress tensor τ_{ij}. A *Newtonian fluid* is one in which there is a linear dependence of τ_{ij} upon the *rate of deformation* $D_{ij} \equiv \frac{1}{2}(\partial v_i/\partial x_j + \partial v_j/\partial x_i)$, that is

$$\tau_{ij} = C_{ijrs} D_{rs}$$

where the coefficients C_{ijrs} may depend upon the local thermodynamic states θ and ρ. Arguments of symmetry and invariance of the stress tensor under translations and rotations lead to (see, for example, Segel and Handelman [5])

$$\tau_{ij} = 2\mu D_{ij} + \lambda \delta_{ij} \operatorname{div} \mathbf{v}$$

where μ and λ are coefficients of viscosity. Generally μ and λ may depend on temperature or density, but here we assume they are constant. The previous assumption forces the stress tensor to have the form

$$\sigma_{ij} = (-p + \lambda \operatorname{div} \mathbf{v})\delta_{ij} + 2\mu D_{ij}$$

and therefore the momentum law (21) becomes

$$\rho \frac{Dv_i}{Dt} = \rho f_i - \frac{\partial}{\partial x_i}(p - (\lambda + \mu)\operatorname{div} \mathbf{v}) + \mu \nabla^2 v_i \tag{35}$$

for $i = 1, 2, 3$. In the incompressible case where $\operatorname{div} \mathbf{v} = 0$, (35) reduces in vector form to

$$\rho \frac{D\mathbf{v}}{Dt} = \rho \mathbf{f} - \operatorname{grad} p + \mu \nabla^2 \mathbf{v} \tag{36}$$

which are the *Navier-Stokes equations*. Exact solutions of (36) in special cases are given in the exercises.

For viscous flows it is generally assumed that the fluid adheres to a rigid boundary. Thus we assume the *adherence boundary condition*

$$\lim_{\mathbf{x} \to \mathbf{x}_b} \mathbf{v}(\mathbf{x}) = \mathbf{v}(\mathbf{x}_b)$$

where $\mathbf{x}_b$ is a point on a rigid boundary and $\mathbf{v}(\mathbf{x}_b)$ is the known velocity of the boundary.

EXERCISES

5.1   Prove the Euler expansion theorem 5.2 when $n = 2$. *Hint*: Recall the rule for differentiating a determinant:

$$\frac{d}{dt} \begin{vmatrix} A & B \\ C & D \end{vmatrix} = \begin{vmatrix} \dfrac{dA}{dt} & B \\ \dfrac{dC}{dt} & D \end{vmatrix} + \begin{vmatrix} A & \dfrac{dB}{dt} \\ C & \dfrac{dD}{dt} \end{vmatrix}$$

5.2   Let $\Delta(\mathbf{h}, t)$ be the Lagrangian density defined by $\Delta(\mathbf{h}, t) = \rho(\mathbf{x}(\mathbf{h}, t), t)$. Show that

$$J(\mathbf{h}, t) = \frac{\Delta(\mathbf{h}, 0)}{\Delta(\mathbf{h}, t)}$$

5.3   Derive the Lagrangian form of the mass conservation law

$$\Delta_t + \Delta \, \mathrm{div}\, \mathbf{v}|_{\mathbf{h}} = 0$$

5.4   Prove that if mass conservation holds then

$$\frac{d}{dt} \int_{\Omega_t} \rho g \, dx = \int_{\Omega_t} \rho \frac{Dg}{Dt} \, dx$$

for any continuously differentiable function g. Show that the scalar function g can be replaced by a vector function $\mathbf{g}$.

5.5   If $\sigma(\mathbf{x}, t; \mathbf{n}) = -p(\mathbf{x}, t)\mathbf{n}$, prove that $\sigma_{ij} = -p(\mathbf{x}, t)\,\delta_{ij}$.

5.6   A two dimensional fluid motion is given by

$$x_1 = h_1 \exp t, \qquad x_2 = h_2 \exp(-t), \qquad t > 0$$

(a) Find $\mathbf{V}$ and $\mathbf{v}$.

(b) Find the streamlines and show they coincide with the particle paths.

(c) If $\rho(x, t) = x_1 x_2$, show that the motion is incompressible.

5.7   The Eulerian velocity of a two dimensional fluid flow is

$$\mathbf{v} = \langle t, x_1 + 1 \rangle, \qquad t > 0$$

Find the streamlines passing through the fixed point (a_0, b_0) and find the particle path of the particle $(h_1, h_2) = (a_0, b_0)$. Does the particle path ever coincide with a streamline at any time t_0?

5.8   A steady flow in two dimensions is defined by

$$\mathbf{v} = \langle 2x_1 + 3x_2, x_1 - x_2 \rangle, \qquad t > 0$$

Find the particle paths. Compute the acceleration measured by an observer located at the fixed point $(3, 4)$.

5.9   A flow is called *potential* on a region D if there exists a continuously differentiable function $\phi(t, \mathbf{x})$ on D for which $\mathbf{v} = \mathbf{grad}\ \phi$, and a flow is called *irrotational* on D if curl $\mathbf{v} = \mathbf{0}$.

(a) Prove that a potential flow is irrotational.

(b) Show that an irrotational flow is not always potential by considering

$$\mathbf{v} = \langle \frac{-y}{x^2 + y^2}, \frac{x}{x^2 + y^2} \rangle \quad \text{on} \quad R^2 - \{0\}$$

(c) Show that if D is simply connected then irrotational flows are potential.

5.10   **(a)** Show that

$$(\mathbf{v} \cdot \mathbf{grad})\mathbf{v} = \tfrac{1}{2}\mathbf{grad}|\mathbf{v}|^2 + \text{curl}(\mathbf{v} \times \mathbf{v})$$

(b) Prove Bernoulli's theorem: For a potential flow of an inviscid fluid, suppose there exists a function $\psi(t, \mathbf{x})$ for which $\mathbf{f} = -\mathbf{grad}\ \psi$. Then for any path C connecting two points $\mathbf{a}$ in $\mathbf{b}$ in a region where $\mathbf{v} = \mathbf{grad}\ \phi$,

$$\left(\phi_t + \frac{1}{2}|\mathbf{v}|^2 + \psi\right)\Big|_{\mathbf{a}}^{\mathbf{b}} + \int_C \frac{dp}{\rho} = 0$$

Hint: Integrate the equation of motion and use (a).

5.11 In a region D where a flow is steady, irrotational, and incompressible with $\rho = $ constant prove that

$$\frac{1}{2}|\mathbf{v}|^2 + \frac{1}{\rho}p + \psi = \text{constant},   (\text{in } D)$$

where $\mathbf{f} = -\mathbf{grad}\ \psi$.

5.12 For a fluid motion with $\sigma_{ij} = \sigma_{ji}$ prove that

$$\frac{dK}{dt} = \int_{\Omega_t} \rho \mathbf{f} \cdot \mathbf{v}\, dx + \int_{\partial\Omega_t} \boldsymbol{\sigma} \cdot \mathbf{v}\, d\tau - \int_{\Omega_t} \sigma_{ij} D_{ij}\, dx$$

where $K(\Omega_t) \equiv \frac{1}{2}\int_{\Omega_t} \rho\mathbf{v} \cdot \mathbf{v}\, dx$ is the kinetic energy of Ω_t. This result is the *energy transport theorem*.

5.13 For an ideal fluid prove that $K(\Omega) = $ constant, where Ω is a fixed region in R^3 and $\mathbf{v}$ is parallel to $\partial\Omega$ (see Exercise 5.12).

5.14 The alternating symbol ε_{ijk} is defined by

$$\varepsilon_{ijk} = \begin{cases} 0, & \text{if any two indices are equal} \\ 1, & \text{if } (ijk) \text{ is an even permutation of } (123) \\ -1, & \text{if } (ijj) \text{ is an odd permutation of } (123) \end{cases}$$

A permutation (ijk) is an even (odd) permutation of (123) if it takes an even (odd) number of switches of adjacent elements to get it to the form (123). For example

$$321 \rightarrow 231 \rightarrow 213 \rightarrow 123$$

so (321) is an odd permutation of (123). Show that if $\mathbf{a} = \langle a_1, a_2, a_3\rangle$ and $\mathbf{b} = \langle b_1, b_2, b_3\rangle$, then the kth component of the cross or wedge product $\mathbf{a} \wedge \mathbf{b}$ is given by

$$(\mathbf{a} \wedge \mathbf{b})_k = \varepsilon_{ijk}a_i b_j,   \text{sum on } i, j = 1, 2, 3$$

5.15 (Conservation of Angular Momentum)   A particle of mass m moving in R^3 with velocity vector $\mathbf{v}$ has linear momentum $\mathbf{p} = m\mathbf{v}$ and angular momentum about the origin $\mathbf{0}$ given by $\mathbf{L} = \mathbf{x} \wedge \mathbf{p}$, where $\mathbf{x}$ is its position vector. If $\mathbf{F}$ is a force on the particle, then $\mathbf{F} = d\mathbf{p}/dt$ by Newton's second law. Show that $\mathbf{N} = d\mathbf{L}/dt$, where $\mathbf{N} = \mathbf{x} \wedge \mathbf{F}$ is the torque (or moment of force) about $\mathbf{0}$. For a continuum argue that one should postulate

$$\frac{d}{dt}\int_{\Omega_t}(\mathbf{x} \wedge \rho\mathbf{v})\,dx = \int_{\Omega_t}(\mathbf{x} \wedge \rho\mathbf{f})\,dx + \int_{\partial\Omega_t}(\mathbf{x} \wedge \boldsymbol{\sigma})\,d\tau \qquad (37)$$

where Ω_t is any material region. Write (37) in component form using the alternating symbol ε_{ijk}. Note that (37) does not account for any internal angular momentum of the fluid particles that would occur in a fluid where the particles are rotating rods; such fluids are called *polar*.

5.16 Prove that (37) of Exercise 5.15 is equivalent to the symmetry of the stress tensor σ_{ij} (i.e., $\sigma_{ij} = \sigma_{ji}$). *Hint:*

$$\frac{d}{dt}\int_{\Omega_t}\varepsilon_{ijk}x_i\rho v_j\,dx = \int_{\Omega_t}\varepsilon_{ijk}x_i\rho\frac{Dv_j}{Dt}\,dx$$

Use the equation of motion and

$$\frac{\partial}{\partial x_l}(x_i\sigma_{lj}) = x_i\frac{\partial}{\partial x_l}\sigma_{lj} + \sigma_{ij}$$

followed by an application of the divergence theorem. This theorem shows that balance of angular momentum is equivalent to symmetry of the stress tensor. Thus, of the nine components of σ_{ij} only six are independent.

5.17 Let $\Gamma \subseteq \Omega_0$ be a simple closed curve and let $\Gamma_t = \phi_t(\Gamma)$, where ϕ_t is an isentropic fluid motion. The *circulation* about Γ_t is defined by the line integral $C_{\Gamma_t} \equiv \int_{\Gamma_t}\mathbf{v}\cdot d\mathbf{l}$. Prove *Kelvin's theorem* that C_{Γ_t} is a constant in time.

5.18 The *vorticity* ω of a flow is defined by $\omega = \operatorname{curl}\mathbf{v}$. Determine the vorticity of the following flows.

(a) $\mathbf{v} = v_0\exp(-x_2^2)\mathbf{j}$, v_0 constant

(b) $\mathbf{v} = v_0\exp(-x_1^2)\mathbf{j}$, v_0 constant

(c) $\mathbf{v} = \omega_0(-x_2\mathbf{i} + x_1\mathbf{j})$, ω_0 constant

Sketch the streamlines in each case.

5.19 If ρ = constant and $\mathbf{f} = \mathbf{grad}\,\psi$ for some ψ, prove that (36) implies the *vorticity equation*

$$\frac{D\boldsymbol{\omega}}{Dt} = (\boldsymbol{\omega} \cdot \mathbf{grad})\mathbf{v} + \frac{\mu}{\rho}\nabla^2\boldsymbol{\omega}$$

5.20 Consider an incompressible viscous flow of constant density ρ_0 under the influence of no body forces governed by the Navier-Stokes equation (36).

(a) If l, ν/l^2, U, and $\rho_0\nu U/l$, where $\nu \equiv \mu/\rho_0$ are length, time, velocity, and pressure scales, show that in dimensionless form the governing equations can be written

$$\text{div}\,\mathbf{v} = 0, \qquad \mathbf{v}_t + \text{Re}(\mathbf{v} \cdot \mathbf{grad})\mathbf{v} = -\mathbf{grad}\,p + \nabla^2\mathbf{v}$$

where $\text{Re} \equiv Ul/\nu$ is a constant called the *Reynolds number*. The constant ν is the kinematic viscosity.

(b) For flows with small Reynolds number, that is, $\text{Re} \ll 1$, show that

$$\nabla^2 p = 0, \qquad \frac{\partial}{\partial t}(\nabla^2\mathbf{v}) = \nabla^2(\nabla^2\mathbf{v})$$

Hint: $\text{curl}(\text{curl}\,\mathbf{v}) = \mathbf{grad}\,\text{div}\,\mathbf{v} - \nabla^2\mathbf{v}$

5.21 (*Plane Couette Flow*) An incompressible viscous fluid of constant density under no body forces is confined to lie between two infinite flat plates at $x_3 = 0$ and $x_3 = d$. The lower plate is stationary and the upper plate is moved at constant velocity U in the x_1 direction. Show that the fluid motion is given by

$$v_1 = \frac{U}{d}x_3, \qquad v_2 = v_3 = 0, \qquad p = p_0 = \text{constant}$$

Show that the stress on the lower plate is $\mu U/d$.

5.22 (*Plane Poiseuille Flow*) An incompressible viscous fluid of constant density under no body forces is confined to lie between two stationary infinite planes $x_3 = -d$ and $x_3 = d$. A flow is forced by a constant pressure gradient $\mathbf{grad}\,p = \langle -C, 0, 0 \rangle, C > 0$. Show that the fluid velocity is given by

$$v_1 = \frac{C}{2\mu}\left(d^2 - x_3^2\right), \qquad v_2 = v_3 = 0$$

Show that the mass flow per unit width in the x_2 direction is given by $2Cd^3\rho/3\mu$.

5.23   Prove Theorems 5.8 and 5.11.

REFERENCES

1. G. B. Whitham, *Linear and Nonlinear Waves*, Wiley-Interscience, New York (1974).
2. P. A. Thompson, *Compressible-Fluid Dynamics*, McGraw-Hill Book Company, New York (1972).
3. J. Smoller, *Shock Waves and Reaction-Diffusion Equations*, Springer-Verlag, New York (1983).
4. R. Courant and K. O. Friedrichs, *Supersonic Flow and Shock Waves*, Wiley-Interscience, New York (1948) (Reprinted by Springer-Verlag, New York (1976)).
5. L. A. Segel and G. H. Handelman, *Mathematics Applied to Continuum Mechanics*, Macmillan Publishing Co., New York (1977).

6

STABILITY AND BIFURCATION

Consider a system of any nature whatsoever that exists in a state S. We say S is stable, in one sense or another, if small perturbations or changes in the system do not drastically affect the state S. For example the solar system currently exists in a time-dependent state in which the planets move about the sun in an orderly fashion. It is known that if a small additional celestial body is introduced into the system, then the original state is not disturbed to any significant degree. We say the original state is stable to small perturbations. Similar questions of stability arise in every physical problem.

Another related concept is that of bifurcation. Bifurcation, or branching, occurs in a system when the state of the system (e.g., an equilibrium state) depends on some parameter and as that parameter varies the state branches or changes to another state at some critical value of the parameter with usually an accompanying change of stability. Of common experience is a stiff rubber rod subject to a force F (a parameter) on one end. The stable state is the lateral deflection of the bar, which is zero if F is small. As F is increased, however, there is a critical value F_c beyond which the equilibrium state is not stable and the rod will buckle with the slightest jar (Fig. 6.1). As F passes through the critical value, the system will branch from one equilibrium state to another. Another example is laminar flow in a pipe. It can be made to persist even at Reynolds numbers above the critical Reynolds number provided the walls of the pipe are smooth and so on. If, however, the slightest perturbation, say a small vibration, is introduced, then there will be transition to turbulent flow, the latter being the stable flow mode above the critical Reynolds number.

In this chapter it is our aim to gain a basic understanding of the ideas underlying stability and bifurcation and to see how these two concepts are

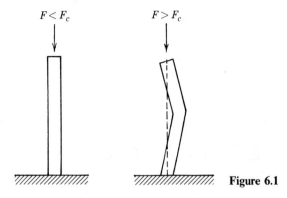

Figure 6.1

connected. In this first section we present several concrete examples that should provide ample motivation of the problems to be considered. The techniques used in these examples are standard methods in applied mathematics that are useful in many physical problems.

6.1 INTUITIVE IDEAS

Stability and Population Dynamics

The notion of stability is introduced in this section in the context of a problem in population dynamics. The simplest model for the dynamics of growth of a population can be stated as follows: The population $p = p(t)$ increases at a rate proportional to the number of individuals in the population. Mathematically this can be expressed as

$$\frac{dp}{dt} = ap, \qquad t > 0 \tag{1}$$

where a is a positive constant called the growth constant. If initially there are p_0 individuals, that is,

$$p(0) = p_0$$

then it follows that

$$p(t) = p_0\exp(at), \qquad t > 0 \tag{2}$$

Equation (1) is known as the Malthusian model of population growth and it predicts that the population will grow exponentially with time. It seems clear that a given population cannot increase exponentially for all time. Eventually, as the population grows competition for food, natural resources, living space, and other factors will limit the growth. Therefore the next approximation would be to include a term in the differential equation that accounts for the competition. The simplest idea is to notice that if there are p individuals in a population, then the average number of encounters of two members per unit time is proportional to p^2. Thus we consider the population model

$$\frac{dp}{dt} = ap - bp^2, \quad t > 0, \quad a, b > 0 \quad (3)$$

called the logistic law of population growth, which contains a linear growth term ap and a nonlinear competition term $-bp^2$. Intuitively if a is large compared to b, then (3) will exhibit the desired features as t increases; if p is not too large then the term $-bp^2$ will be negligible compared to ap and the population will grow exponentially as predicted by Malthusian dynamics (1). As p gets large then the term $-bp^2$ is no longer small and thus it serves to slow down the rate of increase. It is a straightforward exercise to solve the ordinary differential equation (3) by separation of variables to obtain

$$p(t) = \frac{ap_0}{bp_0 + (a - bp_0)\exp(-at)} \quad (4)$$

where p_0 is the initial population. What interests us in this section is an examination of the stability properties of the Equation (3). For general analyses we cannot expect to know the exact solution of the equations so the fact that we know (4) is unimportant.

Let us examine the equilibrium or steady-state solutions of (3). We find them by setting the right side of (3) to zero to get

$$p = 0 \quad (5)$$

and

$$p = \frac{a}{b} \quad (6)$$

Notice that both (5) and (6) are constant solutions of (3) that are independent of time. We focus upon the nonzero solution (6); the standard stability analysis goes as follows. Let $\tilde{p}(t)$ be a perturbation or change in the equilibrium population a/b and let

$$p(t) = \frac{a}{b} + \tilde{p}(t) \quad (7)$$

Since $p(t)$ is governed by (3), we obtain

$$\frac{d}{dt}\left(\frac{a}{b} + \tilde{p}\right) = a\left(\frac{a}{b} + \tilde{p}\right) - b\left(\frac{a}{b} + \tilde{p}\right)^2$$

or

$$\frac{d\tilde{p}}{dt} = -a\tilde{p} - b\tilde{p}^2 \qquad (8)$$

Hence the perturbation $\tilde{p}$ must satisfy (8) exactly. If, however, the perturbation $\tilde{p}$ is small, then the nonlinear term $\tilde{p}^2$ can be neglected compared to $\tilde{p}$ and we obtain the *linearized perturbation equation*

$$\frac{d\tilde{p}}{dt} = -a\tilde{p} \qquad (9)$$

which has the solution

$$\tilde{p} = \text{const} \cdot e^{-at} \qquad (10)$$

Clearly $\tilde{p} \to 0$ as $t \to \infty$ and so the perturbation decays away as $t \to \infty$, which is consistent with the assumption that $\tilde{p}$ was small. Hence we conclude that the equilibrium state a/b is asymptotically stable with respect to small perturbations; by asymptotically stable we mean that the perturbations go to zero as t gets large.

The above conclusions were drawn under the assumption that the perturbation satisfied a linear perturbation equation determined by neglecting the nonlinear terms. The actual perturbation $\tilde{p}$ satisfies the nonlinear equation (8) and in the present case we can carry out a *nonlinear perturbation analysis* by solving (8). For general problems the full nonlinear perturbation equation is usually not solvable and therefore we must resort to a linearized analysis. To solve (8) we separate variables to get

$$\frac{d\tilde{p}}{\tilde{p}(\tilde{p} + b/a)} = -a\,dt$$

Using the partial fraction decomposition

$$\frac{1}{\tilde{p}(\tilde{p} + b/a)} = \frac{a/b}{\tilde{p}} - \frac{a/b}{\tilde{p} + b/a}$$

and integrating gives

$$\frac{a}{b}\ln\tilde{p} - \frac{a}{b}\ln\left(\tilde{p} + \frac{b}{a}\right) = -at + c$$

where c is a constant. Hence

$$\tilde{p}(t) = \frac{(bc/a)\exp(-bt)}{1 - c\exp(-bt)}$$

Therefore the nonlinear analysis shows that $\tilde{p}(t) \to 0$ as $t \to \infty$ regardless of the size of the initial perturbation.

We now investigate the zero solution, that is, the other equilibrium population. As before let

$$p(t) = 0 + \tilde{p}(t)$$

where $\tilde{p}$ is a perturbation. Substituting into (3) gives

$$\frac{d\tilde{p}}{dt} = a\tilde{p} - b\tilde{p}^2 \tag{11}$$

Assuming $\tilde{p}$ is small the linearized perturbation equation is

$$\frac{d\tilde{p}}{dt} = a\tilde{p}$$

which has solution

$$\tilde{p} = \text{const} \cdot e^{at}$$

Therefore the linearized perturbations grow without bound as $t \to \infty$ and the zero solution is unstable with respect to small perturbations. The exact solution to (11) is given by

$$\tilde{p}(t) = \frac{ac}{bc + (a - bc)\exp(-at)}$$

where c is the perturbation at time $t = 0$. A graph of $\tilde{p}(t)$ is shown in Fig. 6.2. Within the nonlinear theory the perturbation grows but remains bounded as $t \to \infty$.

Of course, in the present problem, since the exact solution is known we can deduce the stability properties directly from the solution curves of the equation. Graphs of these curves (4) for different values of p_0 are shown in Fig. 6.3. It is clear that all the integral curves except one (the zero solution) approach the equilibrium solution a/b, and hence it is asymptotically stable

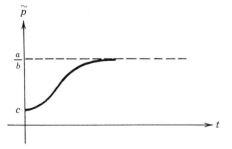

Figure 6.2. Graph of $\tilde{p}(t)$.

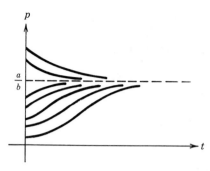

Figure 6.3. Graphs of $p(t)$ for different p_0.

to small perturbations. The zero solution, on the other hand, is not stable, since a small change from the zero state will deviate substantially from that state as the system evolves.

Bifurcation of a Bead on a Hoop

In this section we introduce through an example some of the basic ideas of bifurcation theory. Bifurcation means branching or dividing. For equations containing a parameter and whose solutions depend on that parameter, bifurcation theory is the study of the branching of solutions. That is, we may ask if another solution or solutions branch from a given one, and if so at what point(s) along the given solution does the branching occur. Further, we may inquire into the structure of the solution in a local neighborhood of the point at which bifurcation occurs. Thus bifurcation theory is a study of nonuniqueness, and specifically a study of how the multiplicity of solutions varies with the parameter. Important in this study are the stability properties of the bifurcating solutions. The subject is rapidly developing and it is an area of active research in applied mathematics.

The simplest illustration of bifurcation phenomena occurs in the algebraic equation

$$x^3 - \mu x = 0$$

where μ is a real parameter. As μ varies from negative to positive the number of real solutions jumps from one to three at $\mu = 0$. If we graph the real solutions x as a function of μ, as in Fig. 6.4, we obtain a diagram of the branching or dividing. The point ($x = 0, \mu = 0$) is called a point of criticality or a bifurcation point because the nature of the solution changes there. This type of bifurcation, called the pitchfork bifurcation, occurs in a wide variety of problems.

Another type of bifurcation problem is a differential equation in which a parameter occurs (e.g., the Reynolds number in a hydrodynamic flow). As the parameter varies the equilibrium solution may change and branch from a stable solution to an unstable one, or vice versa, when the parameter reaches a

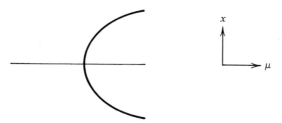

Figure 6.4. Pitchfork bifurcation.

critical value. Again, bifurcation is characterized by a change in the nature of the solution to the problem.

An example will illustrate the basic concepts. Let us consider a bead of mass m that can slide without friction on a circular wire or hoop of radius R. The hoop is constrained to rotate about a vertical diameter with constant angular velocity ω (see Fig. 6.5). It is desired to study the stability of the position of the bead, which is determined by the angle θ, with respect to the angular velocity ω. The governing differential equation is easily found from Newton's second law. The bead's acceleration is $R\ddot{\theta}$ and there are two components of the force on the bead, one due to gravity and the other due to the rotation of the wire. Gravity acts downward with magnitude mg and its tangential component is $mg \sin \theta$ (Fig. 6.6). The force due to the circular motion is $mr\omega^2$ and its tangential component is $mr\omega^2 \cos \theta$; since $r = R \sin \theta$, the tangential component is $mR\omega^2 \cos \theta \sin \theta$ (see Fig. 6.7). Therefore, equating the mass times the acceleration to the total force we obtain

$$mR\ddot{\theta} = mR\omega^2 \cos \theta \sin \theta - mg \sin \theta \tag{12}$$

which is the equation of motion of the bead.

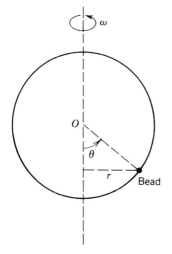

Figure 6.5. Bead on a rotating hoop.

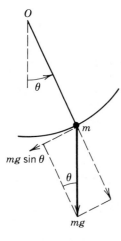

$mg \sin \theta$

mg

Figure 6.6

An equilibrium or steady-state solution occurs when there is no force on the bead, or

$$mR\omega^2\cos \theta \sin \theta - mg \sin \theta = 0$$

or

$$R\omega^2 m \sin \theta \left(\cos \theta - \frac{g}{R\omega^2}\right) = 0$$

Hence the two equilibrium solutions are

$$\theta_0 = 0, \qquad \theta_1 = \arccos \frac{g}{R\omega^2} \tag{13}$$

The first corresponds to the bead remaining at the bottom of the wire and the second corresponds to a positive angular displacement θ_1. In Fig. 6.8 the two equilibrium solutions are sketched as functions of the angular velocity ω. We note that the branches of the equilibrium solutions intersect; we say that the solutions bifurcate and the point $(\sqrt{g/R},0)$ of intersection is called a bifurcation point. Again we encounter the pitchfork bifurcation. If $\omega \leq \sqrt{g/R}$,

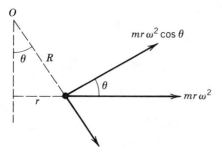

Figure 6.7

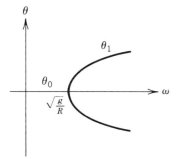

Figure 6.8

then the only equilibrium solution is $\theta_0 = 0$; however, if $\omega > \sqrt{g/R}$, then there are three possible equilibrium solutions.

The question of which solution will be sought out by the system leads to the question of stability. Let us perform a heuristic argument similar to the one made for the population problem. Let $\tilde{\theta}$ be a small perturbation of the solution $\theta_0 = 0$ and write

$$\theta = \theta_0 + \tilde{\theta}$$

Substitution into the differential equation (12) yields

$$mR\ddot{\tilde{\theta}} = mR\omega^2 \cos\tilde{\theta}\sin\tilde{\theta} - mg\sin\tilde{\theta}$$

Using the expansions

$$\sin\tilde{\theta} = \tilde{\theta} + O(\tilde{\theta}^3) \qquad \cos\tilde{\theta} = 1 + O(\tilde{\theta}^2)$$

we get

$$mR\ddot{\tilde{\theta}} = mR\omega^2\big(1 + O(\tilde{\theta}^2)\big)\big(\tilde{\theta} + O(\tilde{\theta}^3)\big) - mg\big(\tilde{\theta} + O(\tilde{\theta}^3)\big)$$
$$= \big(mR\omega^2 - mg\big)\tilde{\theta} + O(\tilde{\theta}^3)$$

Neglecting the higher order terms we obtain the linearized perturbation equation

$$mR\ddot{\tilde{\theta}} = \big(mR\omega^2 - mg\big)\tilde{\theta}$$

or

$$\ddot{\tilde{\theta}} + \left(\frac{g}{R} - \omega^2\right)\tilde{\theta} = 0$$

If $\omega < \sqrt{g/R}$, then the general solution is

$$\tilde{\theta}(t) = A\cos\sqrt{\frac{g}{R} - \omega^2}\,t + B\sin\sqrt{\frac{g}{R} - \omega^2}\,t$$

where A and B are constants. So the perturbations are bounded and θ_0 is stable. On the other hand, if $\omega > \sqrt{g/R}$, then the general solution to the linearized perturbation equation is

$$\tilde{\theta}(t) = A \exp\left(\sqrt{\omega^2 - \frac{g}{R}}\, t\right) + B \exp\left(-\sqrt{\omega^2 - \frac{g}{R}}\, t\right)$$

Consequently there is a mode $\exp(\sqrt{\omega^2 - g/R}\, t)$ that grows exponentially and θ_0 is unstable for $\omega > \sqrt{g/R}$.

Physically one can reason that if ω starts at a small value and is slowly increased, then the bead will remain at the bottom of the hoop ($\theta_0 = 0$) until the critical angular velocity $\omega = \sqrt{g/R}$ is reached. Beyond that the $\theta_0 = 0$ solution is not stable and we expect that the bead will quickly migrate or bifurcate to a branch of the θ_1 curve, that is, jump to a nonzero angle θ_1. This is in fact what occurs. Some equilibrium solutions that bifurcate are stable to small perturbations and some are not. Physically this means that the system favors equilibrium states that are stable. In an unstable state small perturbations, which are almost always present, are not damped out and the system moves to another equilibrium state, ultimately to one that is stable. In this sense every physical system has a mind of its own.

One can also analyze the stability of $\theta_1 = \arccos(g/R\omega^2)$ in the same manner as above. Letting

$$\theta = \arccos \frac{g}{R\omega^2} + \tilde{\theta}$$

where $\tilde{\theta}$ is small, and then substituting into the differential equation, we could retain the lowest order terms in $\tilde{\theta}$ and thus obtain a linearized equation. The calculation would be more formidable in this case. We can accomplish the same result, however, by noting the right side of the differential equation can be expanded in a Taylor series, and from that expression we can identify the lowest order terms. To this end let $f(\theta)$ be defined by

$$f(\theta) = mR\omega^2\cos\theta \sin\theta - mg \sin\theta$$

Then

$$\begin{aligned} f(\theta) &= f(\theta_1) + f'(\theta_1)(\theta - \theta_1) + O\left(|\theta - \theta_1|^2\right) \\ &= f'(\theta_1)(\theta - \theta_1) + O\left(|\theta - \theta_1|^2\right) \\ &= f'(\theta_1)\tilde{\theta} + O(\tilde{\theta}^2) \end{aligned}$$

The linear term in $\tilde{\theta}$ is just $f'(\theta_1)\tilde{\theta}$. In the present case

$$f'(\theta) = mR\omega^2(\cos^2\theta - \sin^2\theta) - mg \cos\theta$$

Consequently, if $f'(\theta_1)$ is denoted by α, then

$$\alpha = mR\omega^2\left[\cos^2\left(\arccos\frac{g}{R\omega^2}\right) - \sin^2\left(\arccos\frac{g}{R\omega^2}\right)\right] - mg\cos\left(\arccos\frac{g}{R\omega^2}\right)$$

$$= -m\frac{(R\omega^2)^2 - g^2}{R\omega^2}$$

$$< 0$$

since $\omega > \sqrt{g/R}$. Hence the linearized perturbation equation is

$$mR\ddot{\tilde{\theta}} - \alpha\tilde{\theta} = 0, \qquad \alpha < 0$$

The general solution is

$$\tilde{\theta}(t) = A\cos\sqrt{\frac{-\alpha}{mR}}\,t + B\sin\sqrt{\frac{-\alpha}{mR}}\,t$$

and the linearized perturbations are bounded; therefore, modulo the correctness of the linearization procedure, the equilibrium solutions on the θ_1 branch are stable.

Stability and Chemotaxis of Amoebae

Another example of a phenomenon that illustrates some of the basic ideas of stability is a model of the motion of amoebae under the influences of diffusion and chemotaxis, the latter being motion induced by variations in the concentration of chemicals produced by the amoebae themselves. This phenomenon is of considerable interest in developmental biology. We follow Lin and Segel [1]. The calculations are reminiscent of those that occur in the study of stability of other fluid mechanical systems.

The essential features of the phenomenon can be described as follows. Slime mold amoebae feed on bacteria in soil or dung and are generally uniformly spatially distributed if the food supply is plentiful. As the food supply becomes depleted, however, the organisms begin to secrete a chemical that acts as an attractant. It is observed that the amoebae move toward the high concentrations of the chemical and form agregates. Many questions remain unanswered about this process, but it is possible to formulate a mathematical model that describes some of the essential features of the behavior of the cellular slime molds. To quantify the process we assume for simplicity that changes take place only in one spatial dimension and that all parameters are constant in any cross section. Figure 6.9 shows the uniform state before the onset of chemotaxis and Fig. 6.10 shows the amoebae just beginning to congregate. The one dimensional assumption is not necessary, but the exposition is simpler. The general case is discussed in a paper by E. F. Keller and L. A. Segel, *J. Theoretical Biology*, **26**, 399–415 (1970).

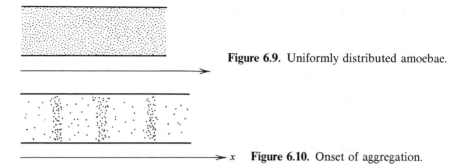

Figure 6.9. Uniformly distributed amoebae.

Figure 6.10. Onset of aggregation.

Let $a(x, t)$ denote the number of amoebae per unit area at position x and at time t and let $c(x, t)$ denote the concentration of the chemical attractant at (x, t) given in mass per unit volume. It is assumed that

(i) There is motion due to the random movement of the amoebae, causing the amoebae to move away from regions of high amoebae concentrations to regions of low amoebae concentrations.

(ii) There is motion of the amoebae toward high concentrations of the chemical.

The formulation of a model equation resembles the calculation in heat conduction problems studied in earlier chapters. We consider an arbitrary region and write a conservation law for the number of amoebae in the region. The conservation law states that the time rate of change of the number of amoebae in $[x_1, x_2]$ must equal the number of amoebae entering the region at $x = x_1$ minus the number of amoebae leaving the region at $x = x_2$. If $\phi(x, t)$ denotes the flux at x at time t, that is, the number of amoebae per unit time per unit area crossing x at time t (here ϕ is positive if motion is to the right or in the positive x direction), then the conservation law can be written

$$\frac{d}{dt} \int_{x_1}^{x_2} a(x, t)\, dx = \phi(x_1, t) - \phi(x_2, t)$$

or using the fundamental theorem of the calculus

$$\frac{d}{dt} \int_{x_1}^{x_2} a(x, t)\, dx = -\int_{x_1}^{x_2} \frac{\partial}{\partial x} \phi(x, t)\, dx$$

The functions are assumed to be smooth so the time derivative can be pulled inside the integral on the left and we may write

$$\int_{x_1}^{x_2} \left(\frac{\partial a}{\partial t}(x, t) + \frac{\partial}{\partial x} \phi(x, t) \right) dx = 0$$

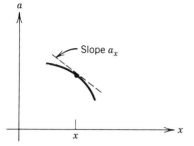

Figure 6.11

Since the interval of integration is arbitrary, it follows that

$$\frac{\partial a}{\partial t} + \frac{\partial \phi}{\partial x} = 0 \tag{14}$$

which is a differential form of the conservation law.

Just as in heat flow a *constitutive assumption* must be made regarding the form of the flux ϕ. From assumptions (i) and (ii) we can assume that ϕ is the sum of a flux ϕ_a due to the diffusion of the amoebae and a flux ϕ_c due to the chemotaxis, that is,

$$\phi = \phi_a + \phi_c \tag{15}$$

It seems clear that ϕ_a should be proportional to the slope or gradient of $a(x, t)$, since under the assumption of diffusion amoebae move from high concentrations of a to low concentrations of a; the steeper the slope, the higher the flux (see Fig. 6.11). Hence we *assume*

$$\phi_a = -k\frac{\partial a}{\partial x}(x, t) \tag{16}$$

where $k > 0$ is a constant of proportionality called the motility. The minus sign occurs because there should be a positive flux (rightward motion) whenever the gradient $\partial a/\partial x$ is negative.

To determine an expression for the flux ϕ_c due to chemotaxis we may argue as previously and deduce that ϕ_c is proportional to the gradient of the chemical concentration $\partial c/\partial x$. As well, if the number of amoebae is doubled for a given gradient, then the flux should be twice as great and therefore ϕ_c should also be proportional to $a(x, t)$. Consequently we assume

$$\phi_c = la(x, t)\frac{\partial c}{\partial x}(x, t) \tag{17}$$

where $l > 0$ is a proportionality constant that measures the strength of the chemotaxis. Note here that there is no minus sign on the right side of (17) because the amoebae move toward higher concentrations of attactant; we say that the amoebae move up the gradient. Combining (14), (15), (16), and (17) gives a nonlinear partial differential equation

$$\frac{\partial a}{\partial t} = \frac{\partial}{\partial x}\left(k\frac{\partial a}{\partial x} - la\frac{\partial c}{\partial x} \right) \tag{18}$$

which expresses the original conservation law for the number of amoebae. The nonlinearity occurs because of the product $a\,\partial c/\partial x$. There are two unknowns, a and c, so another equation is still required.

The additional equation can be found by noticing that c must satisfy a conservation law as well. The time rate of change of the amount of chemical in an interval $x_1 \le x \le x_2$ must equal the flux in minus the flux out plus the rate at which the chemical is created within the interval by secreting amoebae. If $\phi_d(x, t)$ denotes the flux of the chemical via diffusion and $Q(x, t)$ denotes the amount of chemical created at x at time t per unit volume then

$$\frac{d}{dt}\int_{x_1}^{x_2} c(x, t)\, dx = \phi_d(x_1, t) - \phi_d(x_2, t) + \int_{x_1}^{x_2} Q(x, t)\, dx$$

From this conservation law it immediately follows that

$$\frac{\partial c}{\partial t} = -\frac{\partial \phi_d}{\partial x} + Q \tag{19}$$

provided the functions are sufficiently smooth. The random motion of the attractant chemical will be modeled by a proportionality of flux to gradient, as before. Hence

$$\phi_d = -D\frac{\partial c}{\partial x} \tag{20}$$

The source term Q is assumed to be of the form

$$Q = q_1 a - q_2 c \tag{21}$$

Here D is the diffusion constant for the attractant, q_1 is the rate of secretion of the amoebae, and q_2 is the rate of decay of the attractant. Thus from (19) we conclude that

$$\frac{\partial c}{\partial t} = D\frac{\partial^2 c}{\partial x^2} + q_1 a - q_2 c \tag{22}$$

which is the companion equation to (18). Consequently we have two equations in two unknowns $a(x, t)$ and $c(x, t)$.

Equations (18) and (22) clearly have the *uniform solution*

$$a(x, t) = a_0, \qquad c(x, t) = c_0 \tag{23}$$

provided that

$$q_1 a_0 = q_2 c_0 \tag{24}$$

where a_0 and c_0 are constants. This uniform state is an equilibrium solution and we may ask if it is stable to small perturbations or disturbances. That is, if at some time the uniform state is slightly disturbed, does that disturbance die away or does it become more intense with the passage of time? We will now investigate that question and find that onset of aggregation can be identified with an instability of the uniform state.

In the standard stability analysis we let

$$a(x, t) = a_0 + \tilde{a}(x, t), \qquad c(x, t) = c_0 + \tilde{c}(x, t) \tag{25}$$

where $\tilde{a}$ and $\tilde{c}$ are small perturbations or departures from equilibrium. Substitution of (25) into (18) and (22) gives the nonlinear perturbation equations

$$\tilde{a}_t = \frac{\partial}{\partial x}(k\tilde{a}_x - la_0\tilde{c}_x - l\tilde{a}\tilde{c}_x) \tag{26}$$

$$\tilde{c}_t = D\tilde{c}_{xx} + q_1\tilde{a} - q_2\tilde{c} \tag{27}$$

where we have used (24). Now Equation (26) can be linearized owing to the fact that $\tilde{a}\tilde{c}_x$ is quadratic and is hence smaller than the linear terms that contain only first powers of the small quantities. Therefore the *linearized perturbation* equations are

$$\tilde{a}_t = k\tilde{a}_{xx} - la_0\tilde{c}_{xx} \tag{28}$$

$$\tilde{c}_t = D\tilde{c}_{xx} + q_1\tilde{a} - q_2\tilde{c} \tag{29}$$

This pair of coupled partial differential equations can be solved by taking a hint from the solution of the diffusion equation in Chapter 4. We might expect to find solutions of the system (28) and (29), which is diffusionlike, to be of the form

$$\tilde{a} = c_1 e^{\alpha t} e^{i\beta x}$$
$$\tilde{c} = c_2 e^{\alpha t} e^{i\beta x} \tag{30}$$

for some constants c_1, c_2, α, and β. Equations (30) represent a Fourier mode. The question is that if a solution of the form (30) exists must α be positive or negative? If $\alpha > 0$, then there would be a Fourier mode for the perturbations that would grow in time and therefore the uniform state would be unstable to small perturbations.

To determine if solutions of the form (30) exist we substitute (30) into the system (28) and (29) to obtain

$$\left(\alpha + k\beta^2\right)c_1 - la_0\beta^2 c_2 = 0$$
$$-q_1 c_1 + \left(\alpha + D\beta^2 + q_2\right)c_2 = 0 \tag{31}$$

Equations (31) represent a homogeneous system of algebraic equations for c_1 and c_2. From a well-known fact in linear algebra there will be nontrivial solutions of (31) only if the determinant of the coefficient matrix vanishes, or if applied to (31),

$$\alpha^2 + \left(k\beta^2 + D\beta^2 + q_2\right)\alpha + kq_2\beta^2 + kD\beta^4 - q_1 la_0\beta^2 = 0 \tag{32}$$

This quadratic equation for α has discriminant

$$\Delta \equiv \left(k\beta^2 - D\beta^2 - q_2\right)^2 + 4q_1 la_0\beta^2 > 0 \tag{33}$$

and thus the roots α are real. By examining the two roots of the quadratic equation it is easy to see that $\alpha > 0$ if, and only if,

$$kD\beta^2 + q_2 k > q_1 la_0 \tag{34}$$

Consequently (34) represents a necessary and sufficient condition on the constants for stability.

We may ask if an instability is ever possible, that is, if (34) is ever violated. Insight into this question comes when we note that β is the wave number of the given Fourier mode (30). As the wave number β decreases to zero, that is, the wavelength increases, there is a better chance to violate (34). In fact if

$$kq_2 < q_1 la_0 \tag{35}$$

then (34) will always be violated for a range of wave numbers β near zero. Therefore instability will arise from the perturbations that have long wavelengths.

We may interpret condition (35) in a physical context as follows. Condition (35) implies that an instability will occur if the motility k or the decay rate q_2 of the attractant is small or if the rate q_1 of secretion or strength l of the chemotaxis is large. For example, if suddenly in the absence of a food supply attractant is produced and overcomes the stabilizing effects, an instability occurs in the uniform state and aggregates begin to form. Interpreted in this way the onset of aggregation is due to an instability.

The above calculation illustrates many of the features of stability analyses commonly occurring in fluid mechanics. It shows how the simple idea of small perturbations can lead ultimately to far-reaching results and significant insight into the behavior of physical systems whether they are biological, mechanical, or whatever. It also illustrates the importance of instabilities in system. Often there is so much emphasis on a problem being well-posed, one aspect of which is being stable, that the role of instability is overlooked. In the previous example it was the instability that gave rise to the interesting phenomena.

EXERCISES

1.1 Derive (4).

1.2 Determine the dimensions of the constants k, l, D, q_1, and q_2 in the amoebae problem.

1.3 Verify (34).

1.4 Another model of population growth is the Gompertz model

$$\frac{dp}{dt} = rp \ln \frac{k}{p}, \qquad r, k > 0$$

(a) Determine the equilibrium solution and investigate its stability.

(b) Solve the differential equation subject to the initial condition $p(0) = p_0$ and sketch the solution curves for various values of p_0.

(c) In what sense is $p = 0$ an equilibrium population? Is it stable?

1.5 If you have studied Chapter 3, derive the governing differential equation (12) for the bead on a hoop from a variational principle. To this end show that the kinetic and potential energies are $T = \frac{1}{2}mR^2(\dot{\theta}^2 + \omega^2 \sin^2\theta)$ and $V = -mgR \cos \theta$.

6.2 ONE DIMENSIONAL PROBLEMS

Stability

Motivated by the examples in the last section we now investigate the stability and bifurcation of equilibrium solutions of problems in a single dimension. In particular we consider the first order differential equation

$$\frac{du}{dt} = f(\mu, u), \qquad t > 0 \tag{1}$$

where μ is a real parameter and f is a given function having continuous partial derivatives of all order. The unknown function u is a function of the real variable t and also depends on the parameter μ. We say that a constant solution $u = u_0$ of (1) is an *equilibrium solution*. The equilibrium solutions are found by setting

$$f(\mu, u) = 0 \qquad (2)$$

and will in general depend upon the parameter μ. To each point (μ_0, u_0) on the locus (2) there corresponds an equilibrium solution $u = u_0$ at the particular value μ_0 of the parameter. To one value of the parameter may correspond several equilibrium solutions (as in the rotating hoop problem). A graph of the locus (2) in the μu plane is called a *branching* or *bifurcation diagram*. Such a diagram gives a visual representation of the equilibrium solutions at different values of the parameter μ. The intersecting branches of (2) are the *bifurcating solutions* and the intersection points are called *bifurcation points*. In the present context the vertical axis u represents equilibrium solutions; in different contexts some other feature of the solution may be graphed on the vertical axis.

Example 2.1   Consider the differential equation

$$\frac{du}{dt} = (1 - u)(u^2 - \mu), \qquad \mu \in R^1$$

The equilibrium solutions are found by setting $(1 - u)(u^2 - \mu) = 0$, which gives

$$u = 1 \qquad \text{and} \qquad \mu = u^2$$

The branching diagram is drawn in Fig. 6.12. For each $\mu > 0$, $\mu \neq 1$, there are three distinct equilibrium solutions. For $\mu < 0$ there is one equilibrium solution.

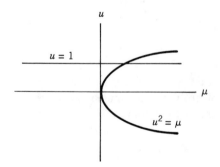

Figure 6.12

For a given, *fixed* value of the parameter μ we can inquire about the stability of any equilibrium solutions that exist for that value of μ. An equilibrium solution u_0 of (1) is *stable* if every solution $u(t)$ of (1) starting sufficiently close to u_0 at $t = 0$ remains close to u_0 for all $t > 0$, that is, for any $\varepsilon > 0$ there exists a $\delta_\varepsilon > 0$ such that

$$\left| u(t) - u_0 \right| < \varepsilon, \qquad t > 0$$

whenever

$$\left| u(0) - u_0 \right| < \delta_\varepsilon$$

If in addition to being stable

$$\lim_{t \to \infty} \left| u(t) - u_0 \right| = 0$$

for every solution $u(t)$ starting sufficiently close to u_0 at $t = 0$, then we say that u_0 is *asymptotically stable*. An equilibrium solution is *unstable* if it is not stable. In these definitions, since μ is fixed, we have omitted in our notation any explicit dependence of the solutions on μ.

Also of interest is how the stability of equilibrium solutions behave as the parameter μ varies. Often an equilibrium solution $u_0(\mu)$ will be stable for $\mu < \mu_c$ and unstable for $\mu \geq \mu_c$, where μ_c is some critical value of μ. Thus as μ is increased slowly, the equilibrium solution of $u_0(\mu)$ becomes unstable at μ_c and the physical system can no longer remain in the state $u_0(\mu)$. After the system becomes unstable at μ_c many resulting behaviors are possible. One is that the system may branch to another stable solution as in the bead problem for $\mu > \mu_c$. This one particular example indicates some of the phenomena that lie at the heart of bifurcation theory.

The restriction to studying differential equations of the type (1) is a considerable one. More generally (1) could be regarded as an equation in some function space (for example, a Hilbert space) and f an operator on that space that depends on a real parameter μ. For example, f could be a linear or nonlinear differential or integral operator on H. Then (1) is an evolution equation governing the motion of u in the space H. Finding equilibrium solutions would amount to solving $f(u, \mu) = 0$, which would be a differential or integral equation. In this case, of course, the problem may be infinite dimensional. With just these vague ideas it is easy to imagine the broad scope of bifurcation and stability theory in analysis. For further discussion in this vein see Stakgold [2].

We now inquire about conditions that guarantee stability or asymptotic stability. The underlying motivation follows the examples in Section 6.1 and can be generally illustrated as follows. Again let μ be fixed. Consider a small perturbation $\tilde{u}(t)$ of an equilibrium solution u_0 of

$$\frac{du}{dt} = f(\mu, u), \qquad t > 0 \tag{3}$$

Letting

$$u(t) = u_0 + \tilde{u}(t) \tag{4}$$

and substituting into the differential equation (3) gives the equation

$$\frac{d\tilde{u}}{dt} = f(\mu, u_0 + \tilde{u}), \qquad t > 0 \tag{5}$$

for the perturbation $\tilde{u}$ (both u_0 and $\tilde{u}$ depend on the fixed value μ). We ask about the longtime behavior of $\tilde{u}$, that is, does it decay or grow as $t \to \infty$? We attempt to answer this question heuristically by linearizing (5). Expanding the right side of (5) in a Taylor series we obtain

$$\frac{d\tilde{u}}{dt} = f(\mu, u_0) + f_u(\mu, u_0)\tilde{u} + \tfrac{1}{2}f_{uu}(\mu, u_0)\tilde{u}^2 + \cdots$$

Since $f(\mu, u_0) = 0$, it follows that the linearized perturbation equation is

$$\frac{d\tilde{u}}{dt} = f_u(\mu, u_0)\tilde{u} \tag{6}$$

The solution is

$$\tilde{u}(t) = Ce^{\alpha t} \tag{7}$$

where α is a real number defined by

$$\alpha \equiv f_u(\mu, u_0) \tag{8}$$

and C is a constant. Therefore if $\alpha < 0$, the small perturbations decay and u_0 is asymptotically stable. On the other hand, if $\alpha > 0$ the perturbations grow and u_0 is unstable. This intuitive but insightful argument forms the basis of a more general theorem for the nonlinear perturbation equation (5). The quantity α defined by (8) is called the *stability indicator*. If $\alpha = 0$, then the previous remarks do not apply.

Before stating and proving the general stability result we first prove a lemma known as *Gronwall's inequality*. It plays a significant role in the systematic study of the theory of ordinary differential equations.

Lemma 2.1   *If $u(t)$ and $v(t)$ are continuous nonnegative functions on $0 \leq t \leq T$ and K is a nonnegative constant, then*

$$u(t) \leq K + \int_0^t v(s)u(s)\, ds, \qquad 0 \leq t \leq T$$

implies

$$u(t) \leq K \exp\left(\int_0^t v(s)\, ds \right), \qquad 0 \leq t \leq T$$

Proof   If $K > 0$, then the given inequality implies

$$\frac{u(t)v(t)}{K + \int_0^t v(s)u(s)\, ds} \leq v(t)$$

Integrating both sides from 0 to t gives

$$\ln\left[K + \int_0^t v(s)u(s)\, ds \right] - \ln K \leq \int_0^t v(s)\, ds$$

which in turn gives

$$u(t) \leq K + \int_0^t v(s)u(s)\, ds \leq K \exp\left(\int_0^t v(s)\, ds \right)$$

If $K = 0$, then $u(t)$ is identically zero and the inequality is trivially true. The reader is asked to supply this argument.

Now we state and prove the basic result.

Theorem 2.1   *Let u_0 be an equilibrium solution of* (3) *and assume that*

$$f(\mu, u_0 + \tilde{u}) = f_u(\mu, u_0)\tilde{u} + R(u_0, \tilde{u})$$

where the remainder term $R(u_0, \tilde{u})$ is $O(\tilde{u}^2)$, that is, $|R(u_0, \tilde{u})| \leq K|\tilde{u}|^2$ for $\tilde{u}$ sufficiently small where K is a positive constant. Then u_0 is asymptotically stable if $\alpha < 0$ and unstable if $\alpha > 0$, where α is given by (8).

Proof   The differential equation (5) for the perturbation $\tilde{u}(t)$ is

$$\frac{d\tilde{u}}{dt} = f_u(\mu, u_0)\tilde{u} + R(u_0, \tilde{u}) \tag{9}$$

(note that R also depends upon the parameter μ, but for brevity it is not included). Multiplying (9) by $\exp(-\alpha t)$ and integrating from 0 to t gives after rearrangement

$$\tilde{u}(t) - \tilde{u}(0)e^{\alpha t} = \int_0^t R(u_0, \tilde{u}(s))e^{\alpha(t-s)}\, ds \tag{10}$$

Therefore

$$\left| \tilde{u}(t) - \tilde{u}(0)e^{\alpha t} \right| \leq \int_0^t \left| R(u_0, \tilde{u}(s)) \right| e^{\alpha(t-s)}\, ds$$

$$\leq K \int_0^t |\tilde{u}(s)|^2 e^{\alpha(t-s)}\, ds$$

Hence

$$\left| \tilde{u}(t) \right| \leq |\tilde{u}(0)| e^{\alpha t} + K \int_0^t |\tilde{u}(s)|^2 e^{\alpha(t-s)}\, ds$$

or

$$e^{-\alpha t} |\tilde{u}(t)| \leq |\tilde{u}(0)| + K \int_0^t |\tilde{u}(s)|^2 e^{-\alpha s}\, ds$$

By Gronwall's inequality in Lemma 2.1

$$e^{-\alpha t}|\tilde{u}(t)| \leq |\tilde{u}(0)| \exp\left(\int_0^t K|\tilde{u}(s)|\, ds \right)$$

Therefore

$$|\tilde{u}(t)| \leq |\tilde{u}(0)| \exp\left(\alpha t + K \int_0^t |\tilde{u}(s)|\, ds \right) \tag{11}$$

Let $\alpha < 0$ and let us assume for the moment that

$$|\tilde{u}(t)| \leq \frac{\eta}{K}, \qquad \text{for} \quad 0 \leq t \leq T \tag{12}$$

where η is chosen so that $\alpha + \eta < 0$, and $T > 0$. Equation (11) implies

$$|\tilde{u}(t)| \leq |\tilde{u}(0)| e^{(\alpha + \eta)t}, \qquad 0 \leq t \leq T \tag{13}$$

Consequently, if (12) holds, then (13) follows for any $T > 0$. We will now show in fact that (12) holds provided $|\tilde{u}(0)|$ is sufficiently small. Let $|\tilde{u}(0)| < \eta/K$. Suppose there is some $t_1 > 0$ such that $|\tilde{u}(t_1)| = \eta/K$, but $|\tilde{u}(t)| < \eta/K$

for all $0 < t < t_1$ (recall that u is continuous). By (13)

$$\left|\tilde{u}(t_1)\right| \leq \left|\tilde{u}(0)\right|e^{(\alpha+\eta)t_1} < \left|\tilde{u}(0)\right| < \frac{\eta}{K}$$

which is a contradiction. Thus (12) must hold for all $t \geq 0$ and hence

$$\left|\tilde{u}(t)\right| \leq \left|\tilde{u}(0)\right|e^{(\alpha+\eta)t}, \qquad t \geq 0$$

This inequality shows that $\tilde{u}(t) \to 0$ as $t \to \infty$ and therefore we infer asymptotic stability.

For the case $\alpha > 0$ assume, by way of contradiction, that u_0 is stable. Let α/K be given. Then there exists a $\delta > 0$ such that $|\tilde{u}(0)| < \delta$ implies

$$\left|\tilde{u}(t)\right| \leq m \equiv \sup_{t>0}\left|\tilde{u}(t)\right| < \frac{\alpha}{K} \qquad \text{for all} \quad t > 0$$

Equation (10) still holds, so multiplying (10) by $\exp(-\alpha t)$ and taking the limit as $t \to \infty$ gives

$$\tilde{u}(0) = -\int_0^\infty R\big(u_0, \tilde{u}(s)\big)e^{-\alpha s}\,ds$$

Then (10) may be written

$$\tilde{u}(t) = -\int_t^\infty R\big(u_0, \tilde{u}(s)\big)e^{\alpha(t-s)}\,ds$$

Therefore

$$\left|\tilde{u}(t)\right| \leq Km^2\int_t^\infty e^{\alpha(t-s)}\,ds = \frac{Km^2}{\alpha}$$

and so

$$m \leq \frac{Km^2}{\alpha}$$

which contradicts the choice of $m < \alpha/K$.

The theorem asserts that if $\alpha = f_u(\mu, u_0) < 0$, then u_0 is asymptotically stable; in other words, the perturbations decay, provided they are small enough initially. If the initial perturbation $\tilde{u}(0)$ were unrestricted, then global stability would result. This is a strong type of stability that would imply that a single equilibrium solution exists and attracts all other solutions of (1). Again we remind the reader that the parameter μ is fixed in the previous discussion.

Example 2.2 Consider the differential equation

$$\frac{du}{dt} = \mu u - u^2$$

The equilibrium solutions are found from

$$\mu u - u^2 = 0$$

and are $u = 0$ and $u = \mu$. In the present case $f_u = \mu - 2u$. For the solution $u = 0$ we have $\alpha = f_u(\mu, 0) = \mu$; hence $u = 0$ is asymptotically stable if $\mu < 0$ and unstable if $\mu > 0$. For the equilibrium solution $u = \mu$ we have $\alpha = f_u(\mu, \mu) = -\mu$, and therefore $u = \mu$ is asymptotically stable if $\mu > 0$ and unstable if $\mu < 0$. (What can be said in the case $\mu = 0$?)

Classification of Bifurcation Points

One of the fundamental questions in bifurcation theory is: As the parameter μ is varied, does there reach a critical value where there is an exchange of stability? For example, at some μ_c does the state that was stable become unstable? We have already observed this phenomenon in the problem of a bead on a rotating hoop; the bifurcation diagram looked schematically like the one shown in Fig. 6.13. For $\mu < \mu_c$ the $u = 0$ solution was stable. But at $\mu = \mu_c$ there was an exchange of stability and the system could no longer exist in the state $u = 0$ for $\mu > \mu_c$; the system tended to a nonzero stable state.

Example 2.2 (Revisited) The branching diagram for

$$\frac{du}{dt} = \mu u - u^2, \qquad \mu \in R^1$$

is shown in Fig. 6.14. We note a different type of stability exchange from that which occurs in Fig. 6.13.

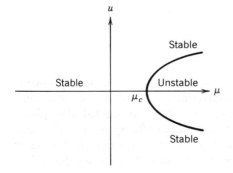

Figure 6.13

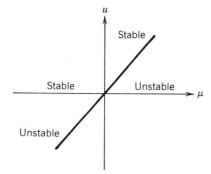

Figure 6.14

Results concerning stability exchange for a given problem depend on the type of bifurcation point involved, the stability of the point, how the branches cross, and so on. To formulate specific results in this direction we classify the points on the branches of solutions to $f(\mu, u) = 0$. Geometrically $z = f(\mu, u)$ represents a surface in R^3 and $f(\mu, u) = 0$ represents the locus of points in the μu plane where the surface intersects the $z = 0$ plane. The intersecting curves in the $z = 0$ plane are the bifurcating solutions. Our approach to bifurcation relates closely to singularity theory, which is a part of a more general analysis; however, many problems in the more general context can be reduced to the simple generic ones that are examined herein. One of the fundamental results used in bifurcation theory and in the classification problem is the implicit function theorem. This theorem answers the question of when the equation $f(\mu, u) = 0$ can be solved locally for one of the variables. The simplest version of this theorem in two dimensions may be stated as follows.

Theorem 2.2   *Let $f(\mu, u)$ be a continuously differentiable function in some open region U of the μu plane containing (μ_0, u_0). If $f_u(\mu_0, u_0) \neq 0$, then there is a rectangle*

$$S: |u - u_0| < a, \qquad |\mu - \mu_0| < b$$

contained in U such that

(i)   *The equation $f(\mu, u) = 0$ has a unique solution $u = u(\mu)$ on S.*
(ii)   *The function $u(\mu)$ is continuously differentiable on $|\mu - \mu_0| < b$ and its derivative is given by*

$$\frac{du}{d\mu} = -\frac{f_\mu(\mu, u(\mu))}{f_u(\mu, u(\mu))} \tag{14}$$

For a proof we refer the reader to any advanced calculus book. We note that the theorem is symmetric in that if $f_\mu(\mu_0, u_0) \neq 0$, then we may solve for μ to obtain $\mu = \mu(u)$. In this case

$$\frac{d\mu}{du} = -\frac{f_u(\mu(u), u)}{f_\mu(u(u), u)} \tag{15}$$

Now let f have continuous partial derivatives up to third order in a neighborhood of a point $P_0 = (\mu_0, u_0)$. The point P_0 is a *regular point* of the locus $f(\mu, u) = 0$, if $f(P_0) = 0$ and if $f_u(P_0) \neq 0$ or $f_\mu(P_0) \neq 0$. In this case the implicit function theorem guarantees that there is a unique curve $u = u(\mu)$ or $\mu = \mu(u)$ passing through P_0. If P_0 is not a regular point, then P_0 is a *singular point* and is characterized by $f(P_0) = f_u(P_0) = f_\mu(P_0) = 0$. Accordingly we may expect unusual behavior at a singular point, since both ratios in (14) and (15) are the indeterminant form $0/0$. Singular points of a locus $f(\mu, u) = 0$ can be studied systematically by assuming further that the second partial derivatives of f do not all vanish simultaneously. Then, letting $P = (\mu, u)$, by Taylor's theorem

$$f(P) = f(P_0) + f_u(P_0)\,\Delta u + f_\mu(P_0)\,\Delta\mu$$
$$+ \tfrac{1}{2}\big(f_{uu}(P_0)\,\Delta u^2 + 2f_{u\mu}(P_0)\,\Delta u\,\Delta\mu + f_{\mu\mu}(P_0)\,\Delta\mu^2 \big)$$
$$+ o\big(\Delta u^2 + \Delta\mu^2\big)$$

where Δu and $\Delta\mu$ are the increments $u - u_0$ and $\mu - \mu_0$, respectively. If P_0 is a singular point and P lies on the locus, then

$$f_{uu}(P_0)\,\Delta u^2 + 2f_{u\mu}(P_0)\,\Delta u\,\Delta\mu + f_{\mu\mu}(P_0)\,\Delta\mu^2 + o\big(\Delta u^2 + \Delta\mu^2\big) = 0$$

In the limit as Δu and $\Delta\mu$ tend to zero,

$$f_{uu}(P_0)\,du^2 + 2f_{u\mu}(P_0)\,du\,d\mu + f_{\mu\mu}(P_0)\,d\mu^2 = 0 \tag{16}$$

which gives a relation between the differentials du and $d\mu$ along any curve $f(\mu, u) = 0$ at P_0. If the *discriminant*

$$D \equiv f_{u\mu}^2(P_0) - f_{uu}(P_0)f_{\mu\mu}(P_0)$$

is positive, then the singular point P_0 is called a *double point*. Solving the quadratic equation (16) gives

$$\frac{du}{d\mu} = -\frac{f_{u\mu}}{f_{uu}} \pm \sqrt{\frac{D}{f_{uu}^2}} \tag{17}$$

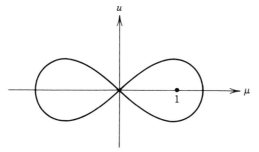

Figure 6.15. A double point.

or

$$\frac{d\mu}{du} = -\frac{f_{u\mu}}{f_{\mu\mu}} \pm \sqrt{\frac{D}{f_{\mu\mu}^2}} \tag{18}$$

Formulas (17) and (18) give the slopes of the tangents to the bifurcating curves at the double point P_0. We shall show in Lemma 2.2 that only two cases are possible when $D > 0$. If $D < 0$, then there are no real tangents and P_0 is called an *isolated point*. If $D = 0$, then at least two curves through P_0 have coincident tangents.

Example 2.3 Consider the lemniscate $f(\mu, u) = (\mu^2 + u^2)^2 - 2(\mu^2 - u^2) = 0$ shown in Fig. 6.15. It is easy to check that $f = f_\mu = f_u = 0$ and $D > 0$ at $(0,0)$. Thus the origin is a double point and two branches pass through the origin with distinct tangents. All of the other points on the lemniscate are regular points.

Example 2.4 The curve $f(\mu, u) \equiv u^3 - \mu^2 = 0$ has a singular point at $(0,0)$ with $D = 0$. In this case two branches have the same vertical tangent at $(0,0)$. Such a point is called a *cusp* (see Fig. 6.16).

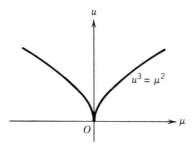

$u^3 = \mu^2$

Figure 6.16. A cusp.

***Example* 2.5**   The locus $f(\mu, u) \equiv u^2 + \mu^2 = 0$ consists of a single point $(0, 0)$. In this case $(0, 0)$ is an isolated singular point with $D < 0$.

A singular point P_0 is a *higher order singular point* if all of the second partials f_{uu}, $f_{u\mu}$, and $f_{\mu\mu}$ vanish at P_0. We shall not investigate this case in which other types of singular behavior occur.

Lemma 2.2   *Let P_0 be a double point of $f(\mu, u) = 0$. Then either*

(i) $f_{\mu\mu}(P_0) \neq 0$ *and the two tangents are given by* (18).
(ii) $f_{\mu\mu}(P_0) = 0$ *and the two tangents are given by*

$$\frac{du}{d\mu} = 0 \quad and \quad \frac{d\mu}{du} = -\frac{f_{uu}}{2f_{u\mu}} \quad at \quad P_0$$

Proof   The quantity $D > 0$ by definition. If $f_{\mu\mu}(P_0) \neq 0$, then two distinct tangents are given by (18). If $f_{\mu\mu}(P_0) = 0$, then $f_{u\mu}(P_0) \neq 0$ and the quadratic (16) becomes

$$du\Big(f_{uu}(P_0)\, du + 2f_{u\mu}(P_0)\, d\mu \Big) = 0$$

Therefore $du/d\mu = 0$ and $d\mu/du = -f_{uu}/2f_{u\mu}$ at P_0. There is, of course, a symmetric version of the lemma when $f_{uu} \neq 0$.

From Lemma 2.2 we infer that two branches with distinct tangents pass through a double point P_0. No more than two branches can pass through a double point, since in that case P_0 would be a higher order singularity.

Exchange of Stability

Because two branches cross at a double point, such points are of interest because a change of stability can occur. In this section we investigate conditions under which such an exchange can take place.

The stability can also change at a regular point. A regular point P_0 is called a *turning point* (with respect to μ) if $f_\mu(P_0) \neq 0$ and $d\mu/du$ changes sign at P_0.

***Example* 2.6**   The function

$$f(\mu, u) = (1 + u^2 - \mu)(\mu^2 - 25u^2)$$

has equilibrium solutions

$$\mu_1 = 1 + u^2, \qquad \mu_2 = 5u, \qquad \mu_3 = -5u$$

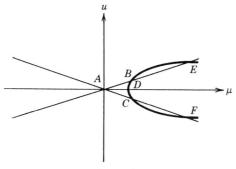

Figure 6.17

The branching diagram is shown in Fig. 6.17. All the points on the branches are regular points except A, B, C, E, and F. The point D is a regular turning point, since $d\mu/du$ changes sign at D. The points A, B, C, E, and F are double points because they are singular points at which there are two distinct tangents.

The following theorem states that the stability must change at a turning point and thus the exchange of stability question at these types of points is easily dismissed.

Theorem 2.3 *Let P_0 be a regular turning point of $f(\mu, u) = 0$. Then equilibrium solutions on one side are stable and on the other side are unstable.*

Proof The proof follows easily from Equation (15), which is equivalent to

$$f_u(\mu(u), u) = -\frac{d\mu}{du}(u)f_\mu(\mu(u), u)$$

or in our previous notation

$$\alpha(u) = -\frac{d\mu}{du}(u)f_\mu(\mu(u), u) \tag{19}$$

where $\alpha(u)$ is the stability indicator. We recall that an equilibrium solution is asymptotically stable if $\alpha < 0$ and unstable if $\alpha > 0$. Since by hypothesis f_μ cannot go through a zero at P_0, (19) implies that α must change sign at P_0 as we move along the branch since $d\mu/du$ changes sign.

We now consider the exchange of stability question at a double point $P_0 = (\mu_0, u_0)$. From Lemma 2.2 there are two cases.

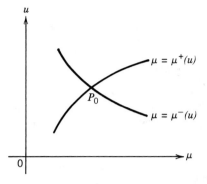

Figure 6.18

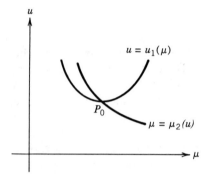

Figure 6.19

Case (i) The two curves $\mu^+(u)$ and $\mu^-(u)$ passing through P_0 have tangents given by (18) (see Fig. 6.18).

Case (ii) One of the curves is $u = u_1(\mu)$ with $du_1/d\mu = 0$ at P_0 and the other is $\mu = \mu_2(u)$ with $d\mu_2/du = -f_{uu}/2f_{u\mu}$ at P_0 (see Fig. 6.19).

We consider Case (i) and let $\alpha^+(u)$ and $\alpha^-(u)$ denote the stability indicators along the curves $\mu^+(u)$ and $\mu^-(u)$, respectively. That is, $\alpha^+(u) \equiv f_u(\mu^+(u), u)$ and $\alpha^-(u) \equiv f_u(\mu^-(u), u)$. Then we have the following theorem for Case (i).

Theorem 2.4 *Let P_0: (μ_0, u_0) be a double point with $f_{\mu\mu}(P_0) \neq 0$. Then*

$$\alpha^+(u) = -\frac{d\mu^+}{du}(u)\left\{\operatorname{sgn} f_{\mu\mu}(P_0)\sqrt{D}\,(u - u_0) + o(|u - u_0|)\right\} \quad (20)$$

$$\alpha^-(u) = \frac{d\mu^-}{du}(u)\left\{\operatorname{sgn} f_{\mu\mu}(P_0)\sqrt{D}\,(u - u_0) + o(|u - u_0|)\right\} \quad (21)$$

where the discriminant D is evaluated at P_0.

Proof From (19) and Taylor's theorem

$$\alpha(u) = -\frac{d\mu}{du}(u)f_\mu(\mu(u), u)$$

$$= -\frac{d\mu}{du}(u)\{ f_\mu(P_0) + f_{\mu u}(P_0)(u - u_0)$$

$$+ f_{\mu\mu}(P_0)(\mu - \mu_0) + o(|u - u_0|)\}$$

$$= -\frac{d\mu}{du}(u)\left\{ f_{\mu u}(P_0) + f_{\mu\mu}(P_0)\frac{\mu - \mu_0}{u - u_0} \right\}(u - u_0) + o(|u - u_0|)$$

But

$$\frac{\mu - \mu_0}{u - u_0} = \frac{d\mu}{du}(u_0) + o(|u - u_0|)$$

and so

$$\alpha(u) = -\frac{d\mu}{du}(u)\left\{ f_{\mu u}(P_0) + f_{\mu\mu}(P_0)\frac{d\mu}{du}(u_0) \right\}(u - u_0) + o(|u - u_0|)$$

But (18) gives (taking the $+$ sign for $\mu^+(u)$)

$$\frac{d\mu^+}{du}(u_0) = -\frac{f_{u\mu}}{f_{\mu\mu}} + \sqrt{\frac{D}{f_{\mu\mu}^2}}, \qquad \text{at } P_0$$

and therefore (20) follows. Equation (21) is obtained similarly.

An analogous theorem can be stated for Case (ii). The proof is left as an exercise.

Theorem 2.5 P_0: (μ_0, u_0) *be a double point with* $f_{\mu\mu}(P_0) = 0$. *Then*

$$\alpha^1(\mu) = \text{sgn} \, f_{u\mu}(P_0)\sqrt{D}\,(\mu - \mu_0) + o(|\mu - \mu_0|) \tag{22}$$

$$\alpha^2(u) = -\text{sgn} \, f_{u\mu}(P_0)\frac{d\mu_2}{du}(u)\{\sqrt{D}\,(u - u_0) + o(|u - u_0|)\} \tag{23}$$

where $\alpha^1(\mu) \equiv f_u(\mu, u_1(\mu))$ *and* $\alpha^2(u) \equiv f_u(\mu_2(u), u)$ *are the stability indicators of* $u_1(\mu)$ *and* $\mu_2(u)$.

The interpretation of Theorem 2.4 in an exchange of stability context is straightforward. When $|u - u_0|$ is small, α^+ and α^- have the same sign if $d\mu^+/du$ and $d\mu^-/du$ have opposite signs, and they have opposite signs if

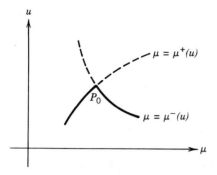

Figure 6.20

$d\mu^+/du$ and $d\mu^-/du$ have the same signs. Consequently the slopes of the curves $\mu^+(u)$ and $\mu^-(u)$ at the double point P_0 indicate how the stability changes.

Example 2.7 Suppose $\mu^+(u)$ and $\mu^-(u)$ are as indicated in Fig. 6.20 with $f_{\mu\mu} < 0$. The solid part of the curves denotes stability ($\alpha < 0$) and the dashed portion denotes unstable solutions. As u passes through u_0, the stability must change along the curves as indicated, consistent with Theorem 2.4.

Example 2.8 We illustrate the preceding concepts on the differential equation

$$\frac{du}{dt} = (\mu - \mu_c)u - u^3$$

where μ_c is a fixed positive constant. The equilibrium solutions are $u_1 = 0$ and $\mu_2 = u^2 + \mu_c$ and a branching diagram is shown in Fig. 6.21. The point P_0: $(\mu_c, 0)$ is a branching point and it is clearly a double point. In fact $f(\mu, u) = (\mu - \mu_c)u - u^3$ and so $f_u = \mu - \mu_c - 3u^2$, $f_{uu} = -6u$, $f_\mu = u$, $f_{\mu\mu} = 0$, and $f_{u\mu} = 1$. Clearly $f_\mu = f_u = 0$ at P_0 and $f_{uu} = f_{\mu\mu} = 0$ at P_0 but $f_{u\mu} \neq 0$. The

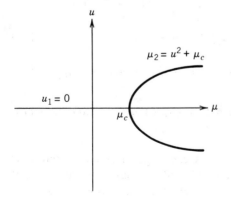

Figure 6.21

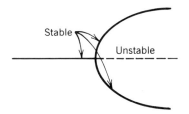

Figure 6.22

discriminant is $D = 1$. By Lemma 2.2 the two distinct tangents are given by $du_1/d\mu = 0$ and $d\mu_2/du = -f_{uu}/2f_{u\mu} = 0$. Theorem 2.5 shows that the stability indicators are

$$\alpha^1(\mu) = \text{sgn}(1)\sqrt{1}\,(\mu - \mu_c) + o(|\mu - \mu_c|)$$

$$\alpha^2(u) = -\text{sgn}(1)\frac{d}{du}(u^2 + \mu_c)\{\sqrt{1}\,(u - 0) + o(|u|)\}$$

or

$$\alpha^1(\mu) = \mu - \mu_c + o(|\mu - \mu_c|)$$

$$\alpha^2(u) = -2u^2 + o(|u|)$$

Since $\alpha^2(u) < 0$, the branch $\mu_2 = u^2 + \mu_c$ is stable; since $\alpha^1(u) > 0$ for $\mu > \mu_c$, the branch $u_1 = 0$ is unstable for $\mu > \mu_0$; and since $\alpha^1(u) < 0$ for $\mu > \mu_c$, the branch $u_1 = 0$ is stable for $\mu < \mu_c$. The stability diagram is shown in Fig. 6.22.

A complete discussion of stability exchange in all cases can be found in Iooss and Joseph [3].

Continuously Stirred Tank Reactor

In this section we set up a model for determining the temperature $\bar{\theta}$ and concentration $\bar{c}$ of a chemically reacting substance in a continuously stirred tank reactor. The reaction takes place in a tank of volume V (see Fig. 6.23) that is stirred continuously in order to maintain uniform temperature and concentration. It is fed by a stream of constant flow velocity q, constant reactant concentration c_{in}, and constant temperature θ_{in}. After mixing and reacting the products are removed at the same volume rate q. We assume that the reaction is first order and irreversible and the reactant disappears at the rate

$$-k\bar{c}e^{-A/\bar{\theta}}$$

where A and k are positive constants. The amount of heat released is assumed

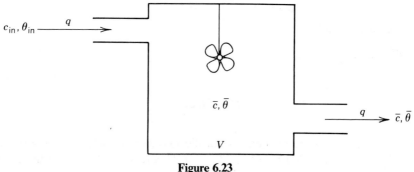

Figure 6.23

to be given by

$$hk\bar{c}e^{-A/\bar{\theta}}$$

where h is a positive constant.

We are able to write a system of two differential equations that govern the concentration $\bar{c}(\bar{t})$ and the temperature $\bar{\theta}(\bar{t})$ of the reactant in the tank. One equation comes from balancing the mass of the reactant and the other arises from conservation of (heat) energy. First, mass balances gives

$$V\frac{d\bar{c}}{d\bar{t}} = qc_{in} - q\bar{c} - Vk\bar{c}e^{-A/\bar{\theta}} \tag{24}$$

This equation states that the rate of change of mass of reactant in the tank equals the mass in, less the mass out, plus the rate at which the mass of the reactant disappears in the chemical reaction. Next, heat balance gives

$$VC\frac{d\bar{\theta}}{d\bar{t}} = qC\theta_{in} - qC\bar{\theta} + hVk\bar{c}e^{-A/\bar{\theta}} \tag{25}$$

where C is the heat capacity of the mixture. These equations may be reduced to dimensionless form by introducing dimensionless variables

$$t = \frac{\bar{t}}{V/q}, \qquad \theta = \frac{\bar{\theta}}{\theta_{in}}, \qquad c = \frac{\bar{c}}{c_{in}}$$

and dimensionless constants

$$\mu = \frac{q}{kV}, \qquad b = \frac{hc_{in}}{C\theta_{in}}, \qquad \gamma = \frac{A}{\theta_{in}}$$

In this case the differential equations (24) and (25) become

$$\frac{dc}{dt} = 1 - c - \frac{c}{\mu}e^{-\gamma/\theta} \tag{26}$$

$$\frac{d\theta}{dt} = 1 - \theta + \frac{bc}{\mu}e^{-\gamma/\theta} \tag{27}$$

This pair of equations may be reduced to a single equation by the following argument. If we multiply (26) by b and add it to (27), we get

$$\frac{d}{dt}(\theta + bc) = 1 + b - (\theta + bc)$$

which integrates to

$$\theta + bc = 1 + b + De^{-t}, \qquad D \text{ constant}$$

Assuming $\theta + bc$ at $t = 0$ is $1 + b$ gives $D = 0$, and then $\theta + bc = 1 + b$. Consequently the heat balance equation becomes

$$\frac{d\theta}{dt} = 1 - \theta + \frac{1 + b - \theta}{\mu}e^{-\gamma/\theta}$$

which is a single differential equation for θ. Introducing $u = \theta - 1$, this equation becomes

$$\frac{du}{dt} = -u + \frac{b - u}{\mu}e^{-\gamma/(u+1)} \tag{28}$$

which is the final form that we shall study.

By definition the parameter μ in (28) can be identified with the flow rate and it will act as the bifurcation parameter in our analysis, since we desire to study equilibrium solutions as a function of flow rate. The equilibrium solutions are given by solutions to the equation

$$\mu u = (b - u)e^{-\gamma/(u+1)} \tag{29}$$

Since (29) cannot be solved in analytic form, we shall resort to graphical techniques. The graph of the function on the right side of (29) has the shape of the curve drawn in Fig. 6.24. Solutions occur at the points where this curve intersects the straight lines μu for different slopes μ. Several such intersection points are shown. For some slopes μ, for example μ_1 and μ_5, there is only one equilibrium solution, while for other slopes there are two (U and Q corresponding to μ_2 and S and W corresponding to μ_4), and for some there are

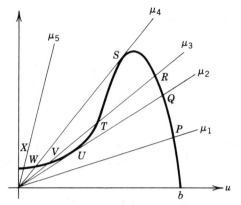

Figure 6.24. Intersections of the graph of $(b - u)\exp(-\gamma/(u + 1))$ with the graph of μu for different values of μ.

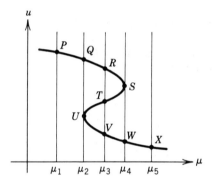

Figure 6.25. Bifurcation diagram.

three (R, T, and V corresponding to μ_3). A graph of the equilibrium solutions versus the parameter μ is shown in Fig. 6.25. The points S and U are regular turning points and therefore by Theorem 2.3 the stability must change at those points.

To determine the stability of each branch of the curve in Fig. 6.25 we note that the right side of the differential equation (28) can be written

$$F(\mu, u) = -u + \frac{1}{\mu}f(u),$$

where $f(u)$ is the function shown in Fig. 6.24. Then

$$F_u(\mu, u) = -1 + \frac{1}{\mu}f'(u)$$

and therefore the condition required for stability of an equilibrium solution u_0 is

$$-1 + \frac{1}{\mu} f'(u_0) < 0$$

On the upper branch of the curve in Fig. 6.25, for example at the point P, we have f' negative and hence the upper branch is asymptotically stable. Therefore, the branch UTS is unstable, and the lower branch $UVWX$ is asymptotically stable.

Physically, if the flow rate μ begins at μ_5 and is slowly decreased, upon reaching the state U there is a sudden jump in temperature to state Q; conversely, if the flow rate is increased from μ_1, upon reaching state S there is a sudden jump to the lower temperature state W. The latter bifurcation is known as *quenching* the reaction.

EXERCISES

2.1 For each of the following differential equations determine the equilibrium solutions and examine their stability

(a) $\dfrac{du}{dt} = au + bu^2, \quad a, b > 0$

(b) $\dfrac{du}{dt} = -2(1 + u^2)^{-1}\arctan u$

(c) $\dfrac{du}{dt} = e^u - 1$

(d) $\dfrac{du}{dt} = u^2(u^2 - 1)$

(e) $\dfrac{du}{dt} = -r(1 - a^{-1}u)(1 - b^{-1}u)u, \quad r, a, b > 0$

2.2 For the following differential equations determine the equilibrium solutions and sketch a branching diagram. Identify the bifurcation points and bifurcating solutions. Investigate the stability of the equilibrium solutions and indicate where an exchange of stability occurs.

(a) $\dfrac{du}{dt} = (u - \mu)(u^2 - \mu)$

(b) $\dfrac{du}{dt} = u(9 - \mu u)(\mu + 2u - u^2)$

2.3 Consider

$$\frac{du}{dt} = u(\mu - u) + \varepsilon$$

where ε is a real number. If $\varepsilon = 0$, show that there are two bifurcating solutions, but if $\varepsilon > 0$, the bifurcation is *broken*, that is, bifurcation no longer occurs.

2.4 Consider the differential equation

$$\frac{du}{dt} = r\left(1 - \frac{u}{k}\right)u - \mu, \qquad r, k > 0$$

Investigate the stability of equilibrium solutions as μ passes through the value $rk/4$.

2.5 Analyze completely the stability and bifurcation properties of the following equations:

(a) $\dfrac{du}{dt} = \mu - u^2$ **(c)** $\dfrac{du}{dt} = \mu^2 u - u^3$

(b) $\dfrac{du}{dt} = \mu u - u^2$ **(d)** $\dfrac{du}{dt} = \mu^2 u + u^3$

2.6 **(a)** Prove (21) in Theorem 2.4

  (b) Prove Theorem 2.5

2.7 Let $P_0 = (u_0, \mu_0)$ be a singular point but not a higher order singular point and let $\alpha(u)$ be given by (19). Prove that if $d\alpha/du = 0$ at $u = u_0$, then P_0 is a double point.

2.8 Discuss the nature of the singular points of the following curves at the origin.

(a) $(\mu^2 - 1)^2 + (u^2 - 1)^2 = 2$ **(d)** $(u - \mu^2)^2 - \mu^5 = 0$

(b) $u^2 - u\mu^2 = 0$ **(e)** $(u^2 - 2\mu^2)^2 - \mu^5 = 0$

(c) $u^2 - \mu^2 = 0$

6.3 TWO DIMENSIONAL PROBLEMS

Phase Plane Phenomena

In preceding sections we examined questions of stability and bifurcation for a single differential equation governing a single dynamic state $u = u(t)$. We now extend the analysis to a system of two simultaneous equations

$$\dot{x} = P(x, y), \qquad \dot{y} = Q(x, y) \tag{1}$$

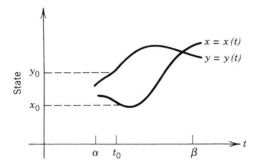

Figure 6.26. Solution representation in state space.

in two unknowns x and y. The overdot denotes d/dt and P and Q are assumed to be given functions that possess continuous partial derivatives of all orders in a domain D of the xy plane. A system of the type (1) in which the independent variable t does not appear in P and Q is said to be *autonomous*. Under the given assumptions on P and Q it is known that there is a unique solution $x = x(t)$, $y = y(t)$ to the initial value problem

$$\dot{x} = P(x, y), \qquad \dot{y} = Q(x, y)$$
$$x(t_0) = x_0, \qquad y(t_0) = y_0 \qquad (2)$$

where t_0 is an instant of time and (x_0, y_0) is in D. The solution is defined in some interval $\alpha < t < \beta$ containing t_0, and it can be graphed in state space as shown in Fig. 6.26. The variables x and y are known as *state variables*. If $x(t)$ and $y(t)$ are not both constant functions, then $x = x(t)$ and $y = y(t)$ define the parametric equations of a curve in the xy plane called the *phase plane*. This curve is called a *path* or *trajectory* of the system (1), and it is graphed schematically in Fig. 6.27. The path is directed in the sense that it is traced out in a certain direction as t increases, as shown by the arrow in Fig. 6.27. Since the solution to the initial value problem (2) is unique, it follows that at most one path passes through each point of the phase plane and all of the paths cover the entire phase plane without intersecting each other.

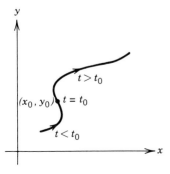

Figure 6.27. Solution representation in phase space.

A constant solution $x(t) = x_0$, $y(t) = y_0$ of (1) is called an *equilibrium* or *steady-state* solution. Such a solution does not define a path in the phase plane but rather a point. Clearly such solutions occur where both P and Q vanish, that is,

$$P(x_0, y_0) = Q(x_0, y_0) = 0 \qquad (3)$$

The points at which (3) holds are called *critical points*. It is evident that no path can pass through a critical point; otherwise, uniqueness would be violated. The totality of all the paths and critical points graphed in the phase plane is called the *phase portrait* of the system (1). Indeed the qualitative behavior of all the paths in the phase plane is determined to a large extent by the location of the critical points and the local behavior of the paths near those points. One can prove the following.

(i) A path cannot approach a critical point in finite time; that is, if a path approaches a critical point, then necessarily $t \to \pm\infty$.

(ii) As $t \to \pm\infty$ a path either approaches a critical point, moves on a closed path, approaches a closed path, or leaves every bounded set.

A closed path corresponds to a periodic solution of (1). These matters are developed further in the sequel.

Example 3.1   Consider the system

$$\dot{x} = y$$
$$\dot{y} = -x \qquad (4)$$

Here $P = y$ and $Q = -x$ and $(0,0)$ is the only critical point corresponding to the equilibrium solution $x(t) = 0$, $y(t) = 0$. To find the paths we note that

$$\ddot{x} + x = 0$$

and so

$$x = x(t) = c_1\cos t + c_2\sin t$$

and

$$y = y(t) = -c_1\sin t + c_2\cos t$$

give the solutions where c_1 and c_2 are constants. We may eliminate the parameter t by squaring and adding to obtain

$$x^2 + y^2 = c_1^2 + c_2^2$$

Therefore the paths are circles centered at the origin and the phase portrait is

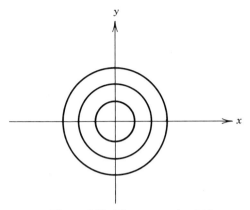

Figure 6.28. Phase portrait of (4).

shown in Fig. 6.28. Alternatively, the differentials dx and dy in (4) satisfy

$$\frac{dy}{dx} = -\frac{x}{y}$$

which can be integrated directly to obtain the paths

$$x^2 + y^2 = \text{constant}$$

In this example all the paths are closed and each corresponds to a periodic solution of period 2π since

$$x(t + 2\pi) = x(t) \qquad \text{and} \qquad y(t + 2\pi) = y(t)$$

Example 3.2   Consider the autonomous system

$$\dot{x} = 2x, \qquad \dot{y} = 3y$$

This system is *uncoupled* and each equation can be solved separately to obtain the solution

$$x = c_1 e^{2t}, \qquad y = c_2 e^{3t}$$

The origin $(0,0)$ is the only critical point. The trajectories are found by integrating

$$\frac{dy}{dx} = \frac{3y}{2x}$$

to get $y^2 = cx^3$, where c is a constant and they are graphed in Fig. 6.29. In this case the critical point is termed a node.

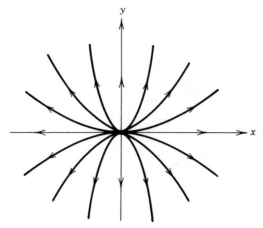

Figure 6.29. A node.

To obtain a physical interpretation of the critical points other than as an equilibrium solution from above, let us consider a particle of mass m moving in one dimension x subject to a force $f(x, \dot{x})$. By Newton's second law the governing equation of motion is

$$m\ddot{x} = f(x, \dot{x}) \tag{5}$$

Equation (5) may be recast as a system of two first order equations by introducing as a new variable the velocity $y = \dot{x}$. Then

$$\dot{x} = y$$
$$\dot{y} = \frac{1}{m} f(x, y) \tag{6}$$

A critical point occurs at $(x_0, 0)$, where $f(x_0, 0) = 0$. In other words, a critical point corresponds to a point where the velocity and the acceleration vanish. Thus the particle is at rest and no forces are acting upon it.

More generally P and Q can be regarded as the components of a velocity vector field $\mathbf{v}(x, y) = \langle P(x, y), Q(x, y) \rangle$ defined in the plane representing the velocity of a two dimensional fluid motion or flow. The paths $x = x(t)$, $y = y(t)$ of the fluid particles are tangent to $\mathbf{v}$ and therefore satisfy (1). The critical points are points where $\mathbf{v} = \mathbf{0}$ or where the fluid particles are at rest. In fluid mechanics these points are called stagnation points.

In principle the critical points of (1) can be found by solving the simultaneous equations $P(x, y) = 0$ and $Q(x, y) = 0$. The paths in phase space can be found by integrating the differential relationship

$$\frac{dy}{dx} = \frac{Q(x, y)}{P(x, y)}.$$

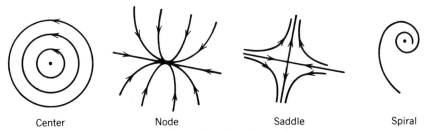

Center   Node   Saddle   Spiral

Figure 6.30. Generic critical points.

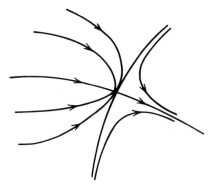

Figure 6.31. Higher order critical point.

A critical point of (1) is said to be *isolated* if there is a neighborhood of the critical point that contains no other critical points. There are four types of isolated critical points that occur frequently; these are a *center*, *node*, *saddle*, and *spiral*. In Fig. 6.30 such critical points are drawn schematically with the local path structure. In some nonlinear systems higher order nongeneric critical points can occur where, for example, a node structure is on one side and a saddle structure is on the other as shown in Fig. 6.31. It is an interesting and important exercise to determine the location and nature of the critical points for a given problem.

We have seen that critical points represent equilibrium solutions of (1). The next important question is that of stability, that is, does the equilibrium solution have some degree of permanence to it when it is subjected to small perturbations? Roughly speaking a critical point is stable if all paths that start sufficiently close to the point remain close to the point. To formulate this mathematically let us suppose that the origin $(0, 0)$ is an isolated critical point of (1). We say that $(0, 0)$ is *stable* if for each $\varepsilon > 0$ there exists a positive number δ_ε such that every path that is inside the circle of radius δ_ε at some time t_0 remains inside the circle of radius ε for all $t > t_0$ (see Fig. 6.32). The critical point is *asymptotically stable* if it is stable and there exists a circle of radius δ_ε such that every path that is inside this circle at $t = t_0$ approaches $(0, 0)$ as $t \to \infty$. If $(0, 0)$ is not stable, then it is *unstable*. We note that a center is stable but not asymptotically stable; a saddle is unstable. A spiral is either

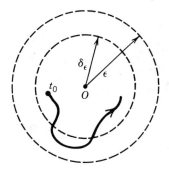

Figure 6.32

asymptotically stable or unstable, depending upon the direction of the paths and similarly for nodes.

To determine analytic criteria for stability we can proceed as follows. Without loss of generality suppose the origin $(0,0)$ is an isolated critical point of (1).[†] This means $x_0(t) = 0$ and $y_0(t) = 0$ is an equilibrium solution. Let

$$x(t) = x_0(t) + \tilde{x}(t)$$
$$y(t) = y_0(t) + \tilde{y}(t)$$

where $\tilde{x}(t)$ and $\tilde{y}(t)$ represent small perturbations of the equilibrium state. Then substituting into (1) gives the perturbation equations

$$\frac{d\tilde{x}}{dt} = P(\tilde{x}, \tilde{y}), \qquad \frac{d\tilde{y}}{dt} = Q(\tilde{x}, \tilde{y})$$

Using Taylor's theorem

$$P(\tilde{x}, \tilde{y}) = P(0,0) + P_x(0,0)\tilde{x} + P_y(0,0)\tilde{y} + O(\tilde{x}^2 + \tilde{y}^2)$$
$$= a\tilde{x} + b\tilde{y} + O(\tilde{x}^2 + \tilde{y}^2)$$

where $a = P_x(0,0)$ and $b = P_y(0,0)$. Similarly

$$Q(\tilde{x}, \tilde{y}) = c\tilde{x} + d\tilde{y} + O(\tilde{x}^2 + \tilde{y}^2)$$

where $c = Q_x(0,0)$ and $d = Q_y(0,0)$. Therefore the linearized perturbation equations are

$$\frac{d\tilde{x}}{dt} = a\tilde{x} + b\tilde{y}$$

$$\frac{d\tilde{y}}{dt} = c\tilde{x} + d\tilde{y} \tag{7}$$

[†] If (x_0, y_0) is a critical point, then the transformation $\bar{x} = x - x_0, \bar{y} = y - y_0$ translates the critical point to the origin.

It seems reasonable that the stability of $(0,0)$ of the nonlinear system (1) is indicated by the stability of $(0,0)$ of the associated linearized system (7). Under suitable conditions this is for the most part true and so we embark upon a classification of the solutions and stability properties of (7).

Linear Systems

The simplest system is the linear system

$$\dot{x} = ax + by$$
$$\dot{y} = cx + dy \tag{8}$$

where a, b, c, and d are constants. We assume that

$$ad - bc \neq 0 \tag{9}$$

Otherwise, the algebraic system

$$ax + by = 0$$
$$cx + dy = 0$$

would have a whole line of nontrivial solutions and $(0,0)$ would not be an isolated critical point. It is easier to examine (8) if we use a matrix formulation. Such is also advantageous in that it generalizes immediately to higher dimensions. Therefore letting

$$\mathbf{u} = \begin{bmatrix} x \\ y \end{bmatrix}, \qquad A = \begin{bmatrix} a & b \\ c & d \end{bmatrix}$$

(8) can be written

$$\frac{d\mathbf{u}}{dt} = A\mathbf{u} \tag{10}$$

Condition (9) is just $\det A \neq 0$. Following our experience with linear equations with constant coefficients we try a solution of (10) of the form

$$\mathbf{u} = \mathbf{v}e^{\lambda t}, \qquad \mathbf{v} \text{ constant} \tag{11}$$

where $\mathbf{v}$ and λ are to be determined. Substituting (11) into (10) gives after simplification

$$A\mathbf{v} = \lambda\mathbf{v} \tag{12}$$

which is the algebraic eigenvalue problem. Consequently λ is an eigenvalue of A and $\mathbf{v}$ is an associated eigenvector. We recall that the eigenvalues of A are found as roots of the equation

$$\det(A - \lambda I) = 0 \tag{13}$$

the corresponding eigenvector(s) are then determined from

$$(A - \lambda I)\mathbf{v} = 0 \tag{14}$$

Since (14) will be a quadratic equation in λ, and hence will have two roots, there are several cases to consider depending on whether the roots λ_1 and λ_2 are real, equal or unequal, of the same or different signs, or are complex, or perhaps purely imaginary. We shall work out in detail one of the cases and only indicate the solution for the others, leaving the details to the reader.

Case (i) $0 < \lambda_1 < \lambda_2$ (λ_1 and λ_2 are real, distinct, and positive). Let $\mathbf{v}_1$ and $\mathbf{v}_2$ be eigenvectors corresponding to λ_1 and λ_2, respectively. Then $\mathbf{v}_1 e^{\lambda_1 t}$ and $\mathbf{v}_2 e^{\lambda_2 t}$ are linearly independent solutions of (10) and the general solution is given by

$$\mathbf{u}(t) = c_1 \mathbf{v}_1 e^{\lambda_1 t} + c_2 \mathbf{v}_2 e^{\lambda_2 t} \tag{15}$$

where c_1 and c_2 are arbitrary constants. If we set $c_2 = 0$ then $\mathbf{u} = c_1 \mathbf{v}_1 e^{\lambda_1 t}$ is a path consisting of the half line l_1^+ if $c_1 > 0$ and the half line l_1^- if $c_1 < 0$ (see Fig. 6.33). Similarly, if $c_1 = 0$, then $\mathbf{u} = c_2 \mathbf{v}_2 e^{\lambda_2 t}$ is a path consisting of the half line l_2^+ if $c_2 > 0$ and l_2^- if $c_2 < 0$. The indicated half lines are in the directions of eigenvectors and their negatives. If $c_1 \neq 0$ and $c_2 \neq 0$, then (15) represents a curved path that enters $(0,0)$ as $t \to -\infty$. To see how the paths enter the origin we note that for large negative t the term $e^{\lambda_1 t}$ dominates $e^{\lambda_2 t}$ since $\lambda_2 > \lambda_1$. Thus

$$\mathbf{u} \approx c_1 \mathbf{v}_1 e^{\lambda_1 t}, \qquad \text{large negative } t$$

and so the curves enter the origin tangent to $\mathbf{v}_1$. For large positive t all the curves (15) go to infinity with slope asymptotic to $\mathbf{v}_2$, since in that case

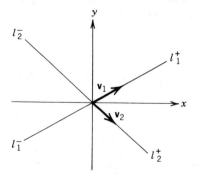

Figure 6.33

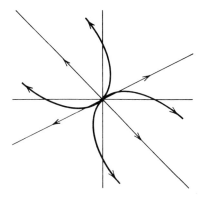

Figure 6.34. Unstable node.

$e^{\lambda_2 t} \gg e^{\lambda_1 t}$. Therefore $(0, 0)$ is an unstable node with phase portrait shown in Fig. 6.34. The lines defined by the eigenvectors $\mathbf{v}_1$ and $\mathbf{v}_2$ are called *separatrices*.

Case (ii) $\lambda_2 < \lambda_1 < 0$ (real, distinct, negative). By a similar argument $(0, 0)$ is a *stable node*. The portrait is the same as in Fig. 6.33 with the direction of the arrows reversed.

Case (iii) $\lambda_1 < 0 < \lambda_2$ (real, distinct, opposite signs). In this case the general solution is again

$$\mathbf{u} = c_1 \mathbf{v}_1 e^{\lambda_1 t} + c_2 \mathbf{v}_2 e^{\lambda_2 t}$$

As in Cases (i) and (ii) the half lines $l_1^+, l_1^-, l_2^+, l_2^-$ defined by the eigenvectors $\mathbf{v}_1$ and $\mathbf{v}_2$ and their negatives are paths. No solution other than $c_1 \mathbf{v}_1 e^{\lambda_1 t}$ can enter the origin at $t \to \infty$ and all solutions are asymptotic to the half lines as $t \to \pm\infty$. The result is the *saddle* shown in Fig. 6.35.

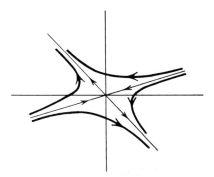

Figure 6.35. Saddle point.

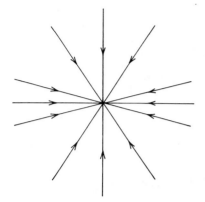

Figure 6.36. Stable node.

Case (iv) $\lambda_1 = \lambda_2 < 0$ (real, equal, negative). In this case there is a single eigenvalue $\lambda \equiv \lambda_1 = \lambda_2$ of multiplicity two and the solution depends upon whether there are one or two linearly independent eigenvectors. (a) If $\mathbf{v}_1$ and $\mathbf{v}_2$ are linearly independent eigenvectors corresponding to λ then the general solution of (10) is

$$\mathbf{u}(t) = c_1 \mathbf{v}_1 e^{\lambda t} + c_2 \mathbf{v}_2 e^{\lambda t}$$
$$= (c_1 \mathbf{v}_1 + c_2 \mathbf{v}_2) e^{\lambda t}$$

Since $\mathbf{v}_1$ and $\mathbf{v}_2$ are independent, every direction is covered and every path is therefore a half line entering the origin (see Fig. 6.36). (b) If $\mathbf{v}$ is the only eigenvector corresponding to λ then $\mathbf{v}e^{\lambda t}$ is a solution of (10). A second linearly independent solution is of the form $(\mathbf{w} + \mathbf{v}t)e^{\lambda t}$ for some vector $\mathbf{w}$ and so the general solution to (10) is

$$\mathbf{u}(t) = (c_1 \mathbf{v}_1 + c_2 \mathbf{w} + c_2 \mathbf{v}t) e^{\lambda t}$$

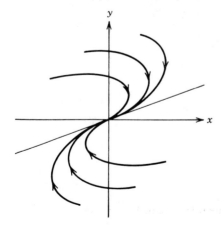

Figure 6.37. Stable node.

For large t we have $\mathbf{u} \approx c_2 t \mathbf{v} e^{\lambda t}$ and so the paths enter the origin with direction $\mathbf{v}$. The phase portrait is shown in Fig. 6.37 (l is the line defined by $\mathbf{v}$) and $(0,0)$ is a *stable node*.

Case (v) $\lambda_1 = \lambda_2 > 0$ (real, equal, positive). This case is exactly like Case (iv) with the directions reversed; $(0,0)$ is an *unstable node*.

Case (vi) $\lambda_1 = \alpha + i\beta$, $\lambda_2 = \alpha - i\beta$ (complex roots). Let $\mathbf{w} + i\mathbf{v}$ be an eigenvector corresponding to λ_1. Then a complex solution of (10) is

$$\left(\mathbf{w} + i\mathbf{v}\right)e^{(\alpha+i\beta)t}$$

or, after expanding using Euler's Formula,

$$e^{\alpha t}\left(\mathbf{w}\cos\beta t - \mathbf{v}\sin\beta t\right) + ie^{\alpha t}\left(\mathbf{w}\sin\beta t + \mathbf{v}\cos\beta t\right)$$

Therefore the general solution can be written as the linear combination of real solutions as

$$\mathbf{u} = c_1 e^{\alpha t}\left(\mathbf{w}\cos\beta t - \mathbf{v}\sin\beta t\right) + c_2 e^{\alpha t}\left(\mathbf{w}\sin\beta t + \mathbf{v}\cos\beta t\right)$$

If $\alpha = 0$, then the two components $x(t)$ and $y(t)$ of $\mathbf{u}$ are periodic of period $2\pi/\beta$. Therefore the paths are closed curves and $(0,0)$ is a *center*. If $\alpha < 0$ the amplitude of $\mathbf{u}$ decreases and the paths are *stable spirals* that wind around the origin; if $\alpha > 0$ the amplitude of u increases and the paths are *unstable spirals* (see Fig. 6.38).

In the previous discussion we have completely characterized the stability properties of the isolated critical point $(0,0)$ of the linear system (10). We summarize the information in the following theorem.

Theorem 3.1 *The critical point $(0,0)$ of the linear system*

$$\frac{d\mathbf{u}}{dt} = A\mathbf{u}, \qquad \det A \neq 0$$

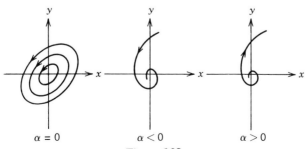

$\alpha = 0$ \qquad\qquad $\alpha < 0$ \qquad\qquad $\alpha > 0$

Figure 6.38

is stable if, and only if, the eigenvalues of A have nonpositive real parts; it is asymptotically stable if, and only if, the eigenvalues have negative real parts.

Example 3.3   Consider the linear system

$$\dot{x} = 3x - 2y$$
$$\dot{y} = 2x - 2y$$

The coefficient matrix

$$\begin{bmatrix} 3 & -2 \\ 2 & -2 \end{bmatrix}$$

has characteristic equation

$$\det\begin{bmatrix} 3 - \lambda & -2 \\ 2 & -2 - \lambda \end{bmatrix} = \lambda^2 - \lambda - 2 = 0$$

and hence the eigenvalues are $\lambda = -1, 2$. This is Case (iii), and the origin is an unstable saddle point. The eigenvectors are found from the linear system

$$\begin{bmatrix} 3 - \lambda & -2 \\ 2 & -2 - \lambda \end{bmatrix}\begin{bmatrix} v_1 \\ v_2 \end{bmatrix} = \begin{bmatrix} 0 \\ 0 \end{bmatrix}$$

When $\lambda = -1$, we have

$$4v_1 - 2v_2 = 0, \qquad 2v_1 - v_2 = 0$$

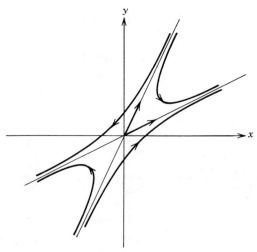

Figure 6.39

and an eigenvector corresponding to $\lambda = -1$ is $[v_1, v_2]^T = [1, 2]^T$. When $\lambda = 2$ the equations become

$$v_1 - 2v_2 = 0, \qquad 2v_1 - 4v_2 = 0$$

which give $[v_1, v_2]^T = [2, 1]^T$. The general solution to the system is

$$\mathbf{u} = c_1 \begin{bmatrix} 1 \\ 2 \end{bmatrix} e^{-t} + c_2 \begin{bmatrix} 2 \\ 1 \end{bmatrix} e^{2t}$$

The eigenvectors define the directions of the separatrices. A phase portrait is shown in Fig. 6.39.

Nonlinear Systems

We now address the question of whether the linear system

$$\dot{x} = ax + by$$
$$\dot{y} = cx + dy \qquad (16)$$

mirrors the qualitative behavior of the nonlinear system

$$\dot{x} = ax + by + f(x, y)$$
$$\dot{y} = cx + dy + g(x, y) \qquad (17)$$

in a neighborhood of a critical point. We assume

(i) That $\det A \neq 0$, where

$$A = \begin{bmatrix} a & b \\ c & d \end{bmatrix}$$

so that the linear system (16) has an isolated critical point at $(0, 0)$.
(ii) That f and g have continuous partial derivatives of all order and

$$\lim_{(x, y) \to (0,0)} \frac{f(x, y)}{\sqrt{x^2 + y^2}} = \lim_{(x, y) \to (0,0)} \frac{g(x, y)}{\sqrt{x^2 + y^2}} = 0$$

Assumption (ii) guarantees that $(0, 0)$ is an isolated critical point of (17). The limit conditions mean that $f = o(\sqrt{x^2 + y^2})$ and $g = o(\sqrt{x^2 + y^2})$, that is, f and g are higher order terms.

We now state the following result of Poincaré, which gives a partial answer to our question. Its proof can be found in advanced treatments of differential equations (see, for example, Birkhoff and Rota [4]).

Theorem 3.2   *Let* $(0,0)$ *be a critical point of the linear system* (16) *and the nonlinear system* (17) *subject to Assumptions* (i) *and* (ii). *For the nonlinear system* $(0,0)$ *is a critical point of the same type as for the linear system in the following cases.*

(i)   *The eigenvalues of A are real, distinct, and have the same sign* (*node*).
(ii)   *The eigenvalues of A are real, distinct, and have opposite signs* (*saddle*).
(iii)   *The eigenvalues of A are complex, but not purely imaginary* (*spiral*).

It is easy to construct examples where the theorem fails to apply.

Example 3.4   Consider the linear system

$$\dot{x} = -y, \qquad \dot{y} = x$$

Here

$$A = \begin{bmatrix} 0 & -1 \\ 1 & 0 \end{bmatrix}$$

has purely imaginary eigenvalues $\pm i$ and hence $(0,0)$ is a center. For the nonlinear system, howevei,

$$\dot{x} = -y - x^3, \qquad \dot{y} = x$$

$(0,0)$ is a spiral (proof?).

The question of stability is answered by the following theorem of Liapunov, which is an immediate consequence of Theorem 3.2.

Theorem 3.3   *Let* $(0,0)$ *be a critical point of* (16) *and* (17) *subject to Assumptions* (i) *and* (ii). *If* $(0,0)$ *is asymptotically stable for* (16), *then it is asymptotically stable for* (17).

Example 3.5   Consider the nonlinear system

$$\dot{x} = -2x + 3y + xy$$
$$\dot{y} = -x + y - 2xy^3$$

which has an isolated critical point at $(0,0)$. Since

$$\frac{xy}{\sqrt{x^2 + y^2}} \to 0, \qquad \frac{2xy^3}{\sqrt{x^2 + y^2}} \to 0$$

as $(x, y) \to (0,0)$, Assumption (ii) is satisfied. The nonsingular matrix

$$A = \begin{bmatrix} -2 & 3 \\ -1 & 1 \end{bmatrix}$$

has eigenvalues $-\frac{1}{2} \pm (\sqrt{3}/2) i$, and so the associated linear system has an asymptotically stable spiral at $(0,0)$. By Theorems 3.2 and 3.3 the nonlinear system also has an asymptotically stable spiral at $(0,0)$.

Another central problem in the theory of nonlinear systems is to determine whether the system admits any closed paths. Such paths are associated with periodic solutions of the differential equations. A solution $x(t)$, $y(t)$ of

$$\begin{aligned} \dot{x} &= P(x, y) \\ \dot{y} &= Q(x, y) \end{aligned} \tag{18}$$

is *periodic* if neither $x(t)$ nor $y(t)$ is constant and there exists a positive number T such that

$$x(t + T) = x(t), \qquad y(t + T) = y(t)$$

for all t. The smallest such T is called the *period*. As we observed in Example 2.1 the linear system

$$\dot{x} = y, \qquad \dot{y} = -x$$

has 2π-periodic solutions

$$\begin{aligned} x(t) &= c_1\cos t + c_2\sin t \\ y(t) &= -c_1\sin t + c_2\cos t \end{aligned}$$

corresponding to the closed paths (circles)

$$x^2 + y^2 = c_1^2 + c_2^2 \tag{19}$$

in the phase plane. For nonlinear systems the existence of closed paths may be difficult to decide. The following negative result, however, due to Bendixson is easy to prove.

Theorem 3.4 *If $P_x + Q_y$ is of one sign in a region of the phase plane, then system* (18) *cannot have a closed path in that region.*

Proof By way of contradiction assume that the region contains a closed path C given by

$$x = x(t), \qquad y = y(t), \qquad 0 \le t \le T$$

and denote the interior of C by R. By Green's theorem in the plane

$$\int_C P\,dy - Q\,dx = \iint_R (P_x + Q_y)\,dx\,dy \neq 0$$

On the other hand

$$\int_C P\,dy - Q\,dx = \int_0^T (PQ - QP)\,dt = 0$$

which is a contradiction.

Another negative criterion is due to Poincaré.

Theorem 3.5   *A closed path of the system* (18) *surrounds at least one critical point of the system.*

Positive criteria that ensure the existence of closed paths for general systems are not abundant, particularly criteria that are practical and easy to apply. For certain special types of equations such criteria do exist. The following fundamental theorem, called the Poincaré-Bendixson theorem, is a general theoretical result.

Theorem 3.6   *Let R be a closed bounded region in the plane containing no critical points of* (18). *If C is a path of* (18) *that lies in R for some t_0 and remains in R for all $t > t_0$, then C is either a closed path or it spirals toward a closed path as $t \to \infty$.*

Theorem 3.6 guarantees in two dimensions that either a path leaves every bounded set as $t \to \pm\infty$, it is a closed curve, or it approaches a critical point or a closed curve. Thus in the plane the only *attractors* are closed curves or critical points. Interestingly enough the Poincaré-Bendixson theorem does not generalize directly to higher dimensions. In three dimensions, for example, there exist *strange attractors* that have neither the character of a point, a curve, or a surface, which act as attractors to all paths.

Bifurcation

We now study systems of equations of the form

$$\dot{x} = P(x, y, \mu), \qquad \dot{y} = Q(x, y, \mu) \tag{20}$$

depending upon a real parameter μ. In this treatment we shall be content to examine several examples of bifurcation phenomena, including Hopf bifurcation, and not develop a general theory as was done for scalar evolution equations in the preceding section. We refer the reader to the texts listed in the bibliography where substantive results of a theoretical nature can be found.

As μ varies in (20), it often occurs that the basic nature or character of the solution changes as indicated, for example, by fundamentally different phase portraits. If this phenomenon occurs at some value μ_{crit}, then we say that μ_{crit} is a bifurcation point. A simple example will illustrate the idea.

Example 3.6   Consider the linear system

$$\dot{x} = x + \mu y, \qquad \dot{y} = x - y$$

where μ is a real number. The eigenvalues λ are found from

$$\det \begin{bmatrix} 1 - \lambda & \mu \\ 1 & -1 - \lambda \end{bmatrix} = 0$$

or

$$\lambda^2 - (1 + \mu) = 0$$

Hence

$$\lambda = \pm \sqrt{1 + \mu}$$

If $\mu > -1$ then the eigenvalues are real and have opposite signs and therefore the origin is a saddle point (Fig. 6.40). If $\mu = -1$, then the system becomes

$$\dot{x} = x - y, \qquad \dot{y} = x - y$$

and there is a line of equilibrium solutions $y = x$. In this case the phase portrait is shown in Fig. 6.41. If $\mu < -1$, then the eigenvalues are purely imaginary and the origin is a center (Fig. 6.42). Consequently as μ decreases the nature of the solution changes at $\mu = -1$; the equilibrium state $(0,0)$

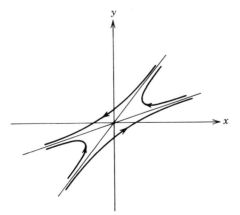

Figure 6.40. $\mu > -1$.

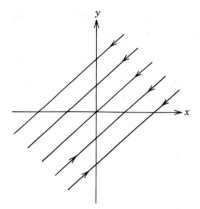

Figure 6.41. $\mu = -1$.

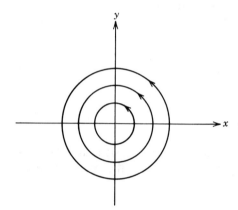

Figure 6.42. $\mu < -1$.

evolves from an unstable saddle to a stable center as μ crosses the critical value. It is interesting to graph the eigenvalues $\lambda = \pm \sqrt{1 + \mu}$ as functions of μ in the complex plane. The two paths in Fig. 6.43 show

$$\lambda_1 = \sqrt{1 + \mu}, \qquad \lambda_2 = -\sqrt{1 + \mu}$$

as μ varies from 0 to -2. At the bifurcation value $\mu = -1$ the two paths intersect and there is an exchange of stability noted by the dashed line (unstable) and the solid line (stable) in the figure.

Example 3.7 (Hopf Bifurcation) In this rather detailed and extensive example we study the system

$$\dot{x} = -y - x\left(x^2 + y^2 - \mu\right)$$
$$\dot{y} = x - y\left(x^2 + y^2 - \mu\right) \qquad (21)$$

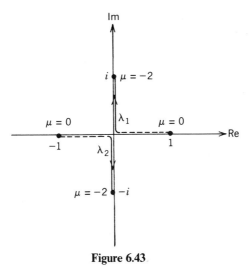

Figure 6.43

where μ is a real parameter. Let us first see whether we can gain some insight from the linearized system

$$\dot{x} = \mu x - y, \qquad \dot{y} = x + \mu y$$

It is easy to see that the eigenvalues λ are given by

$$\lambda = \mu \pm i$$

if $\mu < 0$, then Re $\lambda < 0$, and $(0,0)$ is a stable spiral; if $\mu = 0$, then $\lambda = \pm i$, and $(0,0)$ is a center; if $\mu > 0$, then Re $\lambda > 0$, and $(0,0)$ is an unstable spiral. These phase paths for the linearized system are shown in Fig. 6.44. From results in the preceding sections regarding the relationship between the nonlinear and associated linear system we expect $(0,0)$ to be a stable spiral for the nonlinear system in the case $\mu < 0$ and an unstable spiral when $\mu > 0$. The linearized result does not carry over for $\mu = 0$.

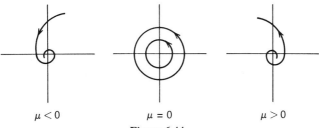

$\mu < 0$ $\mu = 0$ $\mu > 0$

Figure 6.44

It is possible to solve the nonlinear problem directly if we transform to polar coordinates defined by

$$x = r \cos \theta, \qquad y = r \sin \theta$$

Then it is straightforward to derive the two relations

$$x\dot{x} + y\dot{y} = r\dot{r}, \qquad x\dot{y} - y\dot{x} = r^2\dot{\theta}$$

If we multiply the first equation in (21) by x and the second by y and then add, we obtain

$$\dot{r} = r(\mu - r^2) \tag{22}$$

Finally, multiplying the first equation in (21) by $-y$ and the second by x and then adding, we obtain

$$\dot{\theta} = 1 \tag{23}$$

The equivalent system (22) and (23) in polar coordinates may be integrated directly. Clearly

$$\theta = t + t_0 \tag{24}$$

where t_0 is a constant. Now we examine the r equation when $\mu > 0$. By separating variables

$$\frac{dr}{r(r^2 - \mu)} = -dt$$

Using the partial fraction expansion

$$\frac{1}{r(r^2 - \mu)} = \frac{-1/\mu}{r} + \frac{1/2\mu}{r - \sqrt{\mu}} + \frac{1/2\mu}{r + \sqrt{\mu}}$$

and then integrating yields

$$r = \frac{\sqrt{\mu}}{\sqrt{1 + c\exp(-2\mu t)}} \tag{25}$$

where c is a constant of integration. When $c = 0$ we obtain the solution

$$r = \sqrt{\mu}$$
$$\theta = t + t_0$$

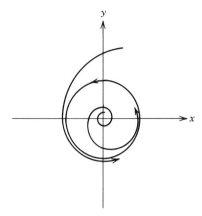

Figure 6.45. $\mu > 0$.

which is a periodic solution whose path is represented by a circle $r = \sqrt{\mu}$ in the phase plane. If $c < 0$, then the solution (25) represents a phase path that spirals toward the circle $r = \sqrt{\mu}$ from the outside, and if $c > 0$, then the solution represents a phase path that spirals toward the circle from the inside. This means that the origin is an unstable spiral as indicated previously by the linear analysis. The phase portrait is shown in Fig. 6.45.

Now consider the case $\mu = 0$. The r equation becomes

$$\frac{dr}{r^3} = -dt$$

and direct integration gives the solution

$$r = \frac{1}{\sqrt{2t + c}}, \qquad \theta = t + t_0$$

which implies that the origin is a stable spiral.

In the case $\mu < 0$, let $k^2 = -\mu > 0$. Then the r equation is

$$\frac{dr}{r(r^2 + k^2)} = -dt \tag{26}$$

The partial fraction expansion is

$$\frac{1}{r(r^2 + k^2)} = \frac{1/k^2}{r} - \frac{(1/k^2)r}{r^2 + k^2}$$

Integrating (26) then gives, after simplification,

$$r^2 = \frac{ck^2 e^{-2k^2 t}}{1 - ce^{-2k^2 t}}$$

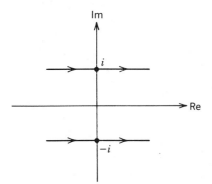

Figure 6.46. Hopf bifurcation.

It is easy to verify that this solution represents spirals that approach the origin at $t \to +\infty$.

Let us review our results. As μ passes through the bifurcation point $\mu = 0$ the origin changes from a stable spiral to an unstable spiral and there appears a new time periodic solution. This type of bifurcation is common and it is an example of what is termed Hopf bifurcation. A graph of the eigenvalues $\lambda = \mu + i$ and $\lambda = \mu - i$ of the linearized system as functions of μ is shown in Fig. 6.46. We note that the complex conjugate pair crosses the imaginary axis at $\mu = 0$; generally this signals the bifurcation to a time periodic solution.

A physical situation where an equilibrium solution bifurcates to a steady convective solution is the classical Bénard problem. A viscous fluid lies between two horizontal planes; the upper plane is maintained at T_1 degrees and the lower plane is held at T_0 degrees with $T_0 > T_1$. For small $\Delta T \equiv T_0 - T_1$ the viscous force is just balanced by gravity and there is an equilibrium solution consisting of no fluid motion and a linear temperature distribution from the lower to upper plane. As ΔT increases, however, the force of buoyancy caused by heating the lower plane comes into play and there is a critical value of ΔT where a convective motion begins and the fluid moves in Bénard cells (see Fig. 6.47). It can be shown that below this critical value the basic equilibrium solution is asymptotically stable, and above the critical value the basic solution is unstable while the cell pattern is asymptotically stable.

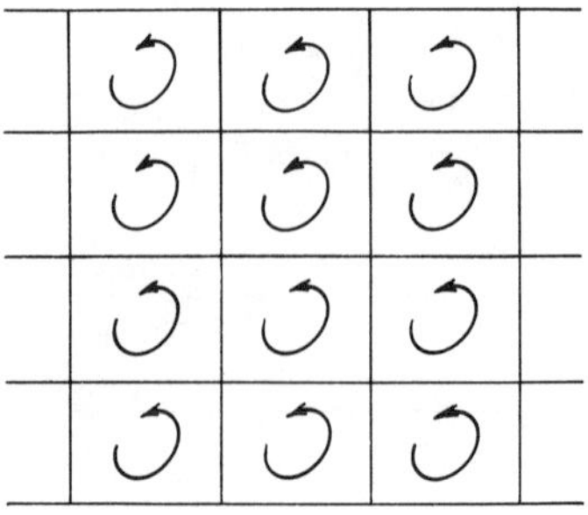

Figure 6.47. Bénard cells.

Example 3.8   In Section 6.1 we discussed the motion of a frictionless bead on
a rotating circular hoop. If R is the radius of the hoop, ω the angular velocity
of rotation, and θ is the angular deflection, then the equation of motion was
found to be

$$\ddot{\theta} = \omega^2 \cos\theta \sin\theta - \frac{g}{R}\sin\theta$$

By introducing the variable $\psi = \dot{\theta}$ the equation can be written as a system of
first order equations

$$\dot{\theta} = \psi$$
$$\dot{\psi} = \omega^2 \cos\theta \sin\theta - \frac{g}{R}\sin\theta \qquad (27)$$

To find the critical points we set the right side of Equations (27) equal to zero
to obtain

$$\theta = 0, \qquad \psi = 0 \qquad (28)$$

and

$$\theta = \arccos\frac{g}{R\omega^2}, \qquad \psi = 0 \qquad (29)$$

We have ignored the critical points that arise because of the periodicity of the
trigometric functions. The critical point $\theta = 0, \psi = 0$ represents the equilibrium position at the bottom of the hoop and the two critical points given by
(29) represent equal angular deflections in the positive and negative directions;
the latter are present only if the inequality

$$\omega > \sqrt{g/R}$$

holds. Therefore as ω increases from zero there appears two additional critical
points at the critical value $\sqrt{g/R}$. In other words two additional critical points
bifurcate from the origin when ω reaches the value $\sqrt{g/R}$.

First we examine the critical point $(0,0)$. Using the expansions $\cos\theta = 1 + O(\theta^2)$ and $\sin\theta = \theta + O(\theta^3)$, we linearize Equations (27) to get

$$\dot{\theta} = \psi, \qquad \dot{\psi} = \left(\omega^2 - \frac{g}{R}\right)\theta$$

The eigenvalues of the coefficient matrix are

$$\lambda_1(\omega) = \sqrt{\omega^2 - \frac{g}{R}}$$

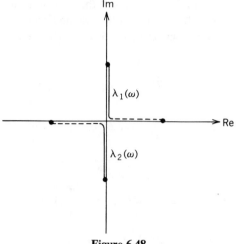

Figure 6.48

and

$$\lambda_2(\omega) = -\sqrt{\omega^2 - \frac{g}{R}}$$

Consequently for the linearized system, if

$$\omega < \sqrt{\frac{g}{R}}$$

then $(0,0)$ is a stable center, and if

$$\omega > \sqrt{\frac{g}{R}}$$

then $(0,0)$ is an unstable saddle. A graph of $\lambda_1(\omega)$ and $\lambda_2(\omega)$ in the complex plane is shown in Fig. 6.48.

Now let $\omega > \sqrt{g/R}$ and consider the critical point (29). The critical point may be translated to the origin by letting

$$\theta = \bar{\theta} + \arccos \frac{g}{R\omega^2}$$

Then equation (27) becomes

$$\dot{\psi} = f\left(\bar{\theta} + \arccos \frac{g}{R\omega^2}\right)$$

where

$$f(\theta) = \omega^2 \cos\theta \sin\theta - \frac{g}{R}\sin\theta.$$

By Taylor's theorem

$$\dot{\psi} = f\left(\arccos\frac{g}{R\omega^2}\right) + f'\left(\arccos\frac{g}{R\omega^2}\right)\bar{\theta} + O(\bar{\theta}^2)$$

$$= f'\left(\arccos\frac{g}{R\omega^2}\right)\bar{\theta} + O(\bar{\theta}^2)$$

so the linearized equation is

$$\dot{\psi} = f'\left(\arccos\frac{g}{R\omega^2}\right)\bar{\theta}$$

It remains to calculate the coefficient $f'(\arccos g/R\omega^2)$. A straightforward differentiation and substitution gives

$$f'(\theta) = \omega^2(\cos^2\theta - \sin^2\theta) - \frac{g}{R}\cos\theta$$

$$f'\left(\arccos\frac{g}{R\omega^2}\right) = -\frac{(R\omega^2)^2 - g^2}{R^2\omega^2}$$

Therefore the linearized system is

$$\dot{\bar{\theta}} = \psi$$

$$\dot{\psi} = -\frac{(R\omega^2)^2 - g^2}{R^2\omega^2}\bar{\theta}$$

and has purely imaginary eigenvalues given by

$$\lambda = \pm\frac{i}{R\omega}\sqrt{(R\omega^2)^2 - g^2}$$

Hence $(\arccos g/R\omega^2, 0)$ is a stable center for the linearized system.

One can show that the nonlinear equations exhibit the same behavior as the linearized system. Thus as ω increases and passes through the critical value $\sqrt{g/R}$, the stable center at the origin bifurcates into an unstable saddle at the origin and two nonzero stable centers located on the θ axis. Schematic phase portraits are sketched in Fig. 6.49. This is an example of *global* bifurcation where the features of the entire phase plane are described.

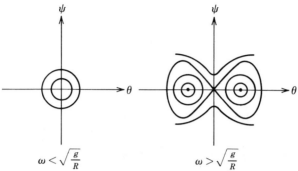

Figure 6.49

EXERCISES

3.1   Sketch phase portraits for the following systems.

(a) $\dot{x} = x - 3y$ (d) $\dot{x} = x + y$
     $\dot{y} = -3x + y$ $\dot{y} = 4x - 2y$

(b) $\dot{x} = -x + y$ (e) $\dot{x} = 3x - 4y$
     $\dot{y} = -x - y$ $\dot{y} = x - y$

(c) $\dot{x} = 4y$
     $\dot{y} = -9x$

3.2   Prove that the critical point $(0,0)$ of the linear system (8) is asymptotically stable if and only if $p > 0$ and $q > 0$, where $p = -(a + d)$ and $q = ad - bc$.

3.3   The equation for a damped harmonic oscillator is

$$m\ddot{x} + a\dot{x} + kx = 0, \qquad m, a, k > 0$$

Write the equation as a system by introducing $y = \dot{x}$ and show that $(0,0)$ is a critical point. Describe the nature and stability of the critical point in the following cases.

(a) $a = 0$ (c) $a^2 - 4km < 0$

(b) $a^2 - 4mk = 0$ (d) $a^2 - 4km > 0$

Interpret the results physically.

3.4   Determine the nature and stability properties of the critical points of the systems:

(a) $\dot{x} = x + y - 2x$
     $\dot{y} = -2x + y + 3y^2$

(b) $\dot{x} = -x - y - 3x^2y$
$\dot{y} = -2x - 4y + y \sin x$

3.5   Consider the system

$$\dot{x} = 4x + 4y - x(x^2 + y^2)$$
$$\dot{y} = -4x + 4y - y(x^2 + y^2)$$

(a) Show there is a closed path in the region $1 \le r \le 3$, where $r^2 = x^2 + y^2$ (use Theorem 3.6).

(b) Find the general solution.

3.6   Determine whether the following systems admit periodic solutions.

(a) $\dot{x} = y$
$\dot{y} = (x^2 + 1)y - x^5$

(b) $\dot{x} = y$
$\dot{y} = y^2 + x^2 + 1$

(c) $\dot{x} = y$
$\dot{y} = 3x^2 - y - y^5$

3.7   Find the values of μ where solutions to the following system bifurcate and examine the stability of the origin in each case.

(a) $\dot{x} = x + \mu y$ **(e)** $\dot{x} = y$
$\dot{y} = \mu x + y$ $\dot{y} = \mu x + x^2$

(b) $\dot{x} = y$ **(f)** $\dot{x} = \mu y + xy$
$\dot{y} = -2x + \mu y$ $\dot{y} = -\mu x + \mu y + x^2 + y^2$

(c) $\dot{x} = x + y$ **(g)** $\dot{x} = y$
$\dot{y} = \mu x + y$ $\dot{y} = -y - \mu + x^2$

(d) $\dot{x} = 2y$ **(h)** $\dot{x} = \mu x - x^2 - 2xy$
$\dot{y} = 2x - \mu y$ $\dot{y} = (\mu - \frac{5}{3})y + xy + y^2$

3.8   Discuss the dependence on the sign of μ of the critical point at the origin of the system

$$\dot{x} = -y + \mu x^3, \qquad \dot{y} = x + \mu y^3$$

3.9   Show that the paths of $\ddot{x} + f(x) = 0$ in the xy phase plane (where $y = \dot{x}$) are closed curves if $f(x)$ is an increasing function and $f(0) = 0$ (*Hint*: Show $\frac{1}{2}y^2 + V(x) = $ constant, where $V(x) = \int_0^x f(u)\, du$.)

3.10 Prove that the origin is an asymptotically stable critical point for the equation

$$\ddot{x} + p(x, \dot{x})\dot{x} + xh(x) = 0$$

if $p(0,0)$ and $h(0)$ are positive.

3.11 Examine the stability of the equilibrium solutions of the system

$$\frac{dx}{dt} = P - y^2 - \nu x, \qquad \frac{dy}{dt} = xy - \nu y$$

where P and ν are positive constants.

6.4 HYDRODYNAMIC STABILITY

Fluid motions provide some of the best and most important examples of stability and instability phenomena encountered in applied problems. In this section, by examining the stability properties of an inviscid fluid confined between two horizontal planes and having a vertical density variation, we illustrate a highly typical calculation common to many problems in fluid dynamics and other branches of continuum mechanics. The techniques involved are fundamental in applied mathematics.

A Layered Fluid

We consider an ideal fluid (inviscid and incompressible) confined between two infinite planes $\bar{z} = 0$ and $\bar{z} = d$ in $\bar{x}\bar{y}\bar{z}$ space, which is under the influence of a constant gravitation field g acting in the negative $\bar{z}$ direction (see Fig. 6.50). We use *barred* variables with anticipation of later introducing unbarred dimensionless quantities for the principal part of the discussion. Let $\bar{t}$ denote time, $\bar{\mathbf{v}}$ the velocity, $\bar{\rho}$ the density, and $\bar{p}$ the pressure. The governing

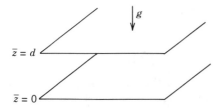

Figure 6.50. Two infinite planes separated by a distance d.

equations are

$$\operatorname{div} \bar{\mathbf{v}} = 0 \tag{1}$$

$$\frac{D\bar{\rho}}{Dt} = 0 \tag{2}$$

$$\bar{\rho}\frac{D\bar{\mathbf{v}}}{Dt} = -\bar{\rho}g\mathbf{k} - \mathbf{grad}\,\bar{p} \tag{3}$$

with boundary conditions

$$\bar{\mathbf{v}} \cdot \mathbf{k} = 0, \qquad \text{at } \bar{z} = 0, d \tag{4}$$

where $\mathbf{k}$ is a unit vector in the $\bar{z}$ direction. Equations (1) and (2) come from the conservation of mass equation and incompressibility condition. Equation (3) is the Euler equation of motion.

The constants in the problem are g and d. It is clear the $\bar{z}$ should be scaled by d and that the velocity component in the $\bar{z}$ direction should be scaled by $\sqrt{dg}$. Time should be scaled by $\sqrt{d/g}$. Even though there are no apparent length scales in the $\bar{x}$ and $\bar{y}$ directions, let us proceed and scale $\bar{x}$ and $\bar{y}$ by d and scale all the velocity components by $\sqrt{dg}$. Hence we introduce dimensionless scaled variables via

$$x = \frac{\bar{x}}{d}, \qquad y = \frac{\bar{y}}{d}, \qquad z = \frac{\bar{z}}{d} \qquad t = \frac{\bar{t}}{\sqrt{d/g}}, \qquad \mathbf{v} = \frac{\bar{\mathbf{v}}}{\sqrt{dg}}$$

To scale density and pressure we select some typical density ρ_0 as yet undetermined. Then

$$\rho = \frac{\bar{\rho}}{\rho_0}, \qquad p = \frac{\bar{p}}{\rho_0 dg}$$

Introducing the scaled variables into the governing equations (1)–(3) and boundary conditions (4) gives

$$\operatorname{div} \mathbf{v} = 0 \tag{5}$$

$$\frac{D\rho}{Dt} = 0 \tag{6}$$

$$\frac{D\mathbf{v}}{Dt} = -\rho\mathbf{k} - \mathbf{grad}\,p \tag{7}$$

and

$$w = 0, \qquad \text{at } z = 0, 1 \tag{8}$$

where $\mathbf{v} = u\mathbf{i} + v\mathbf{j} + w\mathbf{k}$. The undetermined density ρ_0 dropped out of the equations, and the length d and acceleration due the gravity g were scaled out as well.

Our aim is to examine the stability of the *motionless* solution of (5)–(8), that is, the solution

$$\mathbf{v} = \mathbf{0}, \qquad \rho = r(z), \qquad p = -\int r(z)\,dz \qquad (9)$$

where $r(z)$ is a given positive, smooth function of z. We note that (9) is a solution of (5)–(8) for any choice of r (see Exercise 4.1). Because the density varies with the vertical direction z, we call the fluid layered or stratified.

Perturbation Equations

We proceed as in the earlier examples of the perturbation method and consider small perturbations from the equilibrium solution (9). Let

$$\mathbf{v}(t, x, y, z) = \mathbf{0} + \hat{\mathbf{v}}(t, x, y, z) \qquad \rho(t, x, y, z) = r(z) + \hat{\rho}(t, x, y, z)$$

$$p(t, x, y, z) = -\int r(z)\,dz + \hat{p}(t, x, y, z), \qquad (10)$$

where $\hat{\mathbf{v}}$, $\hat{\rho}$, and $\hat{p}$ are small perturbations. Substitution into (5) and (6) gives

$$\hat{u}_x + \hat{v}_y + \hat{w}_z = 0 \qquad (11)$$

$$\hat{\rho}_t + \hat{u}\hat{\rho}_x + \hat{v}\hat{\rho}_y + \hat{w}(r'(z) + \hat{\rho}_z) = 0 \qquad (12)$$

The three momentum equations represented in (7) become

$$\begin{aligned}
(r + \hat{\rho})\hat{u}_t + \langle \hat{u}, \hat{v}, \hat{w} \rangle \cdot \langle \hat{u}_x, \hat{u}_y, \hat{u}_z \rangle &= -\hat{p}_x \\
(r + \hat{\rho})\hat{v}_t + \langle \hat{u}, \hat{v}, \hat{w} \rangle \cdot \langle \hat{v}_x, \hat{v}_y, \hat{v}_z \rangle &= -\hat{p}_y \\
(r + \hat{\rho})\hat{w}_t + \langle \hat{u}, \hat{v}, \hat{w} \rangle \cdot \langle \hat{w}_x, \hat{w}_y, \hat{w}_z \rangle &= -\hat{p}_z - \hat{\rho}
\end{aligned} \qquad (13)$$

Equations (11), (12), and (13) are the *nonlinear perturbation equations* for $\hat{u}$, $\hat{v}$, $\hat{w}$, $\hat{\rho}$, and $\hat{p}$, and it is impossible to solve these equations other than by numerical means. Therefore we opt for a linearized theory and delete all of the products of small terms in (11) through (13) to obtain

$$\begin{aligned}
\hat{u}_x + \hat{v}_y + \hat{w}_z &= 0 \\
\hat{\rho}_t + r'(z)\hat{w} &= 0 \\
r(z)\hat{u}_t + \hat{p}_x &= 0 \\
r(z)\hat{v}_t + \hat{p}_y &= 0 \\
r(z)\hat{w}_t + \hat{p}_z + \hat{\rho} &= 0
\end{aligned} \qquad (14)$$

These are the *linearized perturbation equations*, a system of linear first order partial differential equations with variable coefficients. The boundary condition (8) becomes

$$\hat{w} = 0 \quad \text{at} \quad z = 0, 1 \tag{15}$$

Since the coefficients in the Equations (14) are independent of t, x, and y, we attempt to find a solution of the form

$$
\begin{bmatrix} \hat{u} \\ \hat{v} \\ \hat{w} \\ \hat{p} \\ \hat{\rho} \end{bmatrix} = \begin{bmatrix} U(z) \\ V(z) \\ W(z) \\ P(z) \\ R(z) \end{bmatrix} e^{\sigma t} e^{i(k_1 x + k_2 y)} \tag{16}
$$

where σ is a complex number and $k_1, k_2 \in R^1$. Such solutions are harmonic in x and y with wave numbers k_1 and k_2, respectively. Equation (16) may be regarded as a single Fourier mode. If there exists a single solution of the form (16) with $\text{Re}\,\sigma > 0$, then that mode grows in amplitude as $t \to \infty$ and the equilibrium solution will be unstable. If, however, we show that $\text{Re}\,\sigma < 0$ for all the modes represented by (16), then the perturbations decay and the equilibrium solution (9) will be stable to small perturbations that can be synthesized by superimposing the modes in (16). Actually the latter are quite general, since Fourier analysis guarantees that most reasonably nice functions can be expanded in terms of the fundamental modes represented by (16).

Substituting (16) into the set of equations (14) and the boundary condition (15) gives

$$ik_1 U + ik_2 V + W' = 0 \tag{17}$$

$$\sigma R + r'(z)W = 0 \tag{18}$$

$$r\sigma U = ik_1 P \tag{19}$$

$$r\sigma V = -ik_2 P \tag{20}$$

$$r\sigma W = -P' - R \tag{21}$$

and

$$W(0) = W(1) = 0 \tag{22}$$

where prime denotes differentiation with respect to z. Since the boundary condition is on W, we attempt to obtain one equation for W alone. For other problems it may be quite difficult in practice to eliminate all the variables from one equation but one. In the present case if (19) and (20) are solved for U and

V and then substituted into (17) we obtain

$$\frac{P}{r\sigma}k^2 + W' = 0, \qquad k^2 \equiv k_1^2 + k_2^2$$

Differentiating gives

$$-P' = \frac{\sigma}{k^2}(rW')'$$

But also from (18)

$$R = -\frac{1}{\sigma}r'W$$

Substituting the last two expressions into (21) then gives

$$(r(z)W')' + k^2\left(\frac{1}{\sigma^2}r'(z) - r(z)\right)W = 0, \qquad 0 < z < 1 \qquad (23)$$

with

$$W(0) = W(1) = 0 \qquad (24)$$

Therefore the problem of solving the linearized perturbation equations (14) has been reduced to a single second order ordinary differential equation for $W(z)$. If $r'(z) > 0$, then (23) and (24) is recognized as a regular Sturm-Liouville problem (see Section 4.2) with $\lambda \equiv k^2/\sigma^2$. By Theorem 3.2 in Section 4.2 there are infinite number of eigenvalues $0 \le \lambda_1 < \lambda_2 < \lambda_3 < \cdots$. Therefore if we regard the wave number k as fixed, then there will be infinitely many *positive* values of σ, say $\sigma_1, \sigma_2, \ldots$, with corresponding modes $W_1(z), W_2(z), \ldots$, which are unstable. This is not unexpected, since $r'(z) > 0$ implies that the fluid is top-heavy, meaning that it is more dense above than below.

Rayleigh's Example

A specific example will illustrate the major points. Let $r(z) = \exp z$. Then the Sturm-Liouville problem (23) and (24) becomes

$$W'' + W' + k^2(\sigma^{-2} - 1)W = 0, \qquad 0 < z < 1$$
$$W(0) = W(1) = 0 \qquad (25)$$

The differential equation is linear with constant coefficients. The characteristic equation is

$$m^2 + m + k^2(\sigma^{-2} - 1) = 0$$

and has roots

$$m = -\tfrac{1}{2} \pm \tfrac{1}{2}\sqrt{1 - 4k^2(\sigma^{-2} - 1)}$$

If $1 - 4k^2(\sigma^{-2} - 1) \geq 0$, then it is straightforward to see that no nontrivial solution of (25) exists. On the other hand, if $1 - 4k^2(\sigma^{-2} - 1) < 0$, then the general solution of the differential equation is

$$W(z) = e^{-z/2}(c_1\cos \beta z + c_2\sin \beta z)$$

where

$$\beta = \tfrac{1}{2}\sqrt{4k^2(\sigma^{-2} - 1) - 1}$$

The condition $W(0) = 0$ implies $c_1 = 0$, and $W(1) = 0$ implies $\sin \beta = 0$. Consequently

$$\beta = n\pi, \qquad n = \pm 1, \pm 2, \dots$$

or

$$\sigma^2 = \sigma_n^2 = 4k^2(1 + 4k^2 + 4n^2\pi^2)^{-1}, \qquad n = 1, 2, 3, \dots \tag{26}$$

The eigenfunctions are therefore

$$W_n(z) = e^{-z/2}\sin n\pi z, \qquad n = 1, 2, \dots \tag{27}$$

The remaining perturbations can be calculated from (18) through (20). It is left as an exercise to show

$$
\begin{aligned}
U_n(z) &= \frac{ik_1}{k^2}e^{-z/2}\left(n\pi \cos n\pi z - \frac{1}{2}\sin n\pi z\right) \\
V_n(z) &= \frac{ik_2}{k^2}e^{-z/2}\left(n\pi \cos n\pi z - \frac{1}{2}\sin n\pi z\right) \\
R_n(z) &= -\frac{1}{\sigma_n}e^{z/2}\sin n\pi z \\
P_n(z) &= -\frac{\sigma_n}{k^2}e^{z/2}\left(n\pi \cos n\pi z - \frac{1}{2}\sin n\pi z\right)
\end{aligned}
\tag{28}
$$

for $n = 1, 2, 3, \dots$

We can get a clear picture of the flow pattern by examining two dimensional perturbations and the case $n = 1$, $k = 1$. Then $\hat{v} = 0$, $k_2 = 0$, $k = k_1 = 1$ and

$$\hat{u} = ie^{-z/2}\left(\pi \cos \pi z - \tfrac{1}{2}\sin \pi z\right)e^{\sigma_1 t}e^{ix}$$

$$\hat{w} = e^{-z/2}\sin \pi z\, e^{\sigma_1 t}e^{ix}$$

where $\sigma_1 = 2(5 + 4\pi^2)^{-1/2}$. To obtain real solutions we take the real parts of the previous equations to obtain

$$\hat{u} = -e^{-z/2}\left(\pi \cos \pi z - \tfrac{1}{2} \sin \pi z\right)\sin x\, e^{\sigma_1 t} \tag{29}$$

$$\hat{w} = e^{-z/2}\sin \pi z \cos x\, e^{\sigma_1 t}$$

The solution is periodic in the x direction with period 2π. We can graph the velocity direction field $\langle \hat{u}, \hat{w}\rangle$ for $0 \leq x \leq 2\pi$, $0 \leq z \leq 1$ by noting the following simple facts.

$$\hat{w} > 0 \quad \text{for} \quad 0 < x < \frac{\pi}{2} \quad \text{and} \quad \frac{3\pi}{2} < x < 2\pi$$

$$\hat{w} = 0 \quad \text{for} \quad x = \frac{\pi}{2},\ \frac{3\pi}{2}$$

$$\hat{u} = 0 \quad \text{for} \quad z = \frac{1}{\pi}\arctan 2\pi \approx 0.449$$

$$\hat{u} < 0 \quad \text{for} \quad 0 \leq z < \frac{1}{\pi}\arctan 2\pi, \quad 0 < x < \pi$$

$$\hat{u} > 0 \quad \text{for} \quad \frac{1}{\pi}\arctan 2\pi \leq z \leq 1, \quad \pi < x < 2\pi$$

Figure 6.51 summarizes this information and indicates by an arrow the direction of the velocity field $\langle \hat{u}, \hat{w}\rangle$. Figure 6.52 depicts schematically typical streamlines viewed from the positive y direction where there is no variation. The cell pattern repeats itself every 2π units in the x direction. The fluid circulates along these streamlines with ever increasing speed as time passes.

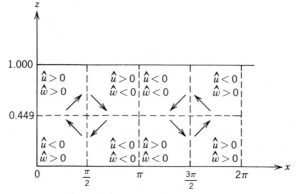

Figure 6.51. Diagram indicating the signs of $\hat{u}$ and $\hat{w}$ and the direction of the velocity vector.

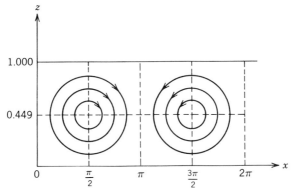

Figure 6.52. Schematic of streamlines of the velocity field $\langle \hat{u}, \hat{w} \rangle$ given by (29) in the case $n = k = 1$ viewed from the y direction.

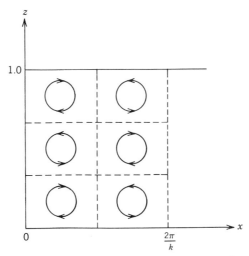

Figure 6.53. Schematic of streamlines for $n = 3$, k arbitrary. In this case three vertical periods appear.

For arbitrary k and n the cell pattern still appears, but with additional cells. Again focusing upon two dimensional perturbations, the case $n = 3$ is drawn in Fig. 6.53. In general the quantity σ_n in $\exp(\sigma_n t)$ determines the growth rate of the perturbations. Since σ_n decreases as n increases, the first mode ($n = 1$) has the fastest growth rate. There is also variation of the growth rate with the wave number k as indicated by (26). It is clear that σ^2 increases as k increases. Hence, perturbations with a larger wave number, or shorter wavelength, grow faster than perturbations with a longer wavelength (see Fig. 6.54).

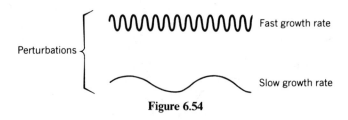

Perturbations

Fast growth rate

Slow growth rate

Figure 6.54

EXERCISES

4.1 Starting with the motionless condition $\mathbf{v} = \mathbf{0}$, derive (9) directly from Equations (5)–(8).

4.2 Fill in the details leading to Equations (28).

4.3 (a) From the Sturm-Liouville system (23) and (24) prove that

$$\frac{1}{\sigma^2}\int_0^1 r'(z)W^2(z)\,dz = \int_0^1 r(z)W^2(z)\,dz + \frac{1}{k^2}\int_0^1 r(z)W'(z)^2\,dz$$

(b) If $r'(z) > 0$ for $0 \le z \le 1$, show that $\sigma^2 > 0$.

(c) If $r'(z) < 0$ for $0 \le z \le 1$, show that σ is purely imaginary.

(d) Interpret (b) and (c) from a stability point of view.

4.4 Investigate the stability problem for a stratified fluid when $r(z) = \exp(-z)$.

4.5 Consider the boundary value problem

$$\frac{1}{R}(u_{xx} + u_{yy}) - u_{xxxx} - u_t - u_{xx}u_y = 0$$

when $u = u(x, y, t)$, $x \in R^1$, $t > 0$, and $0 < y < 1$ with

$$u(x,0,t) = 0, \qquad u(x,1,t) = 1$$

(a) Letting $u = y + \tilde{u}(x, y, t)$, where $u_0 = y$ is an equilibrium solution, determine the boundary conditions and the linearized perturbation equation for $\tilde{u}$.

(b) Assume a solution of the form $\tilde{u} = f(y)\exp(\sigma t)\cos kx$ and show that the fundamental modes are

$$\tilde{u}_n = \exp(\sigma_n t)\cos kx \sin n\pi y$$

where

$$\sigma_n = k^2 - k^4 - \frac{k^2}{R} - \frac{n^2\pi^2}{R}, \qquad n = 1, 2, \ldots$$

(c) Show that if R exceeds $(k^2 + n^2\pi^2)(k^2 - k^4)^{-1}$, then the nth mode grows without bound.

(d) Assuming perturbations of all wave numbers $0 < k < 1$ are present, at what value of R will instability set in?

REFERENCES

1. C. C. Lin and L. A. Segel, *Mathematics Applied to Deterministic Problems in the Natural Sciences*, Macmillan Publishing Co., New York 1974).

2. I. Stakgold, *Green's Functions and Boundary Value Problems*, Wiley-Interscience, New York (1979).

3. G. Iooss and D. D. Joseph, *Elementary Stability and Bifurcation Theory*, Springer-Verlag, New York (1980).

4. G. Birkhoff and G. C. Rota, *Ordinary Differential Equations*, Third Edition, John Wiley and Sons, New York (1978).

7

SIMILARITY METHODS

Nature often exhibits certain symmetries, and the mathematical equations that model nature frequently exhibit these same symmetries. The symmetries in a system give rise to transformations that leave the system invariant or unchanged. For example, a physical system with rotational symmetry may be expected to lead to a mathematical model that is invariant under rotations. If a given system possesses these types of symmetries, it seems reasonable that special properties can be derived. Indeed this is the case, since symmetric equations have a degree of simplicity to them not possessed by more complicated systems.

This chapter contains an introduction to similarity methods in two settings, the calculus of variations and partial differential equations. By a similarity method we understand a technique based on taking advantage of the symmetries or invariance properties in the problem. The method involves determining transformations under which the problem is invariant and then using the transformation to derive special conditions from which the solution can be found or from which the problem can be simplified.

Two examples will illustrate the point. Let us first consider the one dimensional heat equation

$$u_t(x, t) - u_{xx}(x, t) = 0 \tag{1}$$

It is easily checked that if we change variables via

$$\bar{x} = \varepsilon x, \qquad \bar{t} = \varepsilon^2 t, \qquad \bar{u} = u \tag{2}$$

where ε is a real parameter, then an application of the chain rule gives

$$\bar{u}_{\bar{t}} - \bar{u}_{\bar{x}\bar{x}} = \varepsilon^{-2}(u_t - u_{xx})$$

In other words the differential operator that defines the heat equation is unchanged or invariant under the stretching transformation (2), up to a constant factor ε^{-2}. If we now introduce a new independent variable s defined by

$$s = \frac{x}{\sqrt{t}} \tag{3}$$

(note that s is also an invariant of the transformation (2) in that $\bar{s} = s$) and assume that

$$u(x, t) = f(s) \tag{4}$$

for some function f, then upon substitution of (4) into (1) we obtain

$$f''(s) + \frac{s}{2}f'(s) = 0 \tag{5}$$

which is an *ordinary* differential equation for $f(s)$. Therefore the invariance of the partial differential equation leads to the definition of a new, single independent variable with which the partial differential equation can be reduced to an ordinary differential equation. Equation (5) can be solved by straightforward integration to obtain

$$f(s) = c_1 \int_0^s \exp(-z^2) \, dz + c_2$$

where c_1 and c_2 are arbitrary constants. Hence

$$u(x, t) = c_1 \int_0^{x/\sqrt{t}} \exp(-z^2) \, dz + c_2 \tag{6}$$

The class of solutions (6) of Equation (1) are known as similarity solutions and s is the similarity variable. By exploiting the invariance properties of (1) a significant simplification was made. This procedure is known as the similarity method for partial differential equations, and it applies to a wide variety of important problems in science and engineering.

A second example where invariance properties lead to important results is in the calculus of variations. We consider the variational problem

$$J(x) = \int_a^b L(x, \dot{x}) \, dt, \qquad x \in C^2(a, b) \tag{7}$$

where the Lagrangian is independent of explicit dependence upon time t. In such a case the Euler equation $L_x - (d/dt)L_{\dot{x}} = 0$ has a first integral or conservation law of the form

$$L - \dot{x}L_{\dot{x}} = \text{constant} \tag{8}$$

We may reinterpret this result as follows. The independence of t implies that $J(x)$ is invariant under a time translation, or symmetry, given by $\bar{t} = t + \varepsilon$. Hence invariance under time translations leads to a conservation law. A general statement can be made that is the basis of the classical theorem of E. Noether (1882–1935) in the calculus of variations. If the action integral (7) is invariant under certain types of transformations, then conservation laws or first integrals of the Euler equation can be obtained. Consequently symmetries give rise to conservation laws.

These two examples hint of a whole body of theory that can be developed concerning the consequences of invariance in physical problems. Our aim is to formulate the concepts and illustrate the general principles with several practical examples beginning with invariance principles in the calculus of variations. To allow maximum flexibility, however, the two topics of invariant partial differential equations and invariant variational problems are treated independently. Thus the reader may go directly to Section 7.2 to study similarity methods for partial differential equations after studying the paragraphs on local Lie groups that begin the next section.

7.1   INVARIANT VARIATIONAL PROBLEMS

Local Lie Groups

We consider a variational problem defined by the functional

$$J(x) = \int_a^b L(t, x, \dot{x})\, dt, \qquad x \in C^2(a, b) \tag{1}$$

and we assume that the Lagrangian L is four times continuously differentiable on R^3. To keep contact with concrete situations one can regard L as the Lagrange function for a mechanical system and $x = x(t)$ as a generalized coordinate that defines the state of the system at any time t.

By a transformation T of the plane we mean a mapping $T: R^2 \to R^2$ defined by equations

$$T: \bar{t} = \phi(t, x), \qquad \bar{x} = \psi(t, x)$$

where ϕ and ψ are given functions. Geometrically T carries the point (t, x) to another point $(\bar{t}, \bar{x})$ in the same plane with reference to the same coordinate axes. If S is another transformation of the plane defined by the equations

$$S: \bar{t} = \alpha(t, x), \qquad \bar{x} = \beta(t, x)$$

then the composed transformation ST represents the transformation T followed by the transformation S and is defined by

$$ST: \bar{\bar{t}} = \alpha(\phi(t, x), \psi(t, x)), \qquad \bar{\bar{x}} = \beta(\phi(t, x), \psi(t, x))$$

Similarly we may define the composition TS, which represents S followed by T in succession. If the equations defining T are solved for t and x in terms of $\bar{t}$ and $\bar{x}$, then we obtain the inverse transformation T^{-1} represented by

$$T^{-1}: t = \Phi(\bar{t}, \bar{x}), \qquad x = \psi(\bar{t}, \bar{x})$$

Clearly $T^{-1}T$ and TT^{-1} are both the identity transformation I given by the equations $\bar{t} = t$, $\bar{x} = x$.

The types of symmetry transformations under which we study the invariance properties of (1) are transformations of the plane with the additional stipulation that the transformation also depends on a real parameter that we denote by ε. Thus, let ε vary continuously over an open interval $|\varepsilon| < \varepsilon_0$ and consider the collection T_ε of all transformations included in the family defined by

$$T_\varepsilon: \bar{t} = \phi(t, x, \varepsilon), \qquad \bar{x} = \psi(t, x, \varepsilon) \qquad (2)$$

where ϕ and ψ are given analytic functions on $R^2 \times (-\varepsilon_0, \varepsilon_0)$. A particular transformation of the family is obtained by assigning a particular value to the parameter ε. The transformation defined by (2) is said to be a *one parameter family of transformations* on R^2.

For two specific values ε_1 and ε_2 the composed transformation $T_{\varepsilon_2}T_{\varepsilon_1}$ need not be a transformation of the type (2); that is, there does not necessarily exist a value ε_3 of the parameter ε such that $T_{\varepsilon_2}T_{\varepsilon_1} = T_{\varepsilon_3}$.

Example 1.1   Let

$$T_\varepsilon: \bar{t} = \phi(t, x, \varepsilon) \equiv \varepsilon - t, \qquad \bar{x} = \psi(t, x, \varepsilon) = x$$

Then

$$T_{\varepsilon_1}: \bar{t} = \varepsilon_1 - t, \qquad \bar{x} = x$$

$$T_{\varepsilon_2}: \bar{\bar{t}} = \varepsilon_2 - \bar{t}, \qquad \bar{\bar{x}} = \bar{x}$$

Now $\bar{\bar{t}} = \varepsilon_2 - (\varepsilon_1 - t) = \varepsilon_2 - \varepsilon_1 + t$. Thus there does not exist a value ε_3 such that $\bar{\bar{t}} = \varepsilon_3 - t$.

Generally we require (2) to satisfy the *local closure property*, namely if ε_1 and ε_2 are sufficiently small, then there exists a value ε_3 in the interval $|\varepsilon| < \varepsilon_0$ for which $T_{\varepsilon_2} T_{\varepsilon_1} = T_{\varepsilon_3}$. In terms of the functional notation we assume there exists an open subinterval U of $|\varepsilon| < \varepsilon_0$ and an analytic function $g: U \times U \rightarrow (-\varepsilon_0, \varepsilon_0)$, for which $\varepsilon_3 = g(\varepsilon_1, \varepsilon_2)$ and

$$\phi\big(\phi(t, x, \varepsilon_1), \psi(t, x, \varepsilon_1), \varepsilon_2\big) = \phi(t, x, \varepsilon_3)$$
$$\psi\big(\phi(t, x, \varepsilon_1), \psi(t, x, \varepsilon_1), \varepsilon_2\big) = \psi(t, x, \varepsilon_3)$$

We further assume that the family T_ε contains the identity transformation I when $\varepsilon = 0$, that is,

$$\bar{t} = \phi(t, x, 0) = t, \qquad \bar{x} = \psi(t, x, 0) = x$$

Finally, for each ε sufficiently small we require that the transformation T_{ε_1} have an inverse T_{ε_2}. In other words there exists an open subinterval U of $|\varepsilon| < \varepsilon_0$ such that for each ε_1 in U there is a value ε_2 in U for which $T_{\varepsilon_2} T_{\varepsilon_1} = T_{\varepsilon_1} T_{\varepsilon_2} = I$. This means there exists a function $h: U \rightarrow U$, which we assume to be analytic, such that

$$h(0) = g\big(h(\varepsilon), \varepsilon\big) = g\big(\varepsilon, h(\varepsilon)\big) = 0$$

for all ε in U. This condition follows from the composition law $T_\varepsilon T_{h(\varepsilon)} = T_{g(\varepsilon, h(\varepsilon))} = I = T_0$. In general, a one parameter family of transformations T_ε is called a *local Lie group* (Sophus Lie 1842–1899) *of transformations* if it satisfies the local closure property, contains the identity, and inverses exist for small ε. The associative law

$$T_{\varepsilon_1}\big(T_{\varepsilon_2} T_{\varepsilon_3}\big) = \big(T_{\varepsilon_1} T_{\varepsilon_2}\big) T_{\varepsilon_3}$$

holds whenever both sides are defined. If the function g defining the law of composition is $g(\varepsilon_1, \varepsilon_2) = \varepsilon_1 + \varepsilon_2$, then the group is said to be in *canonical form*.

Example 1.2   Consider the one parameter family of transformations defined by

$$T_\varepsilon: \begin{cases} \bar{t} = \phi(t, x, \varepsilon) \equiv t\sqrt{1 - \varepsilon^2} - x\varepsilon \\ \bar{x} = \psi(t, x, \varepsilon) \equiv t\varepsilon + x\sqrt{1 - \varepsilon^2} \end{cases}$$

where $|\varepsilon| < 1$. Calculation shows that $T_{\varepsilon_2} T_{\varepsilon_1}$ is

$$\phi(\bar{t}, \bar{x}, \varepsilon_2) = \bar{t}\sqrt{1 - \varepsilon_2^2} - \bar{x}\varepsilon_2$$
$$= t\Big(\sqrt{1 - \varepsilon_1^2}\sqrt{1 - \varepsilon_2^2} - \varepsilon_1\varepsilon_2\Big) - x\Big(\varepsilon_1\sqrt{1 - \varepsilon_2^2} + \varepsilon_2\sqrt{1 - \varepsilon_1^2}\Big)$$
$$= \phi(t, x, \varepsilon_3)$$

provided

$$\varepsilon_3 \equiv g(\varepsilon_1, \varepsilon_2) = \varepsilon_1\sqrt{1 - \varepsilon_2^2} + \varepsilon_2\sqrt{1 - \varepsilon_1^2}$$

A similar calculation can be made for ψ. Thus local closure holds if $\varepsilon_1, \varepsilon_2 \in U = (-1/\sqrt{2}, 1/\sqrt{2})$. The identity transformation clearly results when $\varepsilon = 0$ and the inverse is

$$t = \bar{t}\sqrt{1 - \varepsilon^2} + \varepsilon\bar{x}, \qquad x = \sqrt{1 - \varepsilon^2}\,\bar{x} - \varepsilon\bar{t}$$

Hence $h(\varepsilon) = -\varepsilon$, which is also analytic on U. Therefore T_ε is a local Lie group.

Other examples of local Lie groups are the following.

(i) *Translation Group*

$$\bar{t} = t + \varepsilon, \qquad \bar{x} = x$$

(ii) *Rotation Group*

$$\bar{t} = t \cos \varepsilon - x \sin \varepsilon$$
$$\bar{x} = t \sin \varepsilon + x \cos \varepsilon$$

(iii) *Stretching Group*

$$\bar{t} = e^{a\varepsilon}t, \qquad \bar{x} = e^{b\varepsilon}x, \qquad a, b \text{ constants}$$

Their properties are left to the exercises.

If T_ε defined by (2) is a local Lie group, then the right sides of (2) can be expanded in a Taylor series about $\varepsilon = 0$ to obtain

$$\bar{t} = \phi(t, x, 0) + \phi_\varepsilon(t, x, 0)\varepsilon + \frac{1}{2!}\phi_{\varepsilon\varepsilon}(t, x, 0)\varepsilon^2 + \cdots$$

$$= t + \tau(t, x)\varepsilon + o(\varepsilon)$$

$$\bar{x} = \psi(t, x, 0) + \psi_\varepsilon(t, x, 0)\varepsilon + \frac{1}{2!}\psi_{\varepsilon\varepsilon}(t, x, 0)\varepsilon^2 + \cdots$$

$$= x + \xi(t, x)\varepsilon + o(\varepsilon)$$

where τ and ξ are defined by

$$\tau(t, x) \equiv \phi_\varepsilon(t, x, 0), \qquad \xi(t, x) \equiv \psi_\varepsilon(t, x, 0) \tag{3}$$

For small ε the so-called *infinitesimal transformation*

$$\bar{t} = t + \varepsilon\tau(t, x) + o(\varepsilon), \qquad \bar{x} = x + \varepsilon\xi(t, x) + o(\varepsilon) \tag{4}$$

approximates T_ε. The quantities τ and ξ are called the *generators* of T_ε and they define the principal linear part of the transformation in ε; that is, the generators are the coefficients of the lowest order terms in ε in the Taylor expansion. Equations (2) give the *global representation* of the group and (4) gives the *infinitesimal representation*.

Example 1.3 Referring to the transformation in Example 1.2 we have

$$\tau(t, x) = \frac{\partial}{\partial \varepsilon} \phi(t, x, \varepsilon)\big|_{\varepsilon=0} = \left(-t\varepsilon(1 - \varepsilon^2)^{-1/2} - x \right)_{\varepsilon=0} = -x$$

$$\xi(t, x) = \frac{\partial}{\partial \varepsilon} \psi(t, x, \varepsilon)\big|_{\varepsilon=0} = \left(t - \varepsilon x(1 - \varepsilon^2)^{-1/2} \right)_{\varepsilon=0} = t$$

Therefore the infinitesimal transformation is

$$\bar{t} = t - \varepsilon x + o(\varepsilon), \qquad \bar{x} = x + \varepsilon t + o(\varepsilon)$$

which approximates T_ε for small ε.

It is clear that the local Lie group (2) uniquely determines the generators τ and ξ by Equations (3). The converse statement is also true: knowledge of the generators τ and ξ completely determines the global form of the local Lie group given by formulas (2). If (t_0, x_0) is a fixed but arbitrary point in R^2, then $\langle \phi(t_0, x_0, \varepsilon), \psi(t_0, x_0, \varepsilon) \rangle$ represents a curve in the parameter ε that passes through (t_0, x_0) at $\varepsilon = 0$. This curve is called the *orbit* of the group through (x_0, t_0). The tangent vector to the curve at (t_0, x_0) is $\langle (d/d\varepsilon)\phi(t_0, x_0, 0), (d/d\varepsilon)\psi(t_0, x_0, 0) \rangle = \langle \tau(t_0, x_0), \xi(t_0, x_0) \rangle$. Therefore the generators τ and ξ define at each point of the plane a vector $\langle \tau, \xi \rangle$ that points in the direction that points are moved by the given transformation. The collection of all orbits of the group are therefore the integral curves of the vector field $\langle \tau, \xi \rangle$. Hence the global representation of the group (2) can be determined by solving the system of differential equations

$$\frac{d\bar{t}}{d\varepsilon} = \tau(\bar{t}, \bar{x}), \qquad \frac{d\bar{x}}{d\varepsilon} = \xi(\bar{t}, \bar{x})$$

or equivalently

$$\frac{d\bar{t}}{\tau(\bar{t}, \bar{x})} = \frac{d\bar{x}}{\xi(\bar{t}, \bar{x})} = d\varepsilon$$

subject to the initial conditions

$$\bar{t} = t, \qquad \bar{x} = x, \qquad \text{at } \varepsilon = 0$$

This system of differential equations is called the *dynamical system* associated with the group.

Example 1.4 We calculate the group whose generators are given by $\tau = t$ and $\xi = -x/2$. The dynamical system is

$$\frac{d\bar{t}}{\bar{t}} = \frac{d\bar{x}}{-\bar{x}/2} = d\varepsilon$$

with $\bar{t} = t$ and $\bar{x} = x$ at $\varepsilon = 0$. Integrating $d\bar{t}/\bar{t} = -2\,d\bar{x}/\bar{x}$ gives

$$\bar{t}\bar{x}^2 = \text{constant} = c_1$$

Applying the initial conditions gives $\bar{t}\bar{x}^2 = c_1 = tx^2$. Integrating $d\bar{t}/\bar{t} = d\varepsilon$ gives $\ln \bar{t} = \varepsilon + c_2$, which after applying the initial conditions yields $c_2 = \ln t$. Hence $\ln \bar{t} = \varepsilon + \ln t$, and so

$$\bar{t} = t \exp(\varepsilon)$$

Thus $\bar{x}^2 = tx^2/\bar{t} = x^2\exp(-\varepsilon)$. The group is therefore given by

$$\bar{t} = \exp(\varepsilon)t, \qquad \bar{x} = \exp(-\varepsilon/2)x$$

which is in canonical form. Notice that renaming the parameter via $\varepsilon' + 1 = \exp(\varepsilon)$ gives the alternate representation

$$\bar{t} = (1 + \varepsilon')t, \qquad \bar{x} = (1 + \varepsilon')^{-1/2}x$$

In passing, we observe that in the variables r and s defined by $s = tx^2$, $r = \ln t$ the group becomes a simple translation

$$\bar{s} = s, \qquad \bar{r} = r + \varepsilon$$

The coordinates s and r are called *canonical coordinates* and it is always possible to represent the group as a translation. This fact is proved in Section 7.3 for groups on R^3.

In Section 7.3 and in the exercises additional properties of local Lie groups are developed. For the present we discuss the invariance properties of variational problems and their consequences.

Invariance of Functionals

Given the variational problem defined by the functional (1), we ask what is meant by (1) being invariant under a local Lie group T_ε represented by (2). We conjecture that the value of (1) ought to be unchanged for all choices of

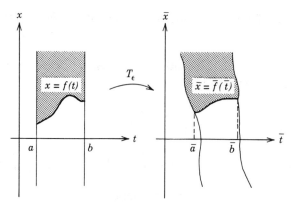

Figure 7.1

$x \in C^2(a, b)$. Therefore we require some notion of the result of applying the transformation T_ε to a given function $x = f(t)$ defined on $[a, b]$. We can visualize the transformation acting on f as shown in Fig. 7.1. Intuitively $x = f(t)$ gets mapped via T_ε to a function $\bar{x} = \bar{f}(\bar{t})$ in $\bar{t}\bar{x}$ space provided ε is sufficiently close to zero. To determine $\bar{f}$ we proceed as follows. If the mapping T_ε is applied to the set of points $(t, f(t))$, we get, as t ranges from a to b,

$$\bar{t} = \phi\big(t, f(t), \varepsilon\big) \tag{5}$$

$$\bar{x} = \psi\big(t, f(t), \varepsilon\big) \tag{6}$$

If ε is sufficiently small, then (5) can be solved for t in terms of $\bar{t}$, a fact we establish in Lemma 1.1. That result can be substituted into (6) to obtain $\bar{x} = \bar{f}(\bar{t})$.

Lemma 1.1 *Let* $f: [a, b] \to R^1$ *be of class* C^2. *Then there exists* $d > 0$ (*depending on* f) *such that if* $|\varepsilon| < \varepsilon_0$ *and* $|\varepsilon| < d$, *then the transformation*

$$\bar{t} = \phi\big(t, f(t), \varepsilon\big)$$

has a unique inverse.

Proof For convenience define $\gamma(t, \varepsilon) = \phi(t, f(t), \varepsilon)$. Now $\gamma(t, 0) = t$, and so $\gamma_t(t_0, 0) = 1$ for all $t_0 \in [a, b]$. We want to show that $\gamma_t > 0$ for all $t \in [a, b]$ and ε sufficiently small. Since γ_t is continuous at $(t_0, 0)$, there is $\delta(t_0) > 0$ and $d(t_0) > 0$ such that $\gamma_t(t, \varepsilon) > 0$ for (t, ε) in $R_{t_0} = (t_0 - \delta(t_0), t_0 + \delta(t_0)) \times (-d(t_0), d(t_0))$ (see Fig. 7.2). Now $\{R(t_0) | t_0 \in [a, b]\}$ covers $[a, b]$, and since $[a, b]$ is compact, there is a finite cover. That is, there is a finite collection of points $t_1, \ldots, t_m$ in $[a, b]$ such that $R_{t_1} \cup \cdots \cup R_{t_m}$ cover $[a, b]$. Now pick $d = \min\{\delta(t_1), \ldots, \delta(t_m)\}$. Then for $|\varepsilon| < d$ and $t \in [a, b]$ we have $\gamma_t(t, \varepsilon) > 0$, and so γ is monotonically increasing. Hence it is invertible.

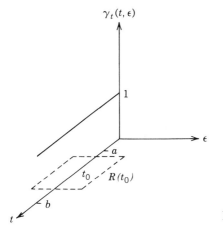

Figure 7.2

Therefore if we are given a transformation and apply it to the curve $x = f(t)$, we obtain (5) and (6). If ε is sufficiently small (5) can be solved for t to obtain

$$t = T(\bar{t}, \varepsilon)$$

Substituting into (6) gives

$$\bar{x} = \psi(T(\bar{t}, \varepsilon), f(T(\bar{t}, \varepsilon)), \varepsilon) \equiv \bar{f}(\bar{t}) \tag{7}$$

which defines $\bar{f}$.

The Lagrange function $L(t, x, \dot{x})$ contains derivatives and if we are to study invariance, then we must determine how derivatives of functions transform under T_ε. The calculation is straightforward. From (7)

$$\bar{f}(\bar{t}) = \psi(t, f(t), \varepsilon)$$

where

$$t = T(\bar{t}, \varepsilon)$$

By the chain rule

$$\frac{d\bar{f}}{d\bar{t}} = \psi_t \frac{dt}{d\bar{t}} + \psi_x \dot{f}(t) \frac{dt}{d\bar{t}}$$

where ψ_t and ψ_x are evaluated at $(t, f(t), \varepsilon)$, $t = T(\bar{t}, \varepsilon)$. But

$$\frac{dt}{d\bar{t}}(\bar{t}) = \frac{1}{(d\bar{t}/dt)(t)} = \frac{1}{(d/dt)\phi(t, f(t), \varepsilon)} = \frac{1}{\phi_t + \phi_x \dot{f}(t)}$$

where $t = T(\bar{t}, \varepsilon)$. Hence the transformed derivative is given by

$$\frac{d\bar{f}}{d\bar{t}} = \frac{\psi_t + \psi_x \dot{f}(t)}{\phi_t + \phi_x \dot{f}(t)}, \qquad t = T(\bar{t}, \varepsilon) \tag{8}$$

Example 1.5   Consider the transformation of Example 1.2 and let $f(t) = mt$. Then applying the transformation to the points $(t, f(t))$ gives

$$\bar{t} = t\sqrt{1 - \varepsilon^2} - mt\varepsilon, \qquad \bar{x} = t\varepsilon + mt\sqrt{1 - \varepsilon^2}$$

Inverting the first equation yields

$$t = \frac{\bar{t}}{\sqrt{1 - \varepsilon^2} - m\varepsilon}$$

Substituting into the second gives

$$\bar{x} = \bar{f}(\bar{t}) \equiv \frac{\varepsilon + m\sqrt{1 - \varepsilon^2}}{\sqrt{1 - \varepsilon^2} - m\varepsilon}\bar{t}$$

Therefore the straight line $x = mt$ with slope m gets rotated into a straight line in $\bar{t}\bar{x}$ space with slope $(\varepsilon + m\sqrt{1 - \varepsilon^2})/(\sqrt{1 - \varepsilon^2} - m\varepsilon)$. The derivative $d\bar{f}/d\bar{t}$ can be computed either directly or by formula (8).

The local Lie group T_ε given by (2) induces naturally a local Lie group $\tilde{T}_\varepsilon$ on $tx\dot{x}$ space defined by the equations

$$\tilde{T}_\varepsilon : \begin{cases} \bar{t} = \phi(t, x, \varepsilon) & \text{(9a)} \\ \bar{x} = \psi(t, x, \varepsilon) & \text{(9b)} \\ \bar{\dot{x}} = \dfrac{\psi_t(t, x, \varepsilon) + \psi_x(t, x, \varepsilon)\dot{x}}{\phi_t(t, x, \varepsilon) + \phi_x(t, x, \varepsilon)\dot{x}} & \text{(9c)} \end{cases}$$

This group is called the *extended group*. Upon expanding the right sides of (9) in a Taylor series about $\varepsilon = 0$, we can obtain the infinitesimal approximation of $\tilde{T}_\varepsilon$ for small ε. This has already been done for (9a) and (9b); to expand (9c) note that

$$\psi_t = \varepsilon\xi_t + o(\varepsilon), \qquad \psi_x = 1 + \varepsilon\xi_x + o(\varepsilon)$$
$$\phi_t = 1 + \varepsilon\tau_t + o(\varepsilon), \qquad \phi_x = \varepsilon\tau_x + o(\varepsilon)$$

Then

$$
\begin{aligned}
\bar{\dot{x}} &= \frac{\varepsilon\xi_t + (1 + \varepsilon\xi_x)\dot{x} + o(\varepsilon)}{1 + \varepsilon\tau_t + \varepsilon\tau_x\dot{x} + o(\varepsilon)} \\
&= \frac{\dot{x} + \varepsilon\dot{\xi} + o(\varepsilon)}{1 + \varepsilon\dot{\tau} + o(\varepsilon)} \\
&= \dot{x} + \left(\dot{\xi} - \dot{x}\dot{\tau}\right)\varepsilon + o(\varepsilon)
\end{aligned}
$$

where $\dot{\xi} \equiv \xi_t + \xi_x\dot{x}$ and $\dot{\tau} \equiv \tau_t + \tau_x\dot{x}$. The lowest order term (the linear term) in ε in the expansion of $\bar{\dot{x}}$ is

$$
\eta \equiv \dot{\xi} - \dot{x}\dot{\tau} \tag{10}
$$

and is the generator of the derivative portion of the transformation. The quantities τ, ξ, and η are called the generators of the extended group. They play an important role in the development of invariance conditions.

Example 1.6 Again considering the transformation of Example 1.2 we have

$$
\phi = t\sqrt{1 - \varepsilon^2} - x\varepsilon, \qquad \psi = t\varepsilon + x\sqrt{1 - \varepsilon^2}
$$

Thus

$$
\psi_x = \phi_t = \sqrt{1 - \varepsilon^2}, \qquad \psi_t = \varepsilon, \quad \phi_x = -\varepsilon
$$

Therefore the extended group is

$$
\begin{aligned}
\bar{t} &= t\sqrt{1 - \varepsilon^2} - x\varepsilon \\
\bar{x} &= t\varepsilon + x\sqrt{1 - \varepsilon^2} \\
\bar{\dot{x}} &= \frac{\varepsilon + \sqrt{1 - \varepsilon^2}\,\dot{x}}{\sqrt{1 - \varepsilon^2} - \varepsilon\dot{x}}
\end{aligned}
$$

and the generators are $\tau = -x$, $\xi = t$ (from Example 1.3) and $\eta = \dot{\xi} - \dot{x}\dot{\tau} = 1 - \dot{x}(-\dot{x}) = 1 + \dot{x}^2$.

We are now in position to define what is meant by the functional (1) being invariant under the symmetry transformation T_ε defined by (2). Intuitively, if $x = x(t)$ is any twice continuously differentiable function on $[a, b]$, then we may compute

$$
J(x) = \int_a^b L\left(t, x(t), \frac{dx}{dt}(t)\right) dt
$$

For sufficiently small ε, Lemma 1.1 allows us to determine $\bar{x} = \bar{x}(\bar{t})$, a twice continuously differentiable function on the interval $[\bar{a}, \bar{b}]$, and we may calculate

$$J(\bar{x}) = \int_{\bar{a}}^{\bar{b}} L\left(\bar{t}, \bar{x}(\bar{t}), \frac{d\bar{x}}{d\bar{t}}(\bar{t})\right) d\bar{t}$$

For invariance we should require that these two integrals be equal, at least up to first order terms in ε, that is, $J(\bar{x}) = J(x) + o(\varepsilon)$. Evidently this is the same as requiring $(d/d\varepsilon)J(\bar{x}) = 0$ at $\varepsilon = 0$. Therefore we make the formal definition.

Definition 1.1   *The functional J is absolutely invariant under the local Lie group T_ε if, and only if,*

$$\frac{d}{d\varepsilon} \int_{\bar{t}_1}^{\bar{t}_2} L\left(\bar{t}, \bar{x}(\bar{t}), \frac{d\bar{x}}{d\bar{t}}(\bar{t})\right) d\bar{t} = 0, \qquad at\ \varepsilon = 0 \qquad (11)$$

for every $x \in C^2[a, b]$ and for every subinterval $[t_1, t_2]$ of $[a, b]$.

The requirement that (11) hold for every subinterval $[t_1, t_2]$ permits the effective removal of the integral from consideration and the focus to be on the Lagrange function L. By changing variables in (11) according to $\bar{t} = \phi(t, x(t), \varepsilon)$ we obtain

$$\int_{\bar{t}_1}^{\bar{t}_2} L\left(\bar{t}, \bar{x}(\bar{t}), \frac{d\bar{x}}{d\bar{t}}(\bar{t})\right) d\bar{t} = \int_{t_1}^{t_2} L\left(\phi, \psi, \frac{\psi_t + \dot{x}\psi_x}{\phi_t + \dot{x}\phi_x}\right) \frac{d\phi}{dt} dt$$

where in the integral on the right $\phi = \phi(t, x(t), \varepsilon)$ and $\psi = \psi(t, x(t), \varepsilon)$. Therefore (11) is equivalent to

$$\frac{d}{d\varepsilon}\left\{ L\left(\phi, \psi, \frac{\psi_t + \dot{x}\psi_x}{\phi_t + \dot{x}\phi_x}\right) \frac{d\phi}{dt}\right\} = 0, \qquad at\ \varepsilon = 0 \qquad (12)$$

Equation (12) is often written as

$$\frac{d}{d\varepsilon}\left\{ L\left(\bar{t}, \bar{x}, \frac{d\bar{x}}{d\bar{t}}\right) \frac{d\bar{t}}{dt}\right\} = 0, \qquad at\ \varepsilon = 0$$

but care must be taken to interpret this as in (12).

Example 1.7   Consider the arclength functional

$$J(x) = \int_a^b \sqrt{1 + \dot{x}^2}\ dt$$

and the local Lie group T_ε discussed in Examples 1.2, 1.3, and 1.6. Equation (12) becomes

$$\frac{d}{d\varepsilon}\left\{\sqrt{1 + \left[\frac{\varepsilon + \dot{x}\sqrt{1 - \varepsilon^2}}{\sqrt{1 - \varepsilon^2} - \varepsilon\dot{x}}\right]^2}\left(\sqrt{1 - \varepsilon^2} - \dot{x}\varepsilon\right)\right\}\Bigg|_{\varepsilon=0} = \frac{d}{d\varepsilon}\{\sqrt{1}\}_{\varepsilon=0} = 0$$

and therefore $J(x)$ is absolutely invariant under T_ε.

The Noether Theorem

We are now postured to develop an important invariance identity that is a necessary and sufficient condition for the functional (1) to be invariant under the symmetry transformation T_ε given by (2). The condition relates the Lagrangian L to the generators τ and ξ of the group and it provides an easy test for invariance.

Theorem 1.1 *The functional $J(x)$ is absolutely invariant under the local Lie group (2) with generators τ and ξ if, and only if,*

$$L_t\tau + L_x\xi + L_{\dot{x}}(\dot{\xi} - \dot{x}\dot{\tau}) + L\dot{\tau} = 0 \tag{13}$$

Proof Carrying out the differentiation on the left side of (12) yields

$$L\frac{d}{d\varepsilon}\left(\frac{d\bar{t}}{dt}\right)\Bigg|_{\varepsilon=0} + \frac{d}{d\varepsilon}L\left(\bar{t}, \bar{x}, \frac{d\bar{x}}{d\bar{t}}\right)\Bigg|_{\varepsilon=0} = 0$$

But

$$\frac{d}{d\varepsilon}\left(\frac{d\bar{t}}{dt}\right)\Bigg|_{\varepsilon=0} = \frac{d}{d\varepsilon}\left(\frac{d}{dt}(t + \varepsilon\tau + o(\varepsilon))\right)\Bigg|_{\varepsilon=0} = \dot{\tau}$$

and the chain rule and (10) imply

$$\frac{d}{d\varepsilon}L\left(\bar{t}, \bar{x}, \frac{d\bar{x}}{d\bar{t}}\right)\Bigg|_{\varepsilon=0} = L_t\tau + L_x\xi + L_{\dot{x}}(\dot{\xi} - \dot{x}\dot{\tau})$$

Putting these results together gives (13) and completes the proof.

Besides providing a check for invariance the identity (13) leads directly to a first integral of the Euler equation.

Theorem 1.2 (E. Noether) *If $J(x)$ is absolutely invariant under the local Lie group T_ε with generators τ and ξ, then*

$$(L - \dot{x}L_{\dot{x}})\tau + L_{\dot{x}}\xi = \text{constant} \tag{14}$$

is a first integral of the Euler equation.

Proof Substitute the identities

$$L_{\dot{x}}\dot{\xi} = \frac{d}{dt}(L_{\dot{x}}\xi) - \xi\frac{d}{dt}L_{\dot{x}}$$

$$L_t\tau = \frac{dL}{dt}\tau - L_x\dot{x}\tau - L_{\dot{x}}\ddot{x}\tau$$

$$L_{\dot{x}}\dot{x}\dot{\tau} + L_{\dot{x}}\ddot{x}\tau = \frac{d}{dt}(L_{\dot{x}}\dot{x}\tau) - \dot{x}\tau\frac{d}{dt}L_{\dot{x}}$$

into (13) and simplify to obtain (14).

The Noether theorem is one of the cornerstone results in the calculus of variations. It states that if a problem represented by a variational integral admits a symmetry, then a conservation law can be written immediately. Thus there is a connection between symmetries and conservation laws in physical systems.

Example 1.8 The functional

$$J(x) = \int_a^b t^2\left(\frac{\dot{x}^2}{2} - \frac{x^6}{6}\right) dt \tag{15}$$

is absolutely invariant under the group T_ε given by $\bar{t} = (1 + \varepsilon)t$, $\bar{x} = (1 + \varepsilon)^{-1/2}x$. This is easily observed by substituting the following quantities into (13) and noting that it is reduced to an identity.

$$L_t = 2t\left(\frac{\dot{x}^2}{2} - \frac{x^6}{6}\right), \qquad L_x = -t^2x^5, \qquad L_{\dot{x}} = t^2\dot{x}$$

$$\tau = t, \qquad \xi = -\frac{x}{2}, \qquad \dot{\tau} = 1, \qquad \dot{\xi} = -\frac{\dot{x}}{2}$$

The Euler equation is

$$\ddot{x} + \frac{2}{t}\dot{x} + x^5 = 0 \tag{16}$$

and is known as the Emden-Fowler equation. By Theorem 1.2 a conservation law or first integral of (16) is

$$\frac{t^3 x^6}{6} + \frac{t^3 \dot{x}^2}{2} + \frac{t^2 x \dot{x}}{2} = \text{constant} \tag{17}$$

The invariance identity (13) may also be used to determine the generators τ and ξ of a group T_ε under which a given functional (1) is absolutely invariant.

Example 1.9   We consider the functional (15) of Example 1.8 and determine the generators $\tau = \tau(t, x)$ and $\xi = \xi(t, x)$ of a group under which (15) is invariant. Equation (13) can be written in expanded form as

$$L_t \tau + L_x \xi + L_{\dot{x}}\left(\xi_t + \dot{x}(\xi_x - \tau_t) - \dot{x}^2 \tau_x \right) + L\left(\tau_t + \dot{x} \tau_x \right) = 0 \tag{18}$$

When L and its partial derivatives are substituted into (18) we obtain an equation of the form

$$A(t, x) + B(t, x)\dot{x} + C(t, x)\dot{x}^2 + D(t, x)\dot{x}^3 = 0$$

which is a polynomial in $\dot{x}$. This equation must hold for all $\dot{x}$ and therefore the coefficients A, B, C, and D must vanish. In this case

$$\frac{x}{3}\tau + t\xi + \frac{tx}{6}\tau_t = 0$$

$$\xi_t - \frac{x^6}{6}\tau_x = 0$$

$$t\tau + t^2 \xi_x - \frac{t^2}{2}\tau_t = 0$$

$$\tau_x = 0$$

Hence $\tau = \tau(t)$, $\xi = \xi(x)$, and these four equations are reduced to two:

$$\frac{x}{3}\tau(t) + t\xi(x) + \frac{tx}{6}\tau'(t) = 0 \tag{19a}$$

$$t\tau(t) + t^2 \xi'(x) - \frac{t^2}{2}\tau'(t) = 0 \tag{19b}$$

Dividing the first equation by xt gives

$$\frac{\tau(t)}{3t} + \frac{\xi(x)}{x} + \frac{\tau'(t)}{6} = 0$$

or

$$\frac{\tau(t)}{3t} + \frac{\tau'(t)}{6} = -\frac{\xi(x)}{x} = k$$

for some constant k. Here we have used the standard separation of variables technique. Hence

$$\xi(x) = -kx$$

and

$$\tau'(t) + \frac{2}{t}\tau(t) = 6k$$

which when integrated gives

$$\tau(t) = 2kt + \frac{c}{t^2}$$

Substituting back into (19b) gives $c = 0$ and k arbitrary. Hence

$$\tau = 2kt, \qquad \xi = -kx, \qquad k \text{ constant}$$

are the required generators. Choosing $k = \frac{1}{2}$ corresponds to the case considered in Example 1.7.

Example 1.10   (Damped Harmonic Oscillator)   We consider (see Fig. 7.3) a harmonic oscillator with restoring force $-kx$ that is emersed in a liquid in such a way that the motion of the mass m is damped by a force proportional to its velocity. Applying Newton's second law to the system we obtain the governing equation of motion

$$m\ddot{x} + a\dot{x} + kx = 0 \tag{20}$$

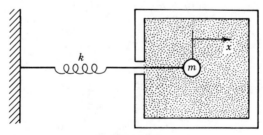

Figure 7.3

where $k > 0$, $a > 0$. A Lagrangian that leads to (20) is (see Example 5.6, Chapter 3)

$$L(t, x, \dot{x}) = \tfrac{1}{2}(m\dot{x}^2 - kx^2)e^{(a/m)t}$$

To determine a first integral of (20) we find generators τ and ξ of a local Lie group under which the functional $J(x) = \int L\,dt$ is absolutely invariant and then we apply Noether's theorem. The invariance identity (13) must hold. For the present case

$$L_t = \frac{1}{2}(m\dot{x}^2 - kx^2)\frac{a}{m}e^{(a/m)t}, \qquad L_x = -kxe^{(a/m)t}, \qquad L_{\dot{x}} = m\dot{x}e^{(a/m)t}$$

Upon substituting these quantities into (13) and expanding the total derivatives $\dot{\xi}$ and $\dot{\tau}$ we obtain

$$\frac{a}{2m}(m\dot{x}^2 - kx^2)\tau - kx\xi + m\dot{x}\left(\frac{\partial\xi}{\partial t} + \frac{\partial\xi}{\partial x}\dot{x} - \dot{x}\frac{\partial\tau}{\partial t} - \dot{x}^2\frac{\partial\tau}{2x}\right)$$

$$+ \frac{1}{2}(m\dot{x}^2 - kx^2)\left(\frac{\partial\tau}{\partial t} + \frac{\partial\tau}{\partial x}\dot{x}\right) = 0$$

When the coefficients of $\dot{x}^0$, $\dot{x}$, $\dot{x}^2$, and $\dot{x}^3$ are collected and equated to zero we obtain the following system of four first order partial differential equations

(i) $\dfrac{a}{2m}\tau x + \xi + \dfrac{x}{2}\dfrac{\partial\tau}{\partial t} = 0$

(ii) $m\dfrac{\partial\xi}{\partial t} - \dfrac{k}{2}\dfrac{\partial\tau}{\partial x}x^2 = 0$

(iii) $\dfrac{a}{2m}\tau + \dfrac{\partial\xi}{\partial x} - \dfrac{1}{2}\dfrac{\partial\tau}{\partial t} = 0$

(iv) $\dfrac{\partial\tau}{\partial x} = 0$ $\qquad\qquad\qquad\qquad\qquad\qquad$ (21)

From (21iv) we conclude that $\tau = \tau(t)$ and consequently from (21ii) it follows that $\xi = \xi(x)$. Therefore these four equations reduce to two ordinary differential equations

(i) $\dfrac{a}{2m}\tau x + \xi + \dfrac{x}{2}\dfrac{d\tau}{dt} = 0$

(ii) $\dfrac{a}{2m}\tau + \dfrac{d\xi}{dx} - \dfrac{1}{2}\dfrac{d\tau}{dt} = 0$ $\qquad\qquad\qquad\qquad$ (22)

If the second equation in (22) is multiplied by x and added to the first equation, then we obtain

$$\frac{a}{m}\tau x + \xi + x\frac{d\xi}{dx} = 0$$

which implies that $\tau = $ constant $= c$. In that case we observe from (22i) that $\xi = -(ac/2m)x$. Upon choosing $c = 1$ we therefore obtain generators

$$\tau = 1, \qquad \xi = -\frac{ax}{2m}$$

of a group under which $J(x)$ is invariant. By Noether's theorem there is a first integral or constant of the motion given by

$$(L - \dot{x}L_{\dot{x}})\tau + L_{\dot{x}}\xi = \text{constant}$$

or

$$(m\dot{x}^2 + kx^2 + ax\dot{x})e^{(a/m)t} = \text{constant}$$

The key to the calculations in Examples 1.9 and 1.10 was to observe that the invariance identity (13) became a polynomial in $\dot{x}$ upon substitution of L and its partial derivatives, thus permitting the coefficients of the powers of $\dot{x}$ to be set to zero. The resulting equations for τ and ξ, the so-called *determining equations of the group*, were then solved to determine the generators.

Yet another important application of the fundamental invariance identity (13) is the characterization or classification of variational problems which have a given symmetry or invariance property. In this case the Lagrangian L is regarded as the unknown in (13).

Example 1.11 Find all variational problems of the form

$$J(x) = \int_a^b L(t, x, \dot{x})\, dt$$

which are invariant under the transformation (dilation)

$$\bar{t} = (1 + \varepsilon)t, \qquad \bar{x} = (1 + \varepsilon)x$$

Here $\tau = t$, $\xi = x$, and (13) becomes

$$L_t t + L_x x = -L$$

which is a first order partial differential equation for L. Its characteristic system is

$$\frac{dt}{t} = \frac{dx}{x} = \frac{d\dot{x}}{0} = \frac{dL}{-L}$$

which has first integrals t/x, $\dot{x}$, and Lx. Therefore

$$L = \frac{1}{x} f\left(\frac{t}{x}, \dot{x}\right)$$

for an arbitrary function f.

Conservation Laws in Mechanics

By the Noether theorem if the action integral

$$J(x) = \int_a^b L(t, x, \dot{x})\, dt$$

is invariant under the local Lie group of transformations T_ε with generators τ and ξ, then a conservation law is given by

$$(L - \dot{x}L_{\dot{x}})\tau + L_{\dot{x}}\xi = \text{constant} \tag{23}$$

In the case of a mechanical system x denotes the generalized coordinate whose trajectory $x = x(t)$ describes the time evolution of the system and satisfies the Lagrange equation

$$L_x - \frac{d}{dt}L_{\dot{x}} = 0$$

We recall from Section 3.5 that the quantities $L_{\dot{x}}$ and $-L + \dot{x}L_{\dot{x}}$ play a special role in the canonical formalism of Hamilton; they are the generalized momentum and Hamiltonian, respectively. In terms of these canonical quantities the conservation law (23) can be written

$$-H\tau + p\xi = \text{constant} \tag{24}$$

where

$$p = L_{\dot{x}} \quad \text{and} \quad H = -L + \dot{x}L_{\dot{x}}$$

In the special case that the group of symmetries is a simple time translation

$$\bar{t} = t + \varepsilon, \qquad \bar{x} = x$$

then the generators are $\tau = 1$ and $\xi = 0$ and (24) becomes

$$H = \text{constant}$$

For systems where H is the total energy we then have the conservation of energy theorem: If J is invariant under a time translation, then the total energy in the system remains constant. Clearly this is equivalent to the

Lagrangian L being independent of t. If the action J is invariant under the spatial translation

$$\bar{t} = t, \qquad \bar{x} = x + \varepsilon$$

then $\tau = 0$, $\xi = 1$ and momentum is conserved, that is,

$$p = \text{constant}$$

These results are easily generalized to systems with n degrees of freedom. Let $x_1, x_2, \ldots, x_n$ be generalized coordinates for a physical system; hence the n functions $x_i = x_i(t)$, $t_0 \le t \le t_1$, $i = 1, \ldots, n$, describe the evolution of the system from time t_0 to time t_1. We assume that there is a Lagrangian $L = L(t, x_1, \ldots, x_n, \dot{x}_1, \ldots, \dot{x}_n)$, depending on the states x_i and their derivatives $\dot{x}_i$ such that the Lagrange equations

$$L_{x_i} - \frac{d}{dt} L_{\dot{x}_i} = 0, \qquad i = 1, \ldots, n \tag{25}$$

for the variational problem

$$J = \int_{t_0}^{t_1} L(t, x_1, \ldots, x_n, \dot{x}_1, \ldots, \dot{x}_n) \, dt \tag{26}$$

coincide with the governing differential equations for the system. As we observed in Chapter 3, for many mechanical systems the Lagrangian L is of the form

$$L = \sum_{i, j=1}^{n} a_{ij}(x_1, \ldots, x_n) \dot{x}_i \dot{x}_j - V(x_1, \ldots, x_n)$$

which is the difference between the kinetic energy (a quadratic form in the generalized velocities $\dot{x}_i$) and the potential energy (which depends only on the coordinates $x_1, \ldots, x_n$). Both Theorems 1.1 and 1.2 are generalized to n functions in the following theorem.

Theorem 1.3 *If $J(x)$ defined by (26) is absolutely invariant under the local Lie group of transformations*

$$\bar{t} = \phi(t, x_1, \ldots, x_n, \varepsilon)$$
$$\bar{x}_i = \psi_i(t, x_1, \ldots, x_n, \varepsilon)$$

with generators

$$\tau = \frac{\partial \phi}{\partial \varepsilon}(t, x_1, \ldots, x_n, 0)$$

$$\xi_i \equiv \frac{\partial \psi_i}{\partial \varepsilon}(t, x_1, \ldots, x_n, 0), \qquad i = 1, \ldots, n$$

then

$$L\tau + \sum_{i=1}^{n} \left\{ L_{x_i}\xi_i + L_{\dot{x}_i}(\dot{\xi}_i - \dot{x}_i\dot{\tau}) \right\} + L\dot{\tau} = 0 \qquad (27)$$

If the Lagrange equations (25) *hold, then*

$$\left(L - \sum_{i=1}^{n} \dot{x}_i L_{\dot{x}_i} \right)\tau + \sum_{i=1}^{n} L_{\dot{x}_i}\xi_i = \text{constant} \qquad (28)$$

In (27) *we have used the notation* $\dot{\tau} = \tau_t + \sum \tau_{x_i}\dot{x}_i$ *and similarly for* $\dot{\xi}_i$.

The proof of Theorem 1.3 is exactly like the proofs of Theorems 1.1 and 1.2. Equation (27) is the invariance identity analogous to (13) and (28) is the conservation law that can be stated in terms of the Hamiltonian

$$H \equiv -L + \sum_{i=1}^{n} \dot{x}_i L_{\dot{x}_i}$$

and the generalized momenta

$$p_i \equiv L_{\dot{x}_i}$$

as

$$-H\tau + \sum_{i=1}^{n} p_i\xi_i = \text{constant}$$

As before, invariance under a time translation leads to an expression of energy conservation in the form $H = \text{constant}$.

A complete discussion and a generalization to multiple integral problems in field theory can be found in Logan [1] or Lovelock and Rund [2].

EXERCISES

1.1 Verify that the following one parameter families of transformations form a local Lie group. Determine the range of the parameter ε and the defining functions g and h.

(a) $\bar{t} = t + \varepsilon$, $\bar{x} = x$

(b) $\bar{t} = t \cos \varepsilon - x \sin \varepsilon$, $\bar{x} = t \sin \varepsilon + x \cos \varepsilon$

(c) $\bar{t} = e^{\varepsilon}t$, $\bar{x} = e^{2\varepsilon}x$

(d) $\bar{t} = t + \varepsilon$, $\bar{x} = \dfrac{tx}{t + \varepsilon}$

1.2   In Exercise 1.1 determine the generators and find the infinitesimal transformation. Find the extended group and its generators.

1.3   For each of the following sets of generators determine the global group (2) and canonical coordinates.

(a) $\tau = t$, $\xi = x$ (c) $\tau = t^2$, $\xi = xt$

(b) $\tau = -x$, $\xi = t$ (d) $\tau = t$, $\xi = 2x + t$

1.4   Verify directly from the definition that

$$J(x) = \int_a^b \sqrt{1 + \dot{x}^2}\, dt$$

is absolutely invariant under the local Lie group of rotations defined in Exercise 1.1b. What about under the group in Exercise 1.1c?

1.5   Directly from the definition show that $J = \int_a^b t\dot{x}^2\, dt$ is not absolutely invariant under the time translation

$$\bar{t} = t + \varepsilon, \qquad \bar{x} = x$$

Use (13) to reach the same conclusion. Show, however, that J is absolutely invariant under the spatial translation

$$\bar{t} = t, \qquad \bar{x} = x + \varepsilon$$

and find a first integral of the Euler equation by Noether's theorem.

1.6   Let

$$J(x) = \int_a^b \left(\dot{x}^2 - (t + 1)x^2 + 2x \right) dt$$

Find the Euler equation. Show that J does not admit any symmetries, that is, show that the only invariance transformation under which J is absolutely invariant is the identity transformation.

1.7   Prove that if the functional $J = \int_a^b L(x, \dot{x})\, dt$ is absolutely invariant under the time dilation $\bar{t} = t + \varepsilon t$, $\bar{x} = x$, then the Lagrangian L must be a homogeneous function of degree one in $\dot{x}$, that is, $L(x, k\dot{x}) = kL(x, \dot{x})$, $k > 0$.

1.8   Consider the action integral in polar coordinates r and θ

$$J(r, \theta) = \int_a^b \left\{ \frac{1}{2}m(\dot{r}^2 + r^2\dot{\theta}^2) + \frac{k}{r} \right\} dt$$

for the motion of a particle of mass m in a central force field, where $k > 0$ is a constant. Find the conservation laws resulting from the two symmetries $\bar{t} = t + \varepsilon$, $\bar{r} = r$, $\bar{\theta} = \theta$ and $\bar{t} = t$, $\bar{r} = r$, $\bar{\theta} = \theta + \varepsilon$. Interpret physically.

1.9   If $J(x) = \int_a^b L(t, x, \dot{x})\, dt$ is absolutely invariant under a local Lie group with generators $\tau = \tau(t)$ and $\xi = \xi(x, t)$ prove that $\int_a^b L_{\dot{x}\dot{x}}(t, x, \dot{x})\, dt$ is absolutely invariant under the same group.

1.10   Consider the local Lie group (2) with generators τ and ξ given by (3). The operator U defined by

$$U \equiv \tau(t, x)\frac{\partial}{\partial t} + \xi(t, x)\frac{\partial}{\partial x}$$

is called the *operator of the group*

(a) If $f(t, x)$ is a given analytic function show that

$$f(\bar{t}, \bar{x}) = f(t, x) + \varepsilon U f + \frac{\varepsilon^2}{2!}U^2 f + \frac{\varepsilon^3}{3!}U^3 f + \cdots$$

where $U^k f$ means $U(U \cdots (Uf) \cdots)$ applied k times. This series is called the *Lie series* for f.

(b) A function $f(t, x)$ is said to be an *invariant* of T_ε if $f(\bar{t}, \bar{x}) = f(t, x)$ for all $|\varepsilon| < \varepsilon_0$. Prove that f is an invariant of T_ε if and only if $Uf = 0$.

(c) Prove that $U = (Ut)(\partial/\partial t) + (Ux)(\partial/\partial x)$.

(d) Under a change of coordinates $s = s(t, x)$, $r = r(t, x)$ show that the operator U can be represented as

$$U_1 = (U_1 s)\frac{\partial}{\partial s} + (U_1 r)\frac{\partial}{\partial r}$$

All of these notions can be generalized to the extended group $\tilde{T}_\varepsilon$.

1.11   Refer to Exercise 1.10.

(a) Find the invariants of the group with generators $\tau = t^2$ and $\xi = x$.

(b) By writing the Lie series for $f \equiv t$ and $f \equiv x$ reconstruct the global representation of the group with generators $\tau = -x$, $\xi = t$.

(c) Illustrate (c) and (d) of Exercise 1.10 using $\tau = -x$ and $\xi = t$ with the change of coordinates $s = \sqrt{t^2 + x^2}$, $r = \arctan(t/x)$.

1.12 The first order ordinary differential equation

$$F(t, x, p) = 0, \qquad p = \dot{x}$$

is said to be *constant conformally invariant* under (2) if there exists a constant k such that

$$\tilde{U}F = kF$$

where

$$\tilde{U} \equiv \tau \frac{\partial}{\partial t} + \xi \frac{\partial}{\partial x} + \eta \frac{\partial}{\partial p}$$

is the operator of the extended group $\tilde{T}_\varepsilon$ and η is given by (10).

(a) If the differential equation is constant conformally invariant prove that T_ε maps solutions to solutions, that is, if $x = g(t)$ is a solution of $F(t, x, p) = 0$, then $\bar{x} = \bar{g}(\bar{t})$ is a solution of $F(\bar{t}, \bar{x}, \bar{p}) = 0$.

(b) Verify that the differential equation $\dot{x}^2 - x - t^2 = 0$ is constant conformally invariant ($k = -2a$) under the group $\bar{t} = e^{a\varepsilon}t$, $\bar{x} = e^{2a\varepsilon}x$.

1.13 (a) Let T_ε be a local Lie group on R^2 given by (2). By computing $(\partial/\partial\varepsilon)(d^2\bar{x}/d\bar{t}^2)$ at $\varepsilon = 0$ find the infinitesimal representation of the twice extended group $\tilde{\tilde{T}}_\varepsilon$, which includes transformation rules for both the first and second derivatives $\dot{x}$ and $\ddot{x}$.
Solution:

$$\frac{\partial}{\partial\varepsilon}\left(\frac{d^2\bar{x}}{d\bar{t}^2}\right)_0 = \dot{\eta} - \ddot{x}\dot{\tau}, \qquad \eta = \dot{\xi} - \dot{x}\dot{\tau}$$

(b) Extend the definition of invariance given in Exercise 1.12 to second order differential equations $F(t, x, \dot{x}, \ddot{x}) = 0$.

1.14 (a) Show that the Poisson-Boltzmann equation

$$\ddot{x} - \frac{1}{t}\dot{x} - e^x = 0$$

is constant conformally invariant under the group $\bar{t} = e^\varepsilon t$, $\bar{x} = x - 2\varepsilon$.

(b) Show that two invariants of the extended group $\bar{t} = e^\varepsilon t$, $\bar{x} = x - 2\varepsilon$, $\bar{p} = e^{-\varepsilon}p$ are given by $u \equiv t^2x$ and $v \equiv tp$, where $p = \dot{x}$.

(c) In the variables u and v show that the Poisson-Boltzmann equation reduces to a first order equation with solution $v = \sqrt{4 + u} - 2$ and thereby determine a first integral.

1.15 Consider the variational problem

$$J(x) = \int_a^b x\left(\dot{x}^3 + \frac{1}{\sqrt{t}}\right) dt$$

(a) Show that the Euler equation is constant conformally invariant under the group $\bar{t} = e^\varepsilon t$, $\bar{x} = e^{5\varepsilon/6}x$.

(b) Show that $J(x)$ is not invariant under the same group. (See J. D. Logan, *J. Phys. A: Math. Gen.* **18**, pp. 2151–2155 (1985), for conditions under which both the variational integral and associated Euler equation are invariant.)

7.2 INVARIANT PARTIAL DIFFERENTIAL EQUATIONS

Self-Similar Solutions

In the introduction we noted that invariance of the heat equation $u_t - ku_{xx} = 0$ led to the definition of a single independent variable $s = x/\sqrt{t}$ with which the heat equation could be reduced to a single ordinary differential equation $f''(s) + (s/2)f'(s) = 0$ for $u(x, t) = f(s)$. If this is a general procedure applicable to many problems, then we shall have a powerful method for solving partial differential equations. Two additional examples of this same type of phenomenon follow.

Example 2.1 Consider Burgers' equation

$$u_t + uu_x = \nu u_{xx}, \qquad \nu > 0$$

Let us attempt to find a traveling wave solution of the form $u(x, t) = f(x - ct)$. Then

$$u_t = -cf'(x - ct), \qquad u_x = f'(x - ct), \qquad u_{xx} = f''(x - ct)$$

Substituting into the partial differential equation gives an ordinary differential equation for f, namely

$$-cf'(s) + f(s)f'(s) = \nu f''(s), \qquad s \equiv x - ct$$

Again the partial differential equation is reduced to an ordinary differential equation.

Example 2.2 Consider the acoustic approximation equations

$$u_t - v_x = 0, \qquad v_t + uu_x = 0 \qquad\qquad (1)$$

where u and v are functions of x and t. Let $s = x/t$ and assume $u = f(s)$ and $v = g(s)$. Then $u_x = f'(s)/t$, $u_t = -xf'(s)/t^2$, $v_x = g'(s)/t$, $v_t = -xg'(s)/t^2$, and the system of partial differential equations (1) becomes

$$-\frac{x}{t^2}f'(s) - \frac{1}{t}g'(s) = 0, \qquad -\frac{x}{t^2}g'(s) + \frac{1}{t}f(s)f'(s) = 0$$

or

$$s^2 f' + sg' = 0 \qquad\qquad (2a)$$
$$-s^2 g' + sf = 0 \qquad\qquad (2b)$$

which is a system of ordinary differential equations for $f(s)$ and $g(s)$. System (2) can be solved by dividing (2b) by s and then adding the result to (2a) to get

$$f'(s^2 + f) = 0$$

Then $f = $ constant or $f(s) = -s^2$. In the latter case $g(s) = 2s^3/3$ and therefore we have obtained a solution

$$u(x, t) = -\frac{x^2}{t^2}, \qquad v(x, t) = \frac{2x^3}{3t^3} \qquad\qquad (3)$$

to the original problem (1). Again the introduction of the new independent variable s, called a similarity variable, permitted a substantial simplification of the original system of partial differential equations and we are able to obtain a solution. For the record we note that the system (1) is invariant, in the following sense, under the stretching transformation

$$\bar{x} = \varepsilon x, \qquad \bar{t} = \varepsilon t, \qquad \bar{u} = u, \qquad \bar{v} = v \qquad\qquad (4)$$

where ε is a real parameter; for

$$\bar{u}_{\bar{x}} = \varepsilon^{-1} u_x, \qquad \bar{u}_{\bar{t}} = \varepsilon^{-1} u_t, \qquad \bar{v}_{\bar{x}} = \varepsilon^{-1} v_x, \qquad \bar{v}_{\bar{t}} = \varepsilon^{-1} v_t$$

and therefore

$$\bar{u}_{\bar{t}} - \bar{v}_{\bar{x}} = \varepsilon^{-1}(u_t - v_x), \qquad \bar{v}_{\bar{t}} + \bar{u}\bar{u}_{\bar{x}} = \varepsilon^{-1}(v_t + uu_x)$$

Hence the operators defining the differential equations (1) are unchanged, up to a constant multiple, upon applying the transformation (4) as a change of variables. Furthermore the similarity variable s has the property $\bar{s} = s$, that is, it is an invariant of (4).

The solution (3) of the system of partial differential equations (1) in Example 2.2 is known as a *similarity* or *self-similar solution*. Similarity

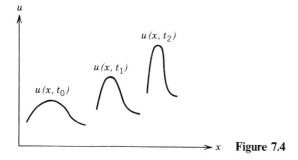

Figure 7.4

solutions are special classes of solutions that possess a type of invariance under certain transformations (stretching, rotations, translations, etc.) of the variables. This means that the self-similar motion of a system having time t and length x as independent variables is one in which the dependent variables or parameters that characterize the system vary in such a way that as time evolves, the spatial variation of these parameters remains geometrically similar. The scale that defines the spatial variation may change with time according to specified rules. Figure 7.4 illustrates three time snapshots of a typical self-similar solution $u(x, t)$ to a partial differential equation. We note the geometric similarity; one curve can be elongated, stretched, translated, etc., into any other member of the class of solutions by the appropriate symmetry transformation.

Although many partial differential equations possess similarity solutions arising from symmetries, arbitrary initial or boundary conditions may break the symmetry. Hence self-similar solutions do not exist for partial differential equations with general auxiliary conditions; too much has been specific and as a result symmetry is lost. One is sometimes satisfied, however, with a self-similar solution to a problem in which the initial conditions are ignored, thereby obtaining a solution that holds after the system has evolved for such a time that the initial conditions no longer affect the motion. In this context similarity solutions are not unlike intermediate asymptotic solutions or long-time solutions to partial differential equations.

Similarity solutions arise because the equations possess invariance properties under certain kinds of transformations. The analytic conditions that the equations admit self-similar solutions lead to the definition of a new single independent variable that is a function of the independent variables in the partial differential equations. With the introduction of this new variable a significant simplification occurs in that the partial differential equation reduces to an ordinary differential equation. The solution of the ordinary differential equation then provides the class of self-similar solutions or motions to the original problem.

The similarity method is an analytic method that in theory is exact. In practice, however, it frequently must be used in conjunction with numerical methods or other approximate techniques to obtain a solution. For some nonlinear partial differential equations the similarity method is the only analytic tool available. And even though we are sometimes unable to solve the exact problem we wish, a similarity solution of a simpler problem with some conditions relaxed cannot only indicate facts relevant to the original problem but can also be useful in itself in providing, for example, an analytic check for computer solutions or other approximation methods.

Similarity and Dimensional Analysis

It is clear from Examples 2.1 and 2.2 that one of the key problems in similarity analysis is to discover the independent variable s, the *similarity variable*, that leads to the reduction of the partial differential equation to an ordinary differential equation. Foremost in the historical development of the subject was the use of dimensional analysis to determine the form of self-similar solutions and the similarity variable. In this section we analyze a thermal diffusion problem to indicate the basic procedures. This is the same problem discussed from a dimensional point of view in Example 2.2 of Chapter 1.

Example 2.3   At $t = 0$ there is an amount of heat energy e located at a point in space. For $t > 0$ the heat is allowed to diffuse outward into a medium of temperature zero characterized by a constant diffusivity k and heat capacity c. If t is time and r the radial distance outward, we wish to determine the temperature $u(r, t)$ at distance r and time t. We notice that because of the spherical symmetry the temperature depends only on the radial distance with no spherical angular dependence. To solve this problem we set up a boundary value problem for $u(r, t)$. The partial differential equation is the spherically symmetric diffusion equation

$$u_t - k\frac{1}{r^2}\frac{\partial}{\partial r}\left(r^2 u_r\right) = 0, \qquad t > 0, \quad r > 0 \tag{5}$$

and the initial and boundary conditions are

$$u(r, 0) = 0, \qquad r > 0 \tag{6}$$

$$\lim_{r \to \infty} u(r, t) = 0, \qquad t > 0 \tag{7}$$

and

$$4\pi c \int_0^\infty r^2 u(r, t)\, dr = e, \qquad t > 0 \tag{8}$$

Condition (6) expresses the fact that at $t = 0$ the temperature of the medium is zero and (7) requires that far away from the source the temperature is zero for all time. Condition (8) states that the total energy present in all of space at any time $t > 0$ equals e, the initial amount of energy present. In this problem there is clearly no mechanism to create or destroy energy.

For a dimensional analysis approach we list the relevant variables

$$t, r, u, e, k, c$$

We expect a relation or physical law of the form

$$g(t, r, u, e, k, c) = 0 \tag{9}$$

From Example 2.2 in Chapter 1 we recall that there are two dimensionless variables that can be formed from the previous list of parameters

$$\pi_1 = \frac{r}{\sqrt{kt}}, \qquad \pi_2 = \frac{uc}{e}(kt)^{3/2}$$

The Buckingham Pi theorem guarantees that (9) is equivalent to a physical law of the form $G(\pi_1, \pi_2) = 0$ or $\pi_2 = f(\pi_1)$. Hence

$$u = \frac{e}{c}(kt)^{-3/2} f\left(\frac{r}{\sqrt{kt}}\right) \tag{10}$$

Equation (10) provides the information we seek; the unknown function $u(r, t)$ is some function f of the single variable

$$s = \frac{r}{\sqrt{kt}}$$

multiplied by a scale factor $e(kt)^{-3/2}/c$ that depends on time. Thus (10) defines both the similarity variable s and the form of the self-similar solutions. To determine the unknown function f we substitute (10) into the original partial differential equation (4). Carrying out the differentiations gives

$$u_r = \frac{e}{c}k^{-2}t^{-2}f'(s), \qquad u_{rr} = \frac{e}{c}k^{-5/2}t^{-5/2}f''(s)$$

and

$$u_t = -\frac{1}{2}\frac{e}{c}k^{-2}t^{-3}rf'(s) - \frac{3}{2}\frac{e}{c}k^{-3/2}t^{-5/2}f(s)$$

Therefore (4) becomes

$$f''(s) + \left(\frac{2}{s} + \frac{s}{2}\right)f'(s) + \frac{3}{2}f(s) = 0 \tag{11}$$

It is easy to see that a bounded solution of (11) is

$$f(s) = A \exp(-s^2/4), \qquad A \text{ constant} \tag{12}$$

Therefore up to a constant the similarity solutions are

$$u(r, t) = \frac{Ae}{c}(kt)^{-3/2}\exp\left(\frac{-r^2}{4kt}\right) \tag{13}$$

For fixed $t > 0$ we have $\lim_{r \to \infty} u(r, t) = 0$, and so the boundary condition (7) will hold. The initial condition (6) holds as well if we interpret (6) in the limiting sense, that is, $\lim_{t \to 0^+} u(r, t) = 0$. The constant A can be evaluated as follows. Upon substituting (13) into (8) and changing variables according to $s = r/\sqrt{kt}$ we obtain

$$4\pi A \int_0^\infty s^2\exp\left(\frac{-s^2}{4}\right)ds = 1$$

The value of the integral is easily found to be $2\sqrt{\pi}$. Thus $A = 1/8\pi^{3/2}$ and

$$u(r, t) = \frac{e/c}{(4\pi kt)^{3/2}}\exp\left(\frac{-r^2}{4kt}\right) \tag{14}$$

which is the solution to the original problem. For a fixed $t > 0$ we have

$$\lim_{r \to 0} u(r, t) = \frac{e/c}{(4\pi kt)^{3/2}}$$

Then as $t \to 0^+$

$$\lim_{t \to 0^+} \frac{e/c}{(4\pi kt)^{3/2}} = +\infty$$

Hence the initial temperature distribution is actually a delta function and we could have formulated the problem in these terms. In Fig. 7.5 several snapshots of the solution are shown for times $0 < t_1 < t_2 < t_3$. All are geometrically

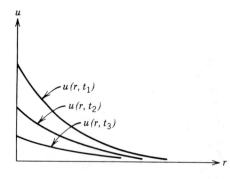

Figure 7.5. Time snapshots of (14).

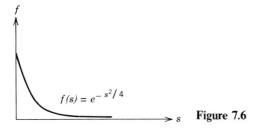

$f(s) = e^{-s^2/4}$

Figure 7.6

similar, which is consistent with our preceding remarks. The temperature distributions in Fig. 7.5 are all of the form

$$u(r, t) = \{\text{time dependent scaling factor}\} \cdot f(s)$$

where $f(s)$ is the basic similarity profile shown in Fig. 7.6. Hence the time snapshot at t_1, say, is formed by taking the graph of $f(s)$, compressing horizontally by a factor of $1/\sqrt{kt_1}$, and then stretching it vertically by the scale factor $(e/c)(4\pi kt_1)^{-3/2}$.

The Method of Stretchings

Similarity solutions arise out of the invariance properties of the partial differential equations under local Lie groups of transformations. In this section we explore this notion in the special case of so-called stretching transformations. Roughly the idea is to find new independent and dependent variables that are multiples of the old ones (i.e., stretched) such that the partial differential equation in the new coordinate system is the same up to some multiple as in the old coordinate system.

Example 2.4 Let us apply this principle to the spherically symmetric heat equation (4) encountered in Example 2.3, which we write as

$$u_t - ku_{rr} - \frac{2k}{r}u_r = 0 \tag{15}$$

We make a change of variables of the form

$$\bar{r} = \varepsilon^a r, \qquad \bar{t} = \varepsilon^b t, \qquad \bar{u} = \varepsilon^c u \tag{16}$$

In other words we stretch the variables. Here ε is a parameter that varies over some open interval containing 1, and a, b, and c are fixed real numbers to be determined. Transformation (16) therefore depends on the parameter ε, and it is a one parameter family of transformations. As ε varies the stretching factor varies for each variable. To determine the form of the partial differential equation in the barred coordinate system we compute the derivatives $\bar{u}_{\bar{t}}$, $\bar{u}_{\bar{r}}$, and $\bar{u}_{\bar{r}\bar{r}}$. By the chain rule

$$\bar{u}_{\bar{t}} = \varepsilon^{c-b} u_t, \qquad \bar{u}_{\bar{r}} = \varepsilon^{c-a} u_r, \qquad \bar{u}_{\bar{r}\bar{r}} = \varepsilon^{c-2a} u_{rr}$$

Then in the barred system

$$\bar{u}_{\bar{t}} - k\bar{u}_{\bar{r}\bar{r}} - \frac{2k}{\bar{r}}\bar{u}_r = \varepsilon^{c-b} u_t - k\varepsilon^{c-2a} u_{rr} - \frac{2k}{r} \varepsilon^{c-2a} u_r$$

If we select a, b, and c via the relation

$$c - b = c - 2a \tag{17}$$

then

$$\bar{u}_{\bar{t}} - k\bar{u}_{\bar{r}\bar{r}} - \frac{2k}{\bar{r}}\bar{u}_r = \varepsilon^{c-b}\left(u_t - ku_{rr} - \frac{2k}{r}u_r \right)$$

Therefore if (17) holds, the differential operator defining the partial differential equation in barred coordinates equals, up to a constant multiple, the operator in unbarred coordinates. In such a case we say that the partial differential equation is constant conformally invariant under (16). A precise definition will come later. Since (17) is a single equation for the three constants a, b, and c, two may be chosen arbitrarily. For definiteness let $a = 1$, $b = 2$, and $c = 3$ so that the stretching transformation (16) becomes

$$\bar{r} = \varepsilon r, \qquad \bar{t} = \varepsilon^2 t, \qquad \bar{u} = \varepsilon^{-3} u \tag{18}$$

Having the partial differential equation invariant under (18) we may pose the question of existence of a similarity variable and self-similar solutions. We noticed in Example 2.2 that the similarity variable s was an invariant of the transformation in that $\bar{s} = s$. Here, from (18) we notice by inspection that $s = r/\sqrt{t}$ is such an invariant since

$$\bar{s} = \frac{\bar{r}}{\sqrt{\bar{t}}} = \frac{\varepsilon r}{\sqrt{\varepsilon^2 t}} = \frac{r}{\sqrt{t}} = s$$

This choice of a similarity variable agrees with the one found by dimensional analysis earlier except for a constant factor k that does not affect the invariance of s. What form will the self-similar solutions take? The key is to look for an expression involving u that is among the invariants of (18). By inspection $ut^{3/2}$ is such an invariant, since

$$\bar{u}\bar{t}^{3/2} = (\varepsilon^{-3}u)(\varepsilon^2 t)^{3/2} = ut^{3/2}$$

Consequently both the quantities s and $ut^{3/2}$ are constants, that is, unchanged, under the action of the transformation (18). It is reasonable to expect therefore that $ut^{3/2}$ is some function of s, that is,

$$ut^{3/2} = f(s)$$

for some f. Hence

$$u = t^{-3/2}f(s), \qquad s = r/\sqrt{t} \qquad (19)$$

This is exactly the same as the earlier result based on dimensional reasoning, up to constant multiples. Therefore by examining the invariants of the stretching transformation we were able to determine both the similarity variable s and the form of the self-similar solutions. As before substitution of (19) into (15) will render the partial differential equation an ordinary differential equation.

We now place this heuristic reasoning on a firm mathematical basis. To fix the idea we consider a single second order partial differential equation for $u = u(x, t)$ given by

$$F(x, t, u, u_x, u_t, u_{xx}, u_{xt}, u_{tt}) = 0, \qquad (x, t) \in D \qquad (20)$$

The transformation of interest is a *one parameter family of stretching transformations* of the form

$$\bar{x} = \varepsilon^a x, \qquad \bar{t} = \varepsilon^b t, \qquad \bar{u} = \varepsilon^c u \qquad (21)$$

where a, b, and c are fixed real constants and ε is a real parameter having values in an open interval I containing $\varepsilon = 1$. If we denote the transformation (21) by T_ε, then $T_\varepsilon\colon R^3 \to R^3$ for each ε in I. The family T_ε contains the identity transformation T_1 on R^3 when $\varepsilon = 1$, and the composition rule $T_{\varepsilon_1}T_{\varepsilon_2} = T_{\varepsilon_1\varepsilon_2}$ holds for all ε_1 and ε_2. The inverse of T_ε is $T_{\varepsilon^{-1}}$. Therefore (21) is a local Lie group of transformations on R^3 with the identity occurring at $\varepsilon = 1$.

Since our discussion involves solutions of (20), we require some notion of the result of applying the transformation T_ε defined by (21), for fixed ε, to a surface in R^3 defined by

$$u = \phi(x, t), \qquad (x, t) \in D \qquad (22)$$

The action of T_ε on ϕ is visualized in Fig. 7.7. The surface (22) gets mapped to a surface in $\bar{x}\bar{t}\bar{u}$ space defined by

$$\bar{u} = \bar{\phi}(\bar{x}, \bar{t}), \qquad (\bar{x}, \bar{t}) \in \bar{D} \qquad (23)$$

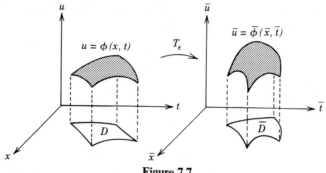

Figure 7.7

where $\overline{D} = \{(\overline{x}, \overline{t}) | \overline{x} = \varepsilon^a x, \overline{t} = \varepsilon^b t, (x, t) \in D\}$ and $\overline{\phi}$ is defined by

$$\overline{\phi}(\overline{x}, \overline{t}) = \varepsilon^c \phi(\varepsilon^{-a} x, \varepsilon^{-b} t) \tag{24}$$

To obtain (24) we applied the transformation (21) to the set of points $(x, t, \phi(x, t))$. We write the transformed partial differential equation as

$$F(\overline{x}, \overline{t}, \overline{u}, \overline{u}_{\overline{x}}, \overline{u}_{\overline{t}}, \overline{u}_{\overline{x}\overline{x}}, \overline{u}_{\overline{x}\overline{t}}, \overline{u}_{\overline{t}\overline{t}}) = 0, \qquad (\overline{x}, \overline{t}) \in \overline{D} \tag{25}$$

with the *same* function F. It is obvious that if $u = \phi(x, t)$ is a solution of (20), then $\overline{u} = \phi(\overline{x}, \overline{t})$ is a solution of (25) since we have just renamed the variables. Notice in general that $\overline{\phi}(\overline{x}, \overline{t})$ is not the same as $\phi(\overline{x}, \overline{t})$.

Example 2.5 Consider a stretching transformation

$$\overline{x} = \varepsilon x, \qquad \overline{t} = \varepsilon^2 t, \qquad \overline{u} = \varepsilon^3 u$$

and let $\phi(x, t) = x^2 - t^2$. Then $\phi(\overline{x}, \overline{t}) = \overline{x}^2 - \overline{t}^2$, whereas from (24) $\overline{\phi}(\overline{x}, \overline{t}) = \varepsilon^3(\varepsilon^{-2}\overline{x}^2 - \varepsilon^{-4}\overline{t}^2) = \varepsilon\overline{x}^2 - \overline{t}^2/\varepsilon$.

Furthermore, if $u = \phi(x, t)$ is a solution of (20), then $\overline{u} = \phi(\overline{x}, \overline{t})$ is not necessarily a solution of (25). If, however, the partial differential equation (20) is invariant, then this latter statement will follow. We make a formal definition motivated by Example 2.4.

Definition 2.1 *The partial differential equation* (20) *is constant conformally invariant under the one parameter family of stretchings* T_ε *defined by* (21) *if, and only if,*

$$F(\overline{x}, \overline{t}, \overline{u}, \overline{u}_{\overline{x}}, \overline{u}_{\overline{t}}, \overline{u}_{\overline{x}\overline{x}}, \overline{u}_{\overline{x}\overline{t}}, \overline{u}_{\overline{t}\overline{t}}) = A(\varepsilon) F(x, t, u, u_x, u_t, u_{xx}, u_{xt}, u_{tt}) \tag{26}$$

for all ε in I, for some function A with $A(1) = 1$. If $A(\varepsilon) \equiv 1$, then we say that (20) *is absolutely invariant.*

The word *conformal* in the definition suggests a stretching or magnification factor. Often we shall just use the word *invariant* when the context is clear.

Example 2.6 Find a stretching transformation under which the wave equation

$$u_{tt} - u_{xx} = 0$$

is invariant. Let

$$\bar{x} = \varepsilon^a x, \qquad \bar{t} = \varepsilon^b t, \qquad \bar{u} = \varepsilon^c u$$

Then

$$\bar{u}_{\bar{t}\bar{t}} - \bar{u}_{\bar{x}\bar{x}} = \varepsilon^{c-2b} u_{tt} - \varepsilon^{c-2a} u_{xx}$$

If $a = b$, then

$$\bar{u}_{\bar{t}\bar{t}} - \bar{u}_{\bar{x}\bar{x}} = \varepsilon^{c-2a}\left(u_{tt} - u_{xx}\right)$$

The conformal factor is $A(\varepsilon) = \varepsilon^{c-2a}$. Hence the wave equation is constant conformally invariant under the given stretching transformation provided $a = b$. Note that a and c may be chosen arbitrarily.

We now formalize the comment preceding Definition 2.1 regarding $\bar{\phi}(\bar{x}, \bar{t})$ being a solution of the transformed partial differential equation (25). The proof follows immediately from Definition 2.1 and Equation (26).

Theorem 2.1 *If the partial differential equation* (20) *is constant conformally invariant under the one parameter family of stretchings T_ε given by* (21) *and if $\phi(x, t)$ is a solution of* (20), *then $\bar{\phi}(\bar{x}, \bar{t})$ defined by* (24) *is a solution of the partial differential equation* (25).

The following example will clear up the question of equality of $\bar{\phi}(\bar{x}, \bar{t})$ and $\phi(\bar{x}, \bar{t})$.

Example 2.7 In Example 2.6 we noted that the wave equation $u_{tt} - u_{xx} = 0$ is invariant under $\bar{x} = \varepsilon x$, $\bar{t} = \varepsilon t$, and $\bar{u} = \varepsilon^2 u$ where we have chosen $a = b = 1$, $c = 2$. The solution $\phi(x, t) = \sin(x - t)$ is transformed to $\bar{\phi}(\bar{x}, \bar{t}) = \varepsilon^2 \sin((\bar{x} - \bar{t})/\varepsilon)$, which is a solution to $\bar{u}_{\bar{t}\bar{t}} - \bar{u}_{\bar{x}\bar{x}} = 0$. But $\phi(\bar{x}, \bar{t}) = \sin(\bar{x} - \bar{t})$ and so $\bar{\phi}(\bar{x}, \bar{t})$ is not generally the same as $\phi(\bar{x}, \bar{t})$. The solution $g(x, t) = (x - t)^2$, however, is transformed to $\bar{g}(\bar{x}, \bar{t}) = (\bar{x} - \bar{t})^2$, which coincides with $g(\bar{x}, \bar{t})$. We note that g is an invariant solution whereas f is not.

Guided by this last example we make the following definition.

Definition 2.2   *A solution $u = \phi(x, t)$, $(x, t) \in D$, of (20) is an invariant solution (or invariant surface) if, and only if,*

$$\phi(\bar{x}, \bar{t}) = \bar{\phi}(\bar{x}, \bar{t}) \tag{27}$$

under the transformation T_ε.

Similarity solutions for partial differential equations are found among the invariant solutions. A condition can be obtained from (27) that is easily used to determine the invariant solutions. To this end we note that (27) is equivalent to

$$\phi(\varepsilon^a x, \varepsilon^b t) = \varepsilon^c \phi(x, t) \tag{28}$$

in terms of the unbarred variables. Since (28) is valid for all ε in I, we may differentiate with respect to ε and then set $\varepsilon = 1$ to obtain the first order partial differential equation

$$ax\phi_x + bt\phi_t = c\phi \tag{29}$$

Equation (29) is called the *invariant surface condition*. The methods of characteristics may be used to determine the general form of ϕ (see Chapter 5). The characteristic system is

$$\frac{dx}{ax} = \frac{dt}{bt} = \frac{d\phi}{c\phi}$$

Integrating the first pair of equations gives

$$x^b t^{-a} = \text{constant}$$

Integrating the second pair gives

$$\phi t^{-c/b} = \text{constant}$$

Therefore the general solution of (29) is

$$\psi\left(\phi t^{-c/b}, x^b t^{-a}\right) = 0$$

for some function ψ. It immediately follows that the invariant surfaces are

$$\phi(x, t) = t^{c/b} f\left(\frac{x^b}{t^a}\right) \tag{30}$$

where f is an arbitrary function. Equation (30) defines the form of possible self-similar solutions to the problem arising from invariance under stretching transformations. The similarity variable s is defined by

$$s = \frac{x^b}{t^a}$$

and so (30) can be written as

$$u = \phi(x, t) = t^{c/b} f(s)$$

It remains to be shown that the original partial differential equation reduces to an ordinary differential equation when (30) is substituted into the partial differential equation. This fact is recorded as a fundamental theorem.

Theorem 2.2 *If the partial differential equation* (20) *is constant conformally invariant under the one parameter family of stretching transformations defined by*

$$\bar{x} = \varepsilon^a x, \qquad \bar{t} = \varepsilon^b t, \qquad \bar{u} = \varepsilon^c u$$

then substitution of the expression

$$u = t^{c/b} f(s) \tag{31}$$

where

$$s = \frac{x^b}{t^a} \tag{32}$$

into (20) *yields an equation of the form*

$$H(s, f, f', f'') = 0$$

which is an ordinary differential equation for f.

Proof The proof of this theorem is straightforward but lengthy for second order equations. Therefore we shall prove the theorem for first order partial differential equations of the form

$$F(x, t, u, p, q) = 0 \tag{33}$$

where $p = u_x$ and $q = u_t$. By invariance under T_ε we mean

$$F(\bar{x}, \bar{t}, \bar{u}, \bar{p}, \bar{q}) = A(\varepsilon) f(x, t, u, p, q) \tag{34}$$

where A is some function, $\bar{p} = \bar{u}_{\bar{x}}$, and $\bar{q} = \bar{u}_{\bar{t}}$. We observe that $\bar{p} = \varepsilon^{c-a} p$

and $\bar{q} = \varepsilon^{c-b}q$. Differentiating (34) with respect to ε and afterward setting $\varepsilon = 1$ yields a first order partial differential equation for the function F, namely

$$axF_x + btF_t + cuF_u + (c-a)pF_p + (c-b)qF_q = A'(1)F$$

This equation can be solved by the method of characteristics to obtain the general form for F. The characteristic system is

$$\frac{dx}{ax} = \frac{dt}{bt} = \frac{du}{cu} = \frac{dp}{(c-a)p} = \frac{dq}{(c-b)q} = \frac{dF}{A'(1)F}$$

Five independent first integrals are

$$\frac{x^b}{t^a}, \qquad ut^{-c/b}, \qquad pt^{(a-c)/b}, \qquad qt^{1-b/c}, \qquad Ft^{-A'(1)/b} \qquad (35)$$

Consequently

$$F = t^{A'(1)/b}G\left(s, ut^{-c/b}, pt^{(a-c)/b}, qt^{1-b/c}\right)$$

for some function G. Now u is given by (31) and it follows that

$$p = bt^{c/b-a}x^{b-1}f'(s)$$

and

$$q = \frac{c}{b}t^{c/b-1}f(s) - ax^bt^{c/b-a-1}f'(s)$$

When u, p, and q are substituted into G Equation (33) becomes

$$G\left(s, f(s), bs^{1-1/b}f'(s), \frac{c}{b}f(s) - asf'(s)\right) = 0$$

or an equation of the form $H(s, f, f') = 0$ and the reduction is complete. It is obvious how to proceed in the second order case. Here, necessarily, $b \neq 0$. If $b = 0$ then the last four first integrals in (35) can be written in terms of x rather than t.

Rather than use Theorem 2.2 as a prescription it is usually better to begin from first principles. We now illustrate the procedure on a nontrivial boundary value problem involving a nonlinear diffusion equation.

Example 2.8   Heat diffuses into a medium $x > 0$ and the temperature $u(x, t)$ of the medium is governed by the nonlinear boundary value problem

$$uu_t - u_{xx} = 0, \qquad x > 0, \quad t > 0 \tag{36}$$

$$u(x, 0) = 0, \qquad x > 0 \tag{37}$$

$$\lim_{x \to \infty} u(x, t) = 0, \qquad t > 0 \tag{38}$$

and

$$u_x(0, t) = -Q, \qquad t > 0 \tag{39}$$

where $Q > 0$ is a fixed constant. We assume that (36)–(39) are already scaled. Conditions (37) and (38) are an initial condition and boundary condition at infinity, respectively. The boundary condition (39) states that heat energy is being injected into the medium at $x = 0$ at a constant rate; recall that u_x is proportional to the flux. The partial differential equation (36) models a situation where the specific heat of the medium increases with temperature.

We assume a stretching transformation

$$\bar{x} = \varepsilon^a x, \qquad \bar{t} = \varepsilon^b t, \qquad \bar{u} = \varepsilon^c u$$

Then

$$\bar{u}\bar{u}_{\bar{t}} - \bar{u}_{\bar{x}\bar{x}} = \varepsilon^{2c-b} uu_t - \varepsilon^{c-2a} u_{xx} = \varepsilon^{2c-b}(uu_t - u_{xx})$$

provided $2c - b = c - 2a$ or $c = b - 2a$. Thus the partial differential equation is invariant under

$$\bar{x} = \varepsilon^a x, \qquad \bar{t} = \varepsilon^b t, \qquad \bar{u} = \varepsilon^{b-2a} u$$

The invariant surface condition is

$$axu_x + btu_t = (b - 2a)u$$

having characteristic system

$$\frac{dx}{ax} = \frac{dt}{bt} = \frac{du}{(b - 2a)u}$$

and first integrals x^b/t^a and $ut^{2a/b-1}$. Therefore letting $s = x^b/t^a$ the invariant surfaces are given by

$$u(x, t) = t^{1-2a/b} f(x^b/t^a)$$

We have been careful not to select particular values of a, b, and c, since flexibility will be required in satisfying the initial and boundary conditions that will entail further restrictions. To satisfy the boundary condition (39) we must have $u_x(x, t) = -Q$ at $x = 0$ or

$$t^{1-(2a/b)-a}bx^{b-1}f(x^b/t^a) = -Q, \qquad \text{at } x = 0$$

So that the left side does not vanish we must choose $b = 1$ to get

$$bt^{1-3a}f(x/t^a) = -Q \qquad \text{at } x = 0$$

But the left side cannot depend on t, so the condition $a = \frac{1}{3}$ is forced. In summary the flux condition at $x = 0$ has fully determined the values of a, b, and c. The stretching transformation is

$$\bar{x} = \varepsilon^{1/3}x, \qquad \bar{t} = \varepsilon t, \qquad \bar{u} = \varepsilon^{1/3}u$$

The similarity variable is

$$s = x/(\sqrt{3}\, t^{1/3}) \tag{40}$$

(we have included a factor of $\sqrt{3}$ to make subsequent calculations easier) and the form of possible self similar solutions is

$$u(x, t) = t^{1/3}f(s) \tag{41}$$

Upon substitution of (41) into the partial differential equation (36) we obtain an ordinary differential equation for the function f,

$$f'' - f(f - sf') = 0, \qquad 0 < s < \infty \tag{42}$$

The flux condition (39) at $x = 0$ forces

$$f'(0) = -\sqrt{3}\, Q \tag{43}$$

It remains to check the implications of the conditions (37) and (38). Note that as $t \to 0$ we have $s \to +\infty$ for a fixed value of x. Hence (37) becomes

$$\lim_{\substack{t \to 0 \\ x > 0}} u(x, t) = \lim_{s \to +\infty} t^{1/3}f(s) = \lim_{\substack{s \to +\infty \\ x > 0}} \frac{x}{\sqrt{3}\, s}f(s) = 0$$

Therefore

$$\lim_{s \to +\infty} f(s) = 0 \tag{44}$$

Finally (38) becomes

$$\lim_{\substack{x \to \infty \\ t > 0}} u(x, t) = \lim_{\substack{s \to \infty \\ t > 0}} t^{1/3}f(s) = 0$$

which leads again to the condition (44). We observe that the three conditions (37), (38), and (39) coalesced into two conditions (43) and (44) on f. This type of coalescence had to occur since f satisfies a second order ordinary differential equation. Therefore we have completed the reduction; the boundary value problem (36)–(39) has been reduced to the solution of an ordinary differential equation (42) subject to boundary conditions (43) and (44). Once solved, the self similar solution will be given by (41).

The Lie Plane

The ordinary differential equation to which we reduced the nonlinear diffusion equation (36) is a formidable equation itself. It cannot be solved analytically and at first it appears that a numerical procedure is not possible since $f(0)$ is not known, only $f'(0)$.

The approach we take, which is common in such problems, is to reduce the problem to a first order ordinary differential equation in so-called Lie variables by taking advantage of its invariance properties in the same manner we exploited the invariance properties of the partial differential equation.

An ordinary differential equation of second order written in the form

$$f'' - G(s, f, f') = 0 \tag{45}$$

is said to be constant conformally invariant under the stretching transformation

$$\bar{s} = \varepsilon s, \qquad \bar{f} = \varepsilon^b f, \qquad \varepsilon \in I \tag{46}$$

if

$$\frac{d^2 \bar{f}}{d\bar{s}^2} - G(\bar{s}, \bar{f}, d\bar{f}/d\bar{s}) = A(\varepsilon)(f'' - G(s, f, f')) \tag{47}$$

for some function A, for all $\varepsilon \in I$. The fundamental reduction theorem was proved by S. Lie in the late 1800s.

Theorem 2.3 *If the second order ordinary differential equation (45) is constant conformally invariant under the stretching transformation defined by (46), then (45) can be reduced to a first order ordinary differential equation of the form*

$$\frac{dv}{du} = \frac{H(u, v) - (b - 1)v}{v - bu} \tag{48}$$

where $u = \phi(s, f)$ and $v = \psi(s, f, f')$ are first integrals of the characteristic system

$$\frac{ds}{s} = \frac{df}{bf} = \frac{df'}{(b - 1)f'} \tag{49}$$

and H is some fixed function depending on G. The quantities u and v are called the Lie variables.

Proof The proof is similar to the proof of Theorem 2.2. Since

$$\frac{d^2\bar{f}}{d\bar{s}^2} - G(\bar{s}, \bar{f}, d\bar{f}/d\bar{s}) = \varepsilon^{b-2}f'' - G(\varepsilon s, \varepsilon^b f, \varepsilon^{b-1}f')$$

for invariance, we must have

$$\varepsilon^{2-b}G(\varepsilon s, \varepsilon^b f, \varepsilon^{b-1}f') = G(s, f, f')$$

for all ε in I. Differentiating with respect to ε at $\varepsilon = 1$ gives a partial differential equation for G, namely

$$sG_s + bfG_f + (b - 1)f'G_{f'} = (b - 2)G$$

The characteristic system is

$$\frac{ds}{s} = \frac{df}{bf} = \frac{df'}{(b - 1)f'} = \frac{dG}{(b - 2)G}$$

with first integrals

$$u \equiv \frac{f}{s^b}, \qquad v \equiv \frac{f'}{s^{b-1}}, \qquad \frac{G}{s^{b-2}}$$

Thus

$$G = s^{b-2}H(u, v)$$

for some function H (note that H is known if G is known). A straightforward calculation of du and dv shows that (48) holds whenever (45) holds.

Example 2.9 Consider the boundary value problem for $f(s)$ obtained in Example 2.8 and given by (42)–(44).

$$f'' - f(f - sf') = 0, \qquad 0 < s < \infty \tag{50}$$

$$f'(0) = -\sqrt{3}\,Q \tag{51}$$

$$\lim_{s \to \infty} f(s) = 0 \tag{52}$$

If we make the stretching transformation

$$\bar{s} = \varepsilon s, \qquad \bar{f} = \varepsilon^b f$$

then

$$\frac{d^2\bar{f}}{d\bar{s}^2} - \bar{f}\left(\bar{f} - \bar{s}\frac{d\bar{f}}{d\bar{s}}\right) = \varepsilon^{b-2}f'' - \varepsilon^{2b}f(f - sf')$$

For invariance $b - 2 = 2b$ or $b = -2$. According to Theorem 2.3 the Lie variables u and v are

$$u = s^2 f, \qquad v = s^3 f' \tag{53}$$

Then

$$\frac{du}{ds} = \frac{v}{s} + \frac{2u}{s}, \qquad \frac{dv}{ds} = \frac{u^2}{2} - \frac{uv}{s} + \frac{3v}{s}$$

and the resulting first order equation is therefore

$$\frac{dv}{du} = \frac{u^2 - uv + 3v}{v + 2u}$$

Actually (53) is not the best choice of Lie variables. Theorem 2.3 requires only that we select them as first integrals of the characteristic system (49). The choice

$$u = s^2 f, \qquad v = s^2(f - sf') \tag{54}$$

leads to the simpler equation

$$\frac{dv}{du} = \frac{v(2 - u)}{3u - v} \tag{55}$$

We have reduced the second order ordinary differential equation (50) to a first order ordinary differential equation (55) in the uv plane (called the *Lie plane*) in terms of the variables u and v defined by (54). A qualitative analysis of the trajectories or integral curves of (55) will give insight into the behavior of the solutions.

First, it is clear that the boundary condition $f'(0) = -\sqrt{3}\,Q$ forces $u = v = 0$ at $s = 0$; this assumes f is finite at $s = 0$. Moreover from (54) we see that $u > 0$ and $v > 0$, since in the latter case we can argue on physical grounds that $-sf' > 0$. Hence the trajectory begins at the origin $(u, v) = (0, 0)$ and goes into the first quadrant of the uv plane. Much information can be obtained by examining the direction field of (55) and its critical points. Easily $dv/du = 0$ along $v = 0$ and $u = 2$ where the direction field is horizontal, and the direction field is vertical along $v = 3u$. The only critical point is $(u, v) = (2, 6)$

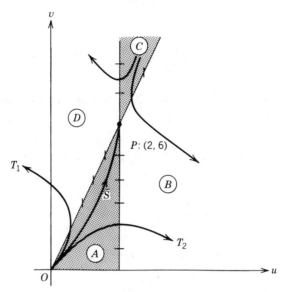

Figure 7.8

where $dv/du = 0/0$. These lines are shown in Fig. 7.8 and the dash marks indicate the slope of the tangent line or the direction field at the given points. One can check that $dv/du > 0$ in Region A, $dv/du < 0$ in Region B, $dv/du > 0$ in Region C, and $dv/du < 0$ in Region D. Therefore the critical point P: $(2, 6)$ must be a saddle point. A trajectory leaving O: $(0,0)$ therefore must leave the quadrant like T_1, cross the line $u = 2$ like T_2, or go into the critical point P along a separatrix S. We reject T_1 on physical grounds, since leaving the quadrant would imply $u < 0$ or $f < 0$, which means the tempera- ture is negative (recall $u = t^{1/3}f(s)$ is the temperature). We reject T_2 because of the following argument. As $u \rightarrow +\infty$ (55) implies $dv/du \sim -v/3$ or $v \sim \exp(-u/3)$, which implies $v \rightarrow 0$. Here we are using the symbol $\sim$ vaguely to mean *behaves like*. But from (54) $v \rightarrow 0$ forces $f - sf' \rightarrow 0$ as $s \rightarrow \infty$. Thus $f \sim$ constant $\cdot s \rightarrow \infty$, which contradicts the boundary condi- tion (52). Consequently the only possibility is that the solution begins at O and goes along the separatrix S into the critical point P as $s \rightarrow \infty$. The actual determination of S must be carried out numerically.

Just from these qualitative features we are able to draw some important conclusions regarding the asymptotic behavior of f and therefore $u(x, t)$ as $s \rightarrow \infty$. We have concluded that $u \rightarrow 2$ and $s \rightarrow +\infty$ and therefore by definition of u

$$f(s) \sim \frac{2}{s^2}, \qquad \text{as } s \rightarrow \infty$$

That is, f goes to zero like $1/s^2$. In terms of the temperature $u(x, t)$ in the originai nonlinear diffusion problem, for large s

$$u(x, t) \sim t^{1/3} \frac{2}{s^2}$$

$$\sim \frac{6t}{x^2}$$

For a fixed time t the temperature falls off like $1/x^2$ as $x \to \infty$.

Further examples involving stretching transformations can be found in Dresner [3] and Ames [4].

EXERCISES

2.1 Show that the wave equation $u_{tt} - u_{xx} = 0$ is constant conformally invariant under $\bar{x} = \varepsilon x$, $\bar{t} = \varepsilon t$, $\bar{u} = u$. Show there exist solutions of the form $u = f(s)$, where $s = x/t$ and where f satisfies the ordinary differential equation

$$(s^2 - 1)f'' + 2sf' = 0$$

By solving this equation show that the self-similar solutions are traveling waves of the form

$$u = c_1 \ln \left| \frac{x - t}{x + t} \right| + c_2$$

2.2 If an ordinary differential equation $y' = F(t, y)$ is absolutely invariant under the stretching transformation $\bar{t} = \varepsilon t$, $\bar{y} = \varepsilon^a y$ prove that it can be reduced to an ordinary differential equation of the form

$$\frac{dx}{x} = \frac{du}{G(u) - au}$$

in which the variables are separated.

2.3 Consider the Emden-Fowler equation

$$\ddot{y} + \frac{2}{t}\dot{y} + y^n = 0, \qquad n > 0, \quad 0 < t < \infty$$

subject to the conditions

$$\dot{y}(0) = 0, \qquad y(t) \text{ finite}$$

(a) When $n = 5$ show that the associated differential equation in Lie variables is integrable and find its solution. Hence find a first integral of the Emden-Fowler equation.

(b) When $n = 3$ find the associated differential equation in Lie variables. By analyzing the Lie plane show that the solution must be oscillatory in this case.

2.4 Consider the nonlinear diffusion equation $u_t = (\partial/\partial x)(k(u)u_x)$, $t > 0$ $0 < x < \infty$, with boundary conditions

$$u(0, t) = u_0, \qquad \lim_{x \to \infty} u(x, t) = u_1$$

and initial condition $u(x, 0) = u_2$, where u_0, u_1, u_2 are constants. Under what conditions on u_0, u_1, u_2 will there be a similarity solution to this problem with similarity variables $s = x^a t^b$ for some choice of a and b?

2.5 The equation $u_y u_{xy} - u_x u_{yy} - u_{yyy} = 0$ occurs in boundary layer theory for flow along a flat plate. Use the method of stretchings to reduce the equation to an ordinary differential equation. Choose the simplest non-trivial transformation.

2.6 Consider the nonlinear partial differential equation

$$uu_t + u_x^2 = 0$$

(a) Find the simplest nontrivial group of stretching transformations under which the equation is invariant.

(b) Find a similarity variable and reduce the partial differential equation to an ordinary differential equation.

(c) Find a class of self-similar solutions to the problem.

2.7 Waves in a one dimensional channel $x > 0$ satisfy the system of nonlinear partial differential equations

$$h_t + (vh)_x = 0$$
$$v_t + vv_x + h_x = 0, \qquad x > 0, \quad t > 0$$

where $h(x, t)$ is the height of the water and $v(x, t)$ is the longitudinal velocity, assumed to be the same for any vertical cross section (see Fig. 7.9). At time $t = 0$ we consider the channel to be dammed at $x = 0$ with h_0 the initial height of the water. At $t > 0$ the dam is removed and we

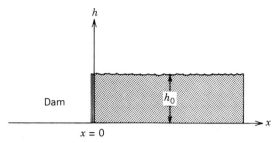

Figure 7.9. Exercise 2.7.

assume

$$\lim_{x \to \infty} h(x, t) = h_0, \qquad t > 0$$

$$\lim_{x \to \infty} v(x, t) = 0, \qquad t > 0$$

Find a similarity solution fitting the requirements just given and sketch typical water height profiles for different times. Compute the speed at which the crest or discontinuity is moving.
Answer:

$$h(x, t) = \left(\frac{2}{3}\sqrt{h_0} - \frac{x}{3t} \right)^2, \qquad v(x, t) = \frac{2x}{3t} - \frac{2}{3}\sqrt{h_0}$$

2.8 Consider the nonlinear diffusion problem

$$u_t - \frac{\partial}{\partial x}(uu_x) = 0, \qquad x > 0, \quad t > 0$$

$$u(x, 0) = 0, \qquad x > 0$$

$$u(\infty, t) = 0, \qquad t > 0$$

$$(uu_x)_{x=0} = -b, \qquad t > 0$$

(a) Use the method of stretchings to reduce the problem to

$$3ff'' + 3f'^2 + 2sf' - f = 0, \qquad 0 < s < \infty$$

$$f(0)f'(0) = -b$$

$$\lim_{s \to \infty} f(s) = 0$$

where $f = f(s)$ and $s = xt^{-2/3}$

(b) Show that $u = f/s^2$ and $v = f'/s$ are Lie variables and reduce the problem to

$$\frac{dv}{du} = \frac{u - 3v^2 - 2v - 3uv}{3u(v - 2u)}$$

in the Lie plane.

(c) Examine the nature of the critical points in the Lie plane and discuss the solution path.

7.3   THE GENERAL SIMILARITY METHOD

Local Lie Groups on R^3

In Section 7.2 the similarity method was developed for partial differential equations invariant under stretching transformations. It seems clear that these concepts can be extended to more general local Lie groups of transformations on R^3. Such transformations include translations, rotations, and other kinds of geometrical mappings. As before, we expect that if a partial differential equation admits such a general symmetry, then it can be reduced to an ordinary differential equation.

Example 3.1   Let ε be a real parameter ranging over some interval containing $\varepsilon = 0$. The one parameter family of transformations

$$\begin{aligned}
\bar{x} &= x \cos \varepsilon - t \sin \varepsilon \\
\bar{t} &= t \cos \varepsilon + x \sin \varepsilon \\
\bar{u} &= u + \varepsilon
\end{aligned} \tag{1}$$

is a rotation by angle ε in the tx plane and a translation by ε in the u direction. Geometrically a point (x, t, u) gets moved along a vertical helix wrapped around the u axis in R^3 as ε varies (see Fig. 7.10). It is observed that setting $\varepsilon = 0$ in (1) reduces it to the identity transformation $\bar{x} = x$, $\bar{t} = t$, $\bar{u} = u$, and replacing ε by $-\varepsilon$ results in an inverse transformation which undoes (1), that is, maps $(\bar{x}, \bar{t}, \bar{u})$ to (x, t, u). Further, if the transformation (1) with $\varepsilon = \varepsilon_1$ is followed by another application of (1) with $\varepsilon = \varepsilon_2$, then the composition is equivalent to a single transformation (1) with $\varepsilon = \varepsilon_1 + \varepsilon_2$. For small ε the right sides of (1) can be expanded in a Taylor series about $\varepsilon = 0$; using

$$\cos \varepsilon = 1 - \frac{\varepsilon^2}{2!} + \cdots, \qquad \sin \varepsilon = \varepsilon - \frac{\varepsilon^3}{3!} + \cdots$$

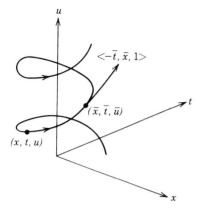

Figure 7.10

we get

$$\bar{x} = x - t\varepsilon + o(\varepsilon)$$
$$\bar{t} = t + x\varepsilon + o(\varepsilon) \tag{2}$$
$$\bar{u} = u + \varepsilon$$

The so-called infinitesimal transformation (2) is an approximation of (1) for small ε. The coefficients $-t$, x, and 1 of ε in (2) are the generators and they define on R^3 a vector field $\langle -t, x, 1 \rangle$ which attaches to each point of R^3 a vector pointing in the direction of the motion of that point under action of (1). Thus the integral curves of the vector field defined by the generators are the helixes that are the actual paths along which (1) moves points.

Now let us consider a general one parameter family of transformations T_ε defined by the equations

$$\begin{aligned} & & \bar{x} = \Phi(x, t, \varepsilon) & & (3a) \\ T_\varepsilon: & & \bar{t} = \Psi(x, t, \varepsilon) & & (3b) \\ & & \bar{u} = \Omega(x, t, u, \varepsilon) & & (3c) \end{aligned}$$

where ε is a real parameter that varies over some open interval $|\varepsilon| < \varepsilon_0$ containing zero, and Φ, Ψ, and Ω are functions analytic on their respective domains. For each ε the transformation T_ε maps R^3 into R^3. When $\varepsilon = 0$ we assume the transformation (3) is the identity transformation I given by

$$\bar{x} = \Phi(x, t, 0) = x$$
$$\bar{t} = \Psi(x, t, 0) = t$$
$$\bar{u} = \Omega(x, t, u, 0) = u$$

It is further assumed that for any ε_1 and ε_2 sufficiently small that there exists an ε_3 in $|\varepsilon| < \varepsilon_0$ such that $T_{\varepsilon_2} T_{\varepsilon_1} = T_{\varepsilon_3}$. In other words the local closure property holds (see Section 7.1). Finally, for each ε sufficiently small we assume that there is an ε_1 such that $T_\varepsilon T_{\varepsilon_1} = T_{\varepsilon_1} T_\varepsilon = I$. The transformation T_{ε_1} is the inverse of T_ε and we usually write $T_{\varepsilon_1} = T_\varepsilon^{-1}$. Hence the one parameter family T_ε defined by (3) is assumed to be a local Lie group of transformations.

In (3) we have taken Φ and Ψ to be independent of u. The more general case of dependence on u is not required in the present analysis. Further generalizations are also possible; for example, local Lie groups depending upon several parameters $\varepsilon^1, \ldots, \varepsilon^n$ can be defined in a similar manner. Using Taylor's Theorem we can expand the right sides of (3) in a power series in ε. For example, (3a) becomes

$$\bar{x} = \Phi(x, t, \varepsilon) = \Phi(x, t, 0) + \frac{\partial}{\partial \varepsilon} \Phi(x, t, 0)\varepsilon + o(\varepsilon)$$

$$= x + \varepsilon\xi(x, t) + o(\varepsilon)$$

where

$$\xi(x, t) \equiv \frac{\partial}{\partial \varepsilon} \Phi(x, t, 0) \tag{4}$$

Similar calculations can be made for (3b) and (3c) to obtain the *infinitesimal transformation*

$$\bar{x} = x + \varepsilon\xi(x, t) + o(\varepsilon)$$
$$\bar{t} = t + \varepsilon\tau(x, t) + o(\varepsilon) \tag{5}$$
$$\bar{u} = u + \varepsilon\omega(x, t, u) + o(\varepsilon)$$

where ξ is defined by (4) and τ and ω are given by

$$\tau(x, t) \equiv \frac{\partial}{\partial \varepsilon} \Psi(x, t, 0), \qquad \omega(x, t, u) \equiv \frac{\partial}{\partial \varepsilon} \Omega(x, t, u, 0) \tag{6}$$

The quantities ξ, τ, and ω are called the *generators* of transformation (3). Viewed as (5) transformation (3) for small ε is just the identity transformation plus a small increment of change whose linear or major part is defined by the quantities $\varepsilon\xi$, $\varepsilon\tau$, and $\varepsilon\omega$.

Example 3.2   (Scale Transformations)   The transformation

$$\bar{x} = (1 + \varepsilon)^a x, \qquad \bar{t} = (1 + \varepsilon)^b t, \qquad \bar{u} = (1 + \varepsilon)^c u \tag{7}$$

is called a *scale* or *stretching transformation*. Here a, b, and c are fixed constants. It is clear that (7) reduces to the identity transformation when

$\varepsilon = 0$. The generators are given by

$$\xi = \frac{\partial}{\partial \varepsilon}\left[(1 + \varepsilon)^a x\right]_{\varepsilon = 0} = ax, \qquad \tau = \frac{\partial}{\partial \varepsilon}\left[(1 + \varepsilon)^b t\right]_{\varepsilon = 0} = bt$$

and

$$\omega = \frac{\partial}{\partial \varepsilon}\left[(1 + \varepsilon)^c u\right]_{\varepsilon = 0} = cu$$

The associated infinitesimal transformation is

$$\bar{x} = x + ax\varepsilon + o(\varepsilon), \qquad \bar{t} = t + bt\varepsilon + o(\varepsilon), \qquad \bar{u} = u + cu\varepsilon + o(\varepsilon)$$

A few words about notation are in order. The scale transformation (7) is precisely the same transformation as studied in Section 7.2 where we wrote

$$\bar{x} = \varepsilon^a x, \qquad \bar{t} = \varepsilon^b t, \qquad \bar{u} = \varepsilon^c u \tag{8}$$

Note that (8) can be obtained from (7) be replacing ε by $\varepsilon - 1$. The reason for the change in notation is that the identity transformation in (7) occurs at $\varepsilon = 0$, whereas in (8) the identity transformation occurs at $\varepsilon = 1$. Clearly the identity transformation can be made to occur at any ε_0 by replacing ε by the translate $\varepsilon - \varepsilon_0$ in (7). From a notational point of view, in the general theory we prefer to study transformations close to $\varepsilon = 0$. Generally, if the identity transformation occurs at $\varepsilon = \varepsilon_0$, then the generators are calculated via $\xi = \Phi_\varepsilon(x, t, \varepsilon_0)$, etc., and the infinitesimal transformation has the form $\bar{x} = x + (\varepsilon - \varepsilon_0)\xi + o(\varepsilon - \varepsilon_0)$, and so on.

The local Lie group T_ε defined by (3) determines by definition a unique set of generators ξ, τ, and ω by formulas (4) and (6). Conversely the generators determine the group. The integral curves of the vector field $\langle \xi, \tau, \omega \rangle$ are the paths along which points are moved by T_ε. Thus the group (3) can be determined by solving the initial value problem

$$\frac{d\bar{x}}{\xi(\bar{x}, \bar{t})} = \frac{d\bar{t}}{\tau(\bar{x}, \bar{t})} = \frac{d\bar{u}}{\omega(\bar{x}, \bar{t}, \bar{u})} = d\varepsilon$$

$$\bar{x} = x, \qquad \bar{t} = t, \qquad \bar{u} = u, \qquad \text{at } \varepsilon = 0$$

Example 3.3 Determine the group with generators

$$\xi = 2x + t, \qquad \tau = t, \qquad \omega = u + 1$$

We form the system

$$\frac{d\bar{x}}{2\bar{x} + \bar{t}} = \frac{d\bar{t}}{\bar{t}} = \frac{d\bar{u}}{\bar{u} + 1} = d\varepsilon$$

subject to the initial conditions $\bar{x} = x$, $\bar{t} = t$, $\bar{u} = u$ at $\varepsilon = 0$. Three first integrals are found to be

$$\frac{\bar{x} + \bar{t}}{\bar{t}^2} = c_1, \qquad \frac{\bar{t}}{\bar{u} + 1} = c_2, \qquad \ln(\bar{u} + 1) = \varepsilon + c_3$$

Applying the initial conditions determines c_1, c_2, and c_3 and we obtain the three equations

$$\frac{\bar{x} + \bar{t}}{\bar{t}^2} = \frac{x + t}{t^2}, \qquad \frac{\bar{t}}{\bar{u} + 1} = \frac{t}{u + 1}, \qquad \ln(\bar{u} + 1) = \varepsilon + \ln(u + 1)$$

The third yields

$$\bar{u} = (u + 1)\exp(\varepsilon) - 1$$

and the second therefore gives

$$\bar{t} = t\exp(\varepsilon)$$

Finally, from the first

$$\bar{x} = (x + t)\exp(2\varepsilon) - t\exp(\varepsilon)$$

and the equations for the group are determined.

Example 3.4   (Canonical Coordinates)   In Example 3.3 the first integrals of the vector field $\langle 2x + t, t, u + 1 \rangle$ define a new set of coordinates in which the group takes a particularly simple form. Let

$$s = \frac{x + t}{t^2}, \qquad r = \ln(u + 1), \qquad w = \frac{t}{u + 1}$$

Then it is clear that in terms of these coordinates the group becomes

$$\bar{s} = s, \qquad \bar{r} = r + \varepsilon, \qquad \bar{w} = w$$

which is a simple translation. The coordinates s, r, and w are called the *canonical coordinates* of the group. The existence of such coordinates for a general group will be shown later. They play an important role in the demonstration that a partial differential equation admitting a symmetry can be reduced to an ordinary differential equation.

As was the case for scale transformations, the transformation T_ε defined by (3) maps for each fixed ε a surface having equation

$$u = g(x, t), \qquad (x, t) \in D$$

in *xtu* space to a surface having equation

$$\bar{u} = \bar{g}(\bar{x}, \bar{t}), \quad (\bar{x}, \bar{t}) \in \overline{D}$$

in $\overline{xtu}$ space where

$$\overline{D} = \{(\bar{x}, \bar{t}) | \bar{x} = \Phi(x, t, \varepsilon), \quad \bar{t} = \Psi(x, t, \varepsilon), \quad (x, t) \in D\}$$

To determine $\bar{g}$ we invert the transformation

$$\bar{x} = \Phi(x, t, \varepsilon), \qquad \bar{t} = \Psi(x, t, \varepsilon)$$

on D to obtain

$$x = x(\bar{x}, \bar{t}), \qquad t = t(\bar{x}, \bar{t})$$

Then $\bar{g}$ is defined by

$$\bar{g}(\bar{x}, \bar{t}) \equiv \Omega(x(\bar{x}, \bar{t}), t(\bar{x}, \bar{t}), g(x(\bar{x}, \bar{t}), t(\bar{x}, \bar{t})), \varepsilon) \tag{9}$$

Invariant Partial Differential Equations

To motivate the definition of invariance of a partial differential equation in the more general case we study two examples.

Example 3.5 Consider the nonlinear wave equation

$$u_t + uu_x = 0$$

and the group of transformations

$$\bar{x} = (x + t)e^{2\varepsilon} - te^{\varepsilon}, \qquad \bar{t} = te^{\varepsilon}, \qquad \bar{u} = (u + 1)e^{\varepsilon} - 1$$

We compute the nonlinear wave operator $u_t + uu_x$ in the barred coordinate system. To accomplish this we invert the *xt* part transformation to obtain

$$x = (\bar{t} + \bar{x})e^{-2\varepsilon} - \bar{t}e^{-\varepsilon}, \qquad t = \bar{t}e^{-\varepsilon}$$

Then by the chain rule

$$\bar{u}_{\bar{t}} = \bar{u}_u u_{\bar{t}} = e^{\varepsilon}(u_t t_{\bar{t}} + u_x x_{\bar{t}}) = u_t + (e^{-\varepsilon} - 1)u_x$$

and

$$\bar{u}_{\bar{x}} = \bar{u}_u u_{\bar{x}} = e^{\varepsilon}(u_t t_{\bar{x}} + u_x x_{\bar{x}}) = e^{-\varepsilon}u_x$$

Therefore after simplification

$$\bar{u}_{\bar{t}} + \bar{u}\bar{u}_{\bar{x}} = u_t + uu_x$$

and the operator defining the partial differential equation is unchanged under the action of the given group.

The absolute invariance of the partial differential equation as occurs in Example 3.5 is too much to demand.

Example 3.6 Consider the diffusion operator $u_t - u_{xx}$ and the group of stretchings

$$\bar{x} = (1 + \varepsilon)x, \qquad \bar{t} = (1 + \varepsilon)^2 t, \qquad \bar{u} = u$$

It is easy to see that

$$\bar{u}_{\bar{t}} - \bar{u}_{\bar{x}\bar{x}} = (1 + \varepsilon)^{-2}(u_t - u_{xx})$$

and therefore invariance occurs up to the conformal factor $(1 + \varepsilon)^{-2}$. The right side of the last equation can be expanded about $\varepsilon = 0$ to obtain

$$\bar{u}_{\bar{t}} - \bar{u}_{\bar{x}\bar{x}} = (1 - 2\varepsilon + \cdots)(u_t - u_{xx})$$
$$= (u_t - u_{xx}) - 2\varepsilon(u_t - u_{xx}) + O(\varepsilon^2)$$

Taking the derivative with respect to ε at $\varepsilon = 0$ gives

$$\frac{d}{d\varepsilon}(\bar{u}_{\bar{t}} - \bar{u}_{\bar{x}\bar{x}})_{\varepsilon=0} = -2(u_t - u_{xx})$$

Hence the derivative of the expression defining the differential equation in barred coordinates equals a constant times the expression in unbarred coordinates. This motivates the following definition.

Consider the first order partial differential equation

$$F(x, t, u, p, q) = 0, \qquad (x, t) \in D \qquad (10)$$

where $p \equiv u_x$ and $q = u_t$. The results will extend in an obvious way to higher order equations.

Definition 3.1 *The partial differential equation* (10) *is constant conformally invariant under the local Lie group* (3) *if, and only if,*

$$\frac{\partial}{\partial \varepsilon} F(\bar{x}, \bar{t}, \bar{u}, \bar{p}, \bar{q})|_{\varepsilon=0} = kF(x, t, u, p, q) \qquad (11)$$

for some constant k. If $k = 0$, the partial differential equation is said to be absolutely invariant. Here $\bar{p} \equiv \bar{u}_{\bar{x}}$ and $\bar{q} \equiv \bar{u}_{\bar{t}}$. In (11) *if it occurs that $k = k(x, t, u, p, q)$, then the partial differential equation is conformally invariant.*

By expanding the left side of (11) we can obtain a condition for invariance in terms of the generators of the local Lie group. Carrying out the differentiation and using the chain rule gives

$$F_x\xi + F_t\tau + F_u\omega + F_p\left(\frac{\partial\bar{p}}{\partial\varepsilon}\right)_{\varepsilon=0} + F_q\left(\frac{\partial\bar{q}}{\partial\varepsilon}\right)_{\varepsilon=0} = kF \tag{12}$$

where ξ, τ, and ω are the generators given by (4) and (6). It remains to compute the two expressions

$$\pi = \pi(x, t, u, p, q) \equiv \left(\frac{\partial\bar{p}}{\partial\varepsilon}\right)_{\varepsilon=0}, \qquad \chi = \chi(x, t, u, p, q) \equiv \left(\frac{\partial\bar{q}}{\partial\varepsilon}\right)_{\varepsilon=0} \tag{13}$$

which entails knowledge of how the derivatives u_x and u_t transform under the group. We record the result in the next theorem.

Theorem 3.1   *If $u = g(x, t)$, then under the transformation T_ε given by (3) the derivatives of $\bar{u} = \bar{g}(\bar{x}, \bar{t})$ are*

$$\bar{p} \equiv \bar{g}_{\bar{x}}(\bar{x}, \bar{t}) = \frac{\bar{t}_t D_x\Omega - \bar{t}_x D_t\Omega}{\det J} \tag{14}$$

$$\bar{q} \equiv \bar{g}_{\bar{t}}(\bar{x}, \bar{t}) = \frac{-\bar{x}_t D_x\Omega + \bar{x}_x D_t\Omega}{\det J} \tag{15}$$

where

$$D_x\Omega \equiv \Omega_x + g_x\Omega_u, \qquad D_t\Omega \equiv \Omega_t + g_t\Omega_u$$

and J is the Jacobi matrix of the transformation $\bar{x} = \Phi(x, t, \varepsilon)$, $\bar{t} = \Psi(x, t, \varepsilon)$ given by

$$J \equiv \begin{pmatrix} \bar{x}_x & \bar{x}_t \\ \bar{t}_x & \bar{t}_t \end{pmatrix}$$

The right sides of (14) and (15) are evaluated at $x = x(\bar{x}, \bar{t})$ and $t = t(\bar{x}, \bar{t})$.

Proof   Differentiating (9) with respect to $\bar{x}$ gives

$$\bar{g}_x = \Omega_x x_{\bar{x}} + \Omega_t t_{\bar{x}} + \Omega_u(g_x x_{\bar{x}} + g_t t_{\bar{x}})$$
$$= x_{\bar{x}} D_x\Omega + t_{\bar{x}} D_t\Omega$$

Similarly $\bar{g}_{\bar{t}} = x_{\bar{t}} D_x\Omega + t_{\bar{t}} D_t\Omega$ and we may write in matrix form

$$(\bar{g}_{\bar{x}} \quad \bar{g}_{\bar{t}}) = (D_x\Omega \quad D_t\Omega)\begin{pmatrix} x_{\bar{x}} & x_{\bar{t}} \\ t_{\bar{x}} & t_{\bar{t}} \end{pmatrix}$$

The matrix on the right is the Jacobi matrix of the inverse transformation $x = x(\bar{x}, \bar{t})$, $t = t(\bar{x}, \bar{t})$ and hence is the inverse of J. Therefore

$$(\bar{g}_{\bar{x}} \quad \bar{g}_{\bar{t}}) = (D_x\Omega \quad D_t\Omega)J^{-1}$$

$$= (D_x\Omega \quad D_t\Omega)\frac{1}{\det J}\begin{pmatrix} \bar{t}_t & -\bar{x}_t \\ -\bar{t}_x & \bar{x}_x \end{pmatrix}$$

Equating components gives formulas (14) and (15) and completes the proof.

Theorem 3.2 *The quantities π and χ defined in* (13) *are given by*

$$\pi = \omega_x + p\omega_u - p\xi_x - q\tau_x \tag{16a}$$

$$\chi = \omega_t + q\omega_u - p\xi_t - q\tau_t \tag{16b}$$

where ξ, τ, and ω are the generators of (3).

Proof We derive the expression for π and leave the calculation of χ for the reader. We could attempt to obtain π directly by calculating $d/d\varepsilon$ at $\varepsilon = 0$ of the expression in (14). It is simpler, however, to expand the right side of (14) in a Taylor series about $\varepsilon = 0$ and read off π as the coefficient of the linear term in ε in the expansion. To this end we note that

$$\bar{t}_t = \frac{\partial}{\partial t}(t + \varepsilon\tau + \cdots) = 1 + \varepsilon\tau_t + o(\varepsilon)$$

$$\bar{t}_x = \frac{\partial}{\partial x}(t + \varepsilon\tau + \cdots) = \varepsilon\tau_x + o(\varepsilon)$$

$$\bar{x}_x = \frac{\partial}{\partial x}(x + \varepsilon\xi + \cdots) = 1 + \varepsilon\xi_x + o(\varepsilon)$$

$$\bar{x}_t = \frac{\partial}{\partial t}(x + \varepsilon\xi + \cdots) = \varepsilon\xi_x + o(\varepsilon)$$

Hence

$$\det J = \bar{x}_x\bar{t}_t - \bar{x}_t\bar{t}_x = 1 + \varepsilon(\xi_x + \tau_t) + o(\varepsilon)$$

and

$$\frac{1}{\det J} = 1 - \varepsilon(\xi_x + \tau_t) + o(\varepsilon)$$

Finally

$$D_x\Omega = \Omega_x + g_x\Omega_u$$

$$= \frac{\partial}{\partial x}(u + \varepsilon\omega + o(\varepsilon)) + g_x\frac{\partial}{\partial u}(u + \varepsilon\omega + o(\varepsilon))$$

$$= g_x + (\omega_x + g_x\omega_u)\varepsilon + o(\varepsilon)$$

Similarly

$$D_t\Omega = g_t + (\omega_t + g_t\omega_u)\varepsilon + o(\varepsilon)$$

Substituting all these expressions into (14) and then collecting terms gives

$$\bar{p} = \bar{g}_{\bar{x}} = g_x + (\omega_x + p\omega_u - p\xi_x - q\tau_x)\varepsilon + o(\varepsilon)$$

Thus $\pi = (\partial\bar{p}/\partial\varepsilon)_{\varepsilon=0}$ is given by (16a), completing the proof.

The local Lie group T_ε on xtu space defined by (3) dictates how functions and their derivatives are transformed (formulas (9), (14), and (15)). Thus T_ε can be extended in a natural way to a one parameter family of transformations $\tilde{T}_\varepsilon$ on $xtupq$ space given by

$$\tilde{T}_\varepsilon: \quad \begin{aligned} \bar{x} &= \Phi(x, t, \varepsilon) \\ \bar{t} &= \Psi(x, t, \varepsilon) \\ \bar{u} &= \Omega(x, t, u, \varepsilon) \\ \bar{p} &= \frac{\Psi_t(\Omega_x + p\Omega_u) - \Psi_x(\Omega_t + q\Omega_u)}{\det J} \\ \bar{q} &= \frac{-\Phi_t(\Omega_x + p\Omega_u) + \Phi_x(\Omega_t + q\Omega_u)}{\det J} \end{aligned}$$

where $\det J = \Phi_x\Psi_t - \Phi_t\Psi_x$. It can be shown that $\tilde{T}_\varepsilon$ is a local Lie group on R^5 and it is called the *extended group*. The quantities ξ, τ, ω, π, and χ are the *generators* of the extended group and represent the principal linear parts of the transformation in infinitesimal form.

$$\begin{aligned} \bar{x} &= x + \varepsilon\xi(x, t) + o(\varepsilon) \\ \bar{t} &= t + \varepsilon\tau(x, t) + o(\varepsilon) \\ \bar{u} &= u + \varepsilon\omega(x, t, u) + o(\varepsilon) \\ \bar{p} &= p + \varepsilon\pi(x, t, u, p, q) + o(\varepsilon) \\ \bar{q} &= q + \varepsilon\chi(x, t, u, p, q) + o(\varepsilon) \end{aligned}$$

The invariance condition (12) can now be expressed as

$$F_x\xi + F_t\tau + F_u\omega + F_p\pi + F_q\chi = kF \tag{17}$$

where π and χ are given in terms of ξ, τ, and ω by (16). The left side of (17) is denoted by $\tilde{U}F$ where $\tilde{U}$ is the *Lie derivative operator* defined by

$$\tilde{U} \equiv \xi\frac{\partial}{\partial x} + \tau\frac{\partial}{\partial t} + \omega\frac{\partial}{\partial u} + \pi\frac{\partial}{\partial p} + \chi\frac{\partial}{\partial q}$$

Thus the condition for invariance of a partial differential equation $F = 0$ is

$$\tilde{U}F = kF \tag{18}$$

In the present context $\tilde{U}F$ is the gradient of $F(x, t, u, p, q)$ in the direction of the extended vector field $\langle \xi, \tau, \omega, \pi, \chi \rangle$. That is

$$\tilde{U}F = \mathbf{grad}\ F \cdot \langle \xi, \tau, \omega, \pi, \chi \rangle$$

The Determination of Symmetries

In the case of stretching transformations it was a simple matter to determine conditions on the transformation that forced invariance on a given partial differential equation. To determine more general symmetries, or general local Lie groups, under which a given partial differential equation is invariant it is convenient to use condition (18) to calculate the group generators ξ, τ, and ω.

Example 3.7   Consider the nonlinear equation $u_t + uu_x = 0$ or

$$q + up = 0 \tag{19}$$

Here $F(x, t, u, p, q) \equiv q + up$ and $F_x = F_t = 0$, $F_u = p$, and $F_q = 1$. The invariance identity (18), or equivalently (17), becomes

$$p\omega + u(\omega_x + \omega_u p - \xi_x p - \tau_x q) + (\omega_t + \omega_u q - \xi_t p - \tau_t q) = k(q + up)$$

where we have substituted (16). This equation may be rewritten as

$$(\omega_t + u\omega_x) + p(\omega + u\omega_u - u\xi_x - \xi_t - ku)$$
$$+ q(\omega_u - u\tau_x - \tau_t - k) = 0$$

We can satisfy this equation by forcing

$$\omega_t + u\omega_x = 0 \tag{20a}$$

$$\omega + u\omega_u - u\xi_x - \xi_t - ku = 0 \tag{20b}$$

$$\omega_u - u\tau_x - \tau_t - k = 0 \tag{20c}$$

which is a system of three coupled first order linear partial differential equations for the generators ξ, τ, and ω. These equations are easily solved. From (20c) we get

$$\omega = a(x, t)u^2 + b(x, t)u + c(x, t)$$

for some functions a, b, and c. Substitution into (20a) forces $a_x = 0$, $c_t = 0$, $a_t = -b_x$, and $b_t = -c_x$. Thus

$$\omega = a(t)u^2 + b(x, t)u + c(x)$$

Substitution into (20c) gives $\tau_x = 2a(t)$ whence

$$\tau = 2a(t)x + g(t)$$

for some function g. Putting ω into (20b) forces $a(t) = 0$, $2b(x, t) = \xi_x + k$, and $\xi_t = c(x)$. Therefore

$$\xi = c(x)t + h(x)$$

for some function h. Also $b_x = 0$, which gives $b = b(t)$, and we have $b'(t) = -c'(x)$, which implies $b(t) = d_0 t + b_0$ and $c(x) = -d_0 x + c_0$ for constants b_0, c_0, and d_0. But $2b(t) = \xi_x + k$ yields $d_0 = 0$ and $h(x) = (2b_0 - k)x + h_0$ for some constant h_0. Summarizing thus far

$$\xi = c_0 t + (2b_0 - k)x + h_0, \qquad \tau = g(t), \qquad \omega = b_0 u + c_0$$

Substituting once again these quantities into (20c) gives $g(t) = (b_0 - k)t + g_0$ for some constant g_0. Therefore the generators of a local Lie group of transformations under which (19) is invariant are given by

$$
\begin{aligned}
\xi &= (2b_0 - k)x + c_0 t + h_0 \\
\tau &= (b_0 - k)t + g_0 \qquad (21) \\
\omega &= b_0 u + c_0
\end{aligned}
$$

for arbitrary constants b_0, c_0, g_0, h_0, and k.

The preceding notions can be generalized to systems of equations in several unknowns. We catalog the result for reference. Let

$$F^{(m)}(x, u, p) = 0, \qquad m = 1, \ldots, n$$

be a system of n partial differential equations, where $x = (x^1, x^2)$, $u = (u^1, \ldots, u^n)$, and $p = (p_i^j) = (\partial u^j / \partial x^i)$. Consider the local Lie group T_ε (written in infinitesimal form)

$$
\begin{aligned}
\bar{x}^i &= x^i + \varepsilon X^i(x) + o(\varepsilon) \\
\bar{u}^j &= u^j + \varepsilon U^j(x, u) + o(\varepsilon)
\end{aligned}
$$

where $i = 1, 2$ and $j = 1, \ldots, n$. The generators P_i^j of the derivative transformations in the extended group $\tilde{T}_\varepsilon$, where

$$\bar{p}_i^j = p_i^j + \varepsilon P_i^j(x, u, p) + o(\varepsilon), \qquad i = 1, 2; \ j = 1, \ldots, n$$

are given in terms of the generators X^i and U^j by

$$P_i^j = \frac{\partial U^j}{\partial x^i} + \sum_{k=1}^{n}\left(\frac{\partial U^j}{\partial u^k}p_i^k\right) - \sum_{l=1}^{2}\frac{\partial X^l}{\partial x^i}p_l^j$$

If the generators X^i depend on u as well, then there is an additional term

$$\sum_{k=1}^{n} \sum_{l=1}^{2} \frac{\partial X^l}{\partial u^k} p_i^l p_i^k$$

on the right side of the last equation for P_i^j. The invariance condition (18) takes the form n equations

$$\tilde{U}F^{(m)} \equiv \sum_{i=1}^{2} F_{x^i}^{(m)} X^i + \sum_{j=1}^{n} F_{u^j}^{(m)} U^j + \sum_{i=1}^{2} \sum_{j=1}^{n} F_{p_i^j}^{(m)} P_i^j = \sum_{r=1}^{n} k_{mr} F^{(r)}$$

for $m = 1, \ldots, n$, where the k_{mr} are n^2 constants. Details of these calculations can be found in Bluman and Cole [5]. A practical application of these procedures for a system of equations in reacting fluids can be found in J. D. Logan and J. J. Pérez (*SIAM J. Appl. Math* **39**(3), 1980, pp. 512–527). An in-depth treatment of the invariance properties of differential equations is given in Olver [6].

Self-Similar Solutions

In summary, we have obtained a condition that is necessary and sufficient for the invariance of the partial differential equation

$$F(x, t, u, p, q) = 0 \tag{22}$$

If the partial differential equation (22) is invariant under (3), what can be said about its solutions? It is clear that if $u = g(x, t)$ is a solution of (22), then $\bar{u} = g(\bar{x}, \bar{t})$ is a solution of

$$F(\bar{x}, \bar{t}, \bar{u}, \bar{p}, \bar{q}) = 0$$

It can be shown that $\bar{u} = \bar{g}(\bar{x}, \bar{t})$ is a solution as well, but in general $g(\bar{x}, \bar{t}) \neq \bar{g}(\bar{x}, \bar{t})$. The following example illuminates these points.

Example 3.8 Consider the nonlinear wave equation $u_t + uu_x = 0$, where $F = q + up$, and the group in Example 3.5 under which the equation is absolutely invariant. Let $g(x, t) = x/t$. One can readily check that g is a solution. Now from (9) one finds that $\bar{g}(\bar{x}, \bar{t}) = \bar{x}/\bar{t}$, and so $g(\bar{x}, \bar{t}) = \bar{g}(\bar{x}, \bar{t})$. On the other hand, let $\phi(x, t) = x(t + 1)^{-1}$. The function ϕ is also a solution and from (9) and the inverse transformation given in Example 3.5

$$\bar{u} = (u + 1)e^{\varepsilon} - 1$$
$$= \left(\frac{x}{t + 1} + 1 \right) e^{\varepsilon} - 1$$
$$= \frac{e^{-\varepsilon}\bar{x} + e^{\varepsilon} - 1}{\bar{t}e^{-\varepsilon} + 1} \equiv \bar{\phi}(\bar{x}, \bar{t})$$

Hence $\phi(\bar{x}, \bar{t}) \neq \bar{\phi}(\bar{x}, \bar{t})$. However, $\bar{\phi}$ is a solution of $\bar{q} + \bar{u}\bar{p} = 0$. Therefore g is an invariant solution and ϕ is not.

Similarity solutions are found among the invariant surfaces, that is, solutions $u = g(x, t)$ for which

$$g(\bar{x}, \bar{t}) = \bar{g}(\bar{x}, \bar{t})$$

After applying (9) we take the derivative with respect to ε at $\varepsilon = 0$ to get

$$g_x(x, t)\Phi_\varepsilon(x, t, 0) + g_t(x, t)\Psi(x, t, 0) = \Omega_\varepsilon(x, t, g, 0)$$

or

$$\xi(x, t)g_x + \tau(x, t)g_t = \omega(x, t, g) \tag{23}$$

Condition (23) is called the *invariant surface condition*. It is a first order partial differential equation in g and it can be solved by the method of characteristics. The characteristic system is

$$\frac{dx}{\xi(x, t)} = \frac{dt}{\tau(x, t)} = \frac{dg}{\omega(x, t, g)} \tag{24}$$

The first pair of equations

$$\frac{dx}{\xi(x, t)} = \frac{dt}{\tau(x, t)}$$

can be integrated to obtain a first integral

$$s \equiv s(x, t) = \text{constant} \tag{25}$$

This is the similarity variable, a fact we shall justify later. Letting $x = X(s, t)$ from (25) the pair

$$\frac{dt}{\tau(X(s, t), t)} = \frac{dg}{\omega(X(x, t), t, g)}$$

can be integrated to obtain another first integral

$$w(s, t, g) = \text{constant} \tag{26}$$

Consequently the general solution of (23) is

$$F(s, w(s, t, g)) = 0 \tag{27}$$

from which we can obtain g and thus candidates for self-similar solutions. The procedure discussed here follows the same pattern of reasoning developed in Section 7.2 for stretching transformations.

Example 3.9   The nonlinear wave equation

$$u_t + uu_x = 0 \tag{28}$$

is absolutely invariant under the local Lie group of transformations with generators

$$\xi = 2x + 2, \qquad \tau = t + 1, \qquad \omega = u$$

The invariant surface condition (23) becomes

$$(2x + 2)u_x + (t + 1)u_t = u \tag{29}$$

where we are using u in lieu of g. The characteristic system is

$$\frac{dx}{2(x + 1)} = \frac{dt}{t + 1} = \frac{du}{u}$$

Integrating the first pair of equations gives first integral

$$s \equiv \frac{(t + 1)^2}{x + 1} = \text{constant} \tag{30}$$

From the second pair

$$\frac{u}{t + 1} = \text{constant}$$

The general solution of (29) is therefore

$$F\left(s, \frac{u}{t + 1}\right) = 0, \qquad F \text{ arbitrary}$$

or

$$u = (t + 1)f(s), \qquad f \text{ arbitrary} \tag{31}$$

Equation (31) where the similarity variable is given by (30) defines the invariant surfaces and candidates for self-similar solutions. When (31) is substituted into the partial differential equation (28) the result is an ordinary differential equation for $f(s)$

$$2sf' + f - s^2ff' = 0$$

This equation can be solved analytically by making the substitution $g = sf$. Then

$$\frac{2 - g}{g(1 - g)} dg = \frac{ds}{s}$$

Integration gives

$$\frac{g^2}{g - 1} = cs, \qquad c \text{ constant}$$

Thus

$$f(s) = \frac{c}{2} \pm \sqrt{\frac{c^2}{4} - \frac{c}{s}}$$

whence

$$u(x, t) = (t + 1)\left[\frac{c}{2} \pm \sqrt{\frac{c^2}{4} - \frac{c(x + 1)}{(t + 1)^2}} \right]$$

These are self-similar motions arising from the invariance of (28) under the given group of transformations.

Without details of the proof we indicate the basis for the reduction of a partial differential equation to an ordinary differential equation. The group T_ε under which the partial differential equation

$$F(x, t, u, p, q) = 0 \tag{32}$$

is invariant defines a preferred coordinate system (canonical coordinates) srw in which the group is a translation $\bar{s} = s$, $\bar{r} = r + \varepsilon$, $\bar{w} = w$. In canonical coordinates the partial differential equation becomes

$$H(s, r, w, w_s, w_r) = 0 \tag{33}$$

and it is invariant under a translation. Thus the independent variable r cannot appear explicitly and (33) will have solutions of the form $w = w(s)$.

Example 3.10 Consider the nonlinear wave equation $u_t + uu_x = 0$ and group of Example 3.9

$$\bar{x} = e^{2\varepsilon}(x + 1) - 1, \qquad \bar{t} = e^\varepsilon(t + 1) - 1, \qquad \bar{u} = e^\varepsilon u$$

Let

$$s = \frac{(t + 1)^2}{x + 1}, \qquad r = \ln(t + 1), \qquad w = \frac{u}{t + 1}$$

It is easily checked that the group is a translation in the *srw* coordinate system. Now $u(x, t) = (t + 1)w(r, s)$ and by the chain rule

$$u_t = (t + 1)[w_s s_t + w_r r_t] + w$$

$$= (t + 1)\left[w_s \frac{2(t + 1)}{x + 1} + w_r \frac{1}{t + 1}\right] + w$$

$$= 2sw_s + w_r + w$$

and

$$u_x = (t + 1)[w_s s_x + w_r r_x] = -\frac{(t + 1)^3}{(x + 1)^2} w_s$$

Hence the partial differential equation $u_t + uu_x = 0$ becomes

$$2sw_s + w_r + w - s^2 ww_s = 0 \tag{34}$$

in the canonical coordinate system. In this equation r does not appear explicitly and so there will be solutions of the form $w = f(s)$ for some function f. Substitution of $w = f(s)$ into (34) yields an ordinary differential equation for f

$$2sf'(s) + f(s) - s^2 f(s)f'(s) = 0$$

This is the same equation found in Example 3.9.

The fact that canonical coordinates can always be chosen is stated in the following theorem.

Theorem 3.3   *Consider the local Lie group (3) with generators ξ, τ, and ω defined by (4) and (6). Then there exists a change of variables*

$$s = s(x, t), \qquad r = r(x, t), \qquad w = w(x, t, u)$$

in which the group T_ε is represented as a translation

$$\bar{s} = s, \qquad \bar{r} = r + \varepsilon, \qquad \bar{w} = w$$

Proof   The vector field $\langle \xi, \tau, \omega \rangle$ defines integral curves along which the group T_ε moves points. The group therefore translates points along these curves that are given as solutions of

$$\frac{d\bar{x}}{\xi(\bar{x}, \bar{t})} = \frac{d\bar{t}}{\tau(\bar{x}, \bar{t})} = \frac{d\bar{u}}{\omega(\bar{x}, \bar{t}, \bar{u})} = d\varepsilon \tag{35}$$

Integrating the first pair of equations yields

$$\bar{s} \equiv s(\bar{x}, \bar{t}) = c_1 \tag{36}$$

where c_1 is a constant. Clearly $\bar{s} = s(\bar{x}, \bar{t}) = s(x, t) = s$. Now consider the pair

$$\frac{d\bar{t}}{\tau(\bar{x}(\bar{t}, c_1), \bar{t})} = d\varepsilon$$

where $\bar{x}(\bar{t}, c_1)$ has been obtained from (36). Integrating gives

$$R(\bar{t}, c_1) = \varepsilon + c_2$$

or

$$R(\bar{t}, s(\bar{x}, \bar{t})) = \varepsilon + c_2$$

Choosing $\bar{r} = R(\bar{t}, s(\bar{x}, \bar{t}))$ we have $\bar{r} = r + \varepsilon$. Finally we integrate

$$\frac{d\bar{t}}{\tau(\bar{x}, \bar{t})} = \frac{d\bar{u}}{\omega(\bar{x}, \bar{t}, \bar{u})}$$

after substituting $\bar{x} = \bar{x}(\bar{t}, c_1)$ to get

$$\bar{w} = W(\bar{x}, \bar{t}, \bar{u}) = c_3$$

Hence $\bar{w} = w$, and the canonical coordinates are given by

$$s = s(x, t), \qquad r = R(t, s(x, t)), \qquad w = W(x, t, u)$$

where s, R, and W are first integrals of $\langle \xi, \tau, \omega \rangle$.

Note that (35) coincides with the characteristic system associated with the invariant surface condition (23). Therefore once a group of symmetries for a given partial differential equation is known, self-similar solutions can be determined from the invariant surface condition as in Example 3.9 or from the canonical coordinates as in Example 3.10.

Example 3.11 Find canonical coordinates for the group

$$\bar{x} = e^{\varepsilon} x, \qquad \bar{t} = e^{2\varepsilon} t, \qquad \bar{u} = u$$

with generators $\xi = x$, $\tau = 2t$, $\omega = 0$. We form

$$\frac{d\bar{x}}{\bar{x}} = \frac{d\bar{t}}{2\bar{t}} = \frac{d\bar{u}}{0} = d\varepsilon$$

Integrating $d\bar{x}/\bar{x} = d\bar{t}/2\bar{t}$ gives

$$\bar{s} \equiv \frac{\bar{x}}{\sqrt{\bar{t}}} = c_1$$

The equation $d\bar{x}/\bar{x} = d\varepsilon$ gives $\ln \bar{x} = \varepsilon + c_2$, where c_2 is a constant. Thus $\bar{r} = \ln \bar{x}$. Finally $\bar{u} = $ constant, so take $\bar{w} = \bar{u}$. Therefore the canonical coordinate system is defined by

$$s = \frac{x}{\sqrt{t}}, \qquad r = \ln x, \qquad w = u$$

To check that the given transformation is a translation in sru coordinates we observe that $\bar{s} = \bar{x}/\sqrt{\bar{t}} = e^\varepsilon x/\sqrt{e^{2\varepsilon}t} = x/\sqrt{t} = s$ and $\bar{r} = \ln \bar{x} = \ln(e^\varepsilon x) = \varepsilon + \ln x = \varepsilon + r$ and $\bar{w} = \bar{u} = u = w$.

The preceding discussion centered around first order partial differential equations and their invariance properties. The extension to second order equations is straightforward with the only complication a mild increase in the difficulty of the calculations. The reader is referred to Bluman and Cole [5] for the details, particularly for the calculation of the generators of a twice extended group.

When initial and boundary conditions are present they must be taken into account as well. Such auxiliary conditions and the curves on which they are defined will be transformed by the group into barred coordinates and must also satisfy invariance requirements. The technique is illustrated in Example 2.8. We take up one additional example that is not atypical of problems encountered in real physical settings.

Example 3.12 A metallic rod of constant cross section initially at rest and occupying the region $x \geq 0$ undergoes longitudinal vibrations. By Exercise 2.2 of Chapter 5 the momentum law in Lagrangian coordinates t and h is

$$V_t - \frac{1}{\rho_0}\Sigma_h = 0$$

where ρ_0 is the initial density, $V(h, t)$ is the velocity of a cross section, and $\Sigma(h, t)$ is the stress. By definition the strain is given by (see Equation (54) of Section 5.2) $E(h, t) = (\partial x/\partial h) - 1$ and so $E_t = V_h$. Now assume a superelastic stress–strain relation $\Sigma = y_0 E^2$, where y_0 is the constant stiffness. Therefore the governing equations can be written

$$V_t - \frac{2y_0}{\rho_0}EE_h = 0, \qquad V_h - E_t = 0 \tag{37}$$

in terms of the velocity and strain. We consider a problem where the back boundary at $x = 0$ is moved at velocity $-t$ beginning at $t = 0$ (this can be accomplished by releasing at $t = 0$ a unit mass permanently attached to the end $x = 0$ of a vertical bar with $x > 0$ upward). The boundary condition is

$$V(0, t) = -t, \qquad t > 0 \tag{38}$$

The motion of the back boundary sends a signal forward into the bar at some nonconstant velocity U. Across this shock front V and E will suffer simple jump discontinuities. The jump conditions can be found by writing the governing partial differential equations (37) in conservation form as

$$V_t + \left(-\frac{y_0}{\rho_0}E^2\right)_h = 0, \qquad E_t + (-V)_h = 0$$

They are (see Section 5.1)

$$[V]U = -\frac{y_0}{\rho_0}[E^2], \qquad [E]U = -[V]$$

Since V and E are zero ahead of the shock we have

$$V(h^-, t)U = -\frac{y_0}{\rho_0}E^2(h^-, t), \qquad E(h^-, t)U = -V(h^-, t)$$

Therefore E and V satisfy the partial differential equations (37) subject to the rear boundary condition (38) and conditions at the shock front that we prefer to write as

$$V^2(h^-, t) = \frac{y_0}{\rho_0}E^3(h^-, t)$$

$$U^2 = \frac{y_0}{\rho_0}E(h^-, t) \tag{39}$$

A space–time diagram is pictured in Fig. 7.11. To determine self-similar solutions we attempt to find a stretching transformation

$$\bar{h} = e^\varepsilon h, \qquad \bar{t} = e^{b\varepsilon}t, \qquad \bar{E} = e^{c\varepsilon}E, \qquad \bar{V} = e^{a\varepsilon}V$$

under which the problem is invariant. Clearly

$$\bar{V}_{\bar{t}} - \frac{2y_0}{\rho_0}\bar{E}\bar{E}_{\bar{h}} = e^{(a-b)\varepsilon}V_t - \frac{2y_0}{\rho_0}e^{(2c-1)\varepsilon}EE_h$$

$$\bar{V}_{\bar{h}} - \bar{E}_{\bar{t}} = e^{(a-1)\varepsilon}V_h - e^{(c-b)\varepsilon}E_t$$

and therefore the partial differential equations (37) are constant conformally

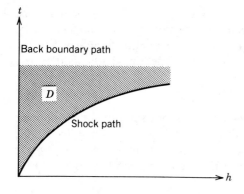

t

Back boundary path

D

Shock path

Figure 7.11. Back boundary and shock in Lagrangian coordinates. D is the domain where the equations are to be solved.

h

invariant provided $a - b = 2c - 1$ and $a - 1 = c - b$ or

$$a = 3(1 - b), \qquad c = 2(1 - b)$$

The group generators are $\xi = h$, $\tau = bt$, $(\partial \bar{E}/\partial \varepsilon)_0 = 2(1 - b)E$ and $(\partial \bar{V}/\partial \varepsilon)_0 = 3(1 - b)V$. The two invariant surface conditions are

$$hE_h + btE_t = 2(1 - b)E$$
$$hV_h + btV_t = 3(1 - b)V$$

with characteristic systems

$$\frac{dh}{h} = \frac{dt}{bt} = \frac{dE}{2(1 - b)E}$$

and

$$\frac{dh}{h} = \frac{dt}{bt} = \frac{dV}{3(1 - b)V}$$

The similarity variable s is found from the integrating $dh/h = dt/bt$ or

$$s \equiv h/t^{1/b} \tag{40}$$

Then it follows that

$$E(h, t) = t^{2(1 - b)/b} f\left(\frac{h}{t^{1/b}} \right)$$

$$V(h, t) = t^{3(1 - b)/b} g\left(\frac{h}{t^{1/b}} \right)$$

for some functions f and g. The rear boundary condition (38) forces $V(0, t) =$

$t^{3(1-b)/b}g(0) = -t$ or $b = \frac{3}{4}$ and $g(0) = -1$. Then

$$E(h,t) = t^{2/3}f\left(\frac{h}{t^{4/3}}\right)$$

$$V(h,t) = tg\left(\frac{h}{t^{4/3}}\right)$$
(41)

The shock path should also be invariant under the given transformation; hence it must be given by $s = h/t^{1/b} = \lambda$ for some constant λ. Thus

$$h = \lambda t^{4/3}$$
(42)

is the shock path and

$$U = \frac{dh}{dt} = \frac{4}{3}\lambda t^{1/3}$$

is the shock velocity. The jump conditions (39) then become

$$g^2(\lambda) = \frac{y_0}{\rho_0}f^3(\lambda)$$
(43)

and

$$\lambda^2 = \frac{9}{16}\left(\frac{y_0}{\rho_0}\right)f(\lambda)$$
(44)

Finally, substituting (41) into (37) we get

$$\frac{2y_0}{\rho_0}ff\dot{} = g - \frac{4}{3}s\dot{g}$$

$$\dot{g} + \frac{4}{3}s\dot{f} = \frac{2}{3}f$$
(45)

where the overdot denotes d/ds, and the reduction is complete. The original boundary value problem has been cast into the pair of ordinary differential equations (45) subject to the conditions (43) and (44) and the condition $g(0) = -1$. The constant λ must be determined as a part of the solution. We can proceed by guessing a value of λ_0 and then use (43) and (44) to compute $f(\lambda_0)$ and $g(\lambda_0)$. The differential equations (45) can then be integrated numerically with initial data $f(\lambda_0)$ and $g(\lambda_0)$ backwards from λ_0 to 0 to determine $\tilde{g}(0)$. Then $\tilde{g}(0)$ can be scaled to -1 to determine λ from

$$\lambda = \frac{3}{4}\sqrt{\frac{y_0}{\rho_0}}f(\lambda)^{1/2} = \frac{3}{4}\left(\frac{y_0}{\rho_0}\right)^{1/3}g^{1/3}(\lambda)$$

The numerical confirmation is requested in the Exercises.

EXERCISES

3.1 Find a system of canonical coordinates for the group

$$\bar{x} = x \cos \varepsilon - t \sin \varepsilon, \qquad \bar{t} = t \cos \varepsilon + x \sin \varepsilon, \qquad \bar{u} = u + \varepsilon$$

3.2 If a partial differential equation in t, x, and u is invariant under a group of the form

$$\bar{t} = \phi(t, x, \varepsilon), \qquad \bar{x} = \psi(t, x, \varepsilon)$$
$$\bar{u} = u + g(x, t)u\varepsilon$$

Show that there will be similarity solutions of the form $u(x, t) = F(x, t)f(s)$, where s is a known function of x and t, f is arbitrary, and F is known.

3.3 Consider the equation $u_{tt} - u_{xx} = 0$.

(a) Determine a group of stretchings $\bar{x} = e^{a\varepsilon}x$, $\bar{t} = e^{b\varepsilon}t$, $\bar{u} = u$ under which the equation is constant conformally invariant.

(b) Show that canonical coordinates are given by

$$s = \frac{x}{t}, \qquad r = \ln t, \qquad w = u$$

(c) In canonical coordinates show that the partial differential equation becomes

$$(1 - s^2)w_{ss} - w_{rr} + 2sw_{sr} + w_r - 2sw_r = 0$$

(d) Find similarity solutions by assuming $w = f(s)$.

3.4 Consider the diffusion equation $u_t - u_{xx} = 0$.

(a) Determine a group of stretchings $\bar{x} = e^{a\varepsilon}x$, $\bar{t} = e^{b\varepsilon}t$, $\bar{u} = u$ under which the equation is constant conformally invariant.

(b) Show that canonical coordinates are given by

$$s = \frac{x}{\sqrt{t}}, \qquad r = \ln x, \qquad w = u$$

(c) In canonical coordinates show that the partial differential equation becomes

$$w_{rr} - w_r + s^2 w_{ss} + 2sw_{rs} + 2sw_s = 0$$

(d) Derive the class of similarity solutions

$$u(x, t) = c_1 \text{erf} \frac{x}{\sqrt{t}} + c_2$$

(e) Show that application of the invariant surface condition leads to the same class of solutions.

3.5 The equation $u_y u_{xy} - u_x u_{yy} - u_{yyy} = 0$ occurs in boundary layer theory for flow along a flat plate.

(a) Verify that the partial differential equation is constant conformally invariant under the group $\bar{x} = (1 - \varepsilon)^2 x$, $\bar{y} = (1 + \varepsilon)y$, $\bar{u} = u$.

(b) Find canonical coordinates.
   Answer:

$$r = \frac{y}{\sqrt{x}}, \qquad s = \frac{1}{2} \ln x, \qquad w = \frac{u}{\sqrt{x}}$$

(c) Show that in canonical coordinates the partial differential equation becomes

$$w_{rrr} + \tfrac{1}{2}ww_{rr} - \tfrac{1}{2}w_r w_{rs} + \tfrac{1}{2}w_{rr}w_s = 0$$

(d) Obtain an ordinary differential equation for solutions of the form $w = f(r)$.

3.6 Perform the numerical integration suggested in Example 3.12. Take $\lambda_0 = 3$ and $y_0/\rho_0 = 1$. Show that $\tilde{g}(0) = -134.5$ and $\lambda = 0.5855$. Sketch velocity and strain profiles and find the equation of the shock path.

3.7 Referring to Example 3.12 show that the differential equations (45) with $y_0/\rho_0 = 1$ can be reduced to

$$\frac{dy}{dx} = \frac{12y + 4x^2 - 18xy}{3y + 8x - 12x^2}$$

when Lie variables $y = g/s^3$, $x = f/s^2$ are introduced. Examine this equation in the Lie plane and show that the shock is represented by the point Q: $(\frac{16}{9}, -\frac{64}{27})$. Show that the solution behind the shock is represented by a curve from Q to infinity.

3.8 In Example 2.2 of Section 7.2 we noted that the nonlinear system of partial differential equations

$$u_t - v_x = 0, \qquad v_t + uu_x = 0$$

is invariant under a stretching transformation. Use the procedure of Example 3.7 to find the generators of a general group $\bar{x} = \Phi(x, t, \varepsilon)$, $\bar{t} = \Psi(x, t, \varepsilon)$, $\bar{u} = \Omega(x, t, u, v, \varepsilon)$, $\bar{v} = \Upsilon(x, t, u, v, \varepsilon)$ under which the system is invariant and reduce it to a system of ordinary differential equations.

REFERENCES

1. J. D. Logan, *Invariant Variational Principles*, Academic Press, Inc., New York (1977).
2. D. Lovelock and H. Rund, *Tensors, Differential Forms, and Variational Principles*, Wiley-Interscience, New York (1975).
3. L. Dresner, *Similarity Solutions of Nonlinear Partial Differential Equations*, Pitman Publishing, Inc., Marshfield, Mass. (1983).
4. W. F. Ames, *Nonlinear Partial Differential Equations in Engineering*, Vol. II, Academic Press, Inc., New York (1972).
5. G. W. Bluman and J. D. Cole, *Similarity Methods for Differential Equations*, Springer-Verlag, New York (1974).
6. P. J. Olver, *Applications of Lie Groups to Differential Equations*, Springer-Verlag, New York (1986).

8

DIFFERENCE METHODS FOR PARTIAL DIFFERENTIAL EQUATIONS

Up to now we have encountered a number of techniques that lead to either an approximate solution or an analytical closed-form solution of a partial differential equation. It would be a mistake to conclude that these techniques are universally applicable; rather, the opposite is true. For the most part solutions can be obtained exactly for only the simplest problems. Even then the analytic form of the solution is often an infinite series or integral representation that requires numerical resolution. For example, Fourier's method of separation of variables leads to a series representation whose coefficients must be calculated numerically. Further, it is generally impossible to sum the infinite series, and so approximate methods must be used in the end. Such analytic techniques, although useful for theoretical discussions, are not always satisfying from a practical point of view.

It is probably not an overstatement to say that almost all partial differential equations that arise in a practical setting are solved numerically on a computer. Since the development of high-speed computing devices the numerical solution of partial differential equations has been in an active state with the invention of new algorithms and the examination of the underlying theory. This area is one of the most active in applied mathematics and it has a great impact on science and engineering because of the ease and efficiency it has shown in solving even the most complicated problems.

In this chapter we develop fundamental numerical algorithms for solving the heat equation, the Laplace equation, the wave equation, and equations represented by hyperbolic systems. The methods are also applicable to other problems, including nonlinear ones. Following the development of some of the algorithms we have included a BASIC program in skeletal form that can easily

489

be put on a microcomputer. The reader is invited to make modifications of the program to improve its efficiency or adapt it to special circumstances. Only slight changes are required to rewrite the program in FORTRAN. Of course, many problems are beyond the capacity of a microcomputer and require very rapid computation and considerable storage. In those cases a large mainframe computer is required. Our purpose here, however, is to communicate the logical structure and indicate the ease with which calculations can be performed to numerically solve partial differential equations.

8.1   FINITE DIFFERENCE METHODS

Discretization

We are primarily interested in solving partial differential equations by a numerical technique called finite differences, although in later sections we briefly examine characteristic methods. The basic idea of the method of finite differences is to cast the continuous problem described by the partial differential equation and auxiliary conditions into a discrete problem that can be solved by a computer in finitely many steps. The discretization is accomplished by restricting the problem to a set of discrete points in the domain of interest and approximating the derivatives in the partial differential equations by difference quotients relating the values of the unknown function at those points. By a systematic procedure we then calculate the unknown function at those discrete points. Consequently a finite difference technique yields a solution only at discrete points in the domain of interest rather than, as we expect for an analytic calculation, a formula or closed-form solution valid at all points of the domain.

We briefly illustrate this discretization process by considering first an ordinary differential equation. Consider the initial value problem

$$\frac{dy}{dx} = f(x, y), \qquad y(0) = y_0 \tag{1}$$

where y_0 is the initial value of $y = y(x)$ at $x = 0$. We wish to determine a solution $y = y(x)$ of (1) in an interval $0 \le x \le 1$. It is well known that a unique solution exists provided $f(x, y)$ satisfies some general regularity conditions. To apply a discrete method the approach is to partition the interval $[0, 1]$ into n subintervals by defining the set of discrete points $x_i = i \, \Delta x$, $i = 0, \ldots, n$, where $\Delta x = 1/n$. The quantity Δx is called the *step size*. At each point x_i we seek to determine a quantity Y_i that is to approximate the exact solution $y(x_i)$ at x_i. From elementary calculus we know the derivative $dy(x_i)/dx$ can be approximated by the difference quotient

$$\frac{dy}{dx}(x_i) \cong \frac{y(x_{i+1}) - y(x_i)}{\Delta x}$$

provided Δx is sufficiently small. Consequently, at x_i the given ordinary differential equation may be approximated by the finite difference relation

$$\frac{Y_{i+1} - Y_i}{\Delta x} = f(x_i, Y_i), \qquad i = 0, 1, \ldots, n - 1$$

or

$$Y_{i+1} = Y_i + \Delta x f(x_i, Y_i)$$

This equation provides an algorithm or *finite difference scheme* for solving the ordinary differential equation. Given $Y_0 = y_0$ the difference equation can be used recursively to compute $Y_1, Y_2, \ldots$, etc., thereby giving an approximate discrete solution to the problem. This method is the classical Euler-Cauchy method.

Now let us consider a simple partial differential equation and perform a similar analysis. One dimensional heat flow in an infinite rod is governed by the partial differential equation

$$u_t - k u_{xx} = 0, \qquad x \in R^1, \quad t > 0$$

where $u = u(x, t)$ is the temperature in the rod at position x at time $t > 0$ and k is the thermal diffusivity, a positive constant characteristic of the material of which the rod is composed. We seek a solution u in the region $R = \{t > 0, -\infty < x < \infty\}$ satisfying the initial condition

$$u(x, 0) = f(x), \qquad x \in R^1$$

where $f(x)$ represents an initial temperature distribution in the bar. We now define in R a two dimensional *lattice* of discrete points (x_j, t_n) by

$$t_n = n \Delta t, \qquad x_j = j \Delta x$$

for $n = 0, 1, 2, \ldots$; $j = 0, \pm 1, \pm 2, \ldots$; the lattice spacings or step sizes Δt and Δx are positive fixed quantities (see Fig. 8.1). At this lattice of points we wish to determine a quantity $U_{j,n}$ that approximates the exact solution $u(x_j, t_n)$. The basic procedure is to replace derivatives by difference quotients and thus reduce the partial differential equation to a finite difference equation for the lattice function $U_{j,n}$. The set of lattice points (x_j, t_n) is also called a *grid* or *mesh*. For the moment let us assume the following finite difference approximations for the derivatives.

$$u_t(x_j, t_n) \cong \frac{u(x_j, t_{n+1}) - u(x_j, t_n)}{\Delta t}$$

$$u_{xx}(x_j, t_n) \cong \frac{u(x_{j+1}, t_n) - 2u(x_j, t_n) + u(x_{j-1}, t_n)}{(\Delta x)^2}$$

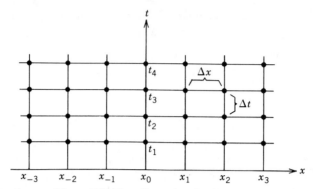

Figure 8.1. Two dimensional grid lattice.

Then the heat equation may be approximated at (x_j, t_n) by

$$\frac{U_{j,n+1} - U_{j,n}}{\Delta t} - k\frac{U_{j+1,n} - 2U_{j,n} + U_{j-1,n}}{\Delta x^2} = 0$$

for $n = 0, 1, 2, \ldots$, and $j = 0, \pm 1, \pm 2, \ldots$. This difference equation can be written in the form

$$U_{j,n+1} = (1 - 2r)U_{j,n} + r(U_{j+1,n} + U_{j-1,n}) \tag{2}$$

where $r = k\,\Delta t/(\Delta x)^2$. Equation (2) relates the values of U at the grid points (x_j, t_{n+1}), (x_j, t_n), (x_{j+1}, t_n), and (x_{j-1}, t_n). These four points form the *computational molecule* of the difference scheme (2). We note that $U_{j,n+1}$ can be calculated with knowledge of $U_{j-1,n}$, $U_{j,n}$, and $U_{j+1,n}$. Consequently, since the values $U_{j,0} = f(x_j)$ are given by the initial values along the $t = 0$ time line, Equation (2) allows us to march forward in time and compute $U_{j,1}, U_{j,2}, \ldots, j = 0, \pm 1, \pm 2, \ldots$, successively, along subsequent time lines.

At this point several questions arise. First, is the algorithm stable? That is, do roundoff errors present in the initial data remain bounded as the scheme propagates those data? Second, does the scheme converge? That is, does the calculated grid function U at (x_j, t_n) approach the exact solution $u(x_j, t_n)$ as the grid spacings Δt and Δx get small? In symbols, does $U_{j,n} \to u(x_j, t_n)$ as $\Delta t \to 0$ and $\Delta x \to 0$? These are two of the important questions that must be asked about any finite difference scheme devised for solving a partial differential equation. Further, we also wish to have bounds on the error in the calculation as well as some notion of the degree of accuracy or consistency of the finite difference approximation to the partial differential equation itself.

We summarize the method of finite differences applied to partial differential equations as follows. To be solved is a partial differential equation on a region

R with auxiliary data on its boundary C. We cover this region with a lattice of discrete points (x_j, t_n), where n and j range over some set of integers. These grid points can be divided into three disjoint sets: the interior grid points R_Δ, the boundary grid points C_Δ, and the exterior points (later we describe exactly how to assign lattice points to a given region, especially regions with irregular boundaries). At the points in $R_\Delta \cup C_\Delta$ we seek an approximation $U_{j,n}$ to $u(x_j, t_n)$, the exact solution. Such an approximation is obtained by setting up a difference equation to replace the given partial differential equation and then finally solving the difference equation over $R_\Delta \cup C_\Delta$.

Finite difference methods applied to the solution of partial differential equations originated in the early part of this century. The advent of large fast computing machines in the 1950s had a significant effect on the methods and algorithms for solving partial differential equations and made the solution of complex scientific and engineering problems possible. The literature since that time has grown extensively and is still in a dynamic state as new methods and ideas are currently being communicated in several important research journals in numerical analysis. The book by Ames [1] can be consulted for references. For additional reading we refer the reader to Richtmyer and Morton [2], Smith [3], Varga [4], and the final chapter of Isaacson and Keller [5]. As stated previously we concentrate here on finite difference methods. There are other techniques for numerically solving partial differential equations, one being the *finite element method*. The interested reader can consult Ames [1] for an introduction and references.

Discrete Approximations for Derivatives

A basic technique for devising algorithms to numerically solve partial differential equations involves making finite difference approximations to the derivatives. Such approximations can often be obtained by Taylor's theorem, which gives both the form and degree of accuracy of the approximation. Taylor's theorem for functions of one variable states that if $F(x)$ has $n + 1$ continuous derivatives in an open interval containing $x = a$, then

$$F(x) = F(a) + F'(a)(x - a) + \frac{F''(a)}{2!}(x - a)^2$$

$$+ \cdots + \frac{F^{(n)}(a)}{n!}(x - a)^n + R_n$$

where the remainder R_n is given by

$$R_n = \frac{F^{(n+1)}(\bar{x})}{(n+1)!}(x - a)^{n+1}$$

where $\bar{x}$ is some number between x and a. Taylor's theorem in several variables is a generalization of this one dimensional expansion. For example, let $u(x, y)$ be a continuous function possessing $n + 1$ continuous partial derivatives in an open circle N centered at (a, b). Then for any (x, y) in N the function u can be expanded as

$$u(x, y) = u(a, b) + u_x(a, b)(x - a) + u_y(a, b)(y - b)$$

$$+ \frac{1}{2!}\left(u_{xx}(a, b)(x - a)^2 + 2u_{xy}(a, b)(x - a)(y - b) \right.$$

$$\left. + u_{yy}(a, b)(y - b)^2 \right)$$

$$+ \cdots + \frac{1}{n!}\left((x - a)\frac{\partial}{\partial x} + (y - b)\frac{\partial}{\partial y} \right)^n u(a, b) + R_n \quad (3)$$

where

$$R_n = \frac{1}{(n + 1)!}\left((x - a)\frac{\partial}{\partial x} + (y - b)\frac{\partial}{\partial y} \right)^{n+1} u(\bar{x}, \bar{y})$$

and $\bar{x} = x + \theta(x - a)$, $\bar{y} = y + \theta(y - b)$ for $0 < \theta < 1$. Thus $(\bar{x}, \bar{y})$ is some point on the line segment connecting (a, b) and (x, y). In (3) we have used x and y as independent variables. In evolution problems the independent variables will be x and t.

To obtain difference formulas for $u_x, u_y, \ldots$, we define a lattice $x_j = j\Delta x$, $y_k = k\,\Delta y$, where j and k range over $0, \pm 1, \pm 2, \ldots$, and where Δx and Δy are the fixed lattice spacings. For convenience denote $u_{j, k} = u(x_j, y_k)$. Then with $n = 2$, $(a, b) = (x_j, y_k)$, and $(x, y) = (x_{j+1}, y_k)$, Equation (3) gives

$$u_{j+1, k} = u_{j, k} + u_x(x_j, y_k)\,\Delta x + \tfrac{1}{2}u_{xx}(\bar{x}_j, y_k)\,\Delta x^2$$

or

$$u_x(x_j, y_k) = \frac{u_{j+1, k} - u_{j, k}}{\Delta x} - \frac{1}{2}u_{xx}(\bar{x}_j, y_k)\,\Delta x \quad (4)$$

where $x_j < \bar{x}_j < x_{j+1}$. Therefore we obtain the *forward difference approximation*

$$u_x(x_j, y_k) \cong \frac{u_{j+1, k} - u_{j, k}}{\Delta x}$$

to the derivative u_x at (x_j, y_k). The term

$$\tau_{j, k} \equiv -\tfrac{1}{2}u_{xx}(\bar{x}_j, y_k)\,\Delta x$$

is called the *local discretization error* or *truncation error* for the approximation. This approximation is said to be first order since the truncation error is bounded by a quantity proportional to Δx to the first power.

It is convenient to recall the *big oh* notation introduced in Chapter 2. A quantity $f(z)$ is said to be of the order z^p if there is a positive constant M such that $|f(z)| \leq M|z|^p$ for all z sufficiently small (that is, $|z| \leq z_0$ for some z_0); we write $f(z) = O(|z|^p)$. In terms of this notation (4) can be written

$$u_x(x_j, y_k) = \frac{u_{j+1, k} - u_{j, k}}{\Delta x} + O(|\Delta x|)$$

The forward difference approximation to u_y at (x_j, y_k) is given similarly by

$$u_y(x_j, y_k) = \frac{u_{j, k+1} - u_{j, k}}{\Delta y} - \frac{1}{2} u_{yy}(x_j, \bar{y}_k) \, \Delta y$$

where $y_k < \bar{y}_k < y_{k+1}$.

Writing out Taylor's formula (3) with $n = 2$, $(a, b) = (x_j, y_k)$, and $(x, y) = (x_{j-1}, y_k)$ gives

$$u_{j-1, k} = u_{j, k} - u_x(x_j, y_k) \, \Delta x + \tfrac{1}{2} u_{xx}(\bar{x}_j, y_k) \, \Delta x^2$$

which yields the *backward difference approximation*

$$u_x(x_j, y_k) = \frac{u_{j, k} - u_{j-1, k}}{\Delta x} + \frac{1}{2} u_{xx}(\bar{x}_j, y_k) \, \Delta x$$

In the same way

$$u_y(x_j, y_k) = \frac{u_{j, k} - u_{j, k-1}}{\Delta y} + \frac{1}{2} u_{yy}(x_j, \bar{y}_k) \, \Delta y$$

Second derivatives may be approximated in a similar fashion. Again from Taylor's formula

$$u_{j+1, k} = u_{j, k} + u_x(x_j, y_k) \, \Delta x + \frac{1}{2} u_{xx}(x_j, y_k) \, \Delta x^2$$

$$+ \frac{1}{3!} u_{xxx}(x_j, y_k) \, \Delta x^3 + \frac{1}{4!} u_{xxxx}(\bar{x}_j, y_k) \, \Delta x^4 \tag{5}$$

and

$$u_{j-1, k} = u_{j, k} - u_x(x_j, y_k) \, \Delta x + \frac{1}{2} u_{xx}(x_j, y_k) \, \Delta x^2$$

$$- \frac{1}{3!} u_{xxx}(x_j, y_k) \, \Delta x^3 + \frac{1}{4!} u_{xxxx}(\bar{\bar{x}}_j, y_k) \, \Delta x^4 \tag{6}$$

where $x_{j-1} < \bar{\bar{x}}_j < x_j < \bar{x}_j < x_{j+1}$. Adding, we obtain the *second order centered difference approximation*

$$u_{xx}(x_j, y_k) = \frac{u_{j+1,k} - 2u_{j,k} + u_{j-1,k}}{\Delta x^2} + \frac{1}{12} u_{xxxx}(\bar{\bar{x}}_j, y_k) \Delta x^2 \quad (7)$$

where $x_{j-1} < \bar{\bar{x}}_j < x_{j+1}$. A similar formula holds for u_{yy}. A second order approximation for first derivatives may be obtained by subtracting (5) from (6) to obtain

$$u_x(x_j, y_k) = \frac{u_{j+1,k} - u_{j-1,k}}{2\Delta x} + O(\Delta x^2)$$

This is the *centered difference approximation* and a similar formula exists for u_y. For mixed partial derivatives it is left as an exercise to show that

$$u_{xy}(x_j, y_k) = \frac{u_{j+1,k+1} - u_{j+1,k-1} + u_{j-1,k-1} - u_{j-1,k+1}}{4\Delta x^2} + O(\Delta x^2) \quad (8)$$

provided that $\Delta x = \Delta y$. Clearly a second order approximation is more accurate than one of the first order, since for small Δx the quantity $(\Delta x)^2$ is smaller than Δx. Discrete approximations with a higher order truncation error do not always lead to more useful and more accurate numerical methods to solve partial differential equations. Some high-order discrete approximations lead to unstable and therefore unusable schemes. Later we indicate examples of this phenomenon.

EXERCISES

1.1 Generate a numerical solution on the interval $0 \le x \le 1$ to the initial value problem

$$y' = xy^2 + 1, \qquad y(0) = 0$$

Use the Euler-Cauchy method with $\Delta x = 0.1$.

1.2 Derive (7) and (8).

1.3 For a spatial grid x_j $(j = 0, \pm 1, \ldots)$ find a second order centered difference approximation for the expression

$$\frac{d}{dx}(f(x)g'(x))$$

involving the points x_{j-1}, x_j, x_{j+1}.

1.4   Write out in full the expression

$$\frac{1}{4!}\left((x-a)\frac{\partial}{\partial x}+(y-b)\frac{\partial}{\partial y}\right)^4 u(a,b)$$

8.2   THE DIFFUSION EQUATION

An Explicit Scheme

In this section we develop the so-called *explicit method* for numerically solving the boundary value problem

$$u_t - ku_{xx} = 0, \qquad 0 < t < T, \quad 0 < x < L \tag{1}$$

$$u(x,0) = f(x), \qquad 0 < x < L \tag{2}$$

$$u(0,t) = g(t), \qquad u(L,t) = h(t), \qquad 0 < t < T \tag{3}$$

where f, g, and h are given functions. By the procedure outlined earlier we introduce a grid in the region $0 \le t \le T$, $0 \le x \le L$ by defining the lattice of points

$$x_j = j\Delta x, \qquad j = 0,1,\ldots,J \tag{4}$$

$$t_n = n\,\Delta t, \qquad n = 0,1,\ldots,N$$

where $\Delta x = L/J$ and $\Delta t = T/N$. The set of interior grid points R_Δ are specified by (x_j, t_n), $j = 1,\ldots,J-1$; $n = 1,\ldots,N-1$, while the boundary grid points C_Δ are specified by (x_0, t_n), (x_J, t_n), $n = 0,\ldots,N$, and (x_j, t_0), (x_j, t_N), $j = 1,\ldots,J-1$. By $U_{j,n}$ we denote a discrete approximation to the exact solution $u(x_j, t_n)$ at (x_j, t_n). Therefore the initial condition (2) can be expressed discretely as

$$U_{j,0} = f(x_j), \qquad j = 0,1,\ldots,J \tag{5}$$

and the boundary conditions (3) are

$$U_{0,n} = g(t_n), \qquad U_{J,n} = h(t_n), \qquad n = 1,2,\ldots,N \tag{6}$$

The problem is to determine U in R_Δ and along the upper boundary. To this end we discretize the partial differential equation (1) at (x_j, t_n) using a forward difference approximation to u_t and a centered difference approximation to u_{xx}. That is

$$u_t \cong \frac{U_{j,n+1} - U_{j,n}}{\Delta t}, \qquad u_{xx} \cong \frac{U_{j+1,n} - 2U_{j,n} + U_{j-1,n}}{\Delta x^2}$$

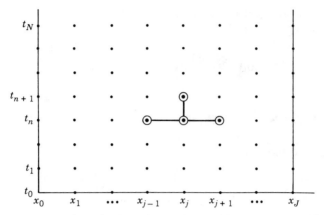

Figure 8.2. Discretization of the boundary value problem (1), (2), (3). The indicated computational molecule relates the points (x_{j-1}, t_n), (x_j, t_n), (x_{j+1}, t_n), and (x_j, t_{n+1}).

Substituting into (1) and rearranging terms we obtain the finite difference equation

$$U_{j,\,n+1} = (1 - 2r)U_{j,\,n} + r\left(U_{j+1,\,n} + U_{j-1,\,n}\right) \tag{7}$$

where

$$r = \frac{k\,\Delta t}{\Delta x^2} \tag{8}$$

and $j = 1, \ldots, J - 1$ and $n = 0, 1, \ldots, N - 1$. The difference scheme described by (5), (6), and (7) is known as an *explicit scheme*, since it permits the direct or explicit calculation of $U_{j,\,n+1}$ from the data at the preceding time step; hence the numerical solution can march forward in time from the initial condition by computing $U_{j,\,n+1}$, $j = 1, \ldots, J - 1$, successively for each $n = 0, 1, \ldots, N - 1$ (Fig. 8.2), thereby filling in the lattice. The values on the side boundaries and lower boundary are given by the boundary conditions and initial conditions (6) and (5), respectively. The *computational molecule* also shown in Fig. 8.2 depicts the points of the lattice that are related by the finite difference scheme (7); it shows how a value at the $(n + 1)$th time step can be calculated from three values at the previous time step and therefore indicates the general flow of the calculation. In this case the numerical calculation simulates the actual physical situation in which the initial temperature distribution is carried forward in time as the diffusion process evolves.

Example 2.1   We indicate how the explicit algorithm can be carried out numerically on a computer. We work in the physical context of studying the evolution of the temperatures in a bar subject to initial and boundary data.

The idea is to store a given temperature distribution in the bar at a specific time step in an array $U(J)$, $J = 0, 1, \ldots, $ NUMXINT, and then compute an array $UN(J)$ over the same range of J that represents the temperature distribution at the next time step using the difference equation (7). At the first step the distribution is just the initial temperature distribution $f(x)$. At the end of each step the data in $UN(J)$ is put back into the array $U(J)$ and the process is repeated; in this manner only two time rows require storage at any step. The variables in the program listing are:

NUMXINT	number of subintervals in which $0 \le x \le L$ is divided
NUMTSTP	number of time steps taken
DELX	Δx
DELT	Δt

The remaining variables are self-explanatory and coincide with the notation in the text. As we shall observe later the constant r must be chosen such that $r \le 0.5$ for stability. The specific problem we solve is

$$u_t - ku_{xx} = 0, \qquad 0 < x < 10, \quad 0 < t < 6$$

$$u(0, t) = u(10, t) = 0, \qquad 0 < t < 6$$

$$u(x, 0) = x(10 - x), \qquad 0 \le x \le 10$$

with $k = 1$. The BASIC program is given in Program Listing 8.1.

```
10    REM--EXPLICIT SCHEME FOR THE HEAT EQUATION--
20    REM--SET THE DISCRETIZATION--
30        DIM U(15),UN(15)
40        K=1: L=10: NUMXINT=10: DELX=L/NUMXINT
50        DELT=.25: NUMTSTP=25: R=K*DELT/DELX^2
60        IF R>.5 THEN GOTO 210
70    REM--DEFINE INITIAL TEMPERATURE--
80        FOR J=0 TO NUMXINT
90        U(J)=J*(10-J): LPRINT USING"##.## ";U(J);:NEXT J:LPRINT
100   REM--COMPUTE TEMPERATURE AT NEXT TIME STEP--
110       FOR N=1 TO NUMTSTP
120         UN(0)=0: UN(NUMXINT)=0
130         FOR J=1 TO NUMXINT-1
140         UN(J)=(1-2*R)*U(J)+R*(U(J-1)+U(J+1))
150         NEXT J
160         REM--PRINT TEMPERATURE PROFILE AND PUT UN(J) INTO U(J)--
170         FOR J=0 TO NUMXINT
180         LPRINT USING"##.## ";UN(J);:U(J)=UN(J): NEXT J: LPRINT
190       NEXT N
200       GOTO 220
210       LPRINT "scheme is not stable"
220       END
```

TABLE 8.1

x	0	1	2	3	4	5	6	7	8	9	10
$t = 0$	0.00	9.00	16.00	21.00	24.00	25.00	24.00	21.00	16.00	9.00	0.00
$t = 1$	0.00	7.54	14.09	19.01	22.00	23.00	22.00	19.01	14.09	7.54	0.00
$t = 2$	0.00	6.66	12.58	17.17	20.04	21.01	20.04	17.17	12.58	6.66	0.00
$t = 3$	0.00	5.97	11.32	15.52	18.19	19.11	18.19	15.52	11.32	5.97	0.00
$t = 4$	0.00	5.38	10.22	14.05	16.49	17.33	16.49	14.05	10.22	5.38	0.00
$t = 5$	0.00	4.86	9.25	12.72	14.94	15.71	14.94	12.72	9.25	4.86	0.00
$t = 6$	0.00	4.40	8.37	11.52	13.54	14.23	13.54	11.52	8.37	4.40	0.00

Table 8.1 shows the discrete temperature values at unit time intervals.

Truncation Error and Convergence

The explicit scheme for the diffusion equation does not always give an accurate approximation to the exact solution. In this section we investigate the convergence of this scheme and set forth some general definitions that will aid in the analysis of all finite difference schemes.

The *truncation error* for a finite difference scheme is a measure of the error that arises from replacing the derivatives in the equation by discrete approximations. In symbolic terms the partial differential equation can be represented in a region R by

$$L(u) = 0, \qquad (x, y) \in R \tag{9}$$

where L is a differential operator. When the derivatives are approximated there results a finite difference equation

$$L_\Delta(U) = 0, \qquad (x, y) \in R_\Delta \tag{10}$$

where L_Δ is a finite difference operator associated with L, and U is a grid function on the lattice R_Δ that approximates the exact solution u. The truncation error is defined to be the difference between the partial differential equation and its difference approximation at the grid points, that is

$$\tau \equiv L(u) - L_\Delta(u), \qquad (x, y) \in R_\Delta \tag{11}$$

Thus τ is grid function on R_Δ. It is important to notice that both terms in (11) involve the exact solution u. Also we remark that (11) defines the truncation error of the difference equation (10) and not the truncation error of the approximate solution. The latter will involve possible errors in the discretization of the boundary and initial conditions and these are not accounted for in (11).

The truncation error for the explicit scheme for the heat equation is according to (11)

$$
\tau_{j,n} = \left[u_t(x_j, t_n) - k u_{xx}(x_j, t_n) \right]
$$
$$
- \left[\frac{u_{j,n+1} - u_{j,n}}{\Delta t} - k \frac{u_{j+1,n} - 2u_{j,n} + u_{j-1,n}}{\Delta x^2} \right] \tag{12}
$$

It easily follows that

$$
\tau_{j,n} = -\frac{1}{2} u_{tt}(x_j, \bar{t}_n) \Delta t - \frac{k}{12} u_{xxxx}(\bar{x}_j, t_n) \Delta x^2 = O(\Delta t) + O(\Delta x^2) \tag{13}
$$

where it is assumed that u is four times continuously differentiable and $t_n < \bar{t}_n < t_{n+1}$, $x_{j-1} < \bar{x}_j < x_{j+1}$. Therefore the truncation error is first order in Δt and second order in Δx and $\tau_{j,n}$ tends to zero as $\Delta t \to 0$ and $\Delta x \to 0$. In general a scheme is called *conditionally consistent* if the truncation error goes to zero provided that some fixed relationship is maintained between Δt and Δx as they tend to zero.

Convergence of a finite difference scheme means that the approximate solution U approaches the true solution u at each fixed point as the grid is refined in such a way that Δt and Δx tend to zero; symbolically

$$
\lim_{\substack{\Delta t \to 0 \\ \Delta x \to 0}} (u - U) = 0 \quad \text{on} \quad R_\Delta \cup C_\Delta
$$

To examine convergence of the explicit scheme for the diffusion equation we define the *error*

$$
e_{j,n} \equiv u_{j,n} - U_{j,n} \quad \text{on} \quad R_\Delta \cup C_\Delta \tag{14}
$$

In general an expression can be obtained for $e_{j,n}$ by subtracting the difference equation (10) from the equation (11) defining the truncation error. In the present case (12) can be rewritten as

$$
u_{j,n+1} = (1 - 2r)u_{j,n} + r(u_{j+1,n} + u_{j-1,n}) - \Delta t \, \tau_{j,n} \tag{15}
$$

where r is the ratio defined by (8). Subtracting (7) from (15) gives a difference equation for the error, namely

$$
e_{j,n+1} = (1 - 2r)e_{j,n} + r(e_{j+1,n} + e_{j-1,n}) - \Delta t \, \tau_{j,n} \tag{16}
$$

For the present assume that $0 < r \le \frac{1}{2}$. Then it follows from (16) that

$$
|e_{j,n+1}| \le (1 - 2r)|e_{j,n}| + r(|e_{j+1,n}| + |e_{j-1,n}|) + \Delta t \, |\tau_{j,n}| \tag{17}
$$

Letting

$$E_n = \max_j |e_{j,n}|, \qquad \hat{\tau} = \max_{n,j} |\tau_{j,n}| \qquad (18)$$

we may rewrite the inequality (17) as

$$|e_{j,n+1}| \le E_n + \Delta t\,\hat{\tau}$$

Taking the maximum over $0 \le j \le J$ then gives

$$E_{n+1} \le E_n + \Delta t\,\hat{\tau} \qquad (19)$$

A repeated application of inequality (19) implies

$$E_n \le E_0 + n\,\Delta t\,\hat{\tau}$$

or

$$E_n \le T\hat{\tau}$$

the last step following from the fact that $E_0 = 0$ and $n\,\Delta t \le T$. Therefore

$$\max_{0 \le j \le J} |e_{j,n}| \le T\hat{\tau} \qquad (20)$$

Since the truncation error $\tau_{j,n}$ and hence $\hat{\tau}$ tends to zero as Δt and Δx tend to zero we see from (20) that

$$\lim_{\substack{\Delta x \to 0 \\ \Delta t \to 0}} |e_{j,n}| = 0 \qquad \text{on} \quad R_\Delta \cup C_\Delta \qquad (21)$$

which shows convergence of the explicit scheme in the case $r \le \frac{1}{2}$. We note from (13) that τ depends on the bounds for u_{tt} and u_{xxxx}.

If $r > \frac{1}{2}$, then (17) does not follow from (16) since $1 - 2r < 0$; hence the proof just given does not go through. By another argument we present later it can be shown that the explicit scheme does not converge when $r > \frac{1}{2}$.

The fact that $r \le \frac{1}{2}$ is required for convergence of the explicit scheme places some limitations on the time step Δt. From the definition of r it is required that

$$\Delta t \le \frac{\Delta x^2}{2k} \qquad (22)$$

This restriction can sometimes impose some difficulties in an actual calculation, since it may force an unreasonably small time step for the problem.

Physically the characteristic time scale L^2/k is a measure of how fast changes are taking place (see Chapter 1) and therefore gives a good notion of the time step needed for the scheme. Finally we point out that the bound for the error given in (20) can be written, using (13), as

$$\max_{0 \le j \le J} |e_{j,n}| \le T\left(\frac{M_1}{2} \Delta t + \frac{M_2}{12} \Delta x^2\right) \tag{23}$$

where $|u_{tt}| \le M_1$ and $|u_{xxxx}| \le M_2$. Equation (23) gives no a priori information, however, since we have no way of evaluating the constants M_1 and M_2 that bound the derivatives of the unknown solution $u(x, t)$.

Implicit Schemes

Whereas an explicit scheme permits the direct calculation of the approximate solution at the next time step an implicit scheme requires an indirect calculation usually involving the solution of a system of equations or some iterative process. In general implicit methods may require more computation but they have an important advantage of being more stable than explicit methods.

In this section we devise an implicit scheme for the problem defined by (1) through (3). The grid is defined in the same manner as (4) and the initial and boundary conditions are defined for the grid function U by (5) and (6). Instead of replacing the time derivative u_t by a forward difference approximation, however, we use a backward difference approximation to obtain

$$\frac{U_{j,n} - U_{j,n-1}}{\Delta t} - k\frac{U_{j+1,n} - 2U_{j,n} + U_{j-1,n}}{\Delta x^2} = 0 \tag{24}$$

or

$$-rU_{j-1,n} + (1 + 2r)U_{j,n} - rU_{j+1,n} = U_{j,n-1} \tag{25}$$

for $n = 1, 2, \ldots, N$ and $j = 1, 2, \ldots, J - 1$. In this case the computational molecule is given in Fig. 8.3, and we notice that this scheme no longer permits a direct calculation of the approximate values forward in time, but rather at each time value t_n Equation (25) relates the grid function at three adjacent points. Therefore for each n Equation (25) represents a system of equations in the $J - 1$ unknowns $U_{1,n}, U_{2,n}, \ldots, U_{J-1,n}$ that can be written in matrix form

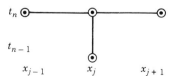

Figure 8.3. Computational molecule for the implicit scheme (25) for the diffusion equation.

as a tridiagonal system

$$
\begin{bmatrix}
1 + 2r & -r & & & & \\
-r & 1 + 2r & -r & & & \\
& -r & 1 + 2r & -r & & \\
& & \ddots & \ddots & \ddots & \\
& & & & & -r \\
& & & & -r & 1 + 2r
\end{bmatrix}
\begin{bmatrix}
U_{1,\,n} \\
U_{2,\,n} \\
U_{3,\,n} \\
\vdots \\
U_{J-2,\,n} \\
U_{J-1,\,n}
\end{bmatrix}
$$

$$
=
\begin{bmatrix}
U_{1,\,n-1} + rU_{0,\,n} \\
U_{2,\,n-1} \\
U_{3,\,n-1} \\
\vdots \\
U_{J-2,\,n-1} \\
U_{J-1,\,n-1} + rU_{J,\,n}
\end{bmatrix}
\tag{26}
$$

Therefore for each $n = 1, \ldots, N$, successively, we must solve the system (26) to obtain U at the lattice points on the nth row.

The solution to a tridiagonal system can be carried out in an easy manner by *forward and back substitution*. We describe this process for an arbitrary tridiagonal system

$$
A\mathbf{x} = \mathbf{f} \tag{27}
$$

or

$$
\begin{bmatrix}
a_1 & c_1 & & & & \\
b_2 & a_2 & c_2 & & & \\
& b_3 & a_3 & c_3 & & \\
& & & \ddots & & \\
& & & & \ddots & \\
& & & & c_{n-1} & \\
& & & & b_n & a_n
\end{bmatrix}
\begin{bmatrix}
x_1 \\
x_2 \\
x_3 \\
\vdots \\
x_{n-1} \\
x_n
\end{bmatrix}
=
\begin{bmatrix}
f_1 \\
f_2 \\
f_3 \\
\vdots \\
f_{n-1} \\
f_n
\end{bmatrix}
\tag{28}
$$

A matrix L is said to be lower triangular if all of its entries above the diagonal are zero. A matrix U is upper triangular if its entries below the diagonal are all zero. If A can be factored into a product of a lower triangular matrix L and an upper triangular matrix U, that is, $A = LU$, then the given system (27) becomes

$$
L\mathbf{g} = \mathbf{f} \tag{29}
$$

where

$$Ux = g \tag{30}$$

The lower triangular system (29) can easily be solved for g by forward substitution, that is solving successively the first through the last equation. The upper triangular system (30) can then be solved to determine the solution x by backward substitution, that is, solving successively the last through the first equation. It is easy to see that the matrix A in (28) can be factored as

$$A = LU = \begin{bmatrix} a'_1 & & & & \\ b_2 & a'_2 & & & \\ & b_3 & a'_3 & & \\ & & & \ddots & \ddots \\ & & & b_n & a'_n \end{bmatrix} \begin{bmatrix} 1 & c'_1 & & & \\ & 1 & c'_2 & & \\ & & 1 & & \\ & & & \ddots & \ddots \\ & & & & c'_{n-1} \\ & & & & 1 \end{bmatrix}$$

where

$$\begin{aligned} a'_1 &= a_1, & c'_1 &= c_1/a'_1 \\ a'_j &= a_j - b_j c'_{j-1}, & j &= 2, \ldots, n \\ c'_j &= \frac{c_j}{a'_j}, & j &= 2, \ldots, n-1 \end{aligned} \tag{31}$$

Then the lower triangular system $Lg = f$ can be solved to obtain

$$g_1 = f_1/a'_1 \tag{32}$$
$$g_j = \frac{f_j - b_j g_{j-1}}{a'_j}, \qquad j = 2, \ldots, n$$

Finally the upper triangular system $Ux = g$ can be easily solved to obtain

$$x_n = g_n \tag{33}$$
$$x_j = g_j - c'_j x_{j+1}, \qquad j = n-1, \ldots, 1$$

Therefore the solution to the tridiagonal system (28) can be accomplished by following steps (31), (32), and (33), provided $a'_j \neq 0$, $j = 1, 2, \ldots, n$. This algorithm can be applied to system (26) at each time step to compute a numerical approximation $U_{j,n}$ to the given problem. This is done in Example 2.2.

The convergence of the implicit scheme can be investigated by the method described in the last section. The truncation error $\tau_{j,n}$ is

$$\tau_{j,n} = u_t(x_j, t_n) - k u_{xx}(x_j, t_n) - \frac{u_{j,n} - u_{j,n-1}}{\Delta t} + k \frac{u_{j+1,n} - 2u_{j,n} + u_{j-1,n}}{\Delta x^2} \tag{34}$$

which gives

$$\tau_{j,n} = \frac{1}{2}u_{tt}(x_j, \bar{t}_n)\Delta t - \frac{k}{12}u_{xxxx}(\bar{x}_j, t_n)\Delta x^2 = O(\Delta t) + O(\Delta x^2) \quad (35)$$

Thus the truncation error is the same as for the explicit scheme. An expression for the error $e_{j,n} = u_{j,n} - U_{j,n}$ can be determined by adding (24) and (34) to get

$$(1 + 2r)e_{j,n} = e_{j,n-1} + r(e_{j+1,n} + e_{j-1,n}) - \Delta t\, \tau_{j,n} \quad (36)$$

Letting

$$E_n = \max_{0 \le j \le J} |e_{j,n}|, \qquad \hat{\tau} = \max_{R_\Delta \cup C_\Delta} |\tau_{j,n}|$$

(36) implies that

$$(1 + 2r)|e_{j,n}| \le E_{n-1} + 2rE_n + \Delta t\, \hat{\tau}$$

Taking the maximum over $0 \le j \le J$ gives

$$E_n \le E_{n-1} + \Delta t\, \hat{\tau}$$

Applying this inequality recursively we conclude that

$$E_n \le E_0 + n\,\Delta t\, \hat{\tau}$$
$$\le T\hat{\tau} \quad (37)$$

the last inequality following from $E_0 = 0$ and $n\,\Delta t \le T$. The inequality (37) shows that

$$|u_{j,n} - U_{j,n}| \le T\hat{\tau} \quad (38)$$

We recall from (35) that $\hat{\tau}$ depends upon the bounds of u_{tt} and u_{xxxx} and moreover $\hat{\tau} \to 0$ as Δt, $\Delta x \to 0$. Therefore the implicit scheme converges for any value of r.

Example 2.2   The implicit scheme is illustrated on the boundary value problem in Example 2.1. The BASIC variables coincide with the notation introduced in the text. Thus the arrays A, B, and C represent a_j, b_j, and c_j, respectively, while AP and CP represent a'_j and c'_j. The arrays F and G represent f and g, respectively, UO is the temperature at the $(n-1)$th step, and U is the temperature at the nth step. The numbers NUMXINT and NUMTSTP are discretization quantities defining the number of subintervals into which the interval $[0, L]$ is divided and the number of time steps to be

```
10 REM--IMPLICIT SCHEME FOR THE HEAT EQUATION--
20 REM--SET UP THE DISCRETIZATION--
30    DIM U(20),A(20),B(20),AP(20),CP(20),F(20),G(20),UO(20),C(20)
40    K=1:L=10:NUMXINT=10:DELX=L/NUMXINT:JJ=NUMXINT-1
50    DELT=.25:NUMTSTP=50:R=K*DELT/DELX^2
60 REM--DEFINE THE MATRIX A--
70    FOR J=1 TO JJ
80    A(J)=1+2*R:NEXT J
90    FOR J=1 TO JJ-1
100   C(J)=-R:NEXT J
110   FOR J=2 TO JJ
120   B(J)=-R:NEXT J
130 REM--DEFINE THE L-U DECOMPOSITION--
140   AP(1)=A(1)
150   CP(1)=C(1)/AP(1)
160   FOR J=2 TO JJ-1
170   AP(J)=A(J)-B(J)*CP(J-1)
180   CP(J)=C(J)/AP(J)
190   NEXT J
200   AP(JJ)=A(JJ)-B(JJ)*CP(JJ-1)
210 REM--SET THE INITIAL TEMPERATURE--
220   FOR J=0 TO NUMXINT
230   UO(J)=J*(10-J)
240   LPRINT USING "##.## ";UO(J);:NEXT J:LPRINT
250 REM--INITIALIZE THE VECTOR F--
260   F(1)=UO(1)
270   FOR J=2 TO JJ-1
280   F(J)=UO(J*DELX):NEXT J
290   F(JJ)=UO(JJ*DELX)
300 REM--BEGIN TIME LOOP--
310   FOR N=1 TO NUMTSTP
320   REM--COMPUTE THE VECTOR G--
330     G(1)=F(1)/AP(1)
340     FOR J=2 TO JJ
350     G(J)=(F(J)-B(J)*G(J-1))/AP(J)
360     NEXT J
370   REM--COMPUTE TEMPERATURE AT NEXT STEP--
380     U(JJ)=G(JJ)
390     FOR J=JJ-1 TO 1 STEP-1
400     U(J)=G(J)-CP(J)*U(J+1)
410     NEXT J
420     U(0)=0:U(NUMXINT)=0
430   REM--PRINT OUT TEMPERATURE PROFILE--
440     FOR J=0 TO NUMXINT
450     LPRINT USING "##.## ";U(J);:NEXT J:LPRINT
460   REM--RENAME THE VECTOR F--
470     F(1)=U(1)
480     FOR J=2 TO JJ-1:F(J)=U(J):NEXT J
490     F(JJ)=U(JJ)
500 REM--END TIME LOOP--
510   NEXT N
520 END
```

taken. Program Listing 8.2 contains the BASIC program and Table 8.2 shows the discrete temperature distributions at unit time intervals.

An obvious extension of the two preceding schemes is to average in some manner the approximations of u_{xx} along the lines t_{n+1} and t_n. Letting λ be a weighting factor we consider the family of schemes for the heat equation given

TABLE 8.2

x	0	1	2	3	4	5	6	7	8	9	10
$t = 0$	0.00	9.00	16.00	21.00	24.00	25.00	24.00	21.00	16.00	9.00	0.00
$t = 1$	0.00	7.65	14.19	19.05	22.01	23.01	22.01	19.05	14.19	7.65	0.00
$t = 2$	0.00	6.75	12.70	17.26	20.10	21.06	20.10	17.26	12.70	6.75	0.00
$t = 3$	0.00	6.05	11.44	15.65	18.30	19.20	18.30	15.65	11.44	6.05	0.00
$t = 4$	0.00	5.45	10.35	14.19	16.64	17.48	16.64	14.19	10.35	5.45	0.00
$t = 5$	0.00	4.93	9.37	12.88	15.12	15.89	15.12	12.88	9.37	4.93	0.00
$t = 6$	0.00	4.47	8.50	11.69	13.73	14.43	13.73	11.69	8.50	4.47	0.00

by

$$\frac{U_{j,n+1} - U_{j,n}}{\Delta t} - k \left[\lambda \frac{U_{j+1,n+1} - 2U_{j,n+1} + U_{j-1,n+1}}{\Delta x^2} \right.$$
$$\left. + (1 - \lambda) \frac{U_{j+1,n} - 2U_{j,n} + U_{j-1,n}}{\Delta x^2} \right] = 0 \quad (39)$$

for $0 \leq \lambda \leq 1$, $j = 1, \ldots, J - 1$, and $n = 0, \ldots, N - 1$. The computational molecule associated with (39) is shown in Fig. 8.4. If $\lambda = 0$, the explicit scheme (7) is recovered. If $\lambda > 0$, then (39) represents an implicit scheme with $\lambda = 1$ giving the scheme (25). For the special case $\lambda = \frac{1}{2}$ the formula is called the *Crank-Nicolson scheme*, after its inventors. In this latter case the truncation error is especially nice. We leave it as an exercise to show that the truncation error for the Crank-Nicolson scheme is $O(\Delta x^2 + \Delta t^2)$.

For the problem defined by (1) through (3), when $\lambda > 0$ the formula (39) represents a system of $J - 1$ equations in $J - 1$ unknowns $U_{1,n}$, $U_{2,n}, \ldots, U_{J-1,n}$ at each time level t_n. The forward and backward substitution algorithm can be applied to solve the resulting tridiagonal system.

Both the implicit and explicit schemes discussed earlier are called *two-level schemes* because they relate the values of the grid function U at two different time levels. By approximating the derivatives in a different manner we can obtain formulas relating the grid function U at three different time levels, thereby obtaining a *three-level formula*. For example, if both derivatives in the heat equation are approximated by centered differences, then we obtain the finite difference equation

$$\frac{U_{j,n+1} - U_{j,n-1}}{2\,\Delta t} - k \frac{U_{j+1,n} - 2U_{j,n} + U_{j-1,n}}{\Delta x^2} = 0$$

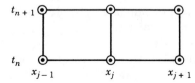

Figure 8.4. Computational molecule for the Crank-Nicolson scheme.

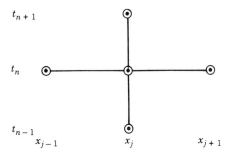

t_{n+1}

t_n

t_{n-1}
x_{j-1} x_j x_{j+1}

Figure 8.5. Computational molecule for the three-level scheme (40).

or

$$U_{j,n+1} = U_{j,n-1} + 2r\left(U_{j+1,n} - 2U_{j,n} + U_{j-1,n}\right) \qquad (40)$$

where $r = k\,\Delta t/\Delta x^2$. The computational molecule is shown in Fig. 8.5. In this case the values of U at the two time levels t_n and t_{n-1} are required to compute U at the time level t_{n+1}. Therefore in order to start the initial value problem values of U must be given along two time steps, $t = 0$ and $t = \Delta t$. Regardless of the method used to obtain the values of U along the time level $t = \Delta t$, however, the scheme (40) does not converge for any value of r. Therefore, even though this scheme appears to be perfectly reasonable it would lead to disastrous results and cannot be used. On the other hand if $U_{j,n}$ in (40) is replaced by the average of the values $U_{j,n+1}$ and $U_{j,n-1}$, then (40) becomes

$$(1 + 2r)U_{j,n+1} + (2r - 1)U_{j,n-1} - 2r\left(U_{j-1,n} + U_{j+1,n}\right) = 0 \qquad (41)$$

This is the *DuFort-Frankel scheme* and it always converges.

General Boundary Conditions

Heretofore we considered only boundary conditions where u itself was specified along an initial curve or boundary curve. Now we indicate the analysis required when boundary conditions involving derivatives occur. We consider the problem

$$u_t - ku_{xx} = 0, \qquad 0 < t < T, \quad 0 < x < L \qquad (42)$$
$$u(x,0) = f(x), \qquad 0 < x < L \qquad (43)$$
$$u(0,t) = g(t), \qquad 0 < t < T \qquad (44)$$
$$u_x(L,t) = h(t), \qquad 0 < t < T \qquad (45)$$

At the right boundary $x = L$ we are no longer given the value of u, but we must compute it subject to the condition (45). In an obvious manner the analysis can be extended to a similar condition at the left boundary. We

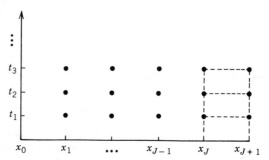

Figure 8.6. Addition of a false boundary to the grid.

discretize the region by defining the grid points

$$x_j = j\Delta x, \qquad j = 0, 1, \ldots, J$$
$$t_n = n\,\Delta t, \qquad n = 0, 1, \ldots, N$$

It is now necessary to compute $U_{j,n}$ for $j = 1, \ldots, J$ and $n = 1, \ldots, N$. For the interior grid points an explicit or implicit formula can be used. To calculate the right boundary points $U_{J,1}, U_{J,2}, \ldots, U_{J,N}$ we create a *false boundary* consisting of the points (x_{J+1}, t_n), $n = 0, 1, 2, \ldots, N$ (see Fig. 8.6). Then (45) can be approximated by the centered difference formula

$$U_{J+1,n} - U_{J-1,n} = 2\,\Delta x\, h(t_n) \tag{46}$$

for $n = 1, 2, \ldots, N$. An explicit algorithm can be devised using (46) in conjunction with (7). Similarly, an explicit scheme can be devised using (46) in conjection with the implicit formula (25) (see Exercise 2.12).

Stability

Earlier we remarked that stability of a numerical algorithm deals with how errors are propagated in a computation using that algorithm. In this section we examine this concept more carefully and state a criterion for stability of a finite difference scheme. For motivation we consider the pure initial value problem for the heat equation

$$u_t - ku_{xx} = 0, \qquad\qquad t > 0, \quad x \in R^1 \tag{47}$$
$$u(x, 0) = \cos \alpha x, \qquad x \in R^1 \tag{48}$$

where α is a positive constant. The solution to this problem is

$$u(x, t) = e^{-k\alpha^2 t}\cos \alpha x \tag{49}$$

and it is clear that u remains bounded as t gets large. On the other hand, introduce the grid $x_j = j \Delta x$, $t_n = n \Delta t$ for $j = 0, \pm 1, \ldots$, and $n = 0, 1, 2, \ldots$, and approximate the partial differential equation by the explicit formula

$$U_{j,n+1} = (1 - 2r)U_{j,n} + r(U_{j+1,n} + U_{j-1,n}) \tag{50}$$

and the initial condition by

$$U_{j,0} = \cos \alpha x_j, \qquad j = 0, \pm 1, \ldots \tag{51}$$

We will now show that for certain values of the ratio $r = k \Delta t / \Delta x^2$ the exact solution to the finite difference equation (50) becomes unbounded as t gets large. Thus the approximate solution cannot converge to the exact solution of the original problem (47) and (48). Looked at differently, if the values $\cos \alpha x_j$ in the initial condition are regarded as errors at the initial time, then we shall see that these errors are propagated without bound using the explicit scheme (50) with $r > \frac{1}{2}$.

We attempt to find a solution of (50) and (51) of the form

$$U_{j,n} = M^n \cos \alpha x_j \tag{52}$$

where M is to be determined. Substituting (52) into (50) we obtain

$$M = 1 - 2r + 2r \cos \alpha \Delta x = 1 - 4r \sin^2 \frac{\alpha \Delta x}{2}$$

Hence the solution of the difference equation (50) subject to (51) is

$$U_{j,n} = \left(1 - 4r \sin^2 \frac{\alpha \Delta x}{2}\right)^n \cos \alpha x_j$$

It is clear that $U_{j,n}$ will be bounded if

$$\left| 1 - 4r \sin^2 \frac{\alpha \Delta x}{2} \right| \le 1$$

or

$$r \le \frac{1}{2 \sin^2 \dfrac{\alpha \Delta x}{2}} \tag{53}$$

The inequality (53) is assured if $r \le \frac{1}{2}$. For $r > \frac{1}{2}$ there will always exist an α for which

$$\left| 1 - 4r \sin^2 \frac{\alpha \Delta x}{2} \right| > 1$$

Therefore the explicit scheme with $r > \frac{1}{2}$ is not stable, as there exists initial conditions that are propagated so that they become unbounded as n increases.

This analysis is a special case of a general method of examining stability due to John von Neumann. The analysis is valid for initial value problems and in general is applicable only for linear equations with constant coefficients. In the general case the method assumes a solution of the difference equation of the form

$$U_{j,\,n} = M^n e^{i\alpha x_j} \tag{54}$$

or in real variables

$$U_{j,\,n} = M^n \cos \alpha x_j \tag{55}$$

If $t = 0$, then (54) or (55) reduces to the initial condition $U_{j,0} = \exp(i\alpha x_j)$ or $U_{j,0} = \cos \alpha x_j$, respectively. The quantity M, called the *magnification factor*, is determined by substitution of (54) or (55) into the finite difference equation. If

$$|M| \le 1 \tag{56}$$

for all α, then the solution (54) or (55) will not grow with time and will remain bounded. Condition (56) is called the *von Neumann stability criterion*. Some insight into this criterion can be gained by noting that if $f(x)$, $x \in R^1$, is a distribution of errors at $t = 0$, then generally $f(x)$ can be resolved into it Fourier modes $e^{i\alpha x}$ via

$$f(x) = \frac{1}{\sqrt{2\pi}} \int_{-\infty}^{\infty} c(\alpha) e^{i\alpha x}\, d\alpha$$

where $c(\alpha)$ is the Fourier transform of f (see (43) and (44) of Section 4.3). In the same way a discrete row of errors at $t = 0$ can be resolved into a finite Fourier series of the form

$$\sum_j c_j e^{i\alpha_j x}$$

The von Neumann analysis focuses upon a single Fourier mode of frequency α representing a row of errors and seeks to determine how that mode is propagated by the difference equation. If there is a single frequency α such that the corresponding mode $M^n e^{i\alpha x}$ grows without bound as $n \to \infty$, then the scheme is unstable. Therefore if the criterion (56) is violated then an instability results. On the other hand, if (56) holds for all α, then every Fourier mode stays bounded. To prove stability in this case one must argue that every initial error configuration can be resolved in the fundamental modes. Fourier analysis guarantees that this can be accomplished for a broad class of functions.

Example 2.3   Consider the three-level scheme (40) for the diffusion equation. Substitution of (54) into (40) yields the quadratic equation

$$M^2 + 8rM \sin^2 \frac{\alpha \,\Delta x}{2} - 1 = 0$$

whose solutions are

$$M = -4r \sin^2 \frac{\alpha \Delta x}{2} \pm \sqrt{1 + 16r^2 \sin^4 \frac{\alpha \Delta x}{2}}$$

Taking the minus sign and expanding the square root term in its binomial series we get

$$M = -4r \sin^2 \frac{\alpha \Delta x}{2} - 1 - 8r^2 \sin^4 \frac{\alpha \Delta x}{2} + O(r^4)$$

It is clear that for any value of $r > 0$ and any value of α we have $|M| > 1$. Thus the scheme (40) is unstable in the von Neumann sense, since there exist initial conditions for which the finite difference solutions $U_{j,n}$ become unbounded as $n \to \infty$.

Matrix Stability Analysis

Another method commonly used to examine the stability of a finite difference scheme is the matrix method. It has the advantage of including boundary conditions and is therefore applicable to a wider range of schemes than the von Neumann method. To illustrate the procedure we examine the explicit scheme (7), which we write as

$$U_{j,n+1} = rU_{j-1,n} + (1 - 2r)U_{j,n} + rU_{j+1,n}$$

with given initial data $U_{j,0}$ ($j = 0, \ldots, J$) and given boundary data $U_{0,n}$ and $U_{J,n}$ ($n = 1, 2, \ldots$). This scheme permits the calculation of the $(n + 1)$th row, which we write as a $J - 1$ vector $\mathbf{U}_{n+1} = [U_{1,n+1}, \ldots, U_{J-1,n+1}]^T$, in terms of the nth row $\mathbf{U}_n = [U_{1,n}, \ldots, U_{J-1,n}]^T$. In matrix form

$$\begin{bmatrix} U_{1,n+1} \\ U_{2,n+1} \\ \vdots \\ U_{J-1,n+1} \end{bmatrix} = \begin{bmatrix} 1 - 2r & r & & & \\ r & 1 - 2r & r & & \\ & \ddots & \ddots & \ddots & \\ & & & & r \\ & & r & & 1 - 2r \end{bmatrix} \begin{bmatrix} U_{1,n} \\ U_{2,n} \\ \vdots \\ U_{J-1,n} \end{bmatrix}$$

$$+ \begin{bmatrix} rU_{0,n} \\ 0 \\ \vdots \\ 0 \\ rU_{J,n} \end{bmatrix}$$

or

$$\mathbf{U}_{n+1} = C\mathbf{U}_n + \mathbf{b}_n, \qquad n = 0, 1, \ldots \tag{57}$$

where C is the coefficient matrix and $\mathbf{b}_n$ is a vector containing the given boundary conditions. Thus the explicit scheme has been represented as an iterative process (57). To simplify the analysis take the boundary conditions $U_{0,n} = 0$, $U_{J,n} = 0$ for $n = 1, 2, \ldots$. Then $\mathbf{b}_n = \mathbf{0}$ and (57) is

$$\mathbf{U}_{n+1} = C\mathbf{U}_n, \qquad n = 0, 1, \ldots$$

Thus $\mathbf{U}_{n+1} = C\mathbf{U}_n = C(C\mathbf{U}_{n-1}) = C^2\mathbf{U}_{n-1} = \cdots = C^{n+1}\mathbf{U}_0$, so that the vector of values at the $(n + 1)$th time step is C^{n+1} applied to the fixed vector $\mathbf{U}_0$ containing the initial data. Now suppose the initial data are a set of errors at time $t = 0$. Then the error at the $(n + 1)$th step is

$$\mathbf{U}_{n+1} = C^{n+1}\mathbf{U}_0$$

and the problem is to determine under what condition does $\mathbf{U}_{n+1}$ remain bounded. This problem is addressed in more detail in Section 8.3, but we now state the result: $\mathbf{U}_{n+1}$ will remain bounded as $n \to \infty$ provided the eigenvalues of C do not exceed unity in modulus. Thus stability of the finite difference scheme is ensured if the eigenvalues of C are in absolute value bounded by 1.

We leave it as an exercise to show that the eigenvalues of C are given by

$$\lambda_j = 1 - 4r \sin^2 \frac{j\pi}{2J}, \qquad j = 1, \ldots, J - 1$$

Therefore the condition for stability of the explicit scheme is

$$\left| 1 - 4r \sin^2 \frac{j\pi}{2J} \right| \le 1, \qquad j = 1, \ldots, J - 1$$

which holds only when $r \le \frac{1}{2}$.

EXERCISES

2.1   Using the explicit formula (7) and initial conditions $U_{0,0} = \varepsilon$, $U_{j,0} = 0$ for $j \ne 0$, calculate by hand $U_{j,n}$ for $n = 1, \ldots, 5$ with $r = 1$ and with $r = \frac{1}{2}$ to determine how the error ε at $(0, 0)$ is propagated by the scheme.

2.2   Prove that the implicit scheme

$$-rU_{j-1, n+1} + (1 + 2r)U_{j, n+1} - rU_{j+1, n+1} = U_{j, n}$$

for the heat equation is von Neumann stable for any $r > 0$.

2.3   The purpose of this exercise is to investigate the numerical instability of the explicit scheme (7). Consider a steel bar that is 2 cm long. The initial temperature in degrees Celsius is given by

$$u = 100x, \qquad 0 \le x \le 1$$
$$u = 100(2 - x), \qquad 1 \le x \le 2$$

Both ends at $x = 0$ and $x = 2$ are maintained at zero degrees. For steel

$$\rho = 7.8 \text{ gm/cm}^3, \qquad c_v = 0.11 \text{ cal/gm} \cdot {}^\circ\text{C},$$
$$K = 0.13 \text{ cal/sec} \cdot \text{cm} \cdot {}^\circ\text{C}$$

With $\Delta x = 0.25$ numerically determine the temperature profile at $t = 0.99$ sec and at $t = 1.98$ sec using the explicit scheme with $r = 0.4$ and $r = 0.6$. Compare with the exact solution

$$u(x, t) = 800 \sum_{n=0}^{\infty} \frac{1}{\pi^2 (2n+1)^2} \cos \frac{(2n+1)\pi(x-1)}{2} e^{-0.3738(2n+1)^2 t}$$

2.4   Redo Exercise 2.3 using the implicit scheme (25).

2.5   Verify that the truncation error for the Crank-Nicolson scheme is $O(\Delta t^2 + \Delta x^2)$.

2.6   Modify the explicit scheme (7) to apply to the equation

$$u_t - ku_{xx} = c^2 u$$

and determine when the scheme is von Neumann stable.

2.7   A hollow tube 20 cm long is initially filled with air containing 2 percent ethyl alcohol vapors. At the bottom of the tube is a pool of alcohol that evaporates into the stagnant gas above. For ethyl alcohol the diffusion coefficient is 0.119 cm²/sec (at 30°C, at which temperature the system is maintained) and the vapor pressure is such that 10 volume percent alcohol in air is present at the surface. At the upper end of the tube the alcohol vapors dissipate to the outside air so that the concentration is essentially zero. Considering only the effects of molecular diffusion numerically determine the concentration of alcohol as a function of time and the distance x measured from the top of the tube (see Fig. 8.7). Use the implicit scheme. What is the long time $(t \to +\infty)$ concentration in the tube?

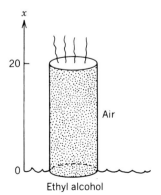

Figure 8.7. Exercise 2.7.

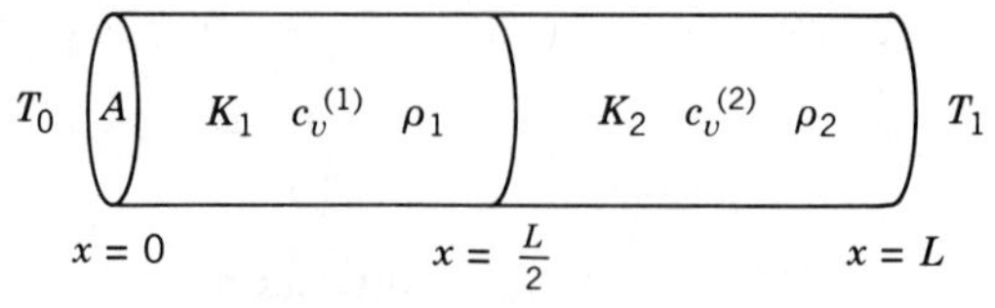

T_0 A K_1 $c_v^{(1)}$ ρ_1 K_2 $c_v^{(2)}$ ρ_2 T_1

$x = 0$ $x = \dfrac{L}{2}$ $x = L$

Figure 8.8. Exercise 2.8.

2.8   A bar of cross-sectional area A is composed of two materials, each homogeneous with constant physical parameters K, c_v, and ρ as shown in Fig. 8.8. The faces at $x = 0$ and $x = L$ are maintained at T_0 and T_1 degrees, respectively. There are no sources and the initial temperature in the bar is $f(x)$ degrees.

(a) What condition(s) must hold at the interface $x = L/2$?

(b) Write a boundary value problem governing the temperature distribution $u(x, t)$ in the bar (use $u = u_1$ for $x < L/2$ and $u = u_2$ for $x > L/2$).

(c) Adapt the implicit scheme (25) to this problem and develop a numerical algorithm to solve heat conduction problems with a material interface.

(d) Take $L = 2$ and $\Delta x = 0.25$ and write explicitly the tridiagonal system that must be solved at each time step.

(e) Use $L = 2$, $\Delta x = 0.25$, $T_0 = 10$, $T_1 = 0$, and $f(x) = 0$ and compute several temperature profiles for various values of t for a bar composed of copper ($0 \le x \le 1$) and cast iron ($1 \le x \le 2$) (see Table 1.1). Determine the long time ($t \to \infty$) temperature distribution.

2.9   It is desired to investigate the detonation properties of the high explosive RX-03-BB at low temperatures. Before an experiment can be performed a pellet of RX-03-BB initially at room temperature ($20°C$) must be cooled down. To accomplish this cool-down dry ice at $-25°C$ is put around a pellet that is 2 cm long and insulated on its lateral sides (see Fig. 8.9). The physical parameters of RX-03-BB are

$$\rho = 1.9 \text{ gm/cm}^3, \qquad c_v = 0.246 \text{ cal/gm} \cdot °C,$$
$$K = 0.002 \text{ cal/cm} \cdot °C \cdot \sec$$

Insulated —
cover

RX-03-BB—

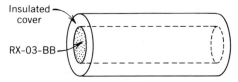

Figure 8.9. Exercise 2.9.

What is the characteristic time for this process? Sketch temperature profiles at $t = 0$, $t = 20$, $t = 60$, $t = 180$, and $t = 550$ sec, and sketch the temperature at the midpoint of the pellet as a function of time. Make a recommendation of what the length of time should be for a cool-down to $-25°C$. (Use $\Delta x = \frac{1}{8}$ cm and $\Delta t = 1$ sec with the explicit scheme.)

2.10 Find the eigenvalues of the $J \times J$ tridiagonal matrix

$$
\begin{bmatrix}
a & b & & & & \\
c & a & b & & & \\
  & c & a & b & & \\
  &   & \ddots & \ddots & \ddots & \\
  &   &   &   &   & b \\
  &   &   &   & c & a
\end{bmatrix}
$$

Solution:

$$
\lambda_j = a + 2b\sqrt{c/b}\,\cos\frac{j\pi}{J + 1}, \qquad j = 1, \ldots, J
$$

2.11 Use the matrix stability method to show that the implicit scheme (25) for the diffusion equation is stable for all $r > 0$.

2.12 Incorporate (46) with the explicit scheme (7) to obtain a finite difference scheme for (42) through (45) with $g = h = 0$ in the form $\mathbf{u}_{n+1} = C\mathbf{u}_n$, where $\mathbf{u}_n = [u_{1,n}, \ldots, u_{J,n}]$. Find the matrix C and perform a matrix stability analysis.

8.3   THE LAPLACE EQUATION

Finite Difference Approximations

One of the fundamental equations of mathematical physics in Laplace's equation

$$
u_{xx} + u_{yy} = 0, \qquad (x, y) \in R \tag{1}
$$

where R is a region in the xy plane. This equation is the prototype of elliptic equations and it arises in electrostatics, steady-state temperatures, hydromechanics, and other areas. As we observed in Example 2.1 in Chapter 4 the initial value problem for Laplace's equation is not well posed. Therefore problems involving Laplace's equation are pure boundary value problems describing equilibrium phenomena. There are two types of boundary conditions that are of interest. The first of these is a condition where u is specified on the boundary ∂R of R, that is

$$u(x, y) = f(x, y), \qquad (x, y) \in \partial R \qquad (2)$$

where $f(x, y)$ is a given function on ∂R. Laplace's equation (1) subject to (2) is called a *Dirichlet problem*.

The second type of boundary condition is one of the form

$$\frac{\partial u}{\partial n}(x, y) = f(x, y), \qquad (x, y) \in \partial R \qquad (3)$$

where $f(x, y)$ is a given function on the boundary. Here $\partial u/\partial n$ denotes the normal derivative, that is, the derivative of u on the boundary in the direction of the outward unit normal n, and so it is given by

$$\frac{\partial u}{\partial n} = \mathbf{grad}\, u \cdot \mathbf{n} \qquad (4)$$

A problem involving (1) subject to (3) is called a *Neumann problem*. Since $\partial u/\partial n$ is proportional to the flux through ∂R the condition describing an *insulated boundary* is

$$\frac{\partial u}{\partial n} = 0, \qquad (x, y) \in \partial R \qquad (5)$$

The numerical solution of problems involving Laplace's equation is carried out in a different manner from the time evolution problems involving the heat equation. Elliptic problems lead to a large-scale linear system of algebraic equations to be solved rather than a time-marching scheme typical of evolution problems.

Consider a Dirichlet problem on the unit square

$$u_{xx} + u_{yy} = 0, \qquad 0 < x < 1, \qquad 0 < y < 1 \qquad (6)$$
$$u(x, 0) = f(x), \qquad u(x, 1) = g(x), \qquad 0 \le x \le 1 \qquad (7)$$
$$u(0, y) = h(y), \qquad u(1, y) = e(y), \qquad 0 \le y \le 1 \qquad (8)$$

where f, g, h, and e are given functions. To illustrate the method we introduce a net with only four interior grid points: $(j\Delta x, k\,\Delta y)$ where $\Delta x = \Delta y = \frac{1}{3}$ and $j, k = 0, 1, 2, 3$ (see Fig. 8.10). We approximate Laplace's equation using centered difference approximations for both second derivatives to obtain

$$\frac{U_{j-1, k} - 2U_{j, k} + U_{j+1, k}}{\Delta x^2} + \frac{U_{j, k-1} - 2U_{j, k} + U_{j, k+1}}{\Delta y^2} = 0$$

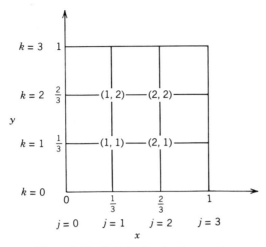

Figure 8.10. Grid for Laplace's equation.

for $j = 1, 2$ and $k = 1, 2$. This simplifies to

$$-\theta_x\left(U_{j-1, k} + U_{j+1, k}\right) + U_{j, k} - \theta_y\left(U_{j, k-1} + U_{j, k+1}\right) = 0 \qquad (9)$$

where the constants θ_x and θ_y are defined by

$$\theta_x = \frac{\Delta y^2}{2\left(\Delta x^2 + \Delta y^2\right)}, \qquad \theta_y = \frac{\Delta x^2}{2\left(\Delta x^2 + \Delta y^2\right)} \qquad (10)$$

The computational molecule consists of the five points shown in Fig. 8.11. When $\Delta x = \Delta y$ we have $\theta_x = \theta_y = \frac{1}{4}$ and (9) reduces to

$$U_{j, k} - \tfrac{1}{4}\left(U_{j-1, k} + U_{j+1, k} + U_{j, k-1} + U_{j, k+1}\right) = 0 \qquad (11)$$

for $j = 1, 2$ and $k = 1, 2$. Thus Equation (11) gives four equations for the four unknowns $U_{1,1}$, $U_{1,2}$, $U_{2,1}$, and $U_{2,2}$. The values of U on the boundary of the grid are given by the boundary conditions (7) and (8). The system of

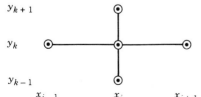

Figure 8.11. Computational molecule for Laplace's equation.

Equations (11) can be written in matrix form as

$$
\begin{bmatrix}
1 & -\frac{1}{4} & -\frac{1}{4} & 0 \\
-\frac{1}{4} & 1 & 0 & -\frac{1}{4} \\
-\frac{1}{4} & 0 & 1 & -\frac{1}{4} \\
0 & -\frac{1}{4} & -\frac{1}{4} & 1
\end{bmatrix}
\begin{bmatrix}
U_{1,1} \\
U_{1,2} \\
U_{2,1} \\
U_{2,2}
\end{bmatrix}
=
\begin{bmatrix}
\frac{1}{4}U_{0,1} + \frac{1}{4}U_{1,0} \\
\frac{1}{4}U_{0,2} + \frac{1}{4}U_{1,3} \\
\frac{1}{4}U_{2,0} + \frac{1}{4}U_{3,1} \\
\frac{1}{4}U_{3,2} + \frac{1}{4}U_{2,3}
\end{bmatrix}
\tag{12}
$$

In the general case when the grid is defined by $x_j = j\Delta x$, $y_k = k\,\Delta y$ for $i = 0, \ldots, J + 1$ and $k = 0, \ldots, K + 1$, where $\Delta x = 1/(J + 1)$ and $\Delta y = 1/(K + 1)$, then equations (9) hold for $i = 1, \ldots, J$ and $k = 1, \ldots, K$, and therefore represent a system of JK equations in the JK unknowns represented by the vector $\mathbf{U} = [U_{1,1}, \ldots, U_{1,K}, U_{2,1}, \ldots, U_{2,K}, \ldots, U_{J,1}, \ldots, U_{J,K}]^T$. This system is $A\mathbf{U} = \mathbf{f}$, where the coefficient matrix A has the form

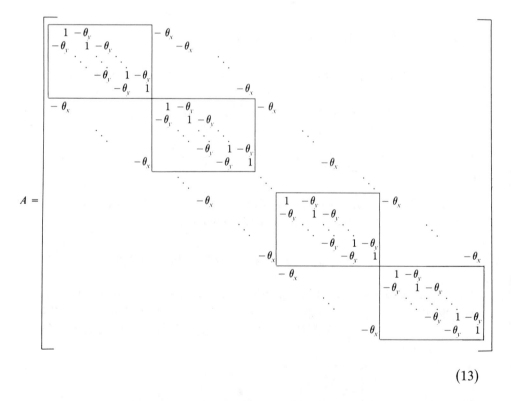

$$\tag{13}$$

and $\mathbf{f}$ is a JK column vector containing the boundary data. It is interesting to note that in a steady-state temperature context (11) states that the temperature $U_{j,k}$ at (x_j, y_k) is the average of the temperatures of the four neighboring points.

Iterative Methods for Linear Systems

The discretization of Laplace's equation subject to a Dirichlet condition leads to a linear system of algebraic equations to be solved for the values of U at the interior grid positions. To obtain (12) we introduced for purpose of illustration only four interior grid points. It seems clear that in most applications we will enforce a much finer lattice structure, since intuition suggests that the finer the grid, the better the discrete approximation to the exact solution. Hence, it is not unusual to divide the intervals $0 \le x \le 1$ and $0 \le y \le 1$ into say 100 subintervals each, thereby giving on the order of 10^4 interior grid points and therefore a linear system with 10^4 unknowns.

Consequently, we focus on large linear systems and methods for their efficient solution. For definiteness we consider the system of equations

$$A\mathbf{x} = \mathbf{f}$$

or

$$
\begin{bmatrix} a_{11} & \cdots & a_{1n} \\ \vdots & & \vdots \\ a_{n1} & \cdots & a_{nn} \end{bmatrix}
\begin{bmatrix} x_1 \\ \vdots \\ x_n \end{bmatrix}
=
\begin{bmatrix} f_1 \\ \vdots \\ f_n \end{bmatrix}
\tag{14}
$$

At first glance it may appear formidable to attempt the solution of (14) when $n = 10{,}000$. From elementary results in numerical analysis it is noted that Gaussian elimination, the standard direct method, requires in the order of n^3 operations to obtain a solution. Hence, if $n = 10{,}000$, then on the order of 10^{12} operations would be required. Therefore we seek more efficient methods, particularly those that take advantage of the sparsity (or large number of zero entries) of the matrix A, as is evident from (13).

A class of methods strikingly suitable for large-scale sparse systems are the iterative methods. The idea is from an initial guess $\mathbf{x}^{(0)}$ to generate recursively a sequence of approximations $\mathbf{x}^{(0)}, \mathbf{x}^{(1)}, \mathbf{x}^{(2)}, \ldots$, that converges to the exact solution. Most iterative schemes involve introducing a *splitting* of A,

$$A = N - P, \qquad \det N \ne 0$$

so that (14) becomes $\mathbf{x} = N^{-1}P\mathbf{x} + N^{-1}\mathbf{f}$. Letting $M = N^{-1}P$ and $\mathbf{b} = N^{-1}\mathbf{f}$ gives a system equivalent to (14), namely

$$\mathbf{x} = M\mathbf{x} + \mathbf{b} \tag{15}$$

with $\mathbf{x}^{(0)}$ an initial approximation we set up the iterative scheme

$$\mathbf{x}^{(\nu)} = M\mathbf{x}^{(\nu-1)} + \mathbf{b}, \qquad \nu = 1, 2, \ldots \tag{16}$$

Questions of convergence can be addressed by defining the error vector $\mathbf{e}^{(\nu)}$ by

$$\mathbf{e}^{(\nu)} = \mathbf{x}^{(\nu)} - \bar{\mathbf{x}}, \qquad \nu = 1, 2, 3, \ldots$$

where $\bar{\mathbf{x}}$ is the exact solution that is not known a priori. It is easy to see that the error at the νth step is

$$\mathbf{e}^{(\nu)} = M^{\nu}\mathbf{e}^{(0)} \tag{17}$$

where $\mathbf{e}^{(0)}$ is initial error in the approximation $\mathbf{x}^{(0)}$ (Exercise 3.2). Thus if the matrix $M = (m_{ij})$ has the property that

$$\lim_{\nu \to \infty} M^{\nu} = 0 \tag{18}$$

then the iterative method (16) will converge for any initial approximation $\mathbf{x}^{(0)}$. A matrix M satisfying equation (18) is called a *convergent matrix*. Unfortunately it is not easy to check condition (18) and so we seek an alternate condition to ensure convergence of the scheme (16).

Before taking up this important topic we examine two splittings of matrix A and the set up of the resulting iterative schemes.

Example 3.1 (The Jacobi Scheme) Take $N \equiv D \equiv \operatorname{diag}(a_{11}, a_{22}, \dots, a_{nn})$ and $P \equiv -L - U$, where L is the triangular part of A below the diagonal and U is the triangular part of A above the diagonal. Then the system (14) becomes

$$\mathbf{x} = D^{-1}(-L - U)\mathbf{x} + D^{-1}\mathbf{f} \equiv M_J\mathbf{x} + \mathbf{b}_J$$

where

$$M_J = (m_{ij}) = \begin{cases} 0, & \text{if } i = j, \\[2mm] -\dfrac{a_{ij}}{a_{ii}}, & \text{if } i \neq j, \end{cases} \qquad \mathbf{b}_J = \left(\dfrac{f_i}{a_{ii}}\right)$$

Then the iterative scheme (16) becomes

$$\mathbf{x}^{(\nu)} = M_J\mathbf{x}^{(\nu-1)} + \mathbf{b}_J \tag{19}$$

or in component form

$$x_i^{(\nu)} = -\frac{1}{a_{ii}}\sum_{j=1}^{n} a_{ij}x_j^{(\nu-1)} + \frac{f_i}{a_{ii}}, \qquad i = 1, \dots, n$$

This is the *Jacobi scheme*. We note that (19) is the same equation we would obtain if we had solved the ith equation in the original system for x_i and then set up the iterative scheme.

Example 3.2 Consider the linear system

$$\begin{bmatrix} 4 & 3 & 0 \\ 3 & 4 & -1 \\ 0 & -1 & 4 \end{bmatrix}\begin{bmatrix} x_1 \\ x_2 \\ x_3 \end{bmatrix} = \begin{bmatrix} 24 \\ 30 \\ -24 \end{bmatrix}$$

We split the matrix A via

$$\begin{bmatrix} 4 & 3 & 0 \\ 3 & 4 & -1 \\ 0 & -1 & 4 \end{bmatrix} = \begin{bmatrix} 4 & 0 & 0 \\ 0 & 4 & 0 \\ 0 & 0 & 4 \end{bmatrix} - \begin{bmatrix} 0 & -3 & 0 \\ -3 & 0 & 1 \\ 0 & 1 & 0 \end{bmatrix}$$

and rewrite the system as

$$\begin{bmatrix} 4 & 0 & 0 \\ 0 & 4 & 0 \\ 0 & 0 & 4 \end{bmatrix}\begin{bmatrix} x_1 \\ x_2 \\ x_3 \end{bmatrix} = \begin{bmatrix} 0 & -3 & 0 \\ -3 & 0 & 1 \\ 0 & 1 & 0 \end{bmatrix}\begin{bmatrix} x_1 \\ x_2 \\ x_3 \end{bmatrix} + \begin{bmatrix} 24 \\ 30 \\ -24 \end{bmatrix}$$

or

$$\begin{bmatrix} x_1 \\ x_2 \\ x_3 \end{bmatrix} = \begin{bmatrix} 0 & -\frac{3}{4} & 0 \\ -\frac{3}{4} & 0 & \frac{1}{4} \\ 0 & \frac{1}{4} & 0 \end{bmatrix}\begin{bmatrix} x_1 \\ x_2 \\ x_3 \end{bmatrix} + \begin{bmatrix} 6 \\ \frac{15}{2} \\ -6 \end{bmatrix}$$

The Jacobi scheme is therefore

$$\begin{bmatrix} x_1^{(\nu)} \\ x_2^{(\nu)} \\ x_3^{(\nu)} \end{bmatrix} = \begin{bmatrix} 0 & -\frac{3}{4} & 0 \\ -\frac{3}{4} & 0 & \frac{1}{4} \\ 0 & \frac{1}{4} & 0 \end{bmatrix}\begin{bmatrix} x_1^{(\nu-1)} \\ x_2^{(\nu-1)} \\ x_3^{(\nu-1)} \end{bmatrix} + \begin{bmatrix} 6 \\ \frac{15}{2} \\ -6 \end{bmatrix} \qquad (20)$$

for $\nu = 1, 2, 3, \ldots$, with $[x_1^{(0)}, x_2^{(0)}, x_3^{(0)}]^T$ given. If for example $[x_1^{(0)}, x_2^{(0)}, x_3^{(0)}]^T = [1, 1, 1]^T$, then the first iterate is

$$\begin{bmatrix} x_1^{(1)} \\ x_2^{(1)} \\ x_3^{(1)} \end{bmatrix} = \begin{bmatrix} 0 & -\frac{3}{4} & 0 \\ -\frac{3}{4} & 0 & \frac{1}{4} \\ 0 & \frac{1}{4} & 0 \end{bmatrix}\begin{bmatrix} 1 \\ 1 \\ 1 \end{bmatrix} + \begin{bmatrix} 6 \\ \frac{15}{2} \\ -6 \end{bmatrix} = \begin{bmatrix} \frac{21}{4} \\ 7 \\ -\frac{23}{4} \end{bmatrix}$$

The next iterate is

$$\begin{bmatrix} x_1^{(2)} \\ x_2^{(2)} \\ x_3^{(2)} \end{bmatrix} = \begin{bmatrix} 0 & -\frac{3}{4} & 0 \\ -\frac{3}{4} & 0 & \frac{1}{4} \\ 0 & \frac{1}{4} & 0 \end{bmatrix}\begin{bmatrix} \frac{21}{4} \\ 7 \\ -\frac{23}{4} \end{bmatrix} + \begin{bmatrix} 6 \\ \frac{15}{2} \\ -6 \end{bmatrix} = \begin{bmatrix} \frac{3}{4} \\ \frac{17}{8} \\ -\frac{17}{4} \end{bmatrix}$$

and so on.

Example 3.3 (The Gauss-Seidel Method)   An obvious improvement in the Jacobi scheme can be realized if at each time a new component of $\mathbf{x}^{(\nu)}$ is computed we use it immediately to compute the next component. Returning to the last example, Equation (20) in component form is

$$
\begin{aligned}
x_1^{(\nu)} &= \qquad -\tfrac{3}{4}x_2^{(\nu-1)} \qquad\qquad +6 \\[4pt]
x_2^{(\nu)} &= -\tfrac{3}{4}x_1^{(\nu-1)} \qquad\qquad +\tfrac{1}{4}x_3^{(\nu-1)} + \tfrac{15}{2} \\[4pt]
x_3^{(\nu)} &= \qquad\qquad \tfrac{1}{4}x_2^{(\nu-1)} \qquad\qquad -6
\end{aligned}
\tag{21}
$$

With the same initial vector $(1,1,1)^T$ we compute $x_1^{(1)}$ as before, namely

$$
x_1^{(1)} = -\tfrac{3}{4}(1) + 6 = \tfrac{21}{4}
$$

But now to compute $x_2^{(1)}$ we use $\tfrac{21}{4}$ or $x_1^{(1)}$ for the first component rather than $x_1^{(0)} = 1$. Hence

$$
x_2^{(1)} = -\tfrac{3}{4}\left(\tfrac{21}{4}\right) + \tfrac{1}{4}(1) + \tfrac{15}{2} = \tfrac{61}{16}
$$

To compute $x_3^{(1)}$ we do not use the values $x_1^{(0)} = 1$ and $x_2^{(0)} = 1$, but rather $x_1^{(1)} = \tfrac{21}{4}$ and $x_2^{(1)} = \tfrac{61}{16}$, respectively. Hence

$$
x_3^{(1)} = \tfrac{1}{4}\left(\tfrac{61}{16}\right) - 6 = -\tfrac{323}{64}
$$

Therefore the first iterate is

$$
\mathbf{x}^{(1)} = \begin{bmatrix} x_1^{(1)} \\ x_2^{(1)} \\ x_3^{(1)} \end{bmatrix} = \begin{bmatrix} \tfrac{21}{4} \\ \tfrac{61}{16} \\ -\tfrac{323}{64} \end{bmatrix}
$$

Rather than use (21) it appears more efficient to use

$$
\begin{aligned}
x_1^{(\nu)} &= \qquad -\tfrac{3}{4}x_2^{(\nu-1)} \qquad\qquad +6 \\[4pt]
x_2^{(\nu)} &= -\tfrac{3}{4}x_1^{(\nu)} \qquad\qquad +\tfrac{1}{4}x_3^{(\nu-1)} + \tfrac{15}{2} \\[4pt]
x_3^{(\nu)} &= \qquad\qquad \tfrac{1}{4}x_2^{(\nu)} \qquad\qquad -6
\end{aligned}
$$

for $\nu = 1, 2, 3, \ldots$. This is the *Gauss-Seidel method*. In general, the Gauss-Seidel method is

$$
x_i^{(\nu)} = -\frac{1}{a_{ii}}\left(\sum_{j=1}^{i-1} a_{ij}x_j^{(\nu)} + \sum_{j=i+1}^{n} a_{ij}x_j^{(\nu-1)} \right) + \frac{f_i}{a_{ii}}
\tag{22}
$$

for $\nu = 1, 2, 3, \ldots$ Equation (22) can be obtained from a splitting of A by defining $N \equiv L + D$ and $P \equiv -U$, and then the system (14) can be written

$$(L + D)\mathbf{x} = -U\mathbf{x} + \mathbf{f}$$

or

$$\mathbf{x} = -(L + D)^{-1}U\mathbf{x} + (L + D)^{-1}\mathbf{f}$$
$$\equiv M_{GS}\mathbf{x} + \mathbf{b}_{GS}$$

where $M_{GS} \equiv -(L + D)^{-1}U$ and $\mathbf{b}_{GS} \equiv (L + D)^{-1}\mathbf{f}$. The Gauss-Seidel method is then

$$\mathbf{x}^{(\nu)} = M_{GS}\mathbf{x}^{(\nu-1)} + \mathbf{b}_{GS}, \qquad \nu = 1, 2, \dots$$

which in component form becomes (22). The Gauss-Seidel method is roughly twice as fast as the Jacobi method. The convergence of these schemes is discussed after the application to the solution of Laplace's equation.

The solution of the system (12) arising from the discretization of Laplace's equation is particularly tailored to application of the Gauss-Seidel method. The system (12) is just the four difference equations

$$U_{j,k} = \tfrac{1}{4}\left(U_{j-1,k} + U_{j+1,k} + U_{j,k-1} + U_{j,k+1}\right) \tag{23}$$

for $k = 1, 2$; $j = 1, 2$, with unknowns $U_{1,1}$, $U_{1,2}$, $U_{2,1}$, and $U_{2,2}$. Upon reflection we observe that the Gauss-Seidel scheme at each step marches through the components $U_{1,1}$, $U_{1,2}$, $U_{2,1}$, and $U_{2,2}$ (say in that order) while at each time updating the approximate vector with the latest computed component. Therefore the Gauss-Seidel scheme corresponding to (23) is

$$U_{j,k}^{(\nu)} = \tfrac{1}{4}\left(U_{j-1,k}^{(\nu)} + U_{j+1,k}^{(\nu-1)} + U_{j,k-1}^{(\nu)} + U_{j,k+1}^{(\nu-1)}\right)$$

for $\nu = 1, 2, 3, \dots$. For example, to compute $U_{2,1}^{(\nu)}$ we may use $U_{1,1}^{(\nu)}$ and $U_{1,2}^{(\nu)}$ (both at the νth step), since we have already computed their values; however, we must use $U_{2,2}^{(\nu-1)}$, since it holds the most recent information regarding $U_{2,2}$. Therefore we can describe in algorithmic form a procedure for solving (23), and hence (12), by the Gauss-Seidel method.

Read in the boundary values $U_{0,0}$, $U_{0,1}$, $U_{0,2}$, $U_{0,3}$, $U_{3,0}$, $U_{3,1}$, $U_{3,2}$, $U_{3,3}$, $U_{1,0}$, $U_{2,0}$, $U_{1,3}$, $U_{2,3}$
Initialize the array $U_{j,k}$ for $j = 1, 2$; $k = 1, 2$.
FOR $\nu = 1$ TO the number of iterations
FOR $j = 1$ TO 2
FOR $k = 1$ TO 2
$\quad U_{j,k} = \tfrac{1}{4}(U_{j-1,k} + U_{j+1,k} + U_{j,k-1} + U_{j,k+1})$
NEXT k
NEXT j
NEXT ν

Some test should be included that terminates the iterations when successive approximations become close in one sense or another, as discussed in the sequel.

Example 3.4 Consider the problem

$$u_{xx} + u_{yy} = 0, \qquad 0 < x, \, y < 1$$

with

$$u(x,0) = \sin \pi x, \qquad 0 < x < 1$$

$$u(x,1) = 0, \qquad 0 < x < 1$$

$$u(0, y) = \sin \pi y, \qquad 0 < y < 1$$

$$u(1, y) = 0, \qquad 0 < y < 1$$

Taking $\Delta x = \Delta y = 0.1$, the Gauss-Seidel method converges to three decimal places in about fifty-five iterations. The BASIC program is given in Program Listing 8.3 and the computed steady-state temperature distribution is shown in Table 8.3.

```
10 REM--GAUSS/SEIDEL FOR LAPLACE'S EQUATION--
20 DIM U(11,11)
30 REM--DISCRETIZE THE DOMAIN--
40    DELX=.1:DELY=.1
50 REM--SET BOUNDARY CONDITIONS--
60    FOR J=0 TO 10
70    U(J,0)=SIN(J*3.14159*DELX)
80    U(J,10)=0
90    NEXT J
100   FOR K=0 TO 10
110   U(0,K)=SIN(K*3.14159*DELY)
120   U(10,K)=0
130   NEXT K
140 REM--INITIALIZE U(J,K) TO ZERO--
150    FOR J=1 TO 9: FOR K=1 TO 9
160    U(J,K)=0
170    NEXT K:NEXT J
180 REM--ITERATE USING GAUSS/SEIDEL--
190    FOR NU=1 TO 75
200    FOR J=1 TO 9: FOR K=1 TO 9
210    U(J,K)=.25*(U(J-1,K)+U(J+1,K)+U(J,K-1)+U(J,K+1))
220    NEXT K: NEXT J
230    NEXT NU
240 REM--PRINT OUT U(J,K)--
250    FOR K=10 TO 0 STEP -1: FOR J=0 TO 10
260    LPRINT USING "#.### ";U(J,K);
270    NEXT J: LPRINT: NEXT K
280 END
```

TABLE 8.3

0.000	0.000	0.000	0.000	0.000	0.000	0.000	0.000	0.000	0.000	0.000
0.309	0.235	0.181	0.143	0.114	0.090	0.070	0.052	0.035	0.017	0.000
0.588	0.448	0.348	0.276	0.221	0.177	0.139	0.104	0.069	0.035	0.000
0.809	0.621	0.488	0.391	0.318	0.259	0.205	0.155	0.104	0.052	0.000
0.951	0.739	0.591	0.484	0.402	0.333	0.269	0.205	0.139	0.070	0.000
1.000	0.793	0.652	0.551	0.473	0.403	0.333	0.259	0.177	0.090	0.000
0.951	0.782	0.673	0.597	0.535	0.473	0.402	0.318	0.221	0.114	0.000
0.809	0.711	0.660	0.628	0.597	0.551	0.484	0.391	0.276	0.143	0.000
0.588	0.595	0.627	0.660	0.673	0.652	0.591	0.488	0.348	0.181	0.000
0.309	0.452	0.595	0.711	0.782	0.793	0.739	0.621	0.448	0.235	0.000
0.000	0.309	0.588	0.809	0.951	1.000	0.951	0.809	0.588	0.309	0.000

(y axis at left, x axis at bottom)

Convergence criteria for iterative schemes can be formulated in terms of matrix norms. First, a *vector norm* on R^n is a rule that associates with each vector $\mathbf{x} = [x_1, \ldots, x_n]^T$ in R^n a real number $\|\mathbf{x}\|$ called the *norm* of $\mathbf{x}$ and that satisfies the conditions

   (i) $\|\mathbf{x}\| \geq 0$ for all $\mathbf{x}$ and $\|\mathbf{x}\| = 0$ if and only if $\mathbf{x} = \mathbf{0}$.

   (ii) $\|c\mathbf{x}\| = |c| \, \|\mathbf{x}\|$ for all $\mathbf{x}$ and real numbers c.

   (iii) $\|\mathbf{x} + \mathbf{y}\| \leq \|\mathbf{x}\| + \|\mathbf{y}\|$ for all vectors $\mathbf{x}$ and $\mathbf{y}$.

The norm of a vector $\mathbf{x}$ is a nonnegative real number that is a measure of the size of $\mathbf{x}$; the reader may be familiar with the following common norms.

   (a) $\displaystyle \|\mathbf{x}\|_1 = \sum_{i=1}^{n} |x_i|$ (1-norm)

   (b) $\displaystyle \|\mathbf{x}\|_2 = \left(\sum_{i=1}^{n} |x_i|^2 \right)^{1/2}$ (2-norm)

   (c) $\displaystyle \|\mathbf{x}\|_\infty = \max_{1 \leq i \leq n} |x_i|$ (infinity norm)

For example if $\mathbf{x} = [1, -2, 3]^T$, then $\|\mathbf{x}\|_1 = 6$, $\|\mathbf{x}\|_2 = \sqrt{14}$, and $\|\mathbf{x}\|_\infty = 3$. It is not difficult to check that norms (a) through (c) each satisfies the conditions (i), (ii), and (iii) for a norm.

   Given a vector norm $\| \cdot \|$ there is an automatic norm induced on the set of $n \times n$ matrices defined by

$$\|A\| = \sup_{\mathbf{x} \neq 0} \frac{\|A\mathbf{x}\|}{\|\mathbf{x}\|} \tag{24}$$

where A is an $n \times n$ matrix. Here $\|A\|$ is called the *induced matrix norm* and it measures the size of matrix A. By definition $\|A\|$ is the largest of the ratios of the size of $A\mathbf{x}$ to the size of $\mathbf{x}$ as $\mathbf{x}$ ranges over all nonzero vectors. We leave, as an exercise, showing that a matrix norm defined by (24) satisfies the conditions

$$\|A\mathbf{x}\| \le \|A\| \, \|\mathbf{x}\| \tag{25}$$

and

$$\|AB\| \le \|A\| \, \|B\| \tag{26}$$

The following theorem provides simple formulas for the matrix norms induced by the vector norms $\|\mathbf{x}\|_1$, $\|\mathbf{x}\|_2$, and $\|\mathbf{x}\|_\infty$.

Theorem 3.1 *Let $A = (a_{ij})$ be an $n \times n$ matrix. Then*

$$\|A\|_1 = \max_{1 \le j \le n} \sum_{i=1}^{n} |a_{ij}| \tag{27}$$

$$\|A\|_2 = \sqrt{\rho(A^T A)} \tag{28}$$

$$\|A\|_\infty = \max_{1 \le i \le n} \sum_{j=1}^{n} |a_{ij}| \tag{29}$$

are the matrix norms induced by the vector norms $\|\mathbf{x}\|_1$, $\|\mathbf{x}\|_2$, and $\|\mathbf{x}\|_\infty$, respectively, and where for any $n \times n$ matrix M the quantity $\rho(M)$ is the spectral radius of M defined by

$$\rho(M) \equiv \max |\lambda_i|$$

where the λ_i are the eigenvalues of M.

Hence the 1-norm $\|A\|_1$ is the maximum absolute column sum of A, and the infinity norm $\|A\|_\infty$ is the maximum absolute row sum of A. The 2-norm $\|A\|_2$, which involves the maximum absolute eigenvalue of $A^T A$, is harder to compute. If A is real and symmetric, then $\|A\|_2 = \rho(A)$, so the 2-norm is just the spectral radius of A.

We prove (29) and leave the proofs of (27) and (28) to the reader. By definition

$$\|A\|_\infty = \sup_{\mathbf{x} \ne 0} \frac{\|A\mathbf{x}\|_\infty}{\|\mathbf{x}\|_\infty}$$

If $A = (a_{ij})$ and $\mathbf{x} = [x_1, \ldots, x_n]^T$, then

$$\frac{\|A\mathbf{x}\|_\infty}{\|\mathbf{x}\|_\infty} = \frac{\displaystyle\max_{1 \le i \le n} \left| \sum_{j=1}^n a_{ij} x_j \right|}{\displaystyle\max_{1 \le k \le n} |x_k|}$$

$$= \max_{1 \le i \le n} \left| \sum_{j=1}^n a_{ij} \frac{x_j}{\displaystyle\max_{1 \le k \le n} |x_k|} \right| \le \max_{1 \le i \le n} \sum_{j=1}^n |a_{ij}|$$

Therefore

$$\|A\|_\infty \le \max_{1 \le i \le n} \sum_{j=1}^n |a_{ij}| \tag{30}$$

To obtain equality we show there exists a vector x for which the supremum is attained. Suppose the maximum on the right side of (30) occurs for the index $i = I$. Then setting $x = [\operatorname{sgn} a_{I1}, \operatorname{sgn} a_{I2}, \ldots, \operatorname{sgn} a_{In}]$ gives

$$\frac{\|A\mathbf{x}\|_\infty}{\|\mathbf{x}\|_\infty} = \frac{\displaystyle\sum_{j=1}^n |a_{Ij}|}{1} = \sum_{j=1}^n |a_{Ij}|$$

Thus (29) follows.

Example 3.5 Let

$$A = \begin{bmatrix} 0 & -1 & 2 \\ 1 & 4 & 6 \\ 2 & -3 & 0 \end{bmatrix}$$

Then $\|A\|_\infty = 11$ and $\|A\|_1 = 8$.

The following theorem shows there is a close relationship between the norm of a matrix M, the spectral radius $\rho(M)$, and M being a convergent matrix.

Theorem 3.2 *The following are equivalent.*

(i) *M is a convergent matrix.*
(ii) *$\rho(M) < 1$.*
(iii) *There exists a matrix norm $\| \cdot \|$ for which $\|M\| < 1$.*

Some portions of the proof of Theorem 3.2 are requested in the exercises. The remaining implications are proved in Isaacson and Keller [5], and are based on the fact that for each $n \times n$ matrix M, and for every $\varepsilon > 0$ a natural norm $\| \cdot \|$ can be found such that $\rho(M) \le \|M\| \le \rho(M) + \varepsilon$.

An approximate value for the number of iterations required to reduce an initial error by a given amount in a convergent scheme (16) can be phrased in terms of the spectral radius. In the 2-norm we have, using (17), (25), and (26),

$$\|\mathbf{e}^{(\nu)}\|_2 \leq \|M\|_2^{\nu}\|\mathbf{e}^{(0)}\|_2$$

where $\|\mathbf{e}^{(0)}\|_2$ is the magnitude of the initial error in the 2-norm. Assuming M is real and symmetric, we have

$$\|\mathbf{e}^{(\nu)}\| \leq \rho(M)^{\nu}\|\mathbf{e}^{(0)}\|_2$$

Therefore if $\rho(M)^{\nu} = 10^{-m}$ or

$$\nu = \frac{m}{-\log\rho(M)}$$

then the initial error is reduced by 10^{-m} in ν iterations. The positive number $-\log\rho(M)$ is called the *rate of convergence*; the closer $\rho(M)$ to unity, the slower the convergence.

In practice it is difficult to compute the spectral radius, and the initial error $\mathbf{e}^{(0)}$ is not known a priori. To determine when the iterative process should stop it is common practice to calculate the residual vector $\mathbf{r}^{(\nu)}$, which is defined as the difference between successive iterates, that is,

$$\mathbf{r}^{(\nu)} \equiv \mathbf{x}^{(\nu+1)} - \mathbf{x}^{(\nu)} = M\mathbf{x}^{(\nu)} + \mathbf{b} - \mathbf{x}^{(\nu)}$$

Thus $\mathbf{r}^{(\nu)}$ is a measure of how well the νth iterate $\mathbf{x}^{(\nu)}$ satisfies the original equation $\mathbf{x} = M\mathbf{x} + \mathbf{b}$; if $\mathbf{r}^{(\nu)} \equiv 0$, then $\mathbf{x}^{(\nu)}$ is the exact solution. If the residual is small, then it may appear that $\mathbf{x}^{(\nu)}$ is near the exact solution $\bar{\mathbf{x}}$. However

$$\mathbf{x}^{(\nu)} - \bar{\mathbf{x}} = M\mathbf{x}^{(\nu-1)} + \mathbf{b} - M\bar{\mathbf{x}} - \mathbf{b} = -M(\mathbf{x}^{(\nu)} - \mathbf{x}^{(\nu-1)}) + M(\mathbf{x}^{(\nu)} - \bar{\mathbf{x}})$$

from which it follows that

$$\|\mathbf{e}^{(\nu)}\| \leq \|M\|\|\mathbf{r}^{(\nu-1)}\| + \|M\|\|\mathbf{e}^{(\nu)}\|$$

Hence

$$\|\mathbf{e}^{(\nu)}\| \leq \frac{\|M\|}{1 - \|M\|}\|\mathbf{r}^{(\nu-1)}\|$$

If $\|M\| \leq \frac{1}{2}$, then $\|\mathbf{e}^{(\nu)}\| \leq \|\mathbf{r}^{(\nu-1)}\|$ and a small residual implies a small error. But if $\frac{1}{2} < M < 1$, then the adjacent iterates can be close without $\mathbf{e}^{(\nu)}$ being small. This fact must be remembered when using a small residual condition to terminate an iterative process.

We now return to the question of convergence of the Jacobi and Gauss-Seidel schemes defined by the iterative processes (19) and (22). For the Jacobi scheme

$$\|M_J\|_\infty = \max_{1 \le i \le n} \sum_{j=1}^n |m_{ij}| = \max_{1 \le i \le n} \sum_{j=1}^n \frac{|a_{ij}|}{|a_{ii}|}$$

From Theorem 3.2 the Jacobi scheme will converge if $\|M_J\|_\infty < 1$, that is, if

$$\sum_{\substack{j=1 \\ j \ne i}}^n |a_{ij}| < |a_{ii}|, \qquad i = 1, \ldots, n \tag{31}$$

A matrix A satisfying (31) is called *strictly diagonally dominant*. Inequality (31) requires the diagonal element in each row to exceed in absolute value the sum of the absolute values of the remaining elements in the row. A is said to be *diagonally dominant* if

$$\sum_{\substack{j=1 \\ j \ne i}}^n |a_{ij}| \le |a_{ii}|, \qquad i = 1, \ldots, n,$$

with strict inequality holding for at least one index i. We have the following theorem.

Theorem 3.3   *If A is strictly diagonally dominant, then the Jacobi and Gauss-Seidel methods converge for any choice of the initial vector $\mathbf{x}^{(0)}$.*

We have proved the previous theorem for the Jacobi scheme. The proof of convergence of the Gauss-Seidel method is left to the reader.

We notice from (13) that the discretization of Laplace's equation does not always lead to a strictly diagonally dominant matrix A, but rather one that is diagonally dominant. Diagonal dominance in itself is not enough to guarantee convergence and another condition is required. It can be formulated in the following manner. With each $n \times n$ matrix $A = (a_{ij})$ we associate a *directed graph* consisting of the n points $\{1, \ldots, n\}$ by drawing an arrow from i to j if $a_{ij} \ne 0$ and no arrow if $a_{ij} = 0$. For example, the directed graph of

$$A = \begin{bmatrix} 0 & 1 & 0 & 1 \\ 1 & 0 & 1 & 0 \\ 0 & 1 & 0 & 1 \\ 1 & 0 & 0 & 1 \end{bmatrix}$$

is shown in Fig. 8.12. This graph has the property that there is a directed path from i to j for all i and j. Such a graph is called *strongly connected*. On the

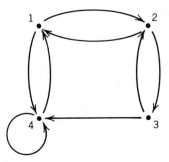

Figure 8.12. Strongly connected graph.

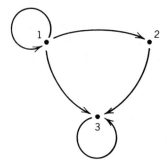

Figure 8.13. Graph not strongly connected.

other hand, the matrix

$$A = \begin{bmatrix} 1 & 1 & 1 \\ 0 & 0 & 1 \\ 0 & 0 & 1 \end{bmatrix}$$

has the graph shown in Fig. 8.13. It is not strongly connected since there is no path from 3 to 2. If a matrix A has a strongly connected graph, then it is called *irreducible*. The fundamental theorem, which we do not prove, is as follows.

Theorem 3.4   *If A is diagonally dominant and irreducible, then*

$$\rho(M_{GS}) < \rho(M_J) < 1$$

and the Jacobi and Gauss-Seidel schemes converge.

Other sufficient conditions can be formulated in terms of positive definiteness (see Varga [4]). Generally the matrix that results from a discretization of Laplace's equation will satisfy the hypotheses of Theorem 3.4, provided the discretization is done properly. As finer and finer grids are introduced, however, the maximum absolute eigenvalue approaches unity and the convergence becomes slower.

It is important to distinguish two different convergent processes in the application of iterative methods to solve the difference equations associated with Laplace's equation. One is the convergence of the iterative process itself and the second is the convergence of the scheme, that is, of $U_{j,k}$ to $u(x_j, y_k)$ as Δx and Δy tend to zero. The iterative process is carried out for fixed values of Δx and Δy; if it converges with a certain accuracy to some value $\hat{U}_{j,k}$ at (x_j, y_k), it is not guaranteed that $\hat{U}_{j,k} - u(x_j, y_k)$ possesses the same degree of smallness.

The Neumann Problem

We now show how to handle boundary conditions of the type $\partial u/\partial n = g$, where g is a given function on the boundary of the region. By way of illustration we consider the following boundary value problem for Laplace's equation on a square R: $0 \le x \le 1, 0 \le y \le 1$.

$$u_{xx} + u_{yy} = 0, \qquad (x, y) \in R \qquad (32)$$

$$u(x, 0) = f(x), \qquad u(x, 1) = g(x), \quad 0 < x < 1 \qquad (33)$$

$$u_x(0, y) = h(y), \qquad u(1, y) = e(y), \quad 0 < y < 1 \qquad (34)$$

The first method is based upon introducing a false boundary along the side(s) where the derivative boundary condition is given, in the present case the left boundary. For simplicity we discretize R by taking $\Delta x = \Delta y = \frac{1}{3}$ (see Fig. 8.14). The values at the grid points on the upper, lower, and right boundaries are given by the boundary conditions. We want to compute the grid function U at the interior points $(1, 1)$, $(1, 2)$, $(2, 1)$, $(2, 2)$, and at the left boundary points $(0, 1)$ and $(0, 2)$. The discretized Laplace's equation

$$U_{j,k} = \tfrac{1}{4}\left(U_{j-1,k} + U_{j+1,k} + U_{j,k-1} + U_{j,k+1}\right) \qquad (35)$$

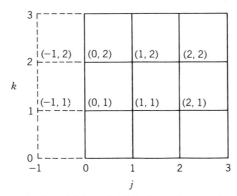

Figure 8.14. Addition of a false boundary to the grid.

$j = 1, 2$; $k = 1, 2$ gives four equations. Two additional equations can be determined by introducing a false boundary (x_{-1}, y_k), $k = 0, 1, 2, 3$ and approximating the left boundary condition by

$$U_{1,k} - U_{-1,k} = 2 \, \Delta x \, h(y_k) \tag{36}$$

for $k = 1, 2$. Substituting (36) into (35) for $j = 0$ gives

$$U_{0,k} - \tfrac{1}{4}(U_{0,k-1} + U_{0,k+1} + 2U_{1,k} - 2\,\Delta x\, h_k) = 0 \tag{37}$$

for $k = 1, 2$. Thus the system for (35) and (37) can be written

$$
\begin{bmatrix}
1 & -\frac{1}{4} & -\frac{1}{2} & & & \\
-\frac{1}{4} & 1 & & -\frac{1}{2} & & \\
-\frac{1}{4} & & 1 & -\frac{1}{4} & -\frac{1}{4} & \\
& -\frac{1}{4} & -\frac{1}{4} & 1 & & -\frac{1}{4} \\
& & -\frac{1}{4} & & 1 & -\frac{1}{4} \\
& & & -\frac{1}{4} & -\frac{1}{4} & 1
\end{bmatrix}
\begin{bmatrix}
U_{0,1} \\ U_{0,2} \\ U_{1,1} \\ U_{1,2} \\ U_{2,1} \\ U_{2,2}
\end{bmatrix}
\begin{bmatrix}
\frac{1}{4}U_{0,0} - \frac{1}{2}\Delta x\, h_1 \\
\frac{1}{4}U_{0,3} - \frac{1}{2}\Delta x\, h_2 \\
\frac{1}{4}U_{1,0} \\
\frac{1}{4}U_{1,3} \\
\frac{1}{4}U_{2,0} + \frac{1}{4}U_{3,1} \\
\frac{1}{4}U_{3,2} + \frac{1}{4}U_{2,3}
\end{bmatrix}
\tag{38}
$$

In practice this system can be solved numerically by the Gauss-Seidel method. After reading in the boundary values and initializing the array $U_{j,k}$, we can proceed as follows.

FOR $\nu = 1$ TO number of iterations
  FOR $k = 1$ TO 2
  $U_{0,k} = \tfrac{1}{4}(U_{0,k-1} + U_{0,k+1} + 2U_{1,k} - 2\,\Delta x\, h_k)$
  NEXT k
  FOR $j = 1$ TO 2
  FOR $k = 1$ TO 2
  $U_{j,k} = \tfrac{1}{4}(U_{j-1,k} + U_{j+1,k} + U_{j,k-1} + U_{j,k+1})$
  NEXT k
  NEXT j
NEXT ν

We now present a general procedure for writing difference equations at points on the boundary of a region where Laplace's equation is satisfied in the interior and $(\partial u / \partial n) = 0$ on the boundary. The method is based upon the divergence theorem in the plane that states that

$$\iint_R \operatorname{div} \mathbf{v} \, dx \, dy = \int_{\partial R} \mathbf{v} \cdot \mathbf{n} \, dl \tag{39}$$

where $\mathbf{n}$ is the outward unit normal to the boundary ∂R of R and $\mathbf{v}$ is a smooth vector field. If $\mathbf{v} = \mathbf{grad}\ u$, where u is a scalar field on R, then (39) gives

$$\iint_R \text{div}\ \mathbf{grad}\ u\ dx\ dy = \int_{\partial R} \mathbf{grad}\ u \cdot \mathbf{n}\ dl$$

or, since $\text{div}(\mathbf{grad}\ u) = u_{xx} + u_{yy} = 0$,

$$\int_{\partial R} \mathbf{grad}\ u \cdot \mathbf{n}\ dl = 0 \tag{40}$$

Condition (40) can be applied to obtain finite difference equations at a variety of boundary points.

To illustrate the method we consider the problem

$$u_{xx} + u_{yy} = 0, \qquad 0 < x < a, \quad 0 < y < b \tag{41}$$

$$u(x, 0) = f(x), \qquad \frac{\partial u}{\partial y}(x, b) = 0, \quad 0 < x < a \tag{42}$$

$$u(a, y) = g(y), \qquad \frac{\partial u}{\partial x}(0, y) = 0, \quad 0 < y < b \tag{43}$$

where f and g are given functions. Physically the solution represents a steady-state temperature distribution in a rectangular plate whose left and upper boundaries are insulated, while the temperature is prescribed on the other two sides. First we discretize the region by defining the grid $x_j = j\Delta x$ and $y_k = k\,\Delta y$ for $j = 0, 1, \ldots, J + 1$ and $k = 0, 1, \ldots, K + 1$, where $\Delta x = a/(J + 1)$ and $\Delta y = b/(K + 1)$ (see Fig. 8.15). Along the lower and right boundaries the values are known. At the interior grid points and at the grid points on the left and upper boundaries, however, we must calculate the

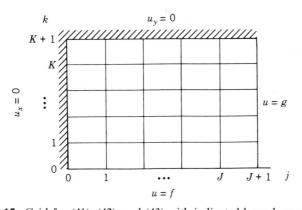

Figure 8.15. Grid for (41), (42), and (43) with indicated boundary conditions.

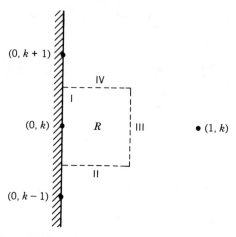

Figure 8.16. Molecule for a left boundary point.

numerical solution U. At the interior points we have (see (9))

$$U_{j,k} = \theta_x(U_{j+1,k} + U_{j-1,k}) + \theta_y(U_{j,k+1} + U_{j,k-1}) \tag{44}$$

for $j = 1, \ldots, J$ and $K = 1, \ldots, K$. Now we obtain additional difference equations at the points on the left and upper boundaries.

First, we consider a left boundary point (x_0, y_k) as shown in Fig. 8.16. Applying (40) to the region R gives

$$\left(\int_I + \int_{II} + \int_{III} + \int_{IV}\right) \mathbf{grad}\, u \cdot \mathbf{n}\, dl = 0 \tag{45}$$

Since $\partial u/\partial x = 0$ along segment I we have

$$\int_I \mathbf{grad}\, u \cdot \mathbf{n}\, dl = 0 \tag{46}$$

Over the path II

$$\int_{II} \mathbf{grad}\, u \cdot \mathbf{n}\, dl = \int_{II} \langle u_x, u_y \rangle \cdot \langle 0, -1 \rangle\, dl$$

$$= -\int_{II} u_y\, dl \cong -\frac{\Delta x}{2}\frac{U_{0,k} - U_{0,k-1}}{\Delta y} \tag{47}$$

Over the path III

$$\int_{III} \mathbf{grad}\, u \cdot \mathbf{n}\, dl = \int_{III} \langle u_x, u_y \rangle \cdot \langle 1, 0 \rangle\, dl = \int_{III} u_x\, dl \cong \frac{U_{1,k} - U_{0,k}}{\Delta x}\Delta y \tag{48}$$

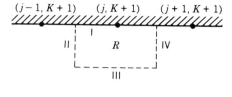

$(j-1, K+1)$ $(j, K+1)$ $(j+1, K+1)$

II   R   IV

III

(j, K)

Figure 8.17. Molecule for an upper boundary point.

Over the path IV

$$\int_{IV} \mathbf{grad}\, u \cdot \mathbf{n}\, dl = \int_{IV} \langle u_x, u_y \rangle \cdot \langle 0, 1 \rangle \, dl = \int_{IV} u_y \, dl \cong \frac{U_{0,k+1} - U_{0,k}}{\Delta y} \cdot \frac{\Delta x}{2} \tag{49}$$

Substituting (46) through (49) into (45) yields

$$U_{0,k} = \frac{\Delta x \, \Delta y}{\Delta x^2 + \Delta y^2} \left[\frac{\Delta x}{2 \Delta y} (U_{0,k+1} + U_{0,k-1}) + \frac{\Delta y}{\Delta x} U_{1,k} \right] \tag{50}$$

for $k = 1, \ldots, K$. Equation (50) is the difference equation at a left boundary point. For an upper boundary point (x_j, y_{K+1}) depicted in Fig. 8.17 condition (40) gives in the same manner

$$U_{j,K+1} = \frac{\Delta x \, \Delta y}{\Delta x^2 + \Delta y^2} \left[\frac{\Delta y}{2 \Delta x} (U_{j-1,K+1} + U_{j+1,K+1}) + \frac{\Delta x}{\Delta y} U_{j,K} \right] \tag{51}$$

for $j = 1, \ldots, J$. Finally we obtain a difference equation for the corner point (x_0, y_{K+1}) (see Fig. 8.18). Applying condition (40) to the region R we have

$$\left(\int_I + \int_{II} + \int_{III} + \int_{IV} \right) \mathbf{grad}\, u \cdot \mathbf{n}\, dl = 0$$

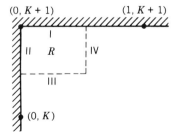

$(0, K+1)$ $(1, K+1)$

I

II   R   IV

III

$(0, K)$

Figure 8.18. Molecule for a corner point.

Clearly $\int_{\mathrm{I}} = \int_{\mathrm{II}} = 0$, since $\partial u/\partial n = 0$ on I and II. For the other two integrals

$$\int_{\mathrm{III}} \mathbf{grad}\; u \cdot \mathbf{n}\; dl = -\int_{\mathrm{III}} u_y\; dl \cong \frac{U_{0,\,K+1} - U_{0,\,K}}{\Delta y} \frac{\Delta x}{2}$$

$$\int_{\mathrm{IV}} \mathbf{grad}\; u \cdot \mathbf{n}\; dl = \int u_x\; dl \cong \frac{U_{1,\,K+1} - U_{0,\,K+1}}{\Delta x} \cdot \frac{\Delta y}{2}$$

Adding gives

$$U_{0,\,K+1} = \frac{2\,\Delta x\,\Delta y}{\Delta x^2 + \Delta y^2}\left[\frac{\Delta x}{2\,\Delta y}U_{0,\,K} + \frac{\Delta y}{2\,\Delta x}U_{1,\,K+1}\right] \tag{52}$$

for the corner point.

Equations (44), (50), (51), and (52) provide $(J + 1)(K + 1)$ equations for the $(J + 1)(K + 1)$ unknowns $U_{j,\,k}$, $j = 0,\dots, J$; $k = 1,\dots, K + 1$. In practice the system can be solved by the Gauss-Seidel method. An outline of a program follows:

```
Read in the boundary values
Initialize the array U_{j,k}
FOR ν = 1 TO number of iterations
    FOR k = 1 TO K
    U_{0,k} = (50)
    NEXT k
    U_{0,K+1} = (52)
    FOR j = 1 TO J
    FOR k = 1 TO K
    U_{j,k} = (44)
    NEXT k
    U_{j,K+1} = (51)
    NEXT j
NEXT ν
```

Example 3.6 Consider

$$u_{xx} + u_{yy} = 0, \qquad\qquad 0 < x, y < 1$$
$$u(x,0) = 36x(1 - x), \qquad 0 < x < 1$$
$$u(1, y) = 0, \qquad\qquad 0 < y < 1$$
$$u_x(0, y) = 0, \qquad\qquad 0 < y < 1$$
$$u_y(x, 1) = 0, \qquad\qquad 0 < x < 1$$

We take $\Delta x = \frac{1}{6}$, $\Delta y = \frac{1}{4}$. The array $U(J, K)$ holds the grid function U at the

```
10 REM--A NEUMANN PROBLEM--
20 REM--DISCRETIZE THE DOMAIN--
30 DIM U(10,5),V(10,5),D(10,5)
40 DELX=1/6:DELY=1/4:KK=3:JJ=5
50 NUMIT=99:S=DELX^2+DELY^2
60 THETAX=.5*(DELY^2)/S:THETAY=.5*(DELX^2)/S
70 REM--SET THE BOUNDARY VALUES--
80      FOR J=0 TO JJ+1
90      U(J,0)=J*(6-J):NEXT J
100     FOR K=1 TO KK+1
110     U(6,K)=0:NEXT K
120 REM--INITIALIZE THE ARRAY U(J,K)--
130     FOR J=0 TO JJ
140     FOR K=1 TO KK+1
150     U(J,K)=1
160     NEXT K:NEXT J
170 REM--SAVE IN V(J,K)--
180     FOR J=0 TO JJ
190     FOR K=1 TO KK+1
200     V(J,K)=U(J,K)
210     NEXT K:NEXT J
220 REM--ITERATE USING GAUSS/SEIDEL--
230     FOR NU=1 TO NUMIT
240        FOR K=1 TO KK
250        U(0,K)=(DELX*DELY/S)*((.5*DELX/DELY)*(U(0,K+1)+U(0,K-1))+(DELY/DELX)
           *U(1,K))
260        NEXT K
270        U(0,KK+1)=((DELX*DELY)/S)*((DELX/DELY)*U(0,KK)+(DELY/DELX)*U(1,KK+1))
280        FOR J=1 TO JJ
290        FOR K=1 TO KK
300        U(J,K)=THETAX*(U(J+1,K)+U(J-1,K))+THETAY*(U(J,K-1)+U(J,K+1))
310        NEXT K
320        U(J,KK+1)=(DELX*DELY/S)*((.5*DELY/DELX)*(U(J-1,KK+1)+U(J+1,KK+1))
           +(DELX/DELY)*U(J,KK))
330        NEXT J
340     REM--COMPUTE ABS(U-V)--
350        FOR J=0 TO JJ:FOR K=1 TO KK+1
360        D(J,K)=ABS(U(J,K)-V(J,K))
370        NEXT K:NEXT J
380     REM--COMPUTE 1-NORM--
390        ONENORM=0
400        FOR J=0 TO JJ:FOR K=1 TO KK+1
410        ONENORM=ONENORM+D(J,K)
420        NEXT K:NEXT J
430     REM--TERMINATION CRITERION--
440        IF ONENORM<.001 THEN GOTO 510
450     REM--SAVE U(J,K) IN V(J,K)--
460        FOR J=0 TO JJ:FOR K=1 TO KK+1
470        V(J,K)=U(J,K)
480        NEXT K:NEXT J
490     NEXT NU
500 GOTO 510
510 LPRINT"The number of iterations is ";NU
520 REM-- PRINT OUT THE TEMPERATURE--
530     LPRINT
540     FOR K=KK+1 TO 0 STEP -1:FOR J=0 TO JJ+1
550     LPRINT USING "#.### ";U(J,K);
560     NEXT J: LPRINT: NEXT K
570 END
```

νth step, while $V(J, K)$ holds the values of U at the $(\nu - 1)$th step. The iteration procedure is terminated when the 1-norm of the residual is less than 0.001 (eighty-seven iterations). The array $D(J, K)$ is the absolute value of the components of the residual, that is, $D(J, K) = |U(J, K) - V(J, K)|$. Program Listing 8.4 contains the BASIC program and Table 8.4 gives the numerical solution.

TABLE 8.4

y 3.026	2.958	2.732	2.315	1.695	0.899	0.000
3.181	3.135	2.947	2.545	1.892	1.014	0.000
3.544	3.630	3.643	3.339	2.598	1.434	0.000
3.517	4.293	5.049	5.119	4.255	2.463	0.000
0.000	5.000	8.000	9.000	8.000	5.000	0.000

$\longrightarrow x$

Accelerating Iterative Schemes

For a given iterative scheme $\mathbf{x}^{(\nu)} = M\mathbf{x}^{(\nu-1)} + \mathbf{b}$, $\nu = 1, 2, \ldots$, it is not possible to compute the error $\mathbf{e}^{(\nu)} = \mathbf{x}^{(\nu)} - \bar{\mathbf{x}}$, since the exact solution $\bar{\mathbf{x}}$ is unknown. Therefore we compute the residual $\mathbf{r}^{(\nu)}$ and terminate the process whenever $\|\mathbf{r}^{(\nu)}\| < \varepsilon$ in some vector norm, where ε is a prescribed error tolerance. The number of iterations required to produce a given degree of accuracy as measured by the residual can be reduced substantially in the following manner. First we compute an approximation $\tilde{\mathbf{x}}^{(\nu)}$ to the νth iterate via

$$\tilde{\mathbf{x}}^{(\nu)} = M\mathbf{x}^{(\nu-1)} + \mathbf{b}$$

Then for $\mathbf{x}^{(\nu)}$ we take $\mathbf{x}^{(\nu-1)}$ plus some fraction ω of the difference between $\mathbf{x}^{(\nu-1)}$ and $\tilde{\mathbf{x}}^{(\nu)}$, that is,

$$\mathbf{x}^{(\nu)} = \mathbf{x}^{(\nu-1)} + \omega\left(\tilde{\mathbf{x}}^{(\nu)} - \mathbf{x}^{(\nu-1)}\right)$$
$$= \omega\tilde{\mathbf{x}}^{(\nu)} + \left(1 - \omega\right)\mathbf{x}^{(\nu-1)}$$

This process can be visualized geometrically. The new iterate $\mathbf{x}^{(\nu)}$ is the convex combination of the previous iterate $\mathbf{x}^{(\nu-1)}$ and the iterate $\tilde{\mathbf{x}}^{(\nu)}$ normally predicted by the iterative scheme. It therefore lies on the line in R^n passing through $\mathbf{x}^{(\nu-1)}$ and $\tilde{\mathbf{x}}^{(\nu)}$. If $0 < \omega < 1$, then $\mathbf{x}^{(\nu)}$ lies between $\mathbf{x}^{(\nu-1)}$ and $\tilde{\mathbf{x}}^{(\nu)}$ and the process is *underrelaxed*; if $\omega > 1$, then the process is *overrelaxed* and amounts to extrapolation since $\mathbf{x}^{(\nu)}$ lies beyond $\tilde{\mathbf{x}}^{(\nu)}$; and if $\omega = 1$, then $\tilde{\mathbf{x}}^{(\nu)} = \mathbf{x}^{(\nu)}$ and the basic iteration scheme results. Figure 8.19 shows the case $\omega > 1$.

When this acceleration process is applied *componentwise* to the Gauss-Seidel scheme by taking

$$\tilde{x}_i^{(\nu)} = -\frac{1}{a_{ii}}\left(\sum_{j=1}^{i-1} a_{ij}x_j^{(\nu)} + \sum_{j=i+1}^{n} a_{ij}x_j^{(\nu-1)}\right) + \frac{f_i}{a_{ii}} \tag{53}$$

$$x_i^{(\nu)} = \omega\tilde{x}_i^{(\nu)} + \left(1 - \omega\right)x_i^{(\nu-1)} \tag{54}$$

it is called *successive overrelaxation* (SOR for short). SOR is a common method for solving linear systems arising from the discretization of Laplace's

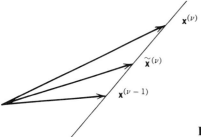

Figure 8.19. SOR with $\omega > 1$.

equation. Equations (53) and (54) can be combined into the single equation

$$x_i^{(\nu)} = \frac{\omega}{a_{ii}}\left(-\sum_{j=1}^{i-1} a_{ij}x_j^{(\nu)} - \sum_{j=i+1}^{n} a_{ij}x_j^{(\nu-1)} + f_i\right) + (1-\omega)x_i^{(\nu-1)} \quad (55)$$

An obvious question is to determine the values of ω for which (55) converges. To answer this question we note that (55) can be written

$$a_{ii}x_i^{(\nu)} = a_{ii}x_i^{(\nu-1)} + \omega\left(f_i - \sum_{j=1}^{i-1} a_{ij}x_j^{(\nu)} - \sum_{j=i+1}^{n} a_{ij}x_j^{(\nu-1)} - a_{ii}x_i^{(\nu-1)}\right)$$

which in matrix form is the same as

$$D\mathbf{x}^{(\nu)} = D\mathbf{x}^{(\nu-1)} + \omega\left(\mathbf{f} - L\mathbf{x}^{(\nu)} - U\mathbf{x}^{(\nu-1)} - D\mathbf{x}^{(\nu-1)}\right)$$

or

$$\mathbf{x}^{(\nu)} = (D + \omega L)^{-1}[(1-\omega)D - \omega U]\mathbf{x}^{(\nu-1)} + \omega(D + \omega L)^{-1}\mathbf{f} \quad (56)$$

where $D = \text{diag}(a_{11}, \ldots, a_{nn})$ and L and U are the lower and upper triangular portions of A, respectively. We recall that the convergence of (56) depends on the properties of the matrix

$$M_\omega \equiv (D + \omega L)^{-1}((1-\omega)D - \omega U)$$

If $\rho(M_\omega) < 1$, then M_ω is a convergent matrix and (56) will be convergent. It can be shown that if A is irreducible and diagonally dominant and if $0 < \omega < 1$, then $\rho(M_\omega) < 1$. This result is of little practical use since in order to obtain fast convergence we must generally take $\omega > 1$. The following theorem shows, however, that ω cannot lie outside the range $0 < \omega < 2$ if the scheme is to converge. A proof can be found in Varga [4].

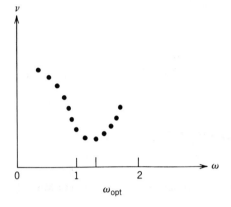

Figure 8.20. Experimental graph of the number of iterations ν required for $\|x^{(\nu)}\| \le 10^{-m}$ versus the relaxation parameter ω.

Theorem 3.5 *If* $\rho(M_\omega) < 1$, *then* $0 < \omega < 2$.

From a practical viewpoint it is helpful to determine the optimum value of ω, that is, the value of ω which gives the fastest convergence as measured by the size of the residual. Often the proper choice of ω can reduce the number of iterations significantly. There are two common methods to experimentally determine ω_{opt}. Both involve solving the homogeneous system

$$Ax = 0$$

which arises from setting the boundary conditions to zero. One is to calculate for different values of ω the number of iterations ν for which $\|x^{(\nu)}\| \le 10^{-m}$ in some vector norm, where m and $x^{(0)}$ are fixed. As a result one obtains a sequence of points (ω, ν) from which a graph can be constructed to choose ω_{opt} (see Fig. 8.20). A second possibility is to perform a fixed number of iterations ν_0 for different values of ω. The ω_{opt} is the value for which $\|x^{(\nu_0)}\|$ is smallest measured in some vector norm. One of these procedures is particularly useful if the problem must be solved many times, for example, with several different sets of boundary conditions. Another method is to choose ω_{opt} as the value that maximizes the convergence rate, that is, minimizes the spectral radius $\rho(M)$. In some simple cases this latter calculation can be done analytically.

Example 3.7 We solve numerically the boundary value problem in Example 3.4 using SOR. The BASIC program is given in Program Listing 8.5 and the numerical solution is the same as in Table 8.3. The number of iterations required to make the 1-norm of the residual less than 0.001 varies significantly with the value of the parameter ω. Table 8.5 shows that a choice of $\omega = 1.6$ reduces the number of iterations to 32 percent of the number required for Gauss-Siedel ($\omega = 1.0$).

```
10    REM--SOR FOR LAPLACE'S EQUATION--
20    DIM U(15,15),V(15,15),D(15,15),UT(15,15)
30    DELX=.1:DELY=.1:JJ=10:KK=10:MAXIT=150
40    INPUT "OMEGA EQUALS",OMEGA
50    REM--SET BOUNDARY CONDITIONS--
60        FOR J=0 TO JJ
70        U(J,0)=SIN(3.14159*J*DELX): U(J,10)=0:NEXT J
80        FOR K=0 TO KK
90        U(0,K)=SIN(3.14159*K*DELY): U(10,K)=0:NEXT K
100   REM--INITIALIZE U AND SAVE IN V--
110       FOR J=1 TO JJ-1: FOR K=1 TO KK-1
120       U(J,K)=.5:V(J,K)=U(J,K):NEXT K:NEXT J
130   REM--BEGIN THE ITERATION
140   FOR NU=1 TO MAXIT
150       FOR J=1 TO JJ-1: FOR K=1 TO KK-1
160       UT(J,K)=.25*(U(J-1,K)+U(J+1,K)+U(J,K-1)+U(J,K+1))
170       U(J,K)=OMEGA*UT(J,K)+(1-OMEGA)*U(J,K)
180       NEXT K: NEXT J
190       REM--INITIATE TERMINATION CRITERION--
200       ONENORM=0
210       FOR J=1 TO JJ-1: FOR K=1 TO KK-1
220       D(J,K)=ABS(U(J,K)-V(J,K)): ONENORM=ONENORM+D(J,K):NEXT K:NEXT J
230       IF ONENORM<.001 THEN GOTO 280
240       REM--SAVE U IN V--
250       FOR J=1 TO JJ-1:FOR K=1 TO KK-1:V(J,K)=U(J,K):NEXT K:NEXT J
260   NEXT NU
270       LPRINT "Does not converge": GOTO 340
280       LPRINT "Number of iterations is ":LPRINT NU: LPRINT
290       LPRINT "OMEGA equals ":LPRINT OMEGA:LPRINT
300       REM--PRINT THE SOLUTION--
310       FOR K=KK TO 0 STEP -1:FOR J=0 TO JJ
320       LPRINT USING"#.### "; U(J,K);
330       NEXT J: LPRINT: NEXT K
340   END
```

TABLE 8.5

ω	1.0	1.1	1.2	1.3	1.4	1.5	1.6	1.7	1.8
Number of Iterations	72	61	51	43	34	25	23	31	46

EXERCISES

3.1 Prove that if $\| \cdot \|$ is a vector norm, then the induced matrix norm is given by

$$\|A\| = \max_{\|\mathbf{x}\|=1} \|A\mathbf{x}\|, \qquad \mathbf{x} \in R^n$$

3.2 Derive (17).

3.3 In R^3 let $\mathbf{x} = [-1, 1, -2]^T$. Find $\|\mathbf{x}\|_1$, $\|\mathbf{x}\|_2$, and $\|\mathbf{x}\|_\infty$.

3.4 For each $\mathbf{x} \in R^n$ prove that $\|\mathbf{x}\|_\infty \le \|\mathbf{x}\|_2 \le \sqrt{n}\,\|\mathbf{x}\|_\infty$.

3.5   Find $\|A\|_\infty$ and $\|A\|_1$ if

$$A = \begin{bmatrix} 1 & 2 & -1 \\ 0 & 3 & -1 \\ 5 & -1 & 1 \end{bmatrix}$$

3.6   Find $\|A\|_2$ if

$$A = \begin{bmatrix} 2 & 1 & 0 \\ 1 & 1 & 1 \\ 0 & 1 & 2 \end{bmatrix}$$

3.7   Find the spectral radii of the matrices:

$$A = \begin{bmatrix} 1 & 0 & 1 \\ 2 & 2 & 1 \\ -1 & 0 & 0 \end{bmatrix} \quad \text{and} \quad A = \begin{bmatrix} 0 & \frac{1}{2} \\ \frac{1}{2} & 0 \end{bmatrix}$$

3.8   Prove that for any induced matrix norm

   (a) $\|A\mathbf{x}\| \le \|A\|\|\mathbf{x}\|$

   (b) $\|AB\| \le \|A\|\|B\|$

   (c) $\rho(A) \le \|A\|$

   (d) $\dfrac{1}{\|A^{-1}\|} \le |\lambda| \le \|A\|$

   where A is nonsingular and λ is an eigenvalue of A.

3.9   Consider the linear system

$$8x_1 - 2x_2 + 3x_3 = 1$$
$$-x_1 + 10x_2 + 2x_3 = 3$$
$$-4x_1 - 5x_2 + 20x_3 = 4$$

Taking $\mathbf{x}^{(0)} = [1,1,1]^T$, find $\mathbf{x}^{(1)}$, $\mathbf{x}^{(2)}$, $\mathbf{x}^{(3)}$, and $\mathbf{x}^{(4)}$ using both the Jacobi and Gauss-Seidel schemes. Find the matrices M_J and M_{GS}.

3.10   In the special case that the $n \times n$ matrix M has n linearly independent eigenvectors $\mathbf{w}_1, \mathbf{w}_2, \ldots, \mathbf{w}_n$ with eigenvalues $|\lambda_1| > |\lambda_2| \ge \cdots \ge |\lambda_n|$, $\rho(M) = |\lambda_1|$, prove that for any choice of $\mathbf{x}^{(0)}$ the iterative scheme

$$\mathbf{x}^{(\nu)} = M\mathbf{x}^{(\nu-1)} + \mathbf{b}, \qquad \nu = 1, 2, \ldots$$

converges if, and only if, $\rho(M) < 1$.

3.11   Show that the 1-norm in R^n

$$\|\mathbf{x}\|_1 = \sum_{i=1}^{n} |x_i|, \qquad \mathbf{x} = [x_1, \ldots, x_n]^{\mathrm{T}}$$

satisfies the conditions of a vector norm.

3.12   Prove (27).

3.13   Prove that if A is strictly diagonally dominant, then the Gauss-Seidel method converges for any choice of the initial vector $\mathbf{x}^{(0)}$.

3.14   Show that the coefficient matrices in (12) and (38) are irreducible and diagonally dominant.

3.15   Show that the truncation error for the finite difference approximation (9) of Laplace's equation is $O(\Delta x^2) + O(\Delta y^2)$.

3.16   A square laminar plate $0 \le x \le \pi, 0 \le y \le \pi$, has temperatures on the boundaries given by

$$u(x, \pi) = \sin x, \qquad u(x, 0) = 0, \quad 0 \le x \le \pi$$
$$u(0, y) = u(\pi, y) = 0, \qquad 0 \le y \le \pi$$

Use Gauss-Seidel with $\Delta x = \Delta y = \pi/10$ to obtain the equilibrium temperature distribution inside the plate. Compare to the exact solution and determine the number of iterations required to obtain three decimal place accuracy.

3.17   A unit square laminar plate is insulated on the left side and has temperatures of 1, 2, and 3 degrees on the top, right, and bottom edges, respectively. Use Gauss-Seidel and grid dimensions compatible with your computer to find a numerical temperature distribution in the plate. Terminate the iterative process when the 1-norm of the residual is less than 0.001.

3.18   Consider the heat conduction problem in Exercise 2.3. Using the implicit scheme for the heat equation, write a program to solve the resulting system that occurs by the Gauss-Seidel method. Use $\Delta x = 0.25$ and $r = 0.6$.

3.19   Consider Laplace's equation over the isosceles triangle shown in Fig. 8.21 with indicated boundary conditions. Write a difference equation that holds at the upper left point P.

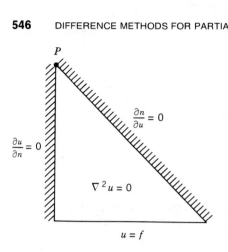

$$u = f$$ **Figure 8.21.** Exercise 3.19.

3.20 Consider the boundary value problem in Example 3.4 with $\Delta x = \Delta y = 0.1$.

(a) Use SOR with $\omega = 1.5$ to find a numerical solution.

(b) By examining the problem with homogeneous boundary conditions determine a value for ω_{opt} using both methods described in the text. Take $m = 3$, $x^{(0)} = [0.1, \ldots, 0.1]^T$, $\nu_0 = 75$. Use the 1-norm.

3.21 Consider Laplace's equation in the unit square with $u = \exp(2\pi x)\sin(2\pi y)$ on the boundary of the square. Take $\Delta x = \Delta y = 0.1$.

(a) Use SOR with $\omega_{opt} = 1.528$ to find a numerical solution. Terminate the process when $\|r^{(\nu)}\|_1 < 0.005$.

(b) Use SOR with $\omega = 1$ (Gauss-Seidel) to find a numerical solution.

(c) Compare the number of iterations required in (a) and (b) and compare the discrete solution to the exact solution.

3.22 Consider Laplace's equation on the region in Fig. 8.22 with the indicated boundary conditions. Take $\Delta x = \Delta y = 1$.

(a) Write difference equations at grid points P, Q, R, S, and T.

(b) Use SOR with $\omega = 1.75$ to obtain a numerical solution.

3.23 Solve the problem in Example 3.6 using SOR and $\omega = 1.6$. Show that the number of iterations required to make the 1-norm of the residual less than 0.001 is about 24 percent of the Gauss-Seidel value.

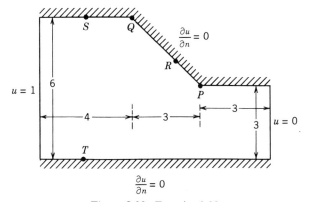

Figure 8.22. Exercise 3.22.

8.4 HYPERBOLIC EQUATIONS

The Wave Equation

The one dimensional wave equation

$$u_{tt} - c^2 u_{xx} = 0, \qquad x \in R^1, \quad t > 0 \tag{1}$$

can be discretized by taking centered difference approximations for u_{tt} and u_{xx} to obtain

$$\frac{U_{j, n+1} - 2U_{j, n} + U_{j, n-1}}{\Delta t^2} - c^2 \frac{U_{j+1, n} - 2U_{j, n} + U_{j-1, n}}{\Delta x^2} = 0 \tag{2}$$

where $x_j = j\Delta x$ and $t_n = n \Delta t$ $(j = 0, \pm 1, \pm 2, \dots; \; n = 0, 1, \dots)$, are grid points introduced in the upper half-plane. Like parabolic equations (2) leads to an explicit marching scheme

$$U_{j, n+1} = 2(1 - r^2)U_{j, n} + r^2\big(U_{j-1, n} + U_{j+1, n}\big) - U_{j, n-1} \tag{3}$$

where

$$r = \frac{c \Delta t}{\Delta x} \tag{4}$$

The computational molecule is shown in Fig. 8.23. Thus in order to implement (3) for the wave equation (1) with initial conditions

$$u(x, 0) = f(x), \tag{5a}$$
$$\qquad\qquad\qquad\qquad x \in R^1$$
$$u_t(x, 0) = g(x), \tag{5b}$$

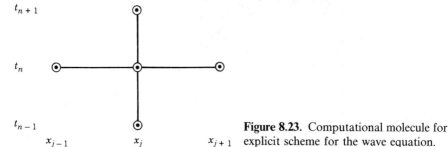

t_{n+1}

t_n

t_{n-1}

x_{j-1} x_j x_{j+1}

Figure 8.23. Computational molecule for explicit scheme for the wave equation.

the values of the grid function $U_{j,n}$ on the first two time lines ($n = 0, 1$) must be given. Equation (5a) gives

$$U_{j,0} = f(x_j) \tag{6}$$

for $j = 0, \pm 1, \ldots$, and (5b) gives

$$\frac{U_{j,1} - U_{j,0}}{\Delta t} = g(x_j)$$

or

$$U_{j,1} = f(x_j) + \Delta t\, g(x_j) \tag{7}$$

for $j = 0, \pm 1, \ldots$. Thus (3), (6), and (7) provide an algorithm for computing $U_{j,n}$ for all grid points in the upper half-plane.

We note that (7) is a first order approximation. A more accurate second order approximation can be obtained as follows. By Taylor's theorem

$$u(x, \Delta t) = u(x,0) + \Delta t\, u_t(x,0) + \frac{\Delta t^2}{2} u_{tt}(x,0) + O(\Delta t^3)$$

But $u_{tt}(x,0) = c^2 u_{xx}(x,0) = c^2 f''(x)$, and therefore

$$u(x, \Delta t) = u(x,0) + \Delta t\, u_t(x,0) + \frac{\Delta t^2 c^2}{2} f''(x) + O(\Delta t^3)$$

So we make the approximation

$$U_{j,1} = f(x_j) + \Delta t\, g(x_j) + \frac{\Delta t^2 c^2}{2} f''(x_j)$$

$$= f(x_j) + \Delta t\, g(x_j) + \frac{r^2}{2}\left(f(x_{j-1}) - 2f(x_j) + f(x_{j+1})\right) \tag{8}$$

and (8) can be used in place of (7). A guiding principle is to approximate boundary data to the same degree of accuracy as the equation itself.

Now we subject the difference equation (3) to the von Neumann stability analysis. Substituting

$$U_{j,n} = M^n \cos \alpha x_j$$

into Equation (3) we obtain

$$M^2 - \left(2 - 4r^2 \sin^2 \frac{\alpha \Delta x}{2}\right) M + 1 = 0$$

which gives

$$M = 1 - 2r^2 \sin^2 \frac{\alpha \Delta x}{2} \pm 2r \sin \frac{\alpha \Delta x}{2} \sqrt{r^2 \sin^2 \frac{\alpha \Delta x}{2} - 1} \qquad (9)$$

There are two cases to consider. If $r \le 1$, then $r^2 \sin^2(\alpha \Delta x/2) - 1 \le 0$ and

$$M = 1 - 2r^2 \sin^2 \frac{\alpha \Delta x}{2} \pm i2r \sin \frac{\alpha \Delta x}{2} \sqrt{1 - r^2 \sin^2 \frac{\alpha \Delta x}{2}}$$

Hence

$$|M| = \left[\left(1 - 2r^2 \sin^2 \frac{\alpha \Delta x}{2}\right)^2 + \left(4r^2 \sin^2 \frac{\alpha \Delta x}{2}\left(1 - r^2 \sin^2 \frac{\alpha \Delta x}{2}\right)\right)\right]^{1/2} = 1$$

If $r > 1$, then choose α such that $\alpha \Delta x = \pi$. Then taking the negative sign in (9) gives

$$M = 1 - 2r^2 \sin^2 \frac{\alpha \Delta x}{2} - 2r \sin \frac{\alpha \Delta x}{2} \sqrt{r^2 \sin^2 \frac{\alpha \Delta x}{2} - 1}$$

$$= 1 - 2r^2 - 2r\sqrt{r^2 - 1} < -1$$

Thus $|M| > 1$. Consequently solutions of the difference equation remain bounded for all α only in the case

$$r = \frac{c \Delta t}{\Delta x} \le 1 \qquad (10)$$

which is the *Courant-Friedrichs-Lewy (CFL) stability condition* for hyperbolic equations. This condition can be understood geometrically by considering a characteristic diagram (see Fig. 8.24). Let (x_j^*, t_n^*) be a grid point. The set of lattice points on which the value of U at (x_j^*, t_n^*) depends is called the

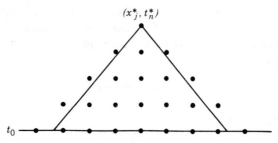

Figure 8.24. The two lines emanating back from (x_j^*, t_n^*) are the characteristics with speed $\pm c$. They enclose the analytical domain of dependence. The grid points indicated form the numerical domain of dependence.

numerical domain of dependence of (x_j^*, t_n^*). We notice that the numerical domain of dependence contains the analytical domain of dependence if the speed the numerical scheme propagates signals exceeds the characteristic speed c, that is,

$$\frac{\Delta x}{\Delta t} \geq c$$

which is the CFL condition. That the numerical domain of dependence should contain the domain of dependence seems reasonable, since otherwise there are initial data in the interval of dependence that the numerical solution would not take into account. Physically the CFL condition means that the numerical solution cannot proceed at a slower rate than the speed of the wave c.

The explicit algorithm (3), (6), and (8) can be implemented in exactly the same manner as the explicit scheme for the diffusion equation given in Program Listing 8.1 (see the Exercises).

We have seen the advantage of stability in implicit schemes for the diffusion equation and the same general observations hold true for the wave equation. Consider the boundary value problem

$$
\begin{aligned}
u_{tt} - c^2 u_{xx} &= 0, &\quad t > 0, &\quad 0 < x < L \\
u(x,0) = f(x), &\quad u_t(x,0) = g(x), &\quad 0 \leq x \leq L \\
u(0, t) = h(t), &\quad u(L, t) = e(t), &\quad t > 0
\end{aligned}
\tag{11}
$$

We define a grid $x_j = j\,\Delta x = 0, 1, \ldots, J + 1$ and $t_n = n\,\Delta t,\ n = 0, 1, 2, \ldots$, where $\Delta x = L/(J + 1)$ and Δt is given. To approximate the partial differential equation we replace u_{tt} by a centered difference formula at (x_j, t_n) and u_{xx} by the average of centered differences at (x_j, t_{n+1}) and (x_j, t_{n-1}). Therefore we obtain a molecule as shown in Fig. 8.25 and the difference

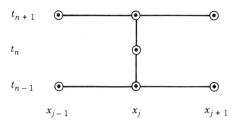

Figure 8.25. Computational molecule for the implicit scheme for the wave equation.

approximation becomes

$$\frac{U_{j,\,n+1} - 2U_{j,\,n} + U_{j,\,n-1}}{\Delta t^2} = \frac{c^2}{2}\left[\frac{U_{j+1,\,n+1} - 2U_{j,\,n+1} + U_{j-1,\,n+1}}{\Delta x^2}\right.$$
$$\left. + \frac{U_{j+1,\,n-1} - 2U_{j,\,n-1} + U_{j-1,\,n-1}}{\Delta x^2}\right]$$

or

$$-r^2 U_{j-1,\,n+1} + 2(1 + r^2)U_{j,\,n+1} - r^2 U_{j+1,\,n+1}$$
$$= r^2\left(U_{j-1,\,n-1} + U_{j+1,\,n-1}\right) + 4U_{j,\,n} - 2(1 + r^2)U_{j,\,n-1} \qquad (12)$$

For each $n \geq 1$ Equation (12) represents a system of J equations in the J unknowns $U_{1,\,n+1}, U_{2,\,n+1}, \ldots, U_{J,\,n+1}$. The boundary conditions are given by

$$U_{0,\,n} = h(t_n), \qquad U_{J+1,\,n} = e(t_n), \qquad n = 1, 2, \ldots \qquad (13)$$

and the initial conditions are given by (6) and (8). As in the case for the diffusion equation the system (12) is tridiagonal with an irreducible diagonally dominant coefficient matrix

$$A = \begin{bmatrix} 2(1 + r^2) & -r^2 & & & & \\ -r^2 & 2(1 + r^2) & -r^2 & & & \\ & -r^2 & 2(1 + r^2) & -r^2 & & \\ & & \ddots & \ddots & \ddots & \\ & & & & & -r^2 \\ & & & & -r^2 & 2(1 + r^2) \end{bmatrix} \qquad (14)$$

Therefore the system (12) may be solved by forward and back substitution or by an iterative method such as the Gauss-Seidel method.

Hyperbolic Systems

Consider a system of first order partial differential equations of the form

$$\frac{\partial u_1}{\partial t} + a_{11} \frac{\partial u_1}{\partial x} + \cdots + a_{1n} \frac{\partial u_n}{\partial x} = 0$$

$$\vdots$$

$$\frac{\partial u_n}{\partial t} + a_{n1} \frac{\partial u_1}{\partial x} + \cdots + a_{nn} \frac{\partial u_n}{\partial x} = 0$$

where the a_{ij} are constants, x and t are independent variables, and $u_1, \ldots, u_n$ are unknown functions. In matrix form with $A = (a_{ij})$ and $\mathbf{u} = [u_1, \ldots, u_n]^T$ the system can be written

$$\mathbf{u}_t + A\mathbf{u}_x = \mathbf{0} \tag{15}$$

We say (15) is *hyperbolic* if the n eigenvalues $\lambda_1, \ldots, \lambda_n$ of A are real and distinct. In this case the equations can be uncoupled as follows. We know there exists a nonsingular matrix P whose columns consist of the normalized eigenvectors of A such that

$$P^{-1}AP = D \equiv \operatorname{diag}(\lambda_1, \ldots, \lambda_n)$$

Multiplying (15) on the left by P^{-1}, we have

$$P^{-1}\mathbf{u}_t + P^{-1}A(PP^{-1})\mathbf{u}_x = \mathbf{0}$$

If $\mathbf{w} = P^{-1}\mathbf{u}$, we obtain the canonical form of (15), namely

$$\mathbf{w}_t + D\mathbf{w}_x = \mathbf{0} \tag{16}$$

The ith equation of (16) is

$$\frac{\partial w_i}{\partial t} + \lambda_i \frac{\partial w_i}{\partial x} = 0$$

which involves only the unknown component w_i. The general solution of (15) is then $\mathbf{u} = P\mathbf{w}$, where $w_i = f_i(x - \lambda_i t)$, $i = 1, \ldots, n$, and the f_i are arbitrary functions.

Example 4.1 The wave equation $u_{tt} - c^2 u_{xx} = 0$ can be written as a system of first order equations by defining $u_1 \equiv u_x$ and $u_2 \equiv u_t$. Then

$$\frac{\partial u_1}{\partial t} - \frac{\partial u_2}{\partial x} = 0, \qquad \frac{\partial u_2}{\partial t} - c^2 \frac{\partial u_1}{\partial x} = 0$$

or

$$\begin{bmatrix} u_1 \\ u_2 \end{bmatrix}_t + \begin{bmatrix} 0 & -1 \\ -c^2 & 0 \end{bmatrix}\begin{bmatrix} u_1 \\ u_2 \end{bmatrix}_x = \begin{bmatrix} 0 \\ 0 \end{bmatrix} \tag{17}$$

The matrix

$$A = \begin{bmatrix} 0 & -1 \\ -c^2 & 0 \end{bmatrix}$$

has eigenvalues $\pm c$ so that (17) is hyperbolic.

Example 4.2   Consider

$$\begin{bmatrix} u_1 \\ u_2 \end{bmatrix}_t + \begin{bmatrix} 4 & -6 \\ 1 & -3 \end{bmatrix}\begin{bmatrix} u_1 \\ u_2 \end{bmatrix} = \begin{bmatrix} 0 \\ 0 \end{bmatrix}$$

The eigenvalues are found from

$$\det\begin{bmatrix} 4 - \lambda & -6 \\ 1 & -3 - \lambda \end{bmatrix} = 0$$

and are $\lambda_1 = -2$ and $\lambda_2 = 3$. Therefore the system is hyperbolic. The eigenvectors are determined from

$$\begin{bmatrix} 4 - \lambda & -6 \\ 1 & -3 - \lambda \end{bmatrix}\begin{bmatrix} r_1 \\ r_2 \end{bmatrix} = \begin{bmatrix} 0 \\ 0 \end{bmatrix}$$

Taking $\lambda = -2$ gives

$$6r_1 - 6r_2 = 0, \qquad r_1 - r_2 = 0$$

and so a normalized eigenvector corresponding to $\lambda_1 = -2$ is $(1/\sqrt{2})[1, 1]^T$ or $[1/\sqrt{2}, 1/\sqrt{2}]^T$. Similarly an eigenvector corresponding to $\lambda_2 = 3$ is $[6/\sqrt{37}, 1/\sqrt{37}]^T$. Hence P and P^{-1} are given by

$$P = \begin{bmatrix} \dfrac{1}{\sqrt{2}} & \dfrac{6}{\sqrt{37}} \\[2mm] \dfrac{1}{\sqrt{2}} & \dfrac{1}{\sqrt{37}} \end{bmatrix}, \qquad P^{-1} = \begin{bmatrix} -\dfrac{\sqrt{2}}{5} & \dfrac{6\sqrt{2}}{5} \\[2mm] \dfrac{\sqrt{37}}{5} & -\dfrac{\sqrt{37}}{5} \end{bmatrix}$$

Taking $\mathbf{w} = P^{-1}\mathbf{u}$ or $\mathbf{u} = P\mathbf{w}$ gives the uncoupled canonical form

$$\begin{bmatrix} w_1 \\ w_2 \end{bmatrix}_t + \begin{bmatrix} -2 & 0 \\ 0 & 3 \end{bmatrix}\begin{bmatrix} w_1 \\ w_2 \end{bmatrix}_x = \begin{bmatrix} 0 \\ 0 \end{bmatrix}$$

or

$$\frac{\partial w_1}{\partial t} - 2\frac{\partial w_1}{\partial x} = 0, \qquad \frac{\partial w_2}{\partial t} + 3\frac{\partial w_2}{\partial x} = 0$$

The preceding argument shows it is sufficient to examine the single equation

$$u_t + \lambda u_x = 0, \qquad x \in R^1, \quad t > 0 \tag{18}$$

We impose the initial condition

$$u(x,0) = f(x), \qquad x \in R^1 \tag{19}$$

and without loss of generality assume $\lambda > 0$. By examining several different numerical schemes for the simple wave equation (18) we illustrate the fundamental issues.

With the standard grid in the upper half-plane, $x_j = j\Delta x$, $t_n = n\,\Delta t$, where $j = 0, \pm 1, \ldots$, and $n = 0, 1, \ldots$, we replace both derivatives in (18) by forward differences to obtain

$$\frac{U_{j,n+1} - U_{j,n}}{\Delta t} + \lambda \frac{U_{j+1,n} - U_{j,n}}{\Delta x} = 0$$

or

$$U_{j,n+1} = (1 + r)U_{j,n} - rU_{j+1,n}, \qquad r = \frac{\lambda\,\Delta t}{\Delta x} \tag{20}$$

The scheme (20) allows us to march forward in time from the initial condition $U_{j,0} = f(x_j)$, $j = 0, \pm 1, \ldots$, and the computational molecule consists of the points (j, n), $(j + 1, n)$, and $(j, n + 1)$. Although this scheme appears completely reasonable further investigation shows that it does not reflect the analytic behavior of (18) and thus can never converge for any value of r. Figure 8.26 shows that the analytic domain of dependence, which consists of a single characteristic line, does not lie in the numerical domain of dependence. Thus convergence is not expected (see Exercise 4.7).

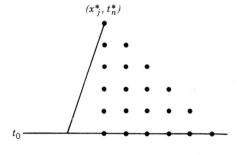

Figure 8.26. Analytical and numerical domain of dependence of (18) and (20), respectively.

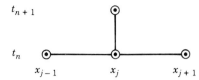

Figure 8.27. Computational molecule for the Lax-Wendroff method.

On the other hand, if (18) is approximated at (x_j, t_n), using a forward difference in time and a backward difference in space we obtain

$$U_{j,n+1} = (1 - r)U_{j,n} + rU_{j-1,n}, \qquad r = \frac{\lambda \, \Delta t}{\Delta x}$$

$$U_{j,0} = f(x_j) \tag{21}$$

with a computational molecule consisting of the points (x_{j-1}, t_n), (x_j, t_n), and (x_j, t_{n+1}). Now it is clear that the numerical domain of dependence contains the analytic domain of dependence provided that $r \leq 1$, which is the CFL condition. One can prove (Exercise 4.8) that if $r \leq 1$, then the scheme (21) is von Neumann stable and convergent to the solution of (18) and (19).

Additional schemes for (18) and (19) are proposed in Exercises 4.9 and 4.10.

One of the most widely used schemes for hyperbolic systems is the *Lax-Wendroff method*. We derive the method for (18). The scheme is explicit so that we attempt to march forward in time by writing a finite difference formula that allows the grid function to be computed at (x_j, t_{n+1}) in terms of the grid function values at (x_{j-1}, t_n), (x_j, t_n) and (x_{j+1}, t_n) (see Fig. 8.27). By Taylor's theorem

$$u_{j,n+1} = u_{j,n} + \Delta t \, u_t(x_j, t_n) + \frac{\Delta t^2}{2} u_{tt}(x_j, t_n) + O(\Delta t^3)$$

But from (18) we note $u_t = -\lambda u_x$ and $u_{tt} = -\lambda u_{xt} = -\lambda u_{tx} = \lambda^2 u_{xx}$. Hence

$$u_{j,n+1} = u_{j,n} - \lambda \, \Delta t \, u_x(x_j, t_n) + \frac{\lambda^2 \, \Delta t^2}{2} u_{xx}(x_j, t_n) + O(\Delta t^3)$$

Replacing u_x and u_{xx} by centered difference approximations

$$u_{j,n+1} = u_{j,n} - \frac{\lambda \, \Delta t}{2 \, \Delta x}(u_{j+1,n} - u_{j-1,n})$$

$$+ \frac{\lambda^2 \, \Delta t^2}{2 \, \Delta x^2}(u_{j+1,n} - 2u_{j,n} + u_{j-1,n}) + O(\Delta t^2 + \Delta x^2)$$

or with $r = \lambda \, \Delta t / \Delta x$

$$u_{j,\,n+1} = (1 - r^2) u_{j,\,n} + \frac{r}{2}(1 + r) u_{j-1,\,n}$$
$$- \frac{r}{2}(1 - r) u_{j+1,\,n} + O(\Delta t^2 + \Delta x^2)$$

Thus we obtain the *Lax-Wendroff formula*

$$U_{j,\,n+1} = (1 - r^2) U_{j,\,n} + \frac{r}{2}(1 + r) U_{j-1,\,n} - \frac{r}{2}(1 - r) U_{j+1,\,n} \qquad (22)$$

which is second order in Δx and Δt.

To examine the von Neumann stability of (22) we substitute $U_{j,\,n} = M^{\,n} \exp(i \alpha x_j)$ to obtain the magnification factor

$$M = 1 - 2r^2 \sin^2 \frac{\alpha \, \Delta x}{2} - ir \sin \alpha \, \Delta x$$

Therefore

$$|M| = \sqrt{1 - 4r^2(1 - r^2) \sin^4 \frac{\alpha \, \Delta x}{2}}$$

If $0 < r \le 1$, then $|M| \le 1$ and the Lax-Wendroff scheme is von Neumann stable. If $r > 1$, then $|M| > 1$ and the scheme is unstable. Hence the stability criterion is the CFL condition

$$\frac{\lambda \, \Delta t}{\Delta x} \le 1$$

The hyperbolic system (15) can be treated similarly. By Taylor's theorem

$$\mathbf{u}_{j,\,n+1} = \mathbf{u}_{j,\,n} + \Delta t \, \mathbf{u}_t(x_j, t_n) + \frac{\Delta t^2}{2} \mathbf{u}_{tt}(x_j, t_n) + O(\Delta t^3)$$

But from the system of Equations (15) we have $\mathbf{u}_t = -A \mathbf{u}_x$ and $\mathbf{u}_{tt} = A^2 \mathbf{u}_{xx}$. Therefore

$$\mathbf{u}_{j,\,n+1} = \mathbf{u}_{j,\,n} - \Delta t \, A \mathbf{u}_x(x_j, t_n) + \frac{\Delta t^2}{2} A^2 \mathbf{u}_{xx}(x_j, t_n) + O(\Delta t^3)$$

Using centered difference approximations for the x derivatives leads to the finite difference equation

$$\mathbf{U}_{j,\,n+1} = (I - B^2) \mathbf{U}_{j,\,n} + \tfrac{1}{2}(B + B^2) \mathbf{U}_{j-1,\,n} - \tfrac{1}{2}(B - B^2) \mathbf{U}_{j+1,\,n} \qquad (23)$$

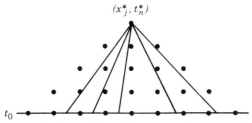

Figure 8.28. Characteristic lines with speeds $\lambda_1, \ldots, \lambda_n$ emanating back from a grid point (x_j^*, t_n^*). The indicated grid points form the numerical domain of dependence of (x_j^*, t_n^*).

where

$$B = \frac{\Delta t}{\Delta x} A$$

This is the Lax-Wendroff formula for hyperbolic systems with constant coefficients. It is a vector equation and can be used with initial starting values $U_{j,0} = f(x_j)$, $j = 0, \pm 1, \ldots$, to numerically solve (15) with initial data $u(x,0) = f(x)$.

We can arrive at a stability condition for (23) without giving a formal argument, but rather appealing to the domain of dependence condition. The characteristics of (15) are $x - \lambda_i t = $ constant, where $\lambda_1, \ldots, \lambda_n$ are the eigenvalues of A. These are sketched in Fig. 8.28. If the numerical domain of dependence is to contain the analytic domain of dependence, then we must have $\Delta x / \Delta t \geq |\lambda_i|$ for $i = 1, 2, \ldots, n$. That is, the speed $\Delta x / \Delta t$ that the numerical scheme progresses must exceed the maximum absolute eigenvalue of A, or

$$\frac{\Delta x}{\Delta t} \geq \max_{1 \leq i \leq n} |\lambda_i| \tag{24}$$

Condition (24) is the stability criterion for the vector Lax-Wendroff method.

Example 4.3   Consider the system in Example 4.2 with eigenvalues $\lambda_1 = -2$ and $\lambda_2 = 3$. Condition (24) forces $\Delta t \leq \Delta x / 3$ for the time step. If initial data $u_1(x,0) = f(x)$, $u_2(x,0) = g(x)$ are given and the solution is sought on an interval I: $t = t^*$, $a \leq x \leq b$, then the initial interval for the Lax-Wendroff difference scheme (23) must be taken large enough to include the numerical interval of dependence.

Boundary value problems require great care for hyperbolic systems since well-posedness must be ensured. For example, consider the system (15) on the region $t > 0$, $0 < x < 1$. If A has only positive eigenvalues, then all of the characteristics are outgoing (to the right). In this case data cannot be pre-scribed on $x = 1$, and the problem

$$\mathbf{u}_t + A\mathbf{u}_x = \mathbf{0}, \quad 0 < x < 1, \quad t > 0$$
$$\mathbf{u}(x,0) = \mathbf{f}(x), \; 0 < x < 1$$
$$\mathbf{u}(1, t) = \mathbf{g}(t), \quad t > 0$$

is well posed. An opposite but similar statement can be made if A has all negative eigenvalues. It is generally the case that if A has only k positive eigenvalues, then only k components of $\mathbf{u}$ can be prescribed along $x = 0$ and $n - k$ components can be prescribed on $x = 1$. Along with initial data $\mathbf{u}(x,0) = \mathbf{f}(x)$ the problem is then well posed. The finite difference approxima-tions used to compute the unknown boundary values must be chosen with extreme care (see [2]).

Equations with variable coefficients where A in (15) is a function of x and t present a greater challenge than constant coefficient equations, since the speed of the characteristics, determined by the eigenvalues $\lambda_i = \lambda_i(x, t)$ ($i = 1, \ldots, n$), vary in space in time. In this case the time step Δt must be controlled for explicit schemes at each time level t_n by requiring that

$$\left| \lambda(x_j, t_n) \right| \frac{\Delta t}{\Delta x} \le 1, \qquad j = 0 \pm 1, \ldots$$

which is the CFL condition. The Lax-Wendroff method can be easily modified to cover equations with variable coefficients (Exercise 4.12).

Characteristic Methods

Since characteristics play a fundamental role in the theory of hyperbolic equations (see Chapter 5), it seems reasonable that a finite difference method that takes advantage of characteristic directions may be quite fruitful. We consider the linear system

$$\mathbf{u}_t + A(x, t)\mathbf{u}_x = \mathbf{0} \tag{25}$$

where $A(x, t)$ is an $n \times n$ matrix and $\mathbf{u} = [u_1, \ldots, u_n]^T$. We say that (25) is *hyperbolic* in a region R of the xt plane if $A(x, t)$ has real distinct eigenvalues $\lambda_i(x, t)$ for each (x, t) in R. In this case there is a matrix $P = P(x, t)$ that diagonalizes A, that is, for which $P^{-1}AP = D \equiv \mathrm{diag}(\lambda_1, \ldots, \lambda_n)$. Letting $\mathbf{u} = P\mathbf{w}$ we obtain

$$\mathbf{u}_t = P\mathbf{w}_t + P_t\mathbf{w}, \qquad \mathbf{u}_x = P\mathbf{w}_x + P_x\mathbf{w}$$

Then (25) becomes

$$P\mathbf{w}_t + P_t\mathbf{w} + AP\mathbf{w}_x + AP_x\mathbf{w} = \mathbf{0}$$

Multiplying on the left by P^{-1} gives the *canonical system*

$$\mathbf{w}_t + D\mathbf{w}_x = C\mathbf{w} \tag{26}$$

where

$$C = -P^{-1}(P_t + AP_x)$$

In (26) the derivative terms are uncoupled. Following the procedure in Chapter 5 we define the characteristic curves as solutions of

$$\frac{dx}{dt} = \lambda_i(x, t), \qquad i = 1, \ldots, n \tag{27}$$

Along these curves (26) reduces to a system of ordinary differential equations

$$\frac{d\mathbf{w}}{dt} = C\mathbf{w} \tag{28}$$

Example 4.4   ($n = 1$)   For the single hyperbolic equation

$$u_t + \lambda(x, t)u_x = 0, \qquad x \in R^1, \quad t > 0$$

with initial condition $u(x, 0) = f(x)$, $x \in R^1$, we have $u = \text{constant}$ on

$$\frac{dx}{dt} = \lambda(x, t) \tag{29}$$

In the upper half-plane introduce the grid $x_j = j\Delta x$, $t_n = n\Delta t$, where $j = 0, \pm 1, \ldots$, and $n = 0, 1, \ldots$. If U denotes the grid function, the goal is to compute U at each lattice point, assuming U is known along the initial time line. Referring to Fig. 8.29 we illustrate how to proceed by indicating how to calculate $U(P)$ if $U(R)$ and $U(S)$ are known. The segment $\overline{QP}$ represents the characteristic emanating backward from P, assuming $\lambda(x, t) > 0$. Approximating (29) we obtain

$$\frac{X(P) - X(Q)}{\Delta t} = \lambda(P)$$

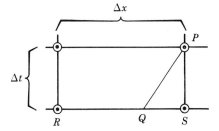

Figure 8.29

which gives the x coordinate of Q. Since $U(P) = U(Q)$ we can find $U(P)$ by interpolating between the values of $U(R)$ and $U(S)$. Using linear interpolation

$$U(P) = U(Q) = \frac{X(Q) - X(R)}{X(S) - X(R)}(U(S) - U(R)) + U(R)$$

By proceeding in this manner the value of U can be computed at each point along a given time line if its values are known along the previous time line. If the point Q falls outside the segment $\overline{RS}$, then the method is unstable, since the domain of dependence condition is not satisfied. Therefore a stability check must be included at each time step to ensure that this condition is met. If it is not, then Δt should be decreased. This method, which involves both a fixed grid and the use of characteristics, is called a hybrid method.

Example 4.5 ($n = 2$) Consider the canonical equations (26) in the case $n = 2$. Along the characteristic

$$\frac{dx}{dt} = \lambda_1(x, t) \tag{30}$$

we have

$$\frac{dw_1}{dt} = c_{11}w_1 + c_{12}w_2 \tag{31}$$

and along

$$\frac{dx}{dt} = \lambda_2(x, t) \tag{32}$$

we have

$$\frac{dw_2}{dt} = c_{21}w_1 + c_{22}w_2 \tag{33}$$

Let (x_j, t_n) be the standard grid in the upper half-plane and W_1 and W_2 be the grid functions corresponding to w_1 and w_2. Assume $\lambda_1 > 0$ and $\lambda_2 < 0$. Referring to Fig. 8.30 we indicate how to compute W_1 and W_2 at P if their values are known at R, S, and T. First approximate (30) by

$$\frac{X(P) - X(Q)}{\Delta t} = \lambda_1(P)$$

from which we can calculate $X(Q)$. Then interpolate to obtain $W_1(Q)$ and $W_2(Q)$ according to

$$W_1(Q) = \frac{X(Q) - X(R)}{X(S) - X(R)}(W_1(S) - W_1(R)) + W_1(R)$$

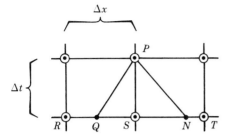

Figure 8.30

and similarly for $W_2(Q)$. Finally $W_1(P)$ can be calculated from the differential equation (31) via the finite approximation

$$\frac{W_1(P) + W_1(Q)}{\Delta t} = c_{11}(Q)W_1(Q) + c_{12}(Q)W_2(Q)$$

Similarly we can calculate the point where the second characteristic through P intersects the previous time line by approximating (32) by

$$\frac{X(P) - X(N)}{\Delta t} = \lambda_2(P)$$

If N lies between S and T, then W_1 and W_2 can be found at $X(N)$ by interpolation according to

$$W_1(N) = \frac{X(N) - X(S)}{X(T) - X(S)}(W_1(T) - W_1(S)) + W_1(S)$$

and similarly for $W_2(N)$. Then $W_2(P)$ can be obtained from the finite difference approximation of (33),

$$\frac{W_2(P) - W_2(N)}{\Delta t} = c_{21}(N)W_1(N) + c_{22}(N)W_2(N)$$

Proceeding in this fashion gives the values of W_1 and W_2 at the grid points on the next time level.

Example 4.6   Consider

$$\begin{array}{ll} u_t + 2tu_x = 0 & 0 \le x < 1, \quad 0 < t < 1 \\ u(x,0) = \sin \pi x, & 0 < x < 1 \\ u(0, t) = 0, & 0 < t < 1 \end{array}$$

We solve this mixed initial boundary value problem numerically on the unit

```
10    REM--HYBRID CHARACTERISTIC SCHEME--
20    REM--SET THE DISCRETIZATION--
30       DIM XP(15),XQ(15),U(15),UN(15)
40       L=1:NUMXINT=10:DELX=L/NUMXINT
50       DELT=.05:NUMTSTP=20
60    REM--DEFINE XP AND INITIAL DATA U--
70       FOR J=0 TO NUMXINT
80       XP(J)=J*DELX: U(J)=SIN(3.14159*J*DELX)
90       LPRINT USING"##.### ";U(J);
100      NEXT J:LPRINT
110   REM--BEGIN TIME STEPS--
120      FOR N=1 TO NUMTSTP
130      REM--SET BOUNDARY VALUE--:UN(0)=0
140      FOR J=1 TO NUMXINT
150         XQ(J)=XP(J)-2*N*DELT^2
160         UN(J)=U(J-1)+(U(J)-U(J-1))*((XQ(J)-XP(J-1))/(XP(J)-XP(J-1)))
170      NEXT J
180      REM--PRINT WAVE PROFILE AND PUT UN(J) INTO U(J)--
190      FOR J=0 TO NUMXINT
200      LPRINT USING "##.### ";UN(J);
210      U(J)=UN(J): NEXT J: LPRINT
220      NEXT N
230      END
```

square using the hybrid scheme of Example 4.4. Since $\lambda(x, t) = 2t$ we have $\max|\lambda| = 2$ and so the CFL condition is $\Delta t \leq 0.5 \Delta x$. The exact solution is $u(x, t) = \sin(\pi(x - t^2))$ for $x > t^2$ and $u(x, t) = 0$ for $x < t^2$. The BASIC program is given in Program Listing 8.6 and the numerical solution is given in Table 8.6

Conservation Laws

Many important equations of mathematical physics can be written in the form of a conservation equation (see Section 5.1)

$$u_t + F(u)_x = 0 \tag{34}$$

or a system of conservation equations

$$\mathbf{u}_t + \mathbf{F}(\mathbf{u})_x = 0 \tag{35}$$

TABLE 8.6

x	0.0	0.2	0.4	0.6	0.8	1.0
$t = 0$	0.000	0.588	0.951	0.951	0.588	0.000
$t = 0.2$	0.000	0.448	0.873	0.967	0.691	0.152
$t = 0.4$	0.000	0.164	0.605	0.910	0.871	0.500
$t = 0.6$	0.000	0.016	0.186	0.566	0.859	0.840
$t = 0.8$	0.000	0.000	0.010	0.103	0.389	0.732
$t = 1.0$	0.000	0.000	0.000	0.001	0.019	0.140

where $\mathbf{u} = [u_1, \ldots, u_n]^\mathrm{T}$ and $\mathbf{F} = [F_1, \ldots, F_n]^\mathrm{T}$. A standard procedure to solve (34) or (35) is the Lax-Wendroff method, which we now develop for (35). The computational molecule is the same as in Fig. 8.27 and by Taylor's theorem

$$\mathbf{u}_{j,\,n+1} = \mathbf{u}_{j,\,n} + \Delta t\, \mathbf{u}_t(x_j, t_n) + \frac{\Delta t^2}{2}\mathbf{u}_{tt}(x_j, t_n) + O(\Delta t^3) \qquad (36)$$

But $\mathbf{u}_t = -\mathbf{F}(\mathbf{u})_x$ and

$$\mathbf{u}_{tt} = -\frac{\partial}{\partial x}\frac{\partial \mathbf{F}}{\partial t} = -\frac{\partial}{\partial x}\left(\frac{\partial \mathbf{F}}{\partial \mathbf{u}}\frac{\partial \mathbf{u}}{\partial t}\right) = \frac{\partial}{\partial x}\left(A(\mathbf{u})\mathbf{F}(\mathbf{u})_x\right)$$

where $A(\mathbf{u}) \equiv \partial \mathbf{F}/\partial \mathbf{u}$ is the Jacobi matrix. Therefore (36) becomes

$$\mathbf{u}_{j,\,n+1} = \mathbf{u}_{j,\,n} - \Delta t\left(\frac{\partial \mathbf{F}}{\partial x}\right)_{j,\,n} + \frac{\Delta t^2}{2}\left[\frac{\partial}{\partial x}\left(A(\mathbf{u})\mathbf{F}(\mathbf{u})_x\right)\right]_{j,\,n} + O(\Delta t^3)$$

where the subscripts j, n mean evaluation at (x_j, t_n). Approximating the spatial derivatives by centered difference formulas gives the second order scheme

$$\mathbf{U}_{j,\,n+1} = \mathbf{U}_{j,\,n} - \frac{\Delta t}{2\,\Delta x}\left(\mathbf{F}_{j+1,\,n} - \mathbf{F}_{j-1,\,n}\right)$$

$$+ \frac{\Delta t^2}{2\,\Delta x^2}\left[A_{j+1/2,\,n}\left(\mathbf{F}_{j+1,\,n} - \mathbf{F}_{j,\,n}\right) + A_{j-1/2,\,n}\left(\mathbf{F}_{j,\,n} + \mathbf{F}_{j-1,\,n}\right)\right]$$

To eliminate the necessity of evaluating A at points other than grid points we can replace $A_{j+1/2,\,n}$ and $A_{j-1/2,\,n}$ by their average values

$$A_{j+1/2,\,n} = \tfrac{1}{2}\left(A_{j+1,\,n} + A_{j,\,n}\right), \qquad A_{j-1/2,\,n} = \tfrac{1}{2}\left(A_{j,\,n} + A_{j-1,\,n}\right)$$

Thus we obtain the scheme

$$\mathbf{U}_{j,\,n+1} = \mathbf{U}_{j,\,n} - \frac{\Delta t}{2\,\Delta x}\left(\mathbf{F}_{j+1,\,n} - \mathbf{F}_{j-1,\,n}\right)$$

$$+ \frac{\Delta t^2}{4\,\Delta x^2}\left[\left(A_{j+1,\,n} + A_{j,\,n}\right)\left(\mathbf{F}_{j+1,\,n} - \mathbf{F}_{j,\,n}\right)\right.$$

$$\left. - \left(A_{j,\,n} + A_{j-1,\,n}\right)\left(\mathbf{F}_{j,\,n} - \mathbf{F}_{j-1,\,n}\right)\right] \qquad (37)$$

which is the Lax-Wendroff formula.

Example 4.7   Consider the scalar conservation law

$$u_t + \left(\tfrac{1}{2}u^2\right)_x = 0$$

Here $F(u) = \frac{1}{2}u^2$ and $A(u) = u$. Then the Lax-Wendroff formula (37) is

$$U_{j,n+1} = U_{j,n} - \frac{\Delta t}{4\Delta x}\left(U_{j+1,n}^2 + U_{j-1,n}^2\right)$$

$$+ \frac{\Delta t^2}{8\Delta x^2}\left[\left(U_{j+1,n} + U_{j,n}\right)\left(U_{j+1,n}^2 - U_{j,n}^2\right)\right.$$

$$\left. - \left(U_{j,n} + U_{j-1,n}\right)\left(U_{j,n}^2 - U_{j-1,n}^2\right)\right]$$

EXERCISES

4.1   Show that the truncation error in the difference approximation (2) to the wave equation is order $O(\Delta x^2) + O(\Delta t^2)$.

4.2   Write a computer program to solve the boundary value problem

$$u_{tt} - 4u_{xx} = 0, \qquad\qquad 0 < x < 1, \quad t > 0$$
$$u(x,0) = \sin 2\pi x, \quad u_t(x,0) = 0, \quad 0 < x < 1$$
$$u(0,t) = u(1,t) = 0, \qquad t > 0$$

using the explicit scheme (3), (6), and (8). Use $\Delta x = 0.1$ and make two runs, one using $\Delta t = 0.025$ and another using $\Delta t = 0.1$. Compare to the exact solution at $t = 1$.

4.3   Show that the matrix A in (14) is irreducible and diagonally dominant.

4.4   Solve the problem in Exercise 4.2 using the implicit scheme (12), (13), (6), and (8). Use $\Delta x = 0.1$ and make two runs, one using $\Delta t = 0.1$ and another using $\Delta t = 0.025$. Compare with Exercise 4.2.

4.5   Determine the truncation error for (12).

4.6   Prove that the implicit scheme (12) is von Neumann stable.

4.7   By example show that the scheme (20) is not convergent.

4.8   (a) Prove that if $r \le 1$, then the scheme (21) is von Neumann stable. Let $U_{j,n} = M^n\exp(i\alpha x_j)$ and show $|M|^2 = 1 - 2r(1 - r)(1 - \cos\alpha\,\Delta x)$.

(b) Prove that if $r \le 1$, then the scheme (21) is convergent. (Find a difference equation for $e_{j,n} \equiv u_{j,n} - U_{j,n}$, and letting $E_n = \max_j|e_{n,j}|$ and $\hat{\tau} = \max_{j,n}|\tau_{j,n}|$, show that $E_{n+1} \le E_n + \hat{\tau}\Delta t$.)

4.9 Consider the initial value problem (18) and (19). Prove that the scheme

$$U_{j,\,n+1} = U_{j,\,n} + \frac{r}{2}\bigl(U_{j-1,\,n} - U_{j+1,\,n}\bigr)$$

determined by taking a forward difference approximation for u_t and a centered difference approximation for u_x is not von Neumann stable for any r.

4.10 If $U_{j,\,n}$ in the scheme in Exercise 4.9 is replaced by the average of $U_{j-1,\,n}$ and $U_{j+1,\,n}$, we obtain the scheme

$$U_{j,\,n+1} = \tfrac{1}{2}(1 + r)U_{j-1,\,n} + \tfrac{1}{2}(1 - r)U_{j+1,\,n}$$

Show that this scheme is von Neumann stable and convergent for $r \le 1$.

4.11 Derive (28).

4.12 Following steps similar to those leading to (23) derive the Lax-Wendroff scheme for the equation $u_t + \lambda(x, t)u_x = 0$.

4.13 Show how the characteristic scheme of Examples 4.4 and 4.5 can be extended to the quasi-linear equation $u_t + \lambda(x, t, u)u_x = 0$.

4.14 Determine the simplest implicit scheme for the equation $u_t + \lambda u_x = 0$ $(\lambda > 0)$ having a computational molecule: $(j, n - 1), (j - 1, n), (j + 1, n), (j, n)$. Find the truncation error and investigate the stability using von Neumann's method.

4.15 **(a)** Reformulate the system of partial differential equations in Example 4.2 on $0 < t < 1, 0 < x < 1$ with auxiliary data

$$
\begin{aligned}
u_1(x,0) &= \sin \pi x, & u_2(x,0) &= \cos(\pi x/2), & 0 &< x < 1 \\
u_2(0, t) &= 1, & u_1(1, t) &= 0, & 0 &< t < 1
\end{aligned}
$$

as a boundary value problem in canonical form.

(b) Solve the problem numerically using the hybrid method of Example 4.5.

4.16 Solve the problem in Example 4.6 by the Lax-Wendroff method.

4.17 Consider the initial value problem

$$
\begin{aligned}
u_t + \bigl(\tfrac{1}{2}u^2\bigr)_x &= 0, & x &\in R^1, \quad t > 0 \\
u(x,0) &= \exp(-x^2), & x &\in R^1
\end{aligned}
$$

Use the Lax-Wendroff formula to advance the initial waveform several time steps until a shock develops.

REFERENCES

1. W. F. Ames, *Numerical Methods for Partial Differential Equations*, 2nd ed., Academic Press, Inc., New York (1977).
2. R. D. Richtmyer and K. W. Morton, *Difference Methods for Initial-Value Problems*, 2nd ed., Wiley-Interscience, New York (1967).
3. G. D. Smith, *Numerical Solution of Partial Differential Equations: Finite Difference Methods*, 3rd ed., Oxford University Press, Oxford (1986).
4. R. S. Varga, *Matrix Iterative Analysis*, Prentice-Hall, Englewood Cliffs, N.J. (1966).
5. E. Isaacson and H. B. Keller, *Analysis of Numerical Methods*, John Wiley & Sons, New York (1966).

INDEX

Abel's integral equation, 223
Acceleration of iterative schemes, 540
Acoustic equations, 283, 441
Action integral, 128
Adiabatic process, 278, 281
Admissible functions, 86
Admissible variations, 96
Advection equation, 231, 234, 554
Approximation:
    difference, 491, 501
    for differential equation, 47
    inner, 57
    outer, 56
    uniformly valid, 46
    uniform singular perturbation, 64
Asymptotic expansion, 48
Asymptotic power series, 48
Asymptotic sequence, 48
Attractor, 394

Bénard problem, 400
Bernoulli's theorem, 336
Bifurcation:
    of bead on hoop, 347, 401
    diagram, 358
    Hopf, 396, 400
    pitchfork, 346
    point, 358
    in tank reactor, 376

Boundary layer:
    over airfoil, 29
    in singular perturbation, 55, 66
    width, 62
Brachistochrone problem, 87, 88, 107
Breaking time, 240, 250
Buckingham pi theorem, 3, 7
    proof, 10
Burgers' equation, 243, 262, 441

$C^n[a,b]$, 92
Canonical coordinates, 423, 468, 480
Catenary, 147 _equation, 195_
Cauchy–Euler method, 491
Cauchy's equations, 329
Cauchy's stress principle, 324, 329
Cauchy's theorem, 326
Characteristics:
    of advection equation, 235
    of hyperbolic systems, 559
    of linear equations, 236
    of nonlinear equations, 237
    for wave equation, 294
Characteristic schemes, 558
Characteristic system, 255
Characteristic time, see Time scale
Chemical reactor, 373
Chemotaxis, 351
Clausius–Duhem inequality, 334

Cole–Hopf transformation, 262
Conductivity, 8, 9, 21, 160
Configuration space, 127
Conservation laws:
  in differential form, 248, 306, 562
  across discontinuity, 251, 314
  in integral form, 249
  for mass, 271, 272, 305, 306, 313, 322
  in mechanics, 128, 435
Consistent scheme, 501
Constraint:
  algebraic, 149
  isoperimetric, 144
Continuity equation, 272, 306, 323
Convection theorem, 320
Convective derivative, 270, 319
Convergence:
  of finite difference scheme, 501
  of Fourier series, 180, 182
  in-the-mean, 182
  pointwise, 45
  uniform, 45
Convergent matrix, 522
Convolution, 189
Coordinates:
  canonical, 423, 468, 480
  cylindrical, 170
  Eulerian, 266
  Lagrangian, 267
  phase, 136, 379
  shock-fixed, 290
  spherical, 171
Courant–Friedrichs–Levy criterion, 549, 555,
    556, 558
Crank–Nicolson scheme, 508
Critical point, 380
  center, 383
  node, 383, 387, 388, 389
  saddle, 383, 387
  spiral, 383, 389

Derivative:
  directional, 97
  discrete approximation of, 494
  of functional, 95, 96
  ordinary, 90, 95
Difference approximation:
  backward, 495
  of boundary conditions, 510, 534
  centered, 496
  forward, 494
Diffusion equation:
  fundamental solution, 158, 194
  initial value problem, 193

linear, 20, 152, 154, 156, 158, 162, 282
nonlinear, 162, 455
similarity solution, 417, 444, 455
spherically symmetric, 444, 447
three dimensional, 168
Diffusivity, 8, 19, 21, 162, 168
Dimensional analysis:
  blast wave problem, 3
  damped oscillator, 71
  diffusion problem, 7, 445
  projectile problem, 23
Dimensional matrix, 6
Dirichlet problem, 174, 518
Discretization error, see Truncation error
Dispersion relation, 245
Dissipation function, 333
Distortion, 285
Divergence theorem, 166, 321, 534
Domain of dependence:
  analytical, 297, 550
  numerical, 550
Duffing equation, 40
DuFort–Frankel scheme, 509

Eigenvalues:
  Freholm integral equation, 210, 212
  Sturm–Liouville problem, 183
  Volterra integral equation, 220
Elliptic equation, 155, 157
Emden–Fowler equation, 143, 461
Energy conservation:
  in diffusion, 19, 160
  in fluids, 278, 280, 332
  in mechanics, 128, 435, 437
Energy integral, 164, 302
Energy transport theorem, 337
Enthalpy, 279, 290
Entropy, 278, 281
Equation of state:
  Abel, 279
  barotropic, 276
  caloric, 281
  ideal gas, 277, 279
  Tait, 279
  thermal, 281
  van der Waals, 279
Equilibrium solution, 358, 380
erf function, 192
Euler equation:
  for fluid, 332
  isoperimetric problem, 146
  multiple integrals, 116, 118
  nth order problem, 112
  second order problem, 112

several functions, 113, 114
simplest problem, 104
Euler expansion formula, 270, 320
Explicit scheme:
  BASIC program, 499
  for diffusion equation, 497
  Lax–Wendroff, 555, 557, 563
  for wave equation, 547
Extremal, 99, 116

False boundary, 510, 530
Fermat's principle, 108
First integral:
  of characteristic system, 255
  for differential equation, 106, 107,
    429
Flow:
  adiabatic, 281
  incompressible, 323
  irrotational, 336
  isentropic, 282
  plane Couette, 339
  plane Poiseuille, 339
  potential, 336
  steady, 318
Fluid motion:
  in one dimension, 267
  in three dimensions, 316
Force:
  body, 273, 323
  on object in fluid, 330
  surface, 273, 324
Fourier:
  coefficients, 180, 185
  cosine series, 181
  series, 180, 185
  sine series, 181
  transform, 192
Fourier's heat law, 19, 161, 167, 281
Fredholm alternative theorem, 209
Froude number, 15
Functional, 86
  continuous, 100
  linear, 100
Fundamental frequencies, 291
Fundamental lemma of calculus of variations,
  see Lagrange's lemma

Gâteaux variation, 96
Gauss–Seidel method, 524
  applied to Laplace's equation, 525
  convergence of, 532
  SOR, 540
Generators of extended group, 472, 475

Graph:
  connected, 531
  directed, 531
Green's function, 194, 229
Green's identity, 167, 173
Green's theorem, 116
Gronwall's inequality, 360
Gurney energy, 15

Hadamard's example, 165
Hamiltonian, 128, 133, 136, 435, 437
Hamilton's equations, 134
Hamilton's principle, 127
Heat capacity, 8
Heat equation, *see* Diffusion equation
Helmholtz equation, 304
Hilbert–Schmidt theorem, 214
Hooke's law, 286, 482
Hugoniot curve, 315
Hyperbolic:
  equation, 155, 157
  system, 552, 558

Ideal fluid, 331
Implicit scheme:
  BASIC program, 507
  Crank–Nicolson, 508
  for diffusion equation, 503
  DuFort–Frankel, 509
  for wave equation, 551
Integral equation:
  Abel's, 223
  convolution type, 221
  Fredholm, 200
  Volterra, 200
Integral surface, 255
Internal energy, 278, 280, 332
Invariance:
  conformal, 440, 450, 470
  of function, 439
  of functional, 428
  of ordinary differential equation, 457, 461
  of partial differential equation, 450, 453, 470
Invariant surface condition, 452, 481
Inverse problem:
  calculus of variations, 137
  in integral equations, 203
  scattering, 298
Inviscid fluid, 331
Isoperimetric problem, 143, 144, 149
Iteration:
  fixed point, 216
  for Fredholm equations, 228
  Gauss–Seidel, 524

Iteration (*Continued*)
  Jacobi, 522
  matrix, 521
  Picard, 217
  rate of convergence, 530
  for Volterra equations, 218

Jacobi method, 522
  convergence of, 532

Kelvin's theorem, 338
Kernel:
  convolution, 221
  resolvent, 216, 221, 228
  separable, 208
  singular, 200, 222
  symmetric, 200, 212
Kinetic energy:
  of fluid, 280, 332, 337
  in generalized coordinates, 127
Korteweg–deVries equation, 246

Lagrange multiplier rule, 143
Lagrange's equations, 128, 435, 436
Lagrange's lemma, 103, 112, 116
Lagrangian, 90, 127, 436
Laplace's equation, 170, 517
  difference approximation, 518, 520
Laplace transforms, 188
  table of, 190   *Laplacian 171*
Lax–Wendroff method, 555, 563
Leibniz' formula, 224, 322
Lennard–Jones potential, 16
Lie derivative, 473
Lie plane, 457
Linear space, 91
Linear system, 385
Local Lie group, 420, 466
  determining equations, 434
  dynamical system of, 423
  extended, 426, 473
  generators of, 422, 466, 473
  global representation, 422
  infinitesimal representation, 422, 466
  in *n* dimensions, 475
  operator of, 439, 473

Mach number, 284
Magnitude:
  of function, 27
  order of, 27
Matching, 62, 63
Material derivative, *see* Convective derivative
Matrix:
  diagonally dominant, 531

  irreducible, 532
  strictly diagonally dominant, 531
  tridiagonal, 504, 517
Matrix stability analysis, 513
Maximum principle, 168
Maxwell's equations, 123, 301
Momentum:
  angular, 338
  balance of, 272, 306, 324
  canonical, 133, 136, 435
  conservation of, 436, 437
  of fluid, 272, 323

Natural boundary conditions, 119
  for multiple integrals, 124
  for simplest problem, 120
Navier–Stokes equations, 335
Neumann problem, 174, 518
  BASIC program, 534, 539
  difference approximation, 534
Neumann series, 219
Newtonian fluid, 334
Noether's theorem, 430, 436
Norm:
  definition, 93, 527
  matrix, 527
  maximum, 95
  strong, 95
  vector, 527
  weak, 94
Normal modes, 291
Normed linear space, 91, 93

O and *o* notation, 44
Orthogonality, 183
Oscillator:
  damped, 71, 72, 138, 432
  linear, 71, 129, 134
  nonlinear, 39
Overlap domain, 62
Overrelaxation, 540

Parabolic equation, 155, 157
Pendulum, 32, 33, 52, 129
Periodic solution, 393
Perturbation:
  regular, 35
  singular, 57
Perturbation series, 35
Perturbation solution, 35
Phase plane, 135, 379
Phase portrait, 380
Phase velocity, 232, 245
Piston problem in gas dynamics, 307
Plateau's problem, 117

Poincaré–Bendixson theorem, 394
Poincaré–Lindstedt method, 32
Point:
  bifurcation, 358
  critical, 380
  double, 366
  high order singular, 368
  isolated, 367
  regular, 366
  singular, 366
  turning, 368
Poisson's equation, 117, 170, 202
Population models: Polar coordinates, 158
  logistics, 343
  Malthusian, 342
Potential energy, 127
Pressure, 276, 331

Quasi-linear equation, general solution, 257

Rankine–Hugoniot conditions, 314
Rate of convergence, 530
Rayleigh line, 315
Rayleigh's example, 410
Reaction–diffusion equation, 168
Residual vector, 49, 530
Reynolds number, 15, 339
Reynold's transport theorem, 322
Riemann invariants, 309

Scaling principle, 26
Scattering, 298
Schrödinger equation, 148
Secular term, 42
Separatrix, 387
Similarity solution, 417, 442
Similarity variable, 444
Soliton, 247, 248
Sound speed, 283, 307
Specific heat:
  constant pressure, 279
  constant volume, 18, 21, 278
Spectral radius, 529
Splitting of matrix, 521
  Gauss–Seidel, 525
  Jacobi, 522
Stability, 165, 359
  asymptotic, 359
  Courant–Friedrichs–Levy criterion, 549
  of critical point, 383
  of difference scheme, 510, 513
  exchange of, 368
  indicator, 360, 371
  of layered fluid, 406
  near double point, 371

near turning point, 369
  von Neumann criterion, 512
Stability theorem:
  for first order equations, 361
  for linear systems, 389
  for nonlinear systems, 392
Stationary, 86, 99
Strain, 285
Streamlines, 318
Stress, 271, 324
Sturm–Liouville problem:
  differential equations, 182
  regular, 182
  variational principle, 149
Summation convention, 328
Superposition principle, 157, 158

Tensor:
  rate of deformation, 334
  stress, 327
  viscous stress, 332
Thermal parameters:
  table of, 21
Thermodynamics:
  first law, 238
  second law, 280
Time scale:
  for diffusion processes, 20, 21
  for general processes, 28
Transformations, *see* Local Lie group
  one parameter family of, 419
  stretching, 449, 466
Tridiagonal system, 504, 517
Truncation error, 495, 500

Unit free, 6
Units, 6

Variation:
  admissible, 96
  of function, 96
  of functional, 96, 97
  Gâteaux, 96
  second, 101
Variational principle, 128
Variational problem, 87
Vibrations of bar, 284, 482
Vorticity, 338

Wave:
  diffusive, 245
  dispersive, 245
  number, 231
  rarefaction, 234
  shock, 233

Wave (*Continued*)
  simple, 311
  traveling, 231
Wave equation:
  in acoustics, 283
  D'Alembert's solution, 293
  in electrodynamics, 301
  finite difference schemes, 547, 551

general solution, 292
nonhomogenous, 303
in spherical symmetry, 304
in vibrations of bar, 287
Well-posed problems, 165
WKB approximation, 305

Young's modulus, 286